Lecture Notes in Artificial Intelligence 838

Subseries of Lecture Notes in Computer Science
Edited by J. G. Carbonell and J. Siekmann

Lecture Notes in Computer Science
Edited by G. Goos and J. Hartmanis

Craig MacNish David Pearce
Luís Moniz Pereira (Eds.)

Logics in Artificial Intelligence

European Workshop JELIA '94
York, UK, September 5-8, 1994
Proceedings

Springer-Verlag
Berlin Heidelberg New York
London Paris Tokyo
Hong Kong Barcelona
Budapest

Series Editors

Jaime G. Carbonell
School of Computer Science, Carnegie Mellon University
Schenley Park, Pittsburgh, PA 15213-3890, USA

Jörg Siekmann
University of Saarland
German Research Center for Artificial Intelligence (DFKI)
Stuhlsatzenhausweg 3, D-66123 Saarbrücken, Germany

Volume Editors

Craig MacNish
Department of Computer Science, University of York
York YO1 5DD, England

David Pearce
LWI, Institut für Philosophie, Freie Universität Berlin
Habelschwerdter Allee 30, D-14195 Berlin, Germany

Luís Moniz Pereira
Departamento de Informatica, Universidade Nova de Lisboa
2825 Monte da Caparica, Portugal

CR Subject Classification (1991): I.2, F.3-4

ISBN 3-540-58332-7 Springer-Verlag Berlin Heidelberg New York
ISBN 0-387-58332-7 Springer-Verlag New York Berlin Heidelberg

CIP data applied for

Printed in Germany

Typesetting: Camera ready by author
SPIN: 10475396 45/3140-543210 - Printed on acid-free paper

Preface

Logics have, for many years, laid claim to providing a formal basis for the study of artificial intelligence. With the depth and maturity of formalisms and methodologies available today, this claim is stronger than ever.

The European Workshops on Logics in AI (or *Journées Européennes sur la Logique en Intelligence Artificielle — JELIA*) began in response to the need for a European forum for discussion of emerging work in this growing field. JELIA'94 follows previous Workshops held in Roscoff, France (1988), Amsterdam, the Netherlands (1990), and Berlin, Germany (1992).

JELIA'94 is taking place in York, England, from 5–8 September, 1994. The Workshop is hosted by the Intelligent Systems Group in the Department of Computer Science at the University of York. Additional sponsorship for the 1994 Workshop is provided by the ESPRIT NOE COMPULOG-NET, ACM SIGART, the Association for Logic Programming—UK Branch (ALP-UK) and the German Informatics Society (GI).

This volume contains the papers selected for presentation at the Workshop along with papers and abstracts from the invited speakers. The increasing importance of the subject was reflected in the submission of 79 papers, from which 24 were selected. Each paper was reviewed by at least 3 referees. We would like to thank all authors and invited speakers for their contributions.

Organisation of the Workshop and the selection of papers was carried out by the Organising Committee consisting of Craig MacNish (Workshop Chair), David Pearce (Programme Co-chair) and Luís Moniz Pereira (Programme Co-chair) together with the Programme Committee consisting additionally of Carlo Cellucci, Luis Farinas del Cerro, Phan Minh Dung, Jan van Eijck, Patrice Enjalbert, Ulrich Furbach, Dov Gabbay, Antony Galton, Michael Gelfond, Vladimir Lifschitz, Victor Marek, Bernhard Nebel, Wolfgang Nejdl and Hans Rott.

Further help in reviewing papers was provided by the additional referees listed overleaf. We would like to extend our thanks to all referees for their valuable assistance.

We would also like to thank the Departments of Computer Science at the University of York, England and Universidade Nova de Lisboa, Portugal, for the provision of facilities, and Antonis Kaniclides and Karen Mews for help with the organisation of the Workshop. Finally we would like to thank Jörg Siekmann, Alfred Hofmann and Springer-Verlag for their assistance in the production of this volume.

June 1994

Craig MacNish
Luís Moniz Pereira
David Pearce

Additional Referees

J. Alferes	Y. Arambulchai	N. Asher
J. Bahsoun	P. Balbiani	P. Baumgartner
B. Bennett	P. Bieber	G. Brewka
K. Broda	H. Bürckert	M. Cayrol
A. Chandrabose	P. Cholewinski	F. Clerin-Debart
F. Cuppens	R. Demolombe	J. Dix
D. Duffy	K. El-Hindi	B. Fronhoefer
C. Gabriella	O. Gasquet	E. Giunchiglia
J. Gooday	M. Grosse	A. Herzig
R. Hirsh	J. Jaspars	Y. Jiang
D. Keen	M. Kerber	C. Kreitz
G. Lakemeyer	R. Letz	R. Li
J. Lonsdale	P. Mancarella	N. McCain
W. Meyer Viol	A. Mikitiuk	L. Monteiro
A. Narayanan	G. Neelakantan Kartha	W. Nutt
C. Paulin-Mohring	H. Prakken	A. Porto
W. Reif	M. Reynolds	M. de Rijke
H. Ruess	K. Schild	C. Schwind
F. Stolzenburg	M. Strecker	M. Truszczynski
H. Turner	I. Urbas	M. Varga von Kibed
Y. Venema	G. Wagner	E. Weydert

Contents

Invited Speaker

Extending Logic Programming

From Carnap's Modal Logic to Autoepistemic Logic*

Georg Gottlob

Institut für Informationssysteme
Technische Universität Wien
Paniglgasse 16, A-1040 Wien, Austria;
Internet: gottlob@vexpert.dbai.tuwien.ac.at

Abstract. In his treatise *Meaning and Necessity*, Carnap introduced an original approach to modal logic which is quite different from the well-known Lewis systems. Recently, it turned out that there are interesting connections between Carnap's modal logic and finite model theory for modal logics. In addition, Carnap's logic has applications in the field of epistemic reasoning. It was also shown that formulas of Carnap's logic are structurally equivalent to trees of **NP** queries, more precisely, of satisfiability tests. In this paper, we give a survey of these results. Moreover, we extend Carnap's logic in a natural way by the possibility of deriving consequences from nonmodal theories and show that the resulting formalism is nonmonotonic. Finally, we explain the relationship between Carnap's logic and autoepistemic logic and show that autoepistemic reasoning corresponds to solving problems equivalent to (possibly cyclic) graphs of interdependent **NP** queries.

1 Carnap's Modal Logic

In 1947 the philosopher and logician Rudolf Carnap published "Meaning and Necessity", a study containing a new approach to semantics and modal logic [3]. In particular, Carnap presents a new modal logic that differs substantially from the former Lewis systems. While Carnap's approach conveys both new semantics for propositional modal sentences and an innovative treatment of individual variables in the first order case (the latter having been fruitfully applied to axiomatizing parts of physics; cf. [1]), we restrict our attention to the propositional part. The presentation of Carnap's logic here differs from the original presentation in [3]. Moreover, we use different (but equivalent) definitions.

Carnap's modal logic, which we will call **C** here[2] extends the classical propositional calculus by a necessity operator □ and a possibility operator ◇. The

* This work was done in the context of the Christian Doppler Laboratory for Expert Systems. Some of the results of the present paper were also presented in [10]. A journal version, integrating the results of [10] and most of the results of the present paper is available from the author.

[2] It is called S_2 in [3] but that name may create confusion with the standard S_2 system of modal logic.

syntax of **C** formulas is the same as that of formulas in standard systems of propositional modal logic (see, e.g. [15] or [4]). Let us write $\models_{\mathbf{C}} \phi$ if formula ϕ is valid in logic **C**. The semantics of **C** is best described by giving a formal definition of $\models_{\mathbf{C}}$. If ϕ is a modal formula, then let *modsub*(ϕ) denote the set of all subformulas of ϕ of the form $\Box\psi$ or $\Diamond\psi$ that occur in ϕ at least at one place not being in the scope of any $\Box$ or $\Diamond$ operator. For instance, if $\phi = (\Diamond\Box(p) \vee \Box(\Box q \wedge \Box r)) \rightarrow \Box r$ then *modsub*(ϕ) contains the three formulas $\Diamond\Box(p)$, $\Box(\Box q \wedge \Box r)$, and $\Box r$.

Let ϕ be a modal formula and let $\psi \in modsub(\phi)$. A *strict* occurrence of ψ in ϕ is an occurrence of ψ which is not in the scope of a modal operator.

Definition 1.1 *Validity in logic* **C** *is inductively defined as follows.*

- *If* $\phi = \Box\psi$ *then* $\models_{\mathbf{C}} \phi$ *iff* $\models_{\mathbf{C}} \psi$.
- *If* $\phi = \Diamond\psi$ *then* $\models_{\mathbf{C}} \phi$ *iff* $\not\models_{\mathbf{C}} \neg\psi$.
- *In all other cases,* $\models_{\mathbf{C}} \phi$ *iff* ϕ^+ *is valid in classical propositional logic, where* ϕ^+ *is the formula resulting from* ϕ *by replacing each strict occurrence of a subformula* $\psi \in modsub(\phi)$ *by* $\top$ *if* $\models_{\mathbf{C}} \psi$ *and by* $\bot$ *otherwise.*

Intuitively, $\Box$ can be considered as a strong provability operator that expresses the metapredicate $\models_{\mathbf{C}}$ within the object language. $\Box\phi$ is valid (and provable) if ϕ is provable. On the other hand, if ϕ is not provable, then $\neg\Box\phi$ is valid (and provable). Accordingly, in Carnap's logic each modal formula ϕ of type $\Box\psi$ or $\Diamond\psi$ is *determinate*; i.e., it either holds that $\models_{\mathbf{C}} \phi$ or that $\models_{\mathbf{C}} \neg\phi$. Examples of valid formulas of **C** are $\Diamond p$, $\Diamond\neg p$, $\neg\Box p$, $\Box q \rightarrow \Box r$. Note that these formulas are not valid in any standard system of modal logic. Familiar substitutivity laws do not hold for **C**. For instance, we have $\models_{\mathbf{C}} \Diamond p$, but not $\models_{\mathbf{C}} \Diamond(q \wedge \neg q)$ even though the second formula is obtained from the first by uniformly substituting the formula $q \wedge \neg q$ for the propositional variable p. Note that in Carnap's logic **C**, as in most other modal logics, a formula of the form $\Diamond\psi$ is in all effects equivalent to the formula $\neg\Box\neg\psi$, so that we may build **C** entirely on the modal operator $\Box$ defining $\Diamond$ as a shorthand for $\neg\Box\neg$. Vice versa, one may construct **C** based on $\Diamond$ and defining $\Box$ as a shorthand for $\neg\Diamond\neg$.

Due to the failure of familiar substitutivity laws, Carnap's modal logic did not give rise to much interest until recently. However, very recently Halpern and Kapron have shown that there is an interesting connection between logic **C** and 0-1-laws for other modal logic.

The 0-1-law for predicate logic [8, 6] states that every property definable without function symbols is either true with probability 1 (almost surely true) or false with probability 1 (almost surely false) on finite structures. Analogous 0-1-laws for various propositional modal logics have recently been established and studied by Halpern and Kapron [11].

Let S be a system of modal logic which is semantically characterized by a corresponding class of Kripke-structures (S-structures). For any modal formula ϕ, let $\nu_n^S(\phi)$ express the ratio between the number of S-structures with state-space $\{1, \ldots, n\}$ satisfying ϕ over the number of *all* S-structures with state space $\{1, \ldots, n\}$. (For a more precise definition of $\nu_n^S(\phi)$ refer to [11].) Let $\nu^S(\phi) = \lim_{n\rightarrow\infty} \nu_n^S(\phi)$. Thus $\nu^S(\phi)$ expresses the asymptotic probability that the formula ϕ is satisfied by an arbitrarily chosen Kripke structure for modal logic S, when the state space of the Kripke structure is finite but grows towards infinity.

The 0-1-law for structure validity in modal logic S holds if for all modal formulas ϕ, we have either $\nu^S(\phi) = 0$ or $\nu^S(\phi) = 1$. We say that a formula ϕ is *almost surely S-valid* if $\nu(\phi) = 1$.

Intuitively, ϕ is almost surely S-valid, if the probability that ϕ is valid in an arbitrarily chosen S Kripke-structure is 1.

Halpern and Kapron [11] have shown that the 0-1-law for structure validity holds for the systems of modal logic **K**, **T**, **S4**, and **S5**. For each of these logics, they have characterized the set of almost sure formulas and have investigated their complexity.

For **S5** it turned out that a formula ϕ is almost surely valid iff ϕ is provable in **S5**.

For **K** and T, the following theorem was shown. Note that the authors of [11] do not explicitly refer to Carnap's logic but independently reconstruct it for ad hoc use.

Proposition 1.1 (Halpern and Kapron [11]) *A propositional modal formula ϕ is almost-surely structure valid in modal logic* **K** *iff ϕ is almost surely structure valid in modal logic* **T** *iff $\models_{\mathbf{C}} \phi$*

The above result gives a new very attractive meaning to Carnap's logic, which was *almost surely* not foreseen by Carnap.

Note that Carnap's logic has a simple and interesting Kripke-type model theory. The sentences true in **C** are precisely those sentences true in the fully connected Kripke structure, where each world corresponds to a finite set of propositional atoms made true, and each such set corresponds to precisely one world. This follows from results of Halpern and Kapron (Theorem 4.8 in [12]) and was independently pointed out to the author by Petr Hàjek and Andreas Herzig (private communications).

It is easy to see that the sentences true in **C** are a superset of the **S5** theorems. This follows, e.g., from our last observation. In particular, it can be seen that those sentences which are true in **C** *for each possible replacement of the propositional variables by formulas* (i.e., those **C** theorems which are stable under uniform substitution) are precisely the **S5** theorems. For example, consider the **S5** axiom $\Diamond A \rightarrow \Box\Diamond A$. As we noted, in **C**, $\Diamond A$ is determinate, i.e. either $\models_{\mathbf{C}} \Diamond A$ or $\models_{\mathbf{C}} \neg\Diamond A$ holds. In the first case, we also have $\models_{\mathbf{C}} \Box\Diamond A$ and in the second case we clearly have $\not\models_{\mathbf{C}} \Box\Diamond A$. Therefore, by Definition 1.1, in the first case, the formula $\Diamond A \rightarrow \Box\Diamond A$ evaluates to true ($\top \rightarrow \top$), and in the second case, it again evaluates to true ($\bot \rightarrow \bot$). This shows that for each formula A, it holds that $\models_{\mathbf{C}} \Diamond A \rightarrow \Box\Diamond A$. In a similar way, we can prove each **S5** theorem based on reasoning by cases in **C**. It follows that **S5** validity can be based entirely on the concept of validity or provability in propositional logic (as is **C**). A reconstruction of **S5**, recurring to such "absolute" principles of validity, was one of Carnap's goals in [3].

2 A Consequence Relation for C

We would like to extend the definition of validity in Carnap's logic to be able to say that a formula ϕ is a *consequence* of a theory T. The latter will be denoted by $T \models_{\mathbf{C}} \phi$. Carnap did not introduce such a consequence relation $T \models_{\mathbf{C}} \phi$ in [3]. The reason may be twofold:

- If T is a modal theory and ϕ a formula, then there is no obvious unique natural meaning to $T \models_{\mathbf{C}} \phi$. Note that there are many possible generalizations (see Section 4).
- If T is a nonmodal theory and ϕ a (possibly modal) formula, then there is a natural meaning to $T \models_{\mathbf{C}} \phi$, however, the corresponding consequence relation is nonmonotonic!

There is in fact a natural generalization of Carnap's logic to cover consequence, in case T is nonmodal. Remember that validity in Carnap's Logic was defined by a recursive reduction to validity in classical propositional logic. Therefore, we can analogously define a Carnap consequence operator on the base of classical consequence, provided our premises are nonmodal.

Definition 2.1 *Let T be a classical propositional theory. Consequence from T in logic* **C** *is inductively defined as follows.*

- *If $\phi = \Box\psi$ then $T \models_{\mathbf{C}} \phi$ iff $T \models_{\mathbf{C}} \psi$.*
- *If $\phi = \Diamond\psi$ then $T \models_{\mathbf{C}} \phi$ iff $T \not\models_{\mathbf{C}} \neg\psi$.*
- *In all other cases, $T \models_{\mathbf{C}} \phi$ iff $T \models \phi[A]$ holds in classical propositional logic, where $\phi[A]$ is the formula resulting from ϕ by replacing each subformula $\psi \in modsub(\phi)$ at each place where it does not occur in the scope of a modal operator by $\top$ if $A \models_{\mathbf{C}} \psi$ and by $\bot$ otherwise.*

The nonmonotonicity of the consequence relation $\models_{\mathbf{C}}$ is easy to see. For example, we have $\models_{\mathbf{C}} p$ but it does not hold that $\neg p \models_{\mathbf{C}} p$. We can show a kind of deduction theorem for **C**. This deduction theorem was stated in a slightly different form in [10] (Theorem 4.3). Let us first give a useful definition.

Definition 2.2 *Let A be a finite nonmodal theory and let ϕ be a (possibly modal) formula. Let $form(A)$ denote the conjunction of all sentences in A ; in case A is empty, $form(A) = \top$ (the constant fot truth). From A and ϕ we construct a modal formula ϕ^A by the following recursive rule. ϕ^A is obtained from ϕ by replacing in ϕ each strict occurrence of a formula $\psi \in modsub(\phi)$ with*

- *$\Box(form(A) \rightarrow \gamma^A)$ if $\psi = \Box\gamma$;*
- *$\Diamond(form(A) \wedge \gamma^A)$ if $\psi = \Diamond\gamma$;*

For example, if $A = \{p, q\}$ and $\phi = (\Box\Box(p \rightarrow r)) \vee \Diamond q$, then $\phi^A = \Box((p \wedge q) \rightarrow \Box((p \wedge q) \rightarrow (p \rightarrow r))) \vee \Diamond((p \wedge q) \wedge q)$.

Of course, the transformation mapping A and ϕ to ϕ^A is feasible in polynomial time.

Theorem 2.1 (Deduction Theorem) *For each finite nonmodal propositional theory A and modal formula* ϕ,

$$A \models_{\mathbf{C}} \phi \text{ iff } \models_{\mathbf{C}} (form(A) \rightarrow \phi^A).$$

PROOF. (Idea). The proof is very similar to the proof of Theorem 4.3 in [10]. The theorem is shown by induction on the modal nesting depth of ϕ by using Definitions 1.1, 2.1, and 2.2. □

Note that $form(A)$ and ϕ^A can be constructed from A and ϕ in polynomial time. Therefore, by our deduction theorem, we have shown that the problem of consequence checking in Logic **C** for nonmodal premises can be polynomially transformed to validity checking in **C**.

3 Logic C and Stable Sets

The concept of stable set has been developed in epistemic logic and in AI in order to formally represent the set of all beliefs that an ideally rational agent with full introspective capabilities (i.e., with the ability to reason about its own beliefs) may adopt. The concept was originally introduced by Stalnaker [30] and later elaborated on by Moore [24] in the context of autoepistemic logic. Formal properties of stable sets are investigated by Halpern and Moses [13] and Marek [18]. A comprehensive survey also containing some interesting new results is given in a chapter (dedicated to stable sets) of a recent book by Marek and Truszczyński [21].

Stable sets are defined in the context of classical logic extended by a modal belief operator L. Throughout this paper, we will not distinguish between the L operator and the modal necessity operator □ and we will use both symbols as synonyms. In a similar fashion, we will use ◇ and $\neg L \neg$ as synonyms.

Informally, if the actual beliefs of an agent are reflected by a stable set T and if ϕ is a formula, then $\phi \in T$ means that the formula ϕ is believed by the agent, while $L\phi \in T$ means that the agent believes that it believes ϕ and so on.

We denote by $\mathcal{L}_L$ the language of propositional logic extended by the modal belief operator L. The L-depth of a formula of $\mathcal{L}_L$ is its maximum level of nesting of L-subformulae. We denote the consequence operator in classical propositional logic by Cn. Stable sets are formally defined as follows.

Definition 3.1 *A theory* $T \subseteq \mathcal{L}_L$ *is stable if* T *satisfies the following three conditions:*

(St1) $Cn(T) = T$; *i.e.,* T *is closed under propositional consequence.*
(St2) *(Positive Introspection) For every* $\phi \in \mathcal{L}_L$, *if* $\phi \in T$ *then* $L\phi \in T$.
(St3) *(Negative Introspection) For every* $\phi \in \mathcal{L}_L$, *if* $\phi \notin T$ *then* $\neg L\phi \in T$.

All sentences involving the belief operator L are called *epistemic sentences*; sentences not involving L, i.e., belonging to the language $\mathcal{L}$ of classical propositional logic are called *objective* or simply nonmodal. The set of all non epistemic sentences of a stable set T is called the ***objective part of*** T and is denoted by $Obj(T)$. Thus, $Obj(T) = T \cap \mathcal{L}$.

It was observed by Moore [24] that stable sets depend on their objective part. Moreover, for each set A of objective sentences, there exists a unique stable set T such that $Obj(T) = Cn(A)$ (see [18]). Let us denote the unique stable set corresponding to an objective theory A by $st(A)$.

The ***membership problem for stable sets*** is formulated as follows. Given a finite objective theory A and a formula $\phi \in \mathcal{L}_L$, is ϕ an element of $st(A)$?

For example, if $A = \{p, q \vee r\}$, then $(LL((\neg Lp) \rightarrow (s \vee Lr))) \wedge ((q \rightarrow (r \rightarrow q)) \rightarrow \neg Lq))$ can be seen to be an element of $st(A)$.

The membership problem for stable sets is an important problem of epistemic reasoning and has been dealt with extensively in the literature. Different algorithms for solving this or tightly related problems have been proposed by Halpern and Moses [13], Marek and Truszczyński [18, 20, 21], and Niemelä [25, 27, 26]. The membership problem for stable sets also plays an essential role in autoepistemic reasoning [20, 26].

The following characterization follows immediately from results of Niemelä [27, 26] and can also be obtained from Theorem 2.1 in [18]. Note that the idea of using recursion on strict subformulas was already present in [13].

Proposition 3.1 *Let A be a set of objective formulas and let ϕ be an epistemic formula. $\phi \in st(A)$ iff $A \models \phi^*$, where ϕ^* is obtained from ϕ by replacing each strict occurrence of $L\gamma \in modsub(\phi)$ in ϕ with $\top$ if $\gamma \in st(A)$ and with $\bot$ if $\gamma \notin st(A)$.*

The above proposition shows that the membership problem for stable sets can be solved by recursion on modal subformulas in a way similar to the entailment problem for Carnap's logic. Actually, these problems turn out to be identical. Recall that we identify $\Box$ with L and $\Diamond$ with $\neg L \neg$. The next theorem follows almost directly from Definition 2.1 and Proposition 3.1.

Theorem 3.1 *Let A be a nonmodal propositional theory and let ϕ be a modal or epistemic formula. Then*

$$\phi \in st(A) \text{ iff } A \models_{\mathbf{C}} \phi.$$

This theorem has a simple corollary:

Theorem 3.2 *If ϕ is a modal formula, then*

$$\models_{\mathbf{C}} \phi \quad \textit{iff} \quad \phi \in st(\emptyset).$$

Based on this corollary, we can give yet another interpretation to Carnap's modal logic: The valid sentences in **C** are exactly those which are adopted by an ideally rational agent with full introspective capabilities in the absence of any information.

By these results have established the polynomial-time equivalence of the Membership problem for stable sets to validity checking in Carnap's logic. This will allows us to give a precise complexity characterization of the former problem (see Section 5).

Remark. Levesque [17] has extended classical logic by a **K** operator for posing epistemic queries to knowledge bases; this operator was further used in [29, 5]. It can be seen that answering an epistemic query to a non-epistemic propositional database in Levesque's setting is precisely equivalent to solving an instance of the membership problem for stable sets. Thus, our complexity result carries over to relevant fragments of Levesque's logic.

4 Generalizations of C and Nonmonotonic Logics

Can we generalize the consequence operator $\models_{\mathbf{C}}$ to cover also modal premises? The answer is yes, but there are several possible generalizations of $\models_{\mathbf{C}}$ and no single one seems to be the most natural candidate. In fact, as the next theorem shows, many well-known nonmonotonic modal logics are generalizations of Carnap's logic. However, there is no agreement upon which nonmonotonic modal logic is the best or the most natural.

Let us first recall the concept of nonmonotonic modal logic. We use the language $\mathcal{L}_L$ defined in Section 3. (Recall, however, that we identified $\Box$ with L, and $\Diamond$ with $\neg L \neg$.)

Definition 4.1 *Let S be a modal logic and let $Cns_S(X)$ denote the set of all S-consequences of a set of formulas X. modal or epistemic logics.*

Let $A \subseteq \mathcal{L}_L$ be a set of (modal or nonmodal) formulas, called the premises.

Δ is an S-expansion of A iff $\Delta = Cns_S(A \cup \{\neg L\phi \mid \phi \notin \Delta\})$.

We write $A \mathrel{\dot{\models}}_S \phi$ if ϕ is an element of each S-expansion of A. This way of reasoning is usually called skeptical (or cautious) nonmonotonic entailment.

Several nonmonotonic modal logics have been studied [23, 22, 21]. With few exceptions, the most relevant examples are included in nonmonotonic, see [19]. **KD45**, i.e., in autoepistemic logic, or in **SW5**. It was shown by Schwarz [32] (see also [19, 21]) that there are no relevant nonmonotonic modal logics properly containing **KD45** or **SW5** because any such logic with reasonable properties would collapse to monotonic **S5** or at least admit multiple expansions of the empty set of premises. The next theorem states that all nonmonotonic logics contained in **KD45** or **SW5** are generalizations of Carnap's logic.

Theorem 4.1 *Let S be a modal logic contained in* **KD45** *or in* **SW5** *and let A be a set of nonmodal propositional formulas. Then, for each modal formula ϕ it holds that*

$$A \mathrel{\dot{\models}}_S \phi \quad \textit{iff} \quad A \models_{\mathbf{C}} \phi.$$

PROOF. Follows directly from our Theorem 3.1, and from Theorem 4.6 in [19] which states that for all nonmonotonic modal logics (with the necessitation rule, which we assume) contained in **KD45** or in **SW5**, it holds that $st(A)$ is the only expansion of A. $\Box$

5 NP Trees and Carnap's Logic

In this section, we define the concept of **NP** trees and review complexity results originally presented in [10]. The connection between **NP** trees and formulas of Carnap's logic is made clear by exhibiting explicit transformations that were not given in [10].

Intuitively, an **NP** tree consists of a tree-like network (circuit) of nodes, each representing a parametric query to an **NP** oracle, such that parts of the query string at node N depend on the oracle answers of N's immediate predecessor nodes. Thus a directed arc $\langle N_1, N_2 \rangle$ in an **NP** tree symbolizes that the oracle query at node N_2 depends on the oracle answer at node N_1. We call N_1 an *input node* for N_2. Note that ther arcs are directed from the leaves towards the root node. An oracle query at a node N can be evaluated as soon as the queries of all input nodes for N have been evaluated. Each **NP** tree has a unique distinguished node called the *result node.* An **NP** tree is positively answered iff the oracle query at its result node is positively answered.

To render our intuitive notion of **NP** tree more precise, we assume in the following definition that the **NP** query at each node is always an instance of SAT (i.e., satisfiability of propositional formulas) and that the oracle query at each node is a SAT instance such that the value of particular propositional variables (the linking variables) may depend on the oracle answers of input nodes to N.

Definition 5.1 *An* **NP** *tree is defined by a triple* $\langle Var, G, F_R \rangle$ *consisting of*

- *a set of propositional variables* $Var = \{v_1, \ldots, v_n\}$, *called the linking variables;*
- *a directed acyclic graph in tree-form* $G = \langle V, E \rangle$, *the nodes* $V = \{F_1, \ldots, F_n\}$ *representing propositional formulas. Different nodes may represent the same formula and we identify the name* F_i *of a node with the name of the formula that is represented by the node. Each propositional formula* F_i, *in addition to zero or more "private" propositional variables not appearing in any other node, contains the following linking variables:* $\{v_j \mid \langle F_j, F_i \rangle \in E\}$.
- *a distinguished terminal node* F_R, *for some* $R \in \{1, \ldots, n\}$, *called the result node.*

For each **NP** *tree the truth value assignment* σ *to the propositional linking variables* $v_1, \ldots, v_n$ *is defined as follows.* $\sigma(v_i)$ *is true if the formula* $F_i^!$ *is satisfiable, where* $F_i^!$ *results from* F_i *by replacing each propositional linking variable* v_j *occurring in* F_i *by* $\top$ *if* $\sigma(v_j) = true$ *and by* $\bot$ *if* $\sigma(v_j) = false$; *otherwise* $\sigma(v_i)$ *is false. (This definition assigns an unambiguous value to each* v_i *representing a bottom node and inductively to each other* v_j*.)*
The result value of an **NP** *tree is equal to* $\sigma(v_R)$.
The problem of deciding whether the result value of an **NP** *tree is* true *is termed* TREES(SAT).

Notation: If N is a node, then the linking variable for N is also referred to as $[N]$. In particular, if $N = F_i$, then $[N] = v_i$.

The following theorem shows that **NP** trees are tightly related to formulas of Carnap's logic.

Theorem 5.1 *There are is a polynomial transformation T mapping each* **C**-*formula ϕ into an* **NP** *tree $T(\phi)$ such that $T(\phi)$ is positively answered iff ϕ is valid in* **C**. *Vice-versa, there is a polynomial transformation frm mapping each* **NP** *tree t into a modal formula $frm(t)$ such that $frm(t)$ is valid in* **C** *iff t is positively answered.*

PROOF. A *strict occurrence* of a subformula ξ in a formula ψ is an occurrence which is not in the scope of a modal operator.

Given a **C** formula ϕ, construct the **NP** tree $T(\phi)$ by the following method.

1.) INITIALIZATION. Let T be the tree obtained as follows.

- Rewrite ϕ by replacing each occurence of $\Box$ by $\neg\Diamond\neg$.
- If $\phi = \neg\Diamond\psi$ then create a result node F_R whose formula is $\neg v_1$ and create a node F_1 with formula ψ. Create an arc $\langle F_1, F_R\rangle$.
- Otherwise create a result node F_R with formula ϕ.

2.) ITERATIVE TREE EXPANSION. WHILE there is a leaf F in T representing a formula ψ containing modal operators, expand F as follows.

FOR each strict occurrence of a subformula $\Diamond\xi \in modsub(\psi)$ in ψ:
- Create a new node F_i;
- Associate formula ξ with F_i,
- Create an arc $\langle F_i, F\rangle$.
- Replace the specific occurrence of $\Diamond\xi$ in ψ with the linking variable v_i.

From Definition 1.1 it is clear that the resulting tree is a positive instance of TREES(SAT) iff $\models_C \phi$. The transformation is obviously feasible in polynomial time.

Conversely, let T be an **NP** tree. Construct a modal formula $frm(T)$ as follows from T.

WHILE T has more than one node DO

Choose a leaf F of T. Let G be the parent node of F and let ψ and ϕ be the formulas associated with F and G, respectively. Note that the the linking variable $[F]$ occurs in ϕ. Eliminate F and replace ϕ in G with the formula $\phi \wedge ([F] \equiv \Diamond\psi)$.

The resulting tree consists of a single node F_R. Let ξ be the formula associated with F_R, and set $frm(T) = \Diamond\xi$.

Again, by Definition 1.1 it is obvious that T is a positive instance of TREES(SAT) iff $\models_C frm(T)$. The transformation is clearly feasible in polynomial time. $\Box$

Let $\mathbf{P}^{\mathbf{NP}}[O(\log n)]$ denote the class of all problems solvable in polynomial time except for a logarithmic number of queries to an oracle in **NP**. This class, which is also denoted by Θ_2^P or $\Delta_2^P[O(\log n)]$, was first defined by Papadimitriou and Zachos [28]. For an overview of several different characterizations, see Wagner [33]. In particular, $\mathbf{P}^{\mathbf{NP}}[O(\log n)]$ is identical to the classes

$\mathbf{LOGSPACE}^{\mathbf{NP}}$ and $\mathbf{P}_{\parallel}^{\mathbf{NP}}$. The latter consists of the decision problems solvable in polynomial time with a single round of parallel queries to an **NP** oracle. Buss and Hay [2] have shown that $\mathbf{P}^{\mathbf{NP}}[O(\log n)]$ is also equal to the class of decision problems solvable in polynomial time with a *constant* number of rounds of parallel queries to an **NP** oracle. Furthermore, it was shown that $\mathbf{P}^{\mathbf{NP}}[O(\log n)]$ coincides with the class of problems reducible by polynomial (or also logspace) truth table reductions to **NP** [14, 33, 2].

In [10], the following theorem was shown by a rather involved proof.

Theorem 5.2 ([10]) TREES(SAT) *is* $\mathbf{P}^{\mathbf{NP}}[O(\log n)]$ *complete.*

By Theorems 5.2 and 5.1, and by the fact that obviously $\mathbf{P}^{\mathbf{NP}}[O(\log n)]$ is closed under polynomial reductions, we get a precise complexity characterization of validity checking in Carnap's modal logic.

Theorem 5.3 ([10]) *Deciding* $\models_{\mathbf{C}} \phi$ *is* $\mathbf{P}^{\mathbf{NP}}[O(\log n)]$ *complete.*

By the deduction theorem for **C**, we get as a corollary that determining whether $A \models_{\mathbf{C}} \phi$ for nonmodal A is $\mathbf{P}^{\mathbf{NP}}[O(\log n)]$ complete. Moreover, by the results in section 3, it follows that the membership problem for stable sets is $\mathbf{P}^{\mathbf{NP}}[O(\log n)]$-complete.

One may consider other graph-theoretical relationships between **NP** queries than trees. For instance, in [10] **NP** dags (acyclic directed graphs) are studied and it is shown that deciding whether an **NP** dag evaluates to true is Δ_2^P complete. $\Delta_2^P = \mathbf{P}^{\mathbf{NP}}$ is the class of all decision problems solvable in polynomial time with access to an oracle in **NP**.

However, it is even possible to go beyond **NP** dags and study general, possibly *cyclic*, **NP** graphs. This is the aim of the next section.

6 NP Graphs and Autoepistemic Logic

Until now we have considered *acyclic* **NP** graphs only. However, it sometimes also makes sense to consider cyclic graphs containing nodes with **NP** oracles. For instance, different nodes of such graphs may represent different agents that have to solve **NP**-problems whose parameters depend on the decisions of other agents. In this section we define the concept of **NP** graphs and study the corresponding complexity problem. We also show how **NP** graphs can be embedded into autoepistemic logic.

The obvious problem with cycles is that they may cause oscillating behavior so that no fixed truth value can be assigned to certain nodes. Note that, of course, the same problem already arises with closed Boolean circuits.

If one deals with cyclic **NP** graphs, it is necessary to find an adequate generalization of the notion of result value of a given **NP** graph. There are several possible approaches to defining such generalizations. Here we will focus on the approach of *stable solutions.* Informally, a stable solution s to an **NP** graph is an equilibrium state described by the truth values each node (or equivalently, each linking variable) takes in this state. An **NP** graph evaluates to *true* iff its result node is assigned *true* in all stable solutions.

Definition 6.1
• **NP** *graphs are defined as* **NP** *trees (see Definition 5.1), except that the directed graph* $G = \langle V, E \rangle$ *may be an arbitrary (possibly cyclic) graph.*
• *A stable solution* s *for* G *is an assignment of a truth value* $s(v_i)$ *to each linking variable* v_i *such that for each node* F_j*, the formula* F_j^s *is satisfiable iff* $s(v_j) = \text{true}$*, where* F_j^s *results from the formula of* F_j *by replacing each linking variable* v_i *with* $\top$ *if* $s(v_i) = true$ *and with* $\bot$ *if* $s(v_i) = false$*.*
• *The decision problem* GRAPHS(SAT) *consists of deciding whether* $s(F_R)$ *is true for each stable solution* s *to a given* **NP** *graph with result node* F_R*.*
• *The class* **GRAPHS(NP)** *consists of all decision problems polynomially transformable to* GRAPHS(SAT).

We now determine the complexity of the decision problem GRAPHS(SAT). The complexity class Σ_2^P (also denoted by $\mathbf{NP}^{\mathbf{NP}}$) is the class of all decision problems solvable by a nondeterministic Turing machine in polynomial time with access to an oracle in **NP**. The class Π_2^P contains all complementary problems to Σ_2^P problems. Σ_2^P and Π_2^P are the central classes of the second level of the Polynomial Hierarchy (see, e.g., [7, 16]).

Theorem 6.1 GRAPHS(SAT) *is* Π_2^P *complete.*

PROOF. *Membership.* An instance $G = \langle V, E \rangle$ of GRAPHS(SAT) is negative iff it has a stable solution s such that $s(F_R) = \mathit{false}$ for the result node F_R. In order to verify that G is a negative instance, it is thus sufficient to guess a truth value assignment s to the nodes in V and check that s is effectively a stable solution. The checking part involves $|V|$ queries to a SAT oracle and is thus in Δ_2^P. Testing if G is a negative instance of GRAPHS(SAT) is thus in $\mathbf{NP}^{\mathbf{NP}} = \Sigma_2^P$. Therefore, GRAPHS(SAT) $\in \Pi_2^P$.
Hardness. Let $Q = \forall p_1 \ldots \forall p_n \exists q_1 \ldots \exists q_m E$ be a quantified propositional formula where E is a propositional formula over propositional variables $p_1, \ldots, p_n$, $q_1, \ldots, q_m$. Deciding whether Q is valid is Π_2^P complete. We transform Q in polynomial time into the **NP** graph G_Q depicted in Figure 1. In this figure, E^* is the propositional formula resulting from E by replacing each variable p_i with the linking variable $[F_i]$. For the rest, the figure is self-explanatory. Obviously, the stable solutions to G_Q exactly correspond to all 2^n possible truth value assignments to the p_i variables. Thus the graph G_Q is a positive instance of GRAPHS(SAT) iff for each such assignment E is satisfiable; i.e, iff Q is valid.
□

Note that an **NP** graph may have no stable solution at all. For a simple example, consider an **NP** graph containing a single node $F_1 = F_R$ labeled by the formula $\neg[F_R]$ such that F_R is connected by a loop to itself (see Figure 2).

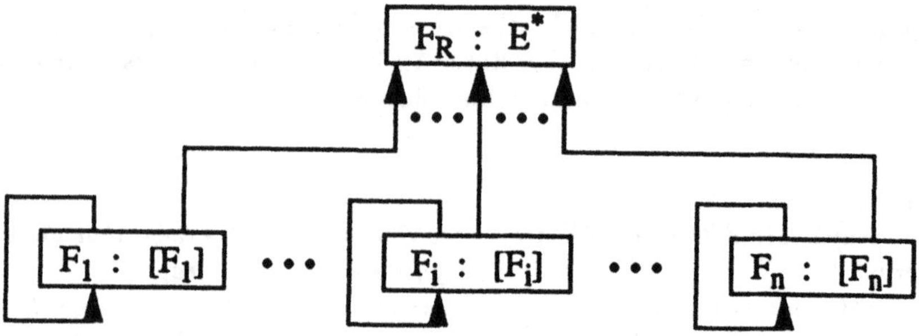

Figure 1: NP-Graph G_Q

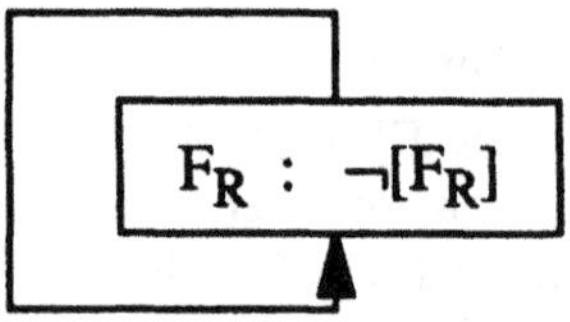

Figure 2: The simplest unstable NP-Graph

The following theorem presents complexity results for related decision problems. We omit the easy proof (which is similar to the dualized version of the proof of the previous theorem).

Theorem 6.2 *Deciding whether an* NP *graph has a stable solution is* Σ_2^P *complete. Deciding whether an* NP *graph has a stable solution which assigns* true *to the result node is* Σ_2^P *complete.*

Remark: According to our definition, NP graphs that admit no stable solution at all are positive instances of GRAPHS(SAT) because the condition on all stable solutions is vacuously fulfilled. This may be considered inappropriate for some applications. However, if necessary, one may change Definition 6.1 and make the additional requirement that each positive instance of GRAPHS(SAT) have at least one stable solution. This modification comes at a slight cost. It is not hard to see that checking whether a given NP graph is a positive instance of the modified problem is complete for the complexity class $\Sigma_2^{\mathbf{P}}(2) = \{L_1 \cap L_2 \mid L_1 \in \Sigma_2^P, L_2 \in \Pi_2^P\}$. Note that this class is the equivalent of the well-known class $\mathbf{D^P}$ at the next higher level of the polynomial-time hierarchy.

In the rest of this section we show that **NP** graphs can be straightforwardly embedded into autoepistemic logic and that the problem GRAPHS(SAT) is basically equivalent to the task of skeptical reasoning in this logic.

Autoepistemic logic (AEL), introduced by Moore [24], is a successful tool for formalizing principles of nonmonotonic reasoning. It was already introfuced in Section 4 as nonmonotonic modal logic **KD45**. Here we will give the original definition of Moore [24].

Recall that this logic is based on the language $\mathcal{L}_L$ of propositional logic extended by a modal belief operator L (or $\Box$), as already introduced in Section 3. In AEL, to each given set $A \subset \mathcal{L}_L$ of initial beliefs corresponds a set of stable expansions, where each expansion is an alternative possible set of total beliefs based on A. Each stable expansion of A is a stable set containing A. If A is an objective set, then the unique stable expansion of A is $st(A)$.

Propositional interpretations are extended to autoepistemic logic by taking autoepistemic formulas of the form $L\phi$ to be atoms (also called epistemic atoms). In particular, the atoms of a formula ψ are all those atoms (propositional or epistemic) which occur *strictly* in ψ, i.e., outside the scope of any L-operator. For example, the atoms of the formula $r \rightarrow (p \wedge LL(q \vee r))$ are r, p, and $LL(q \vee r)$. An interpretation consists of an assignment of truth values to all atomic objective formulas and to all formulas of the form $L\phi$. An interpretation assigns a truth value to each formula of $\mathcal{L}_L$ by the classical rules of truth recursion. The classical consequence relation $\models$ on $\mathcal{L}$ is extended to $\mathcal{L}_L$ as follows. If $A \subseteq \mathcal{L}_L$ and $\phi \in \mathcal{L}_L$ then $A \models \phi$ iff ϕ is true in every interpretation in which A is true. The consequence operation *cons* is defined as usual by $cons(\Delta) = \{\phi \mid \Delta \models \phi\}$ for $\Delta \subseteq \mathcal{L}_L$.

We will use the following notation. If $\Delta \subseteq \mathcal{L}_L$ then $L\Delta = \{L\phi \mid \phi \in \Delta\}$, $\neg\Delta = \{\neg\phi \mid \phi \in \Delta\}$, and $\overline{\Delta} = \mathcal{L}_L - \Delta$.

Formally, a stable expansion of an autoepistemic theory A is defined as follows.

Definition 6.2 ([24]) *Δ is a stable expansion of A iff Δ satisfies the following fixed point equation.*

$$\Delta = cons(A \cup L\Delta \cup \neg L\overline{\Delta}).$$

Skeptical or *cautious* reasoning in AEL is the task of deciding whether a given formula ϕ occurs in all stable expansions of a set A of premises. In the affirmative case we say ϕ is a skeptical consequence of A and write $A \,\dot{\models}\, \phi$

We now specify a transformation mapping each **NP** graph G to an autoepistemic theory $\tilde{G}$.

Definition 6.3 *Let $G = \langle V, E\rangle$ be an* **NP** *graph with node set $V = \{F_1, \ldots, F_n\}$, result node F_R for some $1 \leq R \leq n$, linking variables $v_1 \ldots, v_n$, and edge set E.*
The autoepistemic theory $\tilde{G}$ is defined by $\tilde{G} = \{v_i \equiv \neg L \neg F_i \mid 1 \leq i \leq n\}$.

Theorem 6.3 *Let G be an* **NP** *graph. The stable solutions of G and the stable expansions of $\tilde{G}$ are in a one-to-one correspondence. Moreover, G is a positive instance of* GRAPHS(SAT) *iff $\tilde{G} \not\models v_R$.*

PROOF. Let G be as in Definition 6.3. We first derive some properties of all stable expansions of $\tilde{G}$. Let Δ be an arbitrary stable expansion of $\tilde{G}$. By Def. 6.2 we have:

$$\Delta = cons(\tilde{G} \cup L\Delta \cup \neg L\overline{\Delta}). \qquad (a)$$

We first observe that Δ is *consistent*, since the inconsistent stable set $\mathcal{L}_L$ is not a stable expansion of $\tilde{G}$; to see this, just replace Δ by $\mathcal{L}_L$ in equation (a); the right side evaluates to $cons(\tilde{G} \cup L\mathcal{L}_L)$, which is propositionally consistent (just assign *false* to each v_i and *true* to each atom $L\psi$) and thus different from $\mathcal{L}_L$.

Moreover, note that the only non epistemic propositional atoms occurring in $\tilde{G}$ and thus in the right side of equation (a) are the variables $v_1, \ldots, v_n$. By the Interpolation Theorem it follows that there exists a finite theory Φ built from the propositional variables $v_1, \ldots, v_n$ such that $Obj(\Delta) = cons(\Phi)$. Furthermore, by equation (a), each consistent stable set contains either $L\psi$ or $\neg L\psi$ for each $\psi \in \mathcal{L}_L$. Therefore, due to the formulas $v_i \equiv \neg L \neg F_i$ in $\tilde{G}$, the truth value of each v_i is determined by Δ; i.e., for each v_i, $1 \leq i \leq n$, either $v_i \in \Delta$ or $\neg v_i \in \Delta$. Hence, also, either $v_i \in Obj(\Delta)$ or $\neg v_i \in Obj(\Delta)$, for $1 \leq i \leq n$. Therefore, if $Obj(\Delta) = cons(\Phi)$, then either $\Phi \models v_i$ or $\Phi \models \neg v_i$. From this and our observation that there exists such a Φ built from $v_1, \ldots, v_n$, it follows that there exists a theory $\Phi_\Delta = \{\hat{v}_1, \hat{v}_2, \ldots, \hat{v}_n\}$ where $\hat{v}_i = v_i$ or $\hat{v}_i = \neg v_i$ for $1 \leq i \leq n$ such that $Obj(\Delta) = cons(\Phi_\Delta)$. For such Φ_Δ it furthermore holds that $\Delta = st(\Phi_\Delta)$ because obviously $st(Obj(\Delta)) = \Delta$ and $st(cons(\Phi_\Delta)) = st(\Phi_\Delta)$.

We now show the theorem by proving that each function $s : \{v_1, \ldots, v_n\} \longrightarrow \{true, false\}$ is a stable solution of G iff the set $V_s = \{v_i \mid s(v_i) = true\} \cup \{\neg v_i \mid s(v_i) = false\}$ generates a stable expansion $st(V_s)$ of $\tilde{G}$. (The theorem follows immediately.)

If. Assume $st(V_s)$ is a stable expansion of $\tilde{G}$. We have:

$s(v_i) = true$
iff $v_i \in st(V_s)$
iff $\neg L \neg F_i \in st(V_s)$
iff $\neg F_i \notin st(V_s)$ (by equation (a) and consistency of $st(V_s)$)
iff $V_s \not\models \neg F_i$ (by Prop. 3.1)
iff V_s is consistent with F_i
iff F_i^s is satisfiable. (By the definition of F_i^s, see Def. 6.1.)

Summarizing, F_i^s is satisfiable iff $s(v_i) = true$. Thus s is a stable solution for G.

Only if. Assume s is a stable solution for G. We show that $st(V_s)$ is a stable expansion of $\tilde{G}$ by proving that $st(V_s)$ is a solution to the fixed point equation (a).

We first show that

$$st(V_s) \subseteq cons(\tilde{G} \cup L(st(V_s)) \cup \neg L(\overline{st(V_s)})). \qquad (b)$$

Note that for each consistent stable set S it holds that $S = Obj(S) \cup LS \cup \neg L\overline{S}$. Thus, in particular, $st(V_s) = Obj(st(V_s)) \cup L(st(V_s)) \cup \neg L(\overline{st(V_s)})$. Therefore, since $L(st(V_s)) \cup \neg L(\overline{st(V_s)})$ already appears in the right side of (b), it suffices to show that $Obj(st(V_s))$ is a subset of the right side of (b). Since $Obj(st(V_s)) = cons(V_s)$, it is enough to show that V_s is contained in the right side of (b). This is verified as follows.

- If $v_i \in V_s$ then F_i^s is satisfiable. Hence, $V_s \not\models \neg F_i$, and thus, by Prop. 3.1, it follows that $\neg F_i \notin st(V_s)$, hence $\neg L \neg F_i \in \neg L(\overline{st(V_s)})$. This means that $\neg L \neg F_i$ is an element of the right side of (b). Since the formula $v_i \equiv \neg L \neg F_i$ as an element of $\tilde{G}$ also belongs to the right side of (b), and since this right side is deductively closed, v_i also belongs to it.
- If $\neg v_i \in V_s$ then F_i^s is unsatisfiable. Hence, $V_s \models \neg F_i$, and thus, by Prop. 3.1, it follows that $\neg F_i \in st(V_s)$, hence $L \neg F_i \in L(st(V_s))$. This means that $L \neg F_i$ is an element of the right side of (b). Since the formula $v_i \equiv \neg L \neg F_i$ is an element of $\tilde{G}$, the formula also belongs to the right side of (b), and since this right side is deductively closed, $\neg v_i$ also belongs to it.

Thus V_s is a subset of the right side of (b) and the inclusion (b) is proved.

Let us now consider the other direction:

$$st(V_s) \supseteq cons(\tilde{G} \cup L(st(V_s)) \cup \neg L(\overline{st(V_s)})). \qquad (c)$$

Since $st(V_s) \supseteq L(st(V_s)) \cup \neg L(\overline{st(V_s)})$ and $st(V_s)$ is deductively closed, it is sufficient to see that $st(V_s) \supseteq \tilde{G}$. Indeed, the membership of each formula $v_i \equiv \neg L \neg F_i$ of $\tilde{G}$ in $st(V_s)$ follows from the definition of V_s and Prop. 3.1. □

This theorem shows that AEL is a well-suited device for representing interdependent queries to **NP** oracles and gives us new insights into and better intuitions about this logic. In particular, it fosters interesting new applications for AEL. Recall that we alluded to multi-agent decision making at the beginning of this section. Imagine a group of n agents, each of which has to solve a problem in **NP** whose parameters depend on the decisions of some other agents of the group. This may lead to different stable solutions. The situation is best represented by an **NP** graph G whose nodes stand for the agents and whose edges indicate which agent's decision depends upon which other agent's decision. By our embedding, the problem is thus easily expressible in autoepistemic logic, where each variable v_i now represents the decision of agent i. Deciding whether a certain fact will be true independently of which decision each individual group member makes corresponds to skeptical reasoning in the AEL theory $\tilde{G}$.

Niemelä [27, 26] has shown that skeptical autoepistemic reasoning, i.e., checking whether a formula ϕ belongs to *all* stable expansions of a set of premises A, is in Π_2^P. From this result and the Π_2^P-completeness of GRAPHS(SAT), it follows that we can polynomially transform each skeptical autoepistemic reasoning task to an instance of GRAPHS(SAT), i.e. to an **NP** graph. In this sense, we can say that AEL is not stronger than **NP** graphs. Note, however, that the concrete transformation from AEL to **NP** graphs is somewhat more involved than the embedding in the opposite direction we have presented here. The reason is that AEL allows deep nesting of L-formulas, has particular groundedness requirements, and does not require that the propositional variables occurring in different modal atoms be different.

Finally, let us remark that from our Theorem 6.3 and from Niemelä's results, we immediately get the following.

Theorem 6.4 *Skeptical reasoning in autoepistemic logic is* Π_2^P *complete.*

This theorem was recently shown in [9] via a direct proof. That paper also contains complexity results for other reasoning tasks in AEL and for other nonmonotonic logics.

7 Conclusion and Research Directions

In this paper we have studied various aspects of Carnap's modal logic **C**. In particular, we have established a (partial) deduction theorem for **C**, we have shown the relationship between **C** and reasoning with stable sets, we have shown that several nonmonotonic logics can be conceived as a generalization of **C** , we have outlined the connection between **C** and **NP** trees, leading to a precise complexity characterization for reasoning with **C**. Finally, we have shown that there is a tight connection between **NP** graphs and autoepistemic logic. In particular, we have given a simple embedding scheme for representing **NP** graphs in AEL.

Several questions remain remain for future research. The most interesting two are – in my opinion – :

- As outlined in Section 1, Carnaps modal logic has an important characterization in terms of limit probabilities. Is it possible to extend this characterization by suitable adaptations to at least one of the many nonmonotonic logics that are generalizations of **C**? In particular, is it possible to identify a nonmonotonic logic whose semantics can be explained in terms of (possibly conditional) limit probabilities?
- Is there a nonmonotonic logic which is a simple and natural generalization of Carnap's logic? On one hand, AEL seems to be a good candidate, given its relationship to **NP** graphs. However, other nonmonotonic logics may be close to **C** in other respects.

References

1. A. Bressan. *A General Interpreted Modal Calculus.* Yale University Press, New Haven and London, 1972.

2. S.R. Buss and L. Hay. On Truth-Table Reducibility to SAT. *Information and Computation*, 91:86–102, 1991.
3. R. Carnap. *Meaning and Necessity.* The University of Chicago Press, 1947.
4. B. Chellas. *Modal Logic, an Introduction.* Cambridge Univ. Press, Cambridge, UK, 1980.
5. F.M. Donini, M. Lenzerini, D. Nardi, W. Nutt, and A. Schaerf. Adding Epistemic Operators to Concept Languages. In *Proc. of the 3rd Int. Conf. on Principles of Knowledge Representation and Reasoning KR-92*, pages 342–353, 1992.
6. R. Fagin. Probabilities on Finite Models. *Journal of Symbolic Logic*, 41(1):50–58, 1976.
7. Michael Garey and David S. Johnson. *Computers and Intractability – A Guide to the Theory of NP-Completeness.* W. H. Freeman, New York, 1979.
8. Y.V. Glebskii, D.I. Kogan, M.I. Liogon'kii, and V.A. Talanov. Range and Degree of Realizability of Formulas in the Restricted Predicate Calculus (in Russian). *Kibernetika*, 2:17–28, 1969. English translation in *Cybernetics*, vol.5 (1969), pp.142–154.
9. G. Gottlob. Complexity Results for Nonmonotonic Logics. *The Journal of Logic and Computation*, 2:3:397–425, June 1992.
10. G. Gottlob. NP Trees and Carnap's Modal Logic. In *Proc. 34th Annual Symp. on Foundations of Comp. Sci. (FOCS'93), Palo Alto, CA*, IEEE Computer Soc. Press, pp. 42–51, 1993.
11. J.Y. Halpern and B. Kapron. Zero-One Laws for Modal Logic. In *Proc. of the Seventh Annual IEEE Symposium on Logic in Computer Science*, 1992. Full version: Tech Report RJ 8836 (79218), 1992, IBM Research Division, Almaden Research Center. (See also ref. [12].)
12. J.Y. Halpern and B. Kapron. Zero-One Laws for Modal Logic. To appear in *Annals of Pure and Applied Logic.* (Updated version of [11] which takes into account the results of [10].)
13. J.Y. Halpern and Y. Moses. Towards A Theory of Knowledge and Ignorance: Preliminary Report. In *Proc. Non-Monotonic Reasoning Workshop AAAI*, pages 125–143, 1984. Reprinted in *Logics and Models of Concurrent Systems* K. Apt editor, Springer Verlag, 1985, pp.459–476 .
14. L. Hemachandra. The Strong Exponential Hierarchy Collapses. *Journal of Computer and System Sciences*, 39:299–322, 1989.
15. G.E. Huges and M.J. Cresswell. *An Introduction to Modal Logic.* Methuen London, 1978.
16. David S. Johnson. A Catalog of Complexity Classes. In *Handbook of Theoretical Computer Science*, chapter 2, pages 67–161, Elsevier Science Publishers B.V. (North-Holland), 1990.
17. Hector J. Levesque. Foundations of a Functional Approach to Knowledge Representation. *Artificial Intelligence*, 23(1984),pp.155–212, 1984.
18. W. Marek. Stable Theories in Autoepistemic Logic. *Fundamenta Informaticae*, 12:243–254, 1989.
19. W. Marek, G. Schwarz, and M. Truszczyński. Modal Nonmonotonic Logics: Ranges, Characterization, Computation. *Journal of the ACM*, 40(4):963–990, 1993.
20. W. Marek and M. Truszczyński. Autoepistemic Logic. *Journal of the ACM*, 38(3):588–619, 1991.
21. W. Marek and M. Truszczyński. *Nonmonotonic Logic.* Springer, Berlin 1993.
22. D. McDermott and J. Doyle. Nonmonotonic Logic II. *JACM* 29:1982, pp. 33-57.
23. D. McDermott and J. Doyle. Nonmonotonic Logic I. *Artificial Intelligence* 13:1980, pp. 41–72.

24. R.C. Moore. Semantical Considerations on Nonmonotonic Logics. *Artificial Intelligence*, 25:75–94, 1985.
25. I. Niemelä. Decision Procedure for Autoepistemic Logic. In *Proceedings of 9th Conference on Automated Deduction CADE-88*, pages 676–684, 1988.
26. I. Niemelä. On the Decidability and Complexity of Autoepistemic Reasoning. *Fundamenta Informaticae*, 17:117–155, 1992.
27. I. Niemelä. Towards Automatic Autoepistemic Reasoning. In *Proc. Europ. Workshop on Logics in AI, Amsterdam, September 1990*, Springer, 1991.
28. C.H. Papadimitriou and S. Zachos. Two Remarks on the Power of Counting. In *Proc. 6th GI Conf. on Theoretical Computer Science*, pages 269–276, Springer Verlag, 1983.
29. R. Reiter. On Asking what a Database Knows. In Lloyd, J. W., editor, *Symposium on Computational Logics*, pages 96–113, Springer Verlag, ESPRIT Basic Research Action Series, 1990.
30. R.C. Stalnaker. A Note on Nonmonotonic Modal Logic. 1980. Manuscript, Dept of Philosophy, Cornell Univ.
31. John E. Savage. *The Complexity of Computing*. John Wiley & Sons, 1976.
32. G. F. Shvarts. Autoepistemic Modal Logic of Knowledge. In *Logic Programming and Nonmonotonic Reasoning*, A.Nerode, W. Marek, and V.S. Subrahmanian,eds. MIT Press, Cambridge, Mass., 1991, pp. 260–274.
33. Klaus Wagner. Bounded Query Classes. *SIAM J. Comp.*, 19(5):833–846, 1990.

Compactness Properties of Nonmonotonic Inference Operations

Heinrich Herre
Institute of Computer Science
University of Leipzig
Augustusplatz 10-11, D-04109 Leipzig
e-mail: herre@informatik.uni-leipzig.de

Abstract

The aim of the present paper is to analyse compactness properties of non-monotonic inference operations within the framework of model theory. For this purpose the concepts of a deductive frame and its semantical counterpart, a semantical frame are introduced. Compactness properties play a fundamental in the study of non-monotonic inference, and in the paper several new versions of compactness are studied. It is proved that minimal reasoning in propositional logic is weakly supracompact and that the associated inference operator is uniquely determined by its finitary restriction via an extension operator Δ_4. Furthermore, some generalizations of these results to predicate logic are shown.
keywords : nonmonotonic inference, compactness, model theory

1 Introduction

A. Tarski[Ta56] suggested the study of abstract consequence operations as an important area of mathematical logic. He considered monotonicity and compactness to be properties of all such deductive operations. Given a language L a logic can be described on an abstract level by a functor $C : 2^L \to 2^L$ such that for every set $X \subseteq L$, $C(X)$ is understood to be the set of all the consequences of the set X of presmises. If C is monotonic and compact then C is uniquely determined by its restriction to finite sets $X \subseteq L$. This implies that the finitary part of a deductive operation captures already its relevant properties. The situation changes if the monotonicity requirement is weakened. Today, systems of nonmonotonic inference have won wide acceptance, especially in the area of artificial intelligence, for their ability to treat various forms of commensense reasoning. Logicians and computer scientists have studied inference operations that fail to satisfy monotony, and a broad class of types of nonmonotonic systems is now available.

After a decade of inventing new concrete inference relations there should be an interest to invest more efforts in a global theory to clarify the links between the different approaches. Important steps in this direction are presented by the work of ***Makinson***[Ma89], ***Freund,Lehmann, Makinson*** [FL90], and ***Kraus, Lehmann, Magidor*** [KL90]. In [KL90] the semantics of a nonmonotonic system was defined by cumulative models based on preferential relations. Preferential relations are used to select suitable models from the model class of a knowledge base. It seems, that the previously developed approaches are too narrow to capture all the relevant aspects of nonmonotonic reasoning. Most of the systems are concerned with propositional logic, but in applications of knowledge processing much richer languages are needed.

The present paper offers a broad framework for studying semantical and syntactical aspects of nonmonotonic reasoning. Here, we pursue the strategy to use methods from mathematical logic, in particular from model theory, to contribute to a theoretical foundation of non-monotonic reasoning. Semantics can be investigated in the framework of model theory which is a branch of mathematical logic dealing with the relation between a formal language and its interpretations. This relation is accomplished by giving a basic truth definition, which is the bridge connecting the formal language with its interpretations by means of models. The truth relation can be used to introduce an operator $Mod : 2^L \rightarrow 2^{Int}$ associating to every set $X \subseteq L$ a class $Mod(X)$ of interpretations. We arrive at nonmonotonic model theory if we add an operator $\Phi : 2^L \rightarrow 2^{Int}$ satisfying the condition $\Phi(X) \subseteq Mod(X)$. In this framework the wellknown results in [KL90], [FL93] can be reconstructed, and several new concepts can be formulated which cannot be grasped in previously systems. The interrelations between finitary inference operations and their infinitary extensions are not yet sufficiently elaborated. Here is a need to find new versions of compactness which allow to create tools for constructing new nonmonotonic logics and to characterize known nonmonotonic systems. The present paper is devoted to these topics. Our investigation takes a view which was inspired by the paradigm of abstract model theory [Ba85] and the ideas in [KL90].

Section 2 contains the background concerning pure conditions of nonmonotonic inference operations and notions from monotonic systems, including predicate calculus. In section 3 we propose a general framework for studying relevant aspects of nonmonotonic inference. The main notions are deductive frames and semantical frames. Also, the notion of a predicative semantical frame is introduced which gives a natural generalization of classical model theory to nonmonotonic model theory. In the main section 4 compactness properties are introduced and studied. Several new versions of compactness are investigated, and it is shown that these conditions are satisfied by minimal reasoning in propositional calculus, and a restricted version of predicate calculus. Finally, in section 5 we give a short summary of the results and outline problems for further research.

2 Background

Let L be a nonempty language whose elements are called formulas which are denoted by ϕ, ψ, χ. An operation $C : 2^L \rightarrow 2^L$ is called an *inference operation*, and the pair (L, C_L) is said to be an *inferential system*. The operation C_L represents the notion of logical inference. An inferential system (L, C_L) is a *closure system* and C_L a closure operation if it satisfies the following conditions:

$$
\begin{array}{ll}
X \subseteq C_L(X) & \text{(inclusion)} \\
C_L(C_L(X)) = C_L(X) & \text{(idempotence)} \\
X \subseteq Y \Rightarrow C_L(X) \subseteq C_L(Y) & \text{(monotony)}
\end{array}
$$

An inference operation C_L satisfies *compactness* if $\phi \in C_L(X)$ implies the existence of a finite subset $Y \subseteq X$ such that $\phi \in C_L(Y)$. A closure system (L, C_L) is a *deductive system* if C_L satisfies compactness; then C_L is said to be a *deductive operation*. A set $X \subseteq L$ is closed under C_L if $C_L(X) = X$. (L, Cn) denotes an inferential system based on classical logic; the natural example for such a system is classical propositional logic with standard semantics. An inference system (L, C) is said to be *partial* if C is not necessarily defined for all $X \subseteq L$; let $dom(C)$ be the domain of C, and $Fin(L) = \{X : X \subseteq L \text{ and } X \text{ is finite }\}$. A partial inference operation C is said to be *finitary* iff $dom(C) = Fin(L)$. $F \downarrow X$ denotes the restriction of a function F to the set $X \subseteq dom(F)$; define $C_f = C \downarrow Fin(L)$. The partial relation $\leq$ between inference operations C_1, C_2 is defined as follows: $C_1 \leq C_2 \Leftrightarrow \forall X(C_1(X) \subseteq C_2(X))$.

A semantics for a closure system (L, C_L) can be defined by a *logical system*. A *logical system* $(L, M, \models)$ is determined by a language L , a set (or class) M whose elements are called *worlds* and a *relation of satisfaction* $\models \subseteq M \times L$ between worlds and formulas. Given a logical system $(L, M, \models)$, we introduce the following notions. Let $X \subseteq L$, $Mod^{\models}(X) = \{m : m \in M \text{ and } m \models X\}$, where $m \models X$ iff for every $\phi \in X : m \models \phi$. Let $K \subseteq M$, then $Th^{\models}(K) = \{\phi : \phi \in L \text{ and } K \models \phi\}$, where $K \models \phi$ iff for all $m \in K : m \models \phi$. $C^{\models}(X) = \{\phi : Mod^{\models}(X) \subseteq Mod^{\models}(\phi)\}$, $X \models \phi$ iff $\phi \in C^{\models}(X)$. Obviously, $(L, C^{\models})$ is a closure system. The deductive system (L, C_L) is *correct* (*complete*) with respect to the logical system $(L, M, \models)$ iff $C_L(X) \subseteq C^{\models}(X)$ $(C_L(X) = C^{\models}(X))$. If $C^{\models}$ satisfies compactness then $C^{\models}$ can be defined by a system of derivation rules [Ra73].

The investigation of general properties of inference operations $C : 2^L \rightarrow 2^L$ that do not satisfy monotony is well-established. In [Ma 93] this field is surveyed and properties of many nonmonotonic inference operations are studied that have been proposed in the literature. A condition on inference operations is said to be *pure* if it concerns the operation alone without regard to its interaction with classical consequence operation and truth-functional or other connectives. The most important pure conditions are the following.

$$X \subseteq Y \subseteq C(X) \Rightarrow C(Y) \subseteq C(X) \quad \text{(Cut)}$$
$$X \subseteq Y \subseteq C(X) \Rightarrow C(X) \subseteq C(Y) \quad \text{(Cautious Monotony)}$$
$$X \subseteq Y \subseteq C(X) \Rightarrow C(X) = C(Y) \quad \text{(Cumulativity)}$$
$$X \subseteq C(Y) \text{ and } Y \subseteq C(X) \Rightarrow C(X) = C(Y) \quad \text{(Reciprocity)}$$

There exist certain formal interrelations between these conditions. Given inclusion, idempotence is a special case of cut. An inference operation C is cumulative iff C satisfies inclusion, cut and cautious monotony. Given inclusion, cumulativity is equivalent to reciprocity.

We conclude this section with an overview about notions from logic that are used in the following. A ***signature*** σ is a set of relational, functional and constant symbols. Let L be a first order language of a certain signature, $Const(L)$ the set of the constant symbols, $Rel(L)$ the set of the relational symbols and $Fun(L)$ the set of the function symbols of L. $Fm(L)$ is the set of formulas of L and a set $X \subseteq Fm(L)$ of formulas is called a ***theory***. A prenex formula is said to be ***universal*** if it contains no existential quantifier, a universal theory is a set of universal formulas.

The ***signature*** of a theory P, $sign(P)$, is the set containing the relational, functional and constant symbols appearing in P. $L(P)$ is the first order language associated to P. $B(P)$ denotes the ***Herbrand base*** of P, i.e. the set of all ground atoms of the signature of P. $U(P)$ is the set of all variable-free terms of $sign(P)$. We use the notion of an ***interpretation*** for P as a relational structure $\mathcal{A}$ which associates a declarative meaning to the symbols of $sign(P)$. A structure $\mathcal{A}$ is a ***model*** for P if every formula ϕ in P is true in $\mathcal{A}$, denoted by $\mathcal{A} \models \phi$. Let $Mod(P)$ be the class of all models of P in the language $L(P)$.A ***Herbrand structure*** for P in the language L is one for which the universe equals $U(P), L(P) \subseteq L$, and the function symbols have their canonical interpretation. A ***Herbrand model*** of P is a Herbrand structure in the language $L(P)$ being a model of P. Herbrand structures of the signature $sign(P)$ are determined by subsets $I \subseteq B(P)$. We identify Herbrand structures $\mathcal{A}$ with their associated set $I_{\mathcal{A}} \subseteq B(P)$. For a class of $\mathcal{K}$ of structures let $Th(\mathcal{K}) = \{F : F \in L(\mathcal{K}) \text{ and } \mathcal{K} \models F\}$. A Herbrand model I is said to be ***minimal*** for S if I is a model of S and no proper subsets of I represents a model of S.

3 Deductive frames

Alongside the three conditions of cut, cautious monotony and cumulativity ***Makinson*** [Ma93] emphasizes two mixed conditions of inference: ***supraclassicality*** and ***distributivity***. Supraclassicality is the condition that C extends the usual consequence operation Cn of classical logic, ie. $Cn(X) \subseteq C(X)$ for all $X \subseteq L$. He claims that if C behaves logically it has to be supraclassical and should satisfy certain additional absorption properties. This point of view seems to be too narrow: there are nonmonotonic systems having a logical interpretation but fail these properties. [1] However, there is a challenge behind *Makinsons's* claim, namely the problem whether any reasonable nonmonotonic system behaves logically with respect to a suitable (not necessarily classical) logic. To make this point precise we introduce several definitions.

Definition 1 *1. A system $\mathcal{IF} = (L, C_L, C)$ is said to be a inferential frame iff the following conditions are satisfied:*

(a) L is a language.

(b) C_L is a deductive inference operation on L, i.e. C_L satisfies inclusion , idempotence, monotony and compactness.

(c) C satisfies $C_L(X) \subseteq C(X)$ (supradeductivity) [2].

2. C satisfies left absorption iff $C_L(C(X)) = C(X)$; and C satisfies congruence or right absorption iff $C_L(X) = C_L(Y) \Rightarrow C(X) = C(Y)$. C satisfies full absorption if C satisfies left absorption and congruence.

3. An inferential frame $\mathcal{DF} = (L, C_L, C)$ is said to be a deductive frame if it satisfies full absorption. In this case C is said to be logical over C_L, and (L, C_L) is a deductive basis for C.

If C_L is not assumed to be compact then the system (L, C_L, C) is called a *weak inferential frame*, and analogously, a *weak deductive frame.*

Definition 2 *$\mathcal{LF} = (L, M, \models, \Phi)$ is a semantical frame iff*

1. $(L, M, \models)$ is a logical system.

2. $\Phi : 2^L \to 2^M$ is a functor such that $\Phi(X) \subseteq Mod^{\models}(X)$. Let $C_\Phi(X) = Th^{\models}(\Phi(X))$.

3. $\mathcal{LF}$ is said to be compact if $C^{\models}$ satisfies compactness.

C_Φ satisfies extension and left absorption, and hence $(L, C^{\models}, C_\Phi)$ is an inferential frame. We summarize further properties. Let (L, C_L, C) be a deductive frame. C satisfies *loop* if for any $n \in \omega$: $X_1 \subseteq C(X_2)$, $X_2 \subseteq C(X_3)$, ... $X_{n-1} \subseteq C(X_n)$, $X_n \subseteq C(X_1)$ implies $C(X_1) = C(X_n)$. C is *distributive* iff C is cumulative and satisfies the condition $C(X) \cap C(Y) \subseteq C(C_L(X) \cap C_L(Y))$. C satisfies *conditionalization* iff C is cumulative and satisfies the condition $C(X \cup Y) \subseteq C_L(X \cup C(Y))$. C is *rational* iff C satisfies conditionalization, cumulativity and the following condition: $X \subseteq Y$ and $C_L(C(X) \cup Y) \neq L \Rightarrow C(X) \subseteq C(Y)$.
In the following we consider two concrete families of semantical frames.

[1] *Pearce* [Pe93] analysed this point in detail. Systems failing supraclassicality include general systems of nonmonotonic reasoning extending standard nonclassical logic, eg. intuitionistic logic, and systems of logic programming which embody certain constructive forms of reasoning.

[2] *Pearce*[Pe93] discussed nonmonotonic extensions of the constructive inference rrelation $\vdash_N$, and introduced the notion of *supraconstructivity*. Our notion of *supradeductivity* is more general because it is not connected to a particular logic, and thus seems to cover all nonmonotonic systems having a logical interpretation. A similar view is also presented in [FL93].

Definition 3 *$\mathcal{LF} = (L, M, \models, \Phi)$ is a propositional semantical frame if the language L is built up from propositional variables and contains at least the functors $\wedge, \vee, \neg$. The semantics depends on the type of logic (classical logic, modal logic, temporal logic ,many-valued logics and partial logics).*

The language of a *predicative semantical frame* is constructed from a signature. Signatures are denotes by σ, τ. The class of τ-structures is denoted by $Str(\tau)$. A set $\mathcal{T}$ of signatures is closed if $\sigma \subseteq \tau, \tau \in \mathcal{T}$ implies $\sigma \in \mathcal{T}$ and $\sigma, \tau \in \mathcal{T}$ implies $\tau \cup \sigma \in \mathcal{T}$.

Definition 4 *A predicative semantical frame $\mathcal{LF} = (L, \mathcal{T}, \models_L, \Phi)$ is defined by following conditions:*

1. *$\mathcal{T}$ is a closed set of signatures.*
2. *L is a mapping defined on $\mathcal{T}$,$L(\tau)$ is a class of formulas.*
3. *$\models_{L(\tau)}$ is a relation $\subseteq L(\tau) \times Str(\tau)$, $\models_L = \bigcup_{\tau \in \mathcal{T}} \models_{L(\tau)}$.*
4. *$\tau \subseteq \sigma \Rightarrow L(\tau) \subseteq L(\sigma)$.*
5. *Isomorphism property: Let $\mathcal{A}, \mathcal{B} \in Str(\tau)$, $\phi \in L(\tau)$, then $\mathcal{A} \models_L \phi$ and $\mathcal{A} \simeq \mathcal{B}$ implies $\mathcal{B} \models_L \phi$.*
6. *Reduct property: $\mathcal{A} \in Str(\sigma), \phi \in L(\tau), \tau \subseteq \sigma$ and $\mathcal{A} \models_L \phi$ then $\mathcal{A} \downarrow \tau \models_L \phi$. Furthermore, if $\mathcal{A} \models_L \phi$ then $\phi \in L(\tau_{\mathcal{A}})$.*
7. *Renaming property. Let $F : \tau \rightarrow \sigma$ a renaming. Then for each $\phi \in L(\tau)$ there is a sentence ϕ^f from $L(\tau)$ such that $\mathcal{A} \models_L \phi$ iff $\mathcal{A}^f \models_L \phi^f$.*
8. *Φ is a functor defined on $\mathcal{T}$ such that for $\tau \in \mathcal{T}$ $\Phi(\tau)$ is a function from $2^{L(\tau)}$ in $2^{Str(\tau)}$ satisfying the condition $\Phi(\tau)(X) \subseteq Mod(X)$ for every $X \subseteq L(\tau)$.*

Let $\mathcal{K}$ be a function defined on $\mathcal{T}$ such that for every $\tau \in \mathcal{T}$ $\mathcal{K}(\tau) \subseteq Str(\tau)$. $\mathcal{K}$ is said to be *reductive* if for $\sigma \subseteq \tau$ the condition $\mathcal{K}(\sigma) = \{\mathcal{A} \downarrow \sigma : \mathcal{A} \in \mathcal{K}(\tau)\}$ is fulfilled. Φ is said to be *contextually independent* if there is a reductive function $\mathcal{K}$ such that for every set $X \subseteq L(\tau)$: $\Phi(X) = Mod(X) \cap \mathcal{K}(\tau)$. A predicative logical frame is said to be an *abstract logic* if the functor Φ is contextually independent. This notion coincides with the concept of *abstract logic* introduced in [KV93]. The following proposition is obvious.

Proposition 1 *Let $\mathcal{F} = (L, \mathcal{T}, \models_L, \Phi)$ a predicative semantical frame such that Φ is contextually independent. Then C_Φ is monotonic.*

The following proposition can be proved using Lindenbaum's standard technique, [Li91], [Di94].

Proposition 2 *Let (L, C_L, C) be an inferential frame satisfying left absorption. Then there exists a logical frame $(L, M, \models, \Phi)$ such that $C^{\models} = C_L$ and $C = C_\Phi$.*

Finally, we discuss the question whether a given inference operation C posesses a deductive base (L, C_0) which is maximal below C. In [Di94] such a maximal element is considered as the adequate underlying logic for C.

Proposition 3 (Di94) *(1)Let (L, C) be an inferential system satisfying inclusion. Then the set $\{C_0 : C_0 \leq C$ and (L, C_0, C) is an inferential frame satisfying left absorption $\}$ has a greatest element.*
(2)There are inferential systems (L, C) satisfying inclusion such that the set $\{C_0 : C_0 \leq C$ and (L, C_0, C) is a deductive frame$\}$ has no greatest element.

Proposition 4 (Di94) *If the inferential system (L, C) is cumulative then the set $\{C_0 : C_0 \leq C$ and (L, C_0, C) is a deductive frame $\}$ has a greatest element.*

4 Compactness Properties

Firstly, we given an overview about compactness properties which were studied in the literature. It is reasonable to distinguish two cases.

(1) C is monotonic. Then let $\Delta_0(C_f) = \bigcup_{Y \in Fin(X)} C_f(Y)$. Δ_0 can be considered as an operator extending finitary inference operation to infinitary ones, and if C is monotonic then $\Delta_0(C_f) \leq C$. If C is monotonic and compact then $\Delta_0(C_f) = C$, ie. C is uniquely defined by its finitary restriction via Δ_0. In case C is not compact $\Delta_0(C)$ gives an approximation of C from below. If C represents first order logic then $\Delta_0(C_f) = C$; if C is second order logic then $\Delta_0(C_f) \neq C$.

(2) C does not satisfy monotony. Then there is no operator Δ allowing to reconstruct every cumulative inference operation from its finitary restriction. However, one could seek for a family of such operators generating a class of inference operations from its finitary elements. Thus, it seems an interesting task to investigate such operators in a general setting. For this purpose the following notions are introduced. [3]

Definition 5 *Let (L, C_L) be a deductive system. $\mathcal{D}(L, C_L) = \{C : (L, C_L, C)$ is a deductive frame $\}$; $\mathcal{D}_f(L, C_L) = \{C : C$ is finitary and (L, C_L, C) is a deductive frame $\}$; $\mathcal{I}(L, C_L) = \{C : (L, C_L, C)$ is an inferential frame $\}$.*

1. *A functor $\Delta : \mathcal{D}_f \rightarrow \mathcal{I}(L, C_L)$ is said to be an extension operator if for every $C \in \mathcal{D}(L, C_L)$ the conditions $dom(\Delta(C)) = 2^L$ and $\Delta(C) \downarrow Fin(L) = C$ are satisfied. Δ is called deductive if $im(\Delta) \subseteq \mathcal{D}(L, C_L)$.*

2. *An inference operation $C : 2^L \rightarrow 2^L$ is Δ-compact iff $C \subseteq \Delta(C_f)$; C is completely Δ-compact iff $C = \Delta(C_f)$.*

[3] *Freund/Lehmann* [FL93] ask whether there is a reasonable notion of *pseudo-compactness* for which any finitary operation has a unique pseudo-compact extension to an infinitary one. It seems that there is no such notion. However, one should be more liberal and admit a well-behaved family of extension operators solving this problem.

3. *Let P be a property on deductive frames and $\mathcal{D}(P)$ the class of all deductive frames satisfying P. An extension operator Δ preserves the property P if for every $C \in dom(\Delta) \cap \mathcal{D}(P)$ the condition $\Delta(C) \in \mathcal{D}(P)$ is fulfilled.*

Abstract compactness properties can be expressed by conditions *compcond*($C_L, C, Fin(L)$) depending on C, C_L and the finite subsets of the language L. Important compactness properties are summarized in the following definition.

Definition 6 *Let (L, C_L, C) be a deductive frame.*

1. *C is weakly compact iff for every $X \subseteq L$, $\phi \in C(X)$ there is a finite subset $A \subseteq C_L(X)$ such that $\phi \in C(A)$.*

2. *C is supracompact iff for every $X \subseteq L$, $\phi \in C(X)$ there is a finite subset $A \subseteq X$ such that $\phi \in C(A \cup B)$ for every subset $B \subseteq C(A)$.*

3. *C is finitary supracompact iff for every $X \subseteq L$, $\phi \in C(X)$ there is a finite set $A \subseteq X$ such that for every finite $B \subseteq C_L(X)$ the condition $\phi \in C(A \cup B)$ is satisfied.*

4. *C is weakly supracompact iff for every $X \subseteq L$, $\phi \in C(X)$ and every finite $A \subseteq C_L(X)$ there is a finite set B, $A \subseteq B \subseteq C_L(X)$ such that $\phi \in C(B)$.*

The notion of supracompactness was introduced by *Freund*[91]; finitary supracompactness was studied in [FL93]. We collect some examples of extension operators, denoted by $\Delta_i, i = 1,2,3,4$. Let (L, C_L, C) be a deductive frame. Then the transform of C is defined as follows: $C^t(X) = \{\phi \in L$: there is a finite set $A \subseteq X$ such that for all $Y : A \subseteq Y \subseteq C(X) : \phi \in C(Y)\}$. It is easily to see that C^t is compact and $C_f = (C^t)_f$.

Definition 7 *Let (L, C_L, C) be a deductive frame, and F a finitary and cumulative inference operation which is logical over (L, C_L).*

1. *$\Delta_1(F)(X) = F(A)$, if there is a finite subset $A \subseteq C_L(X)$ such that $C_L(X) \subseteq F(A)$, and $\Delta_1(F)(X) = C_L(X)$ o.w.*

2. *$\Delta_2(F)(X) = \{\phi \in L$: there is a finite subset $A \subseteq X$ such that for every finite $B \subseteq C_L(X); \phi \in F(A \cup B)\}$.*

3. *Let $C_0 = \Delta_2(F)$, $C_{i+1} = (C_i)^t$, and $\Delta_3(F) = \bigcap_{i<\omega} C_i$.*

4. *$\Delta_4(F)(X) = \{\phi$: for every finite $A \subseteq C_L(X)$ there is a finite B such that $A \subseteq B \subseteq C_L(X)$ and $\phi \in F(B)\}$.*

The extension operators $\Delta_1, \Delta_2, \Delta_3$ were studied in [FL 93]. The deductive extension operator Δ_1 was invented by *Makinson* and has the following interesting property. Let (L, C_L, F) be a cumulative and finitary deductive frame. Then $\Delta_1(F)$ is the smallest extension of F such that $(L, C_L, \Delta_1(F))$ is a cumulative deductive frame. Unfortunately, all these three operators do not cover one of the most natural nonmonotonic systems: *minimal reasoning.* Minimal reasoning is not Δ_i-compact $i \in \{1,2,3\}$ and is neither supracompact nor finitary supracompact. The operator Δ_4 was introduced and presented in [He93]. In the following we show that the extension operator Δ_4 is suitable for analysing minimal reasoning in propositional and predicate logic.

Let $\mathcal{L}(Prop) = (L(Var), M, \models)$ the classical propositional logic with standard semantics, ie. $M = 2^{Var}$. The language $L(Var)$ is generated from the set Var of propositional variables by the functors $\{\wedge, \vee, \neg\}$. For a set $X \subseteq L(Var)$ let $var(X)$ be the set of all variables appearing in X. For $V \subseteq Var$ let $L(V)$ be the sublanguage of $L(Var)$ generated by V. Let $Mod_V(X)\{w \downarrow V : w \in Mod(X)\}$. As usual for $v, w \in M; v \leq w \Leftrightarrow (\forall p \in Var)(v(p) \leq w(p))$. If v, w are partial functions on Var then $v \sqsubseteq w$ iff $dom(v) \subseteq dom(w)$ and $v(p) = w(p)$ for every $p \in dom(v)$. A model v of X is minimal iff there is no model $w \in Mod(X)$ such that $w < v$: $Min(X)$ denotes the set of all minimal models of X. The inference operation C_m of minimal reasoning is defined by $C_m(X) = Th(Min(X))$, and let $X \models_m \phi$ iff $\phi \in C_m(X)$. The following three propositions are wellknown.

Proposition 5 *Let X be a set $\subseteq L(Var)$ and $V \subseteq Var$ a finite subset ov $var(X)$. Then there is a finite subset $B \subseteq C_L(X)$, $var(B) = V$ such that $Mod_V(X) = Mod_V(B)$.*

Proposition 6 *If $\phi \in L(V)$ then $X \models \phi$ iff $Mod_V(X) \subseteq Mod(\{\phi\})$.*

Proposition 7 *Every $v \in Mod(X)$ is extension of a minimal model of X.*

The following example by *J. Dietrich* shows that the inference operation of minimal reasoning C_m is not compact. Let $X = \{p_1 \wedge \ldots \wedge p_i \wedge (p_{i+1} \vee p_0) : 1 \leq i < \omega\}$. Then $X \models \neg(p_0 \leftrightarrow p_1)$. If $m \in Min(X)$ then $m \models \neg p_0, m \models p_1$, hence $m \models \neg(p_0 \leftrightarrow p_1)$. For every finite subset $X_f \subseteq X$ holds $X_f \not\models_{min} \neg(p_0 \leftrightarrow p_1)$. Using an idea of *J. Dietrich* one can prove the following

Proposition 8 *The propositional semantical frame $\mathcal{LF} = (\mathcal{L}(Prop), \Phi_m)$, where $\Phi_m(X) = Min(X)$, is weakly supracompact.*

Proof Let $X \models_m \phi$, $A \subseteq C_L(X)$, A finite and $var(A) \cup var(\phi) = \{p_1, \ldots, p_s\} = V$. By prop. 5 there is a finite subset $B \subseteq C_L(X)$ such that $var(B) = \{p_1, \ldots, p_s\}$ and $Mod_V(X) = Mod_V(A \cup B)$. Let $w \in Min_V(A \cup B)$, then $w \in 2^V$. w can be extended to a model $d \in Mod(X)$. By prop. 7 there is a minimal model $e \in Min(X)$ such that $e \leq d$. By assumption $e \models \phi$. It is $e \downarrow V = w$, since w is minimal for $A \cup B$. By prop.6: $w \models \phi \Leftrightarrow e \models \phi$, hence $A \cup B \models_m \phi$. □

Proposition 9 *Let $X \subseteq L(Var)$ and $\phi \in L(Var)$. Following conditions are equivalent:*

1. $X \models_m \phi$,
2. *for every finite subset $A \subseteq C_L(X)$ there exists a finite subset $B \subseteq C_L(X)$ such that $var(B) \subseteq var(A)$ and $A \cup B \models_m \phi$.*

Proof: The implication (1) → (2) follows immediately from prop.8.

We show (2) → (1). Let $var(\phi) = \{q_1, \ldots, q_m\}$ and for any finite subset $A \subseteq C_L(X)$ there is a $B \subseteq C_L(X)$ such that $var(B) \subseteq var(A)$ and $A \cup B \models_m \phi$. Using prop.5 one can construct an infinite sequence $A_1, A_2, \ldots, A_n, \ldots$ of finite sets $A_i \subseteq C_L(X)$ such that $var(A_i) \subseteq var(A_{i+1})$, $var(\bigcup_{i\in\omega} A_i) = var(X)$, and $Mod_{var(A_i)}(A_i) = Mod_{var(A_i)}(X)$. Let $v(i) =_{df} var(A_i)$. Firstly, we show the following.

1. $C_L(\bigcup_{i\in\omega} A_i) = C_L(X)$.
2. for all $j > i$: $Mod_{v(i)}(A_i) = Mod_{v(i)}(A_j)$.

2.: It is $Mod_{v(i)}(X) \subseteq Mod_{v(i)}(A_j)$, since $X \models A_j$. On the other hand, $Mod_{v(j)}(X) \subseteq Mod_{v(i)}(A_i)$, since $X \models A_j$.

1.: It is sufficient to show: $C_L(X) \subseteq C_L(\bigcup_{i\in\omega} A_i)$. Let be $X \models \phi$, and $var(\phi) = \{q_1, \ldots, q_m\}$. By construction there is a $j < \omega$ such that $var(\phi) \subseteq v(j)$; hence, by prop.6 and since $Mod_{v(j)}(A_j) = Mod_{v(j)}(X)$, finally $A_j \models \phi$.

By assumption for every A_i there is a $B_i \subseteq C_L(X)$ such that $var(B_i) \subseteq var(A_i)$ and $A_i \cup B_i \models_m \phi$. Let be $D_i = A_i \cup B_i$; denote $va(i) =_{df} var(D_i)$. Obviously, the sequence $D_1, D_2, \ldots$ satisfies the following conditions:

1. $D_i \models_m \phi$;
2. $var(\phi) \subseteq D_i$ and $va(i) \subseteq va(i+1)$, and $var(\bigcup D_i) = var(X)$;
3. $Mod_{va(i)}(D_i) = Mod_{va(i)}(X)$, and hence $Mod_{va(i)}(D_i) = Mod_{va(i)}(D_j)$, $j > i$.
4. $C_L(\bigcup D_i) = C_L(X)$.

We prove that $X \models_m \phi$. Let $v \in Min(X)$. It is sufficient to show that there is a $i < \omega$ such that $v \downarrow va(i)$ is minimal for D_i (since $v \downarrow var(D_i) \models \phi$ implies $v \models \phi$ and $D_i \models_m \phi$). This is proved by contradiction.

Assume $v \in Min(X)$, but $v \downarrow var(D_i)$ is not minimal for D_i for every $i < \omega$. Let $v_i = v \downarrow var(D_i)$ and $\Delta(i) = \{\rho : \rho \in Mod_{va(i)}(D_i)$ and ρ is minimal and $\rho < v_i$. Then $\Delta(i) \neq \emptyset$ for every $i < \omega$. Every $\lambda \in \Delta(i)$ can be extended to a model $\sigma \in \Delta(i+1)$. This follows from the condition $Mod_{va(i)}(D_i) = Mod_{va(i)}(D_{i+1})$. Let $\lambda \sqsubseteq \sigma$, $\sigma \in Mod(D_{i+1})$ and σ is minimal in D_{i+1}. Then $\sigma \downarrow var(D_i)$ is a model of D_i. It cannot be $\sigma \downarrow var(D_i) < \lambda$, hence $\lambda = \sigma \downarrow var(D_i)$. Since $\sigma \downarrow var(D_i) < v_i$ it follows $\sigma < v_{i+1}$, hence $\sigma \in \Delta(i+1)$.

We define now a sequence $\rho_1 \sqsubseteq \rho_2 \sqsubseteq \ldots \sqsubseteq \rho_n \sqsubseteq \ldots$ and let $\rho = \bigcup \rho_i$. Then (1) $\rho \models X$ and (2) $\rho < v$.

Condition (2) is clear from the construction. We show (1). It is sufficient to prove the following: if $X \models \psi$, then $\rho \models \psi$, (hence $\rho \models X$). Let $X \models \psi$, then $\bigcup D_i \models \psi$. Then there is a finite subset $Y \subseteq \bigcup D_i$, such that $Y \models \psi$. From this follows the existence of $j < i$ such that $D_j \models \psi$. Since $\rho \downarrow var(D_j) \in Mod_{va(j)}(D_j)$ it follows $\rho \downarrow var(D_j) \models \psi$ and because of $\rho \downarrow var(D_j) \sqsubseteq \rho$, finally $\rho \models \psi$. This gives a contradiction. □

Proposition 10 (Corollary) *Let C be the inference operator of minimal reasoning in classical propositional logic. Then C is completely Δ_4-compact, ie. $C = \Delta_4(C_f)$.*

Proof: Let $C_1 = \Delta_4(C_f)$: We prove that for every subset $X \subseteq L : C(X) = C_1(X)$.

(1) $C(X) \subseteq C_1(X)$. Let $\phi \in C(X)$, then for every finite subset $A \subseteq Cn(X)$ there is a finite set $B \subseteq Cn(X)$ such that $\phi \in C(A \cup B)$. This implies the condition $\phi \in C_1(X)$ by proposition 9 and the construction of Δ_4.

(2) $C_1(X) \subseteq C(X)$. Let $\phi \in C_1(X)$. By definition: for every finite $A \subseteq Cn(X)$ there is a finite subset $B \subseteq Cn(X)$ such that $A \subseteq B$ and $\phi \in C_f(B)$. Since B is finite, then $C_f(B) = C(B)$. This implies by prop.9 $\phi \in C(X)$. □

Now we investigate the problem whether these propositions can be generalized to first order predicate logic. Let $\mathcal{L}(Pred) = (L, \bigcup_{\tau \in \mathcal{T}} Str(\tau), \models_L)$ be the logical system corresponding to first order predicate logic over the signatures from $\mathcal{T}$. $\mathcal{LF}(Pred) = (\mathcal{L}(Pred), \Phi_m)$ is the predicatively logical frame determined by $\mathcal{L}(Pred)$ and the operator Φ_m which selects minimal models, ie. $\Phi_m(X) = \{\mathcal{B} : \mathcal{B} \text{ is a minimal model of } X\}$. The minimality of a model is defined as follows. $\mathcal{B} = (U, (R_i)_{i \in I}, (f_j)_{j \in J})$ is a minimal model of X if $\mathcal{B} \models X$ and for every $\mathcal{C} = (U, (R_i^*)_{i \in I}, (f_j)_{j \in J})$: if $R_i^* \subseteq R_i$ for all $i \in I$ and there is an $i \in I$ such that $R_i^* \neq R_i$ then $\mathcal{B} \not\models X$.

Proposition 11 *The predicative semantical frame $\mathcal{LF}(Pred) = (\mathcal{L}(Pred), \Phi_m)$ is not weakly compact.*

Proof: Let σ the following signature: $\sigma = (\cup, \cap, -, I(x), 0, 1)$, where $\cup, \cap$ are binary operation symbols, $-$ is a unary operation symbol, I is a unary predicate, and $0, 1$ are constants. Let $Ax(BA)$ be a finite set of sentences axiomatizing the theory of atomic Boolean algebras. We introduce following definitions and formulas:

- $at(x) \leftrightarrow x$ *is an atom*
- $atomic(x) \leftrightarrow x = sup(y : y \leq x \wedge at(y))$.
- $\phi_1 := \forall xy(I(x) \wedge I(y) \rightarrow I(x \cup y)) \wedge \forall xy(I(x) \rightarrow I(x \cap y))$
- $\phi_2 := \forall x(at(x) \rightarrow I(X))$

- $a(i) \leftrightarrow$ *there exist at least i many atoms*
- $a(\infty) = \{a(i) : i \in \omega\}$

Obviously, all these conditions can be formalized in first order predicate logic. Let T be the following theory: $T = \{\phi_1, \phi_2\} \cup Ax(BA) \cup a(\infty)$. Let ϕ be the following sentence: $\phi_1 \wedge \phi_2 \wedge \bigwedge\{Ax(BA)\} \wedge \forall x \exists y (I(x) \rightarrow (x < y \wedge I(y))$.

Then $T \models_m \psi$. Let $\mathcal{A} = (U, \cup, \cap, -, 0, 1, I)$ be a minimal model of T. Then $\mathcal{A}$ is an atomic Boolean algebra having an infinite number of atoms and I is ideal. Since I is a minimal predicate I is a proper ideal and hence I has no maximal element. From this follows that $\forall x \exists y (I(x) \rightarrow (x < y \wedge I(y))$.

We show that for every finite subset $S_0 \subseteq T^{\models}$ holds $S_0 \not\models_m \psi$. Let $S_0 \subseteq T^{\models}$ be a finite subset; then there is a finite subset $T_0 \subseteq T$ (by compactness) such that $T_0 \models S_0$. From this follows that there is a number n such that $S_0 \not\models a(j)$ for every $j \geq n$. We show that S_0 has a finite model. Let $a(j_0) \in T_0$, j_0 maximal. Then there is a Boolean algebra with j_0 many atoms, and since $\mathcal{B}$ is atomic, is finite. From $T_0 \models S_0$ follows $\mathcal{B} \models S_0$. Let $\mathcal{B}_1$ be a minimal model of S_0: obviously, $\mathcal{B}_1 \not\models \psi$. □

If we restrict T and ψ in a suitable way, then weak compactness is true. Let T be a universal theory of signature σ, ϕ a quantifier free sentence; $B(T)$ the set of all atomic sentences. Then ϕ is a Boolean combination of sentences from $B(T)$: let $At(\phi)$ the set of all atomic sentences appearing in ϕ. A model I of T can be represented by a subset $I \subseteq B(T)$ such that $I \models T$. For a set $P \subseteq B(\sigma)$ and $I \subseteq B(\sigma)$ let $I \downarrow P =_{df} I \cap P$.

Proposition 12 *Let $I \subseteq B(\sigma)$, ϕ a P-formula, ie. $at(\phi) \subseteq P$: then $I \models \phi$ if and only if $I \cap P \models \phi$.*

Proof: Let $at(\phi) = \{A_1, \ldots, A_k\}$. We prove this condition inductively on the complexity of ϕ. 1.: ϕ is atomic; then $\phi := A$. Then $I \models A$ iff $A \in I$ and since $A \in P$ then $A \in A \cap P$.
2.: $\phi := \phi_1 \wedge \phi_2$, and $at(\phi) \subseteq P$. Let $I \models \phi_1 \wedge \phi_2$, then $I \models \phi_1, \phi_2$ and, using induction hypothesis $I \cap P \models \phi_1, \phi_2$, hence $I \cap P \models \phi$. And if $I \cap P \models \phi_1 \wedge \phi_2$, then $I \cap P \models \phi_1, \phi_2$ and by i.h. $I \models \phi_1 \wedge \phi_2$. Let $\phi := \neg\psi$. Assume $I \models \neg\psi$ then $I \not\models \psi$ iff $I \cap P \not\models$ (by i.h.), and hence $I \cap P \models \neg\psi$.
3.:Let $\phi := \phi_1 \vee \phi_2$. If $I \models \phi_1 \vee \phi_2$ then $I \models \phi_1$ or $I \models \phi_2$. If $I \models \phi_1$ then by i.h. $I \cap P \models \phi_1$ and hence $I \cap P \models \phi_1 \vee \phi_2$. Let $I \cap P \models \phi_1 \vee \phi_2$, then $I \cap P \models \phi_1$ or $I \cap P \models \phi_2$, by i.h. $I \models \phi_1$ or $I \models \phi_2$, hence $I \models \phi_1 \vee \phi_2$. □.

For every finite subset $P \subseteq B(\sigma)$ and universal theory T let $Mod_P(T) = \{I \cap P : I \in Mod(T)\}$. The following proposition is obvious.

Proposition 13 *Let T be a universal theory, $P \subseteq B(\sigma)$ a finite set. Then there is a quantifier free sentence ϕ such that $at(\phi) \subseteq P$, and $Mod_P(\phi) = Mod_P(T)$.*

Proposition 14 *Let T be a set of universal sentences, and ϕ a quantifier free sentence. Then the following holds: if $T \models_m \phi$ then there exists a finite subset $A \subseteq Th(T)$ such that $A \models_m \phi$.*

Proof: Let $at(\phi) = P$. Then there is a sentence ψ such that $at(\psi) = P$ and $Mod_P(\psi) = Mod_P(T)$. Obviously, $T \models \psi$. We show $\{\psi\} \models_m \phi$. Let I be a partial minimal model of ψ. Then by $Mod_P(\psi) = Mod_P(T)$ I can be extended to a two valued model J of T. J is an extension of a minimal model J_0. Then $J_0 \downarrow P = I$ (this follows from the fact that J is an extension of I, if $J_0 \downarrow P < I$ then this contradicts the minimality of I (because $J_0P \in Mod_P(\psi)$). Since J_0 is minimal, by assumption $J_0 \models \phi$ and since $at(\phi) = P$ finally $I \models \phi$. Altogether, $\{\psi\} \models_m \phi$. □.

5 Conclusions and Problems

We have presented a framework for studying nonmonotonic inference operations. This framework is based on the notion of a deductive frame, representing the syntactical level, and on the notion of a semantical frame, describing the semantics of a nonmonotonic system. The deductive basis which represents a monotonic logic and the nonmonotonic inference operation are considered as equally important. This abstract handling of monotonic logic is even more general than the approach in [FL93]. Then we introduced several new versions of compactness allowing to analyse such natural nonmonotonic systems as minimal reasoning in propositional and predicate calculus; the previously invented concepts of compactness do not cover these systems. We conclude this section with a collection of open problems.

Representation problems
Which cumulative inference systems (L, C) can be expanded to deductive frames (L, C_L, C) satisfying distributivity, conditionalization, rationality? Which properties are satisfied by natural nonmonotonic systems? Find a characterisation theorem for minimal reasoning in propositional calculus. The paradigm of such a kind of theorem is given by classical propositional logic: Let (L, C) be a deductive base, $L = L(\wedge, \vee, \neg)$ the language of propositional calculus. Assume that C satisfies inclusion, monotony, idempotence, the deduction theorem, and $C(\emptyset) =$ set of all tautologies. Then C is classical propositional logic.

Compactness properties.
Find natural construction Δ which extends finitary i.o. to infinitary ones. Of importance is a notion of *rule compactness*. One may try to associate for every deductive base (L, C_L, C) a system $\mathcal{R}$ of finitary nonmonotonic derivation rules. Then define an an inference operation $C_{\mathcal{R}}$ based on these rules. Let (L, C) be cumulative. Does there exist a deductive base (L, C_L) such that (L, C_L, C) is a deductive frame and C is Δ-compact for a certain extension operator Δ?

Application to logic programs.
By a refinement of the abstract theory of nonmonotonic inference operations one can investigate nonmonotonic operators, as for example, introduced in [Dx92]. Then there is the open problem of finding an adequate underlying logic, say for the well-founded semantics of normal logic programs.

References

[Ba85] Barwise,J.,S.Feferman : Model-Theoretic Logics, Springer-Verlag, 1985

[Bl91] Bell,J.: Pragmatic Logics, in: Proc. of 2nd. Int. Conference of Knowledge Representation and Reasoning, Cambridge MA, 1991

[Di94] Dietrich,J.: Deductive Bases of Nonmonotonic Inference Operations; NTZ Report, Universit"a Leipzig, 1994

[DH94] Dietrich,J., Herre,H.: Outline of Nonmonotonic Model Theory; NTZ Report, Universit"at Leipzig, 1994

[FL90] Freund,M., Lehmann,D., Makinson,D.: Canonical Extensions to the Infinite Case of Finitary Nonmonotonic Inference Operations; Arbeitspapiere der GMD 443, 1990

[Dx91] Dix,J.: Nichtmonotones Schliessen und dessen Anwendung auf Semantiken logischer Programme, Doctoral Dissertation, Karlsruhe, 1992

[Fr91] Freund,M.: Supracompact inference operations, LNCS vol.543, 59-73 (1991)

[FL93] Freund,M. D. Lehmann: Nonmontonic inference operations; Preprint, Hebrew University of Jerusalem, 1993

[He91] Herre,H.: Nonmonotonic Reasoning and Logic Programs NIL'91, LNCS vol. 543, p. 38-58

[He93] Herre,H.:Contributions to nonmonotonic model theory, Workshop on Nonclassical Logics in Computer Science, Dagstuhl-Berichte (ed. Marek,V.,A. Nerode, P.H.Schmitt), 1993

[KL90] Kraus,S., D. Lehmann, M. Magidor: Nonmonotonic Reasoning, Preferential models and cumulative logics; A.I. 44 (1990), 167 - 207

[KV93] Kolaitis,P.G., J.A.Väänänen: Generalized Quantifiers and Pebble Games on Finite Structures, Preprint, University of Helsinki, 1993

[Ld69] Lindström,P.: On extensions of elementary logic, Theoria, 35, 1-11 (1969)

[Li91] Lindström,S.:A semantic approach to nonmonotonic reasoning: inference operations and choice; Dept. of Philosophy, Uppsala University, Preprint, 1991

[Ma89] Makinson,D.: General Theory of Cumulative Inference,in: Reinfrank,M.(Ed.) Non-monotonic Reasoning,LNAI vol. 346 Berlin Springer-Verlag, 1989, 1-18

[Ma93] Makinson,D.: General Patterns in Nonmonotonic Reasoning; in: D. Gabbay (ed.) Handbook of Logic in Artificial Intelligence and Logic Programming, Oxford University Press, 1993

[Pe93] Pearce,D.: Remarks on Monmonotonicity and Supraclassicality, Preprint,Berlin, 1993

[Ra79] Rautenberg,W.: Klassische und nichtklassische Aussagenlogik, Vieweg, 1979

[Ta56] Tarski,A.: Logic, Semantics, Metamathematics. Papers from 1923 -1938. Clarendon Press, Oxford, 1956

[Th89] Thiele,H.: Monotones und nichtmonotones Schliessen; in: Grabowski,J., Jantke,H.-J., H. Thiele: Grundlagen der Kuenstlichen Intelligenz, 80 - 160, Akademie-Verlag, Berlin 1989

Around a Powerful Property of Circumscriptions

Yves Moinard[1] and Raymond Rolland[2]

[1] IRISA, Campus de Beaulieu, 35042 RENNES-Cedex, FRANCE, tel.: (33) 99 84 73 13, E-mail: moinard@irisa.fr

[2] IRMAR, Campus de Beaulieu, 35042 RENNES-Cedex, FRANCE, tel.: (33) 99 28 60 19, E-mail: Raymond.Rolland@univ-rennes1.fr

Abstract. The notion of preferential entailment has emerged as a generalization of circumscription. Here we study only "classical" preferential entailment, i.e. the underlying logic is classical. Also, as we want to apply these results to the various notions of circumscriptions which have been defined in the literature, we examine which preferential entailments can be considered as a kind of circumscription. Among the results given are a precise study of cumulativity (with respect to well-foundedness), even for preferential entailments based on non transitive relations, and the isolation of a property called "reverse monotony", which is a fundamental characteristic of any preferential entailment associated to a circumscription. We prove that the main properties of circumscriptions can be considered as corollaries of reverse monotony. We introduce also a new property of circumscriptions, "disjunctive coherence", which is not a consequence of reverse monotony.

1 Introduction

We begin with a reminder about preferential entailment. We need preferential relations which enjoy none of the properties which are generally considered as necessary. For example, we need to consider relations which are not transitive or which are neither antireflexive nor antisymetrical if we want to encompass the various existing circumscriptions. We precise the results about cumulativity, reasoning by cases, and related topics. Circumscriptions are "syntactic" by nature (even if they do posess a semantics, which is precisely given by preferential entailment). Thus, a preference relation naturally associated to a circumscription cannot be defined from any relation between models: it must be somehow coherent with the syntax. We investigate a powerful property of circumscription, reverse monotony, and we introduce a new property: disjunctive coherence. We show that idempotence, the principle of deduction, the ability to reason by cases, and transitive cumulativity are all corollaries of reverse monotony. Also, we provide a few examples with predicate and pointwise circumscriptions. These examples are chosen to prove rigourously some counter-properties of circumscriptions.

2 Preference Relations

We refer the reader to [BS88], [Sho88], [Mak88] and [KLM90].

Notations 2.1 $\mathcal{L}$ denotes a logical language, generally first order, but some incursions in second order will be precised when necessary. $\mathcal{T}$ is a theory (i.e. a set of formulas) in $\mathcal{L}$. If Φ is a formula (resp. $\mathcal{T}'$ a theory) in $\mathcal{L}$, and μ an interpretation over $\mathcal{L}$, we note as usual $\mu \models \mathcal{T}$ if μ is a model of $\mathcal{T}$ and $\mathcal{T} \models \Phi$ (resp. $\mathcal{T} \models \mathcal{T}'$) if any model of $\mathcal{T}$ is a model of Φ (resp. $\mathcal{T}'$). We note $Th(\mathcal{T})$ for the entailment closure of $\mathcal{T}$. $\mathcal{L}$ also contains the equality symbol "=" and an infinite number of individual variables. $\equiv$ is a metasymbol meaning "equivalent to". If μ is an interpretation over $\mathcal{L}$, if P is a predicate symbol, Φ a formula, f a function symbol, all of arity k, in $\mathcal{L}$, then:
- The set D_μ denotes the domain of μ.
- The subset $|P|_\mu$ (or $|\Phi|_\mu$) of D_μ^k denotes the extension of P (or Φ) in μ.
- The application f_μ from D_μ^k to D_μ denotes the interpretation of f in μ.
- $Th(\mu)$ denotes the (complete) theory made of all the formulas in $\mathcal{L}$ which are true in μ.

Our models and interpretations are *normal* (= interpreted as identity). $\mathcal{L}_\mu$, the *language of* μ, adds to $\mathcal{L}$ a name for each element in D_μ.

Considering some *preferred* models among the models of a theory $\mathcal{T}$ is useful in knowledge representation if we want to deal with incomplete information or rules with exceptions. Here are the precise definitions involved:

Definition 2.2 A binary relation $\prec$ among the interpretations over $\mathcal{L}$ is a *preference relation*. A model μ of a theory $\mathcal{T}$ is *a model of* $\mathcal{T}$ *minimal for* $\prec$ (or *is minimal for* $(\mathcal{T}, \prec)$) iff there exists no model ν of $\mathcal{T}$ with $\nu \prec \mu$; in this case we note $\mu \models_\prec \mathcal{T}$. If any model of $\mathcal{T}$ minimal for $\prec$ is a model of ϕ (resp. $\mathcal{T}'$), we note $\mathcal{T} \models_\prec \phi$ (resp. $\mathcal{T} \models_\prec \mathcal{T}'$).

We call *preferential entailment* the relation $\models_\prec$ between a theory $\mathcal{T}$ and a formula ϕ. Each preference relation $\prec$ gives rise to one preferential entailment $\models_\prec$, but each preferential entailment may be associated to various preference relations. If some preference relation $\prec$ associated to $\models_\prec$ is antireflexive, antisymetrical and transitive, $\models_\prec$ is an *ordered preferential entailment.*

We do not require antireflexivity nor transitivity for preference relations.

Notations 2.3 For any preference relation $\prec$, we note $\overline{\prec}$ for its transitive closure and $\prec_\mathcal{T}$ for the restriction of $\prec$ to the class of the models of some theory $\mathcal{T}$. A model of $\mathcal{T}$ may be minimal for $\prec$ and not for $\overline{\prec}$.

An important notion for circumscription is the so-called *well-foundedness*. This notion appears with a lot of different names, and some differences, in the literature. Here are the notions useful for circumscriptions.

Definition 2.4 A theory $\mathcal{T}$ is *well-founded for a preference relation* $\prec$ iff for any model μ of $\mathcal{T}$ not minimal for $\prec$, there exists ν minimal for $(\mathcal{T}, \prec)$ with $\nu \prec \mu$.

Theorem 2.1 [Moi92] If $\mathcal{T}$ is well-founded for $\prec$ and if $\mathcal{T} \models_{\prec} \phi$, then $\mathcal{T} \cup \{\phi\}$ is well-founded for $\prec$.

For an arbitrary ϕ, $\mathcal{T}$ may be well-founded while $\mathcal{T} \cup \{\phi\}$ is not (see example 6.1). Theorem 2.1 explains why the results given in [Mak88] or [KLM90] for their stronger notion of well-foundedness for *any theory* are also true with the more reasonable condition of well-foundedness *of a given* $\mathcal{T}$.

Definition 2.5 A theory $\mathcal{T}$ is *well-behaved for* $\prec$ iff for any sentence ϕ we have: *if* $\mathcal{T} \models_{\prec} \phi$, *then* the models minimal for $(\mathcal{T} \cup \{\phi\}, \prec)$ are the models minimal for $(\mathcal{T}, \prec)$.

Lemma 2.2 [MR94] If $\mathcal{T}$ is well-founded for $\prec$, then $\mathcal{T}$ is well-behaved.

It is rare that a theory is well-founded for a non transitive $\prec$. Some circumscriptions give naturally rise to non transitive relations, e.g. strong circumscription [Moi92]. The relation naturally arising from pointwise circumscription [Lif88] is *antitransitive*: if $\mu_1 \prec \mu_2$ and $\mu_2 \prec \mu_3$ then $\mu_1 \not\prec \mu_3$. Also, some notions of common sense reasoning can be rendered by a preferential entailment only if the associated $\prec$ is not transitive [Ryc90]. Thus it is important to give a milder definition in this case. We need two new definitions as the most natural one (the first one) does not satisfy theorems 2.1 and 4.7:

Definition 2.6 A theory $\mathcal{T}$ is *weakly well-founded for* $\prec$ iff for any model μ of $\mathcal{T}$ not minimal for $(\mathcal{T}, \prec)$, there exists ν minimal for $(\mathcal{T}, \prec)$ with $\nu \overline{\prec_{\mathcal{T}}} \mu$.

Definition 2.7 A theory $\mathcal{T}$ is *mildly well-founded for* $\prec$ iff it is weakly well-founded and for any sentence ϕ such that $\mathcal{T} \models_{\prec} \phi$ and any models μ and ν minimal for $(\mathcal{T} \cup \phi, \prec)$ we do *not* have $\nu \overline{\prec_{\mathcal{T}}} \mu$.

Property 2.3 [MR94] $\mathcal{T}$ is mildly well-founded iff $\mathcal{T}$ is weakly well-founded and well-behaved.

Remarks 2.1 • If $\prec$ is transitive then the three kinds of well-foundedness coincide.
• Any well-founded theory is mildly well-founded, and any mildly well-founded theory is weakly well-founded, but the converses are not guaranteed.

Definition 2.4 is the classical meaning of the expression "well-founded" in the literature about circumscription: see e.g. [EMR85, Lif86]. [Mak88] or [KLM90] define similar notions, using the terms *stoppered* or *smooth*. In mathematical texts, "well-founded" ([CK73, p.150]) prevents the existence of infinitely decreasing chains, which is not the case here. *Bounded* in [Sho88] corresponds to this "mathematical meaning". If the preference relation $\prec$ is transitive, boundedness implies well-foundedness. If $\prec$ is not transitive, boundedness only implies weak well-foundedness (converse false, see example 6.5); boundedness implies neither well-foundedness nor mild well-foundedness.

Note that we have given a *"unitary version"*, we could also give the stronger "general – or infinitary – version" of well-behaveness (in the finite propositionnal case, these two notions coincide):

Definitions 2.8 [MR94] A theory $\mathcal{T}$ is *well-behaved for $\prec$ (infinitary version)* iff for any theory $\mathcal{T}'$ in $\mathcal{L}$, we have: if ($\mathcal{T} \models_{\prec} \mathcal{T}'$ and $\mathcal{T}' \models \mathcal{T}$) then a model is minimal for $(\mathcal{T}, \prec)$ iff it is minimal for $(\mathcal{T}', \prec)$.
Thanks to property 2.3, there is a corresponding *"general"* – or *"infinitary"* – version of mild well-foundedness.

3 Circumscriptions

We precise now what we call a *circumscription*, introduced [McC80, McC86] in order to express some commonsense problems formally.

Definitions 3.1 A process which, to any theory $\mathcal{T}$, closed for entailment, associates a new theory $\overline{\mathcal{T}} = Th(\mathcal{T} \cup \mathcal{T}^c)$ for some theory $\mathcal{T}^c$ is what we call a *pre-circumscription.*

If $\mathcal{T}$ is any theory in $\mathcal{L}$, we define $\overline{\mathcal{T}}$ as: $\overline{Th(\mathcal{T})}$.

A *"circumscription"* is a pre-circumscription for which there is an associated preference relation $\prec$ such that, for any theory $\mathcal{T}$ in $\mathcal{L}$ and any interpretation μ over $\mathcal{L}$: $\mu \models \overline{\mathcal{T}}$ iff $\mu \models_{\prec} \mathcal{T}$. $\mathbf{Circ_g}(\mathcal{T})$ denotes some circumscription of $\mathcal{T}$. □

Thus, for any circumscription $\mathbf{Circ_g}$, there exist relations $\prec$ such that for any theory $\mathcal{T}$ and any formula Φ in $\mathcal{L}$: $\mathbf{Circ_g}(\mathcal{T}) \models \Phi$ iff $\mathcal{T} \models_{\prec} \Phi$.

Conversely, to any preference relation $\prec$ and to any theory $\mathcal{T}$, we may associate the unique set $\mathcal{T}'$ whose elements are all the formulas ϕ in $\mathcal{L}$ which are true in all the models of $\mathcal{T}$ minimal for $\prec$. Clearly, $\mathcal{T}' \models \mathcal{T}$, thus, with our definitions, any preferential entailment is a pre-circumscription. The converse is false, as proved by the examples 5.1 and 5.2 below. Note that any model minimal for $(\mathcal{T}, \prec)$ is a model of $\mathcal{T}'$, but that the converse is not guaranteed, i.e., a preference relation $\prec$ is not necessarily "coherent" with respect to the syntax:

Definition 3.2 A relation $\prec$ is *definability preserving* [1] iff for any first order theory $\mathcal{T}$, the class $\mathcal{M}_{\prec}(\mathcal{T})$ of all the models minimal for $(\mathcal{T}, \prec)$ is the class of all the models of some theory $\mathcal{T}'$ in $\mathcal{L}$ (clearly $\mathcal{T}' \models \mathcal{T}$).

This notion has a meaning also in a second order language, but this second order version is different from the first order version (see example 6.3 below).

Clearly, with our definitions, any preferential entailment associated to a preference relation which is definability preserving is a circumscription, and conversely, any circumscription is a preferential entailment to which we may associate a preference relation which is definability preserving. [Sch92] gives an example of a preferential entailment which is not definability preserving. As this example is given in a framework which is different (the kind of preferential entailment found in e.g. [KLM90]), and also as it is of great importance, let us recall it here:

[1] In earlier texts, we have called this property *"coherence"*. At that time, we were not aware of Schlechta's work. Schlechta has independantly introduced the same notion in a related context. In order to unify the terminology, we adopt here the well fitted, and previously published, name given in [Sch92].

Example 3.1 (Example 1.3-1 in [Sch92]) $\mathcal{L}$ is the propositional language for which the set of propositional symbols is the infinite set $\mathcal{P} = \{p_i \ / i \in \mathbb{N}\}$. As we are in the propositional case, the set of all the interpretations over $\mathcal{L}$ may be assimilated to the set of the subsets of $\mathcal{P}$. The context will make clear when we consider p_i as a propositional symbol, member of an interpretation, or when p_i is considered as a formula in $\mathcal{L}$, member of some theory $\mathcal{T}$. μ_0 and μ_1 are two different interpretations over $\mathcal{L}$, e.g. $\mu_0 = \mathcal{P}$, $\mu_1 = \{p_i \ / i \geq 1\}$. $\prec$ is defined by:

$\mu \prec \nu$ iff $\mu = \mu_1$ and $\nu = \mu_0$.

The models minimal for $(Th(\emptyset), \prec)$ are all the models over $\mathcal{L}$, except μ_0. But, the set of all the formulas true in all these models is the set $Th(\emptyset)$ of all the tautologies in $\mathcal{L}$, thus, it has the set of all the interpretations over $\mathcal{L}$ as set of models. This set contains μ_0, which is not minimal for $(Th(\emptyset), \prec)$. □

The preference relations $\prec$ "naturally associated" to the circumscriptions defined in the literature are definability preserving (however, see the end of subsection 6.1). It seems likely that they have more properties. Till now, we are unable to fully characterize these properties. One could think that the seemingly benign following property would be a good candidate:

Definition 3.3 A preference relation $\prec$ is *"semi-compatible with elementary equivalence"* iff, for any interpretations μ, ν and μ', if $\nu \prec \mu$ and if μ and μ' are elementarily equivalent, (noted $\mu \equiv_{el} \mu'$, meaning that for any formula ϕ in $\mathcal{L}$, $\mu \models \phi$ iff $\mu' \models \phi$), then there exists ν' such that $\nu \equiv_{el} \nu'$ and $\nu' \prec \mu'$.

Again, note that this definition might apply to a first order language, as well as to a second order language, but that these two notions differ.

Unfortunately, the relation naturally associated to the best known circumscription, namely first order predicate circumscription, is not always semi-compatible with elementary equivalence (see example 6.2). All the preference relations naturally associated to a circumscription are *"semi-compatible with isomorphism"* (definition 3.3 where we replace "elementarily equivalent" by "isomorphic"), but this is not surprising. Here we mean the "naturally associated $\prec$" (see section 6). For any circumscription, we can find some associated $\prec$ which respects definition 3.3, but we have much better:

Definition 3.4 $\prec$ is *compatible with elementary equivalence* iff for any interpretations, if $\mu \equiv_{el} \mu'$ and $\nu \equiv_{el} \nu'$, then $\mu \prec \nu$ iff $\mu' \prec \nu'$.

Note that Schlechta's example also shows that compatibility with elementary equivalence (which is trivial in the propositional case) does not imply definability preservation. [Moi93] shows that for any circumscription, we can find a preference relation $\prec'$ associated to it which is compatible with elementary equivalence and definability preserving. But this relation is far from being the "natural one".

4 General Properties of Preferential Entailment

We recall here several well-known results (see [MR94] for more details).

Theorem 4.1 For any formula ϕ in $\mathcal{L}$, if $\mathcal{T} \models \phi$ then $\mathcal{T} \models_{\prec} \phi$, i.e.: $\mathcal{T} \models_{\prec} \mathcal{T}$.
This is called *"reflexivity"* e.g. in [KLM90].

Theorem 4.2 If $\mathcal{T} \models \bot$ then $\mathcal{T} \models_{\prec} \bot$.
If $\mathcal{T}$ is (weakly) well-founded for $\prec$, the converse holds.

Theorem 4.3 [Mak88] *Idempotence* of circumscription: Circumscribing a circumscribed theory does not add anything: $\mathbf{Circ_g}(\mathbf{Circ_g}(\mathcal{T})) \equiv \mathbf{Circ_g}(\mathcal{T})$.
This property is true for any preferential entailment (precisely to any pre-circumscription defined from a preferential entailment).

Theorem 4.4 If $\mathcal{T} \models_{\prec} \phi$ and $\mathcal{T} \models_{\prec} \psi$ then $\mathcal{T} \models_{\prec} \phi \wedge \psi$. This is called the *"And rule"* e. g. in [KLM90].

Theorem 4.5 (cf *observation 5* in [Mak88]) Any preferential entailment is *transitivitively cumulative*: if $\mathcal{T} \models_{\prec} \phi$ and $\mathcal{T} \cup \phi \models_{\prec} \psi$ then $\mathcal{T} \models_{\prec} \psi$.

The inference defined by circumscription is *non monotonic*, i.e.: $\mathbf{Circ_g}(\mathcal{T} \cup \psi) \not\models \mathbf{Circ_g}(\mathcal{T})$. In terms of preferential entailment: for some sentences ϕ and ψ, we may have $\mathcal{T} \models_{\prec} \phi$ and $\mathcal{T} \cup \psi \not\models_{\prec} \phi$. However, one weakening of monotony remains true:

Definitions 4.1 A preferential entailment has the property of *cumulative monotony for* $\mathcal{T}$ iff, for any sentences ψ, ϕ in $\mathcal{L}$: if $\mathcal{T} \models_{\prec} \phi$ and $\mathcal{T} \models_{\prec} \psi$ then $(\mathcal{T} \cup \{\phi\}) \models_{\prec} \psi$. This is *"restricted" or "cautious" monotony* of [Gab85, KLM90]. An inference relation is *cumulative* iff it is transitively *and* monotonically cumulative: for preferential entailment, cumulative monotony and cumulativity are synonymous (theorem 4.5).

Theorem 4.6 A preferential entailment is cumulative for $\mathcal{T}$ iff $\mathcal{T}$ is well-behaved for $\prec$.

Theorem 4.7 (cf *observation 6* in [Mak88]) If $\mathcal{T}$ is (mildly) well-founded for $\prec$, then $\models_{\prec}$ is cumulative for $\mathcal{T}$.

Theorem 4.8 Any preferential entailment *allows to reason by cases*: for any theory $\mathcal{T}$ and any sentences ϕ_1, ϕ_2 and ψ in $\mathcal{L}$, *if* $\mathcal{T} \cup \{\phi_1\} \models_{\prec} \psi$ *and* $\mathcal{T} \cup \{\phi_2\} \models_{\prec} \psi$, *then* $\mathcal{T} \cup \{\phi_1 \vee \phi_2\} \models_{\prec} \psi$ (see e.g. [Som90, KLM90]).

Preferential entailments respect the deduction principle (see e.g. [Sho88]) also called e.g. in [KLM90], the "hard part" of the deduction theorem:

Theorem 4.9 For any ϕ and ψ, if $(\mathcal{T} \cup \{\phi\}) \models_{\prec} \psi$ then $\mathcal{T} \models_{\prec} (\phi \Rightarrow \psi)$.

5 A Powerful Property of Circumscriptions

Theorem 5.1 For any circumscription, for any theories $\mathcal{T}$ and $\mathcal{T}$', and for any sentence ϕ in $\mathcal{L}$ we have:

(RM1) $(\mathbf{Circ_g}(\mathcal{T}) \cup \{\phi\}) \models \mathbf{Circ_g}(\mathcal{T} \cup \{\phi\})$.

(**RM**) $(\mathbf{Circ_g}(\mathcal{T}) \cup \mathcal{T}') \models \mathbf{Circ_g}(\mathcal{T} \cup \mathcal{T}')$.

One extreme case is when $\mathcal{T}$ is empty: $\mathbf{Circ_g}(\emptyset) \cup \mathcal{T} \models \mathbf{Circ_g}(\mathcal{T})$.

We call this property *"reverse monotony"* (RM for short, RM1 being the finitary version) because monotony (plus reflexivity) is $\mathbf{Circ_g}(\mathcal{T} \cup \{\phi\}) \models (\mathbf{Circ_g}(\mathcal{T}) \cup \{\phi\})$ (RM) shows that, the more axioms are introduced into the circumscribed theory, the less new results (besides the original axioms) the circumscription produces. With $\mathbf{Circ_g}(\mathcal{T}) \models \mathcal{T}$ (reflexivity), RM puts limits to the circumscription. Note that this property has appeared in [FL90] and in [Sch92] under the name "infinite conditionalization". This name suggests that it was only considered as a variant of the deduction principle. As we think that this property is fundamental in this context, we think that it deserves a proper name, corresponding to its meaning. Note that the context of [FL90] and [Sch92] is different: the preferential entailment considered is the complicated notion defined in [KLM90], which uses a preference relation between sets of models, instead of models as here. In order to encompass all the circumscriptions, we do not need such a notion (in which (DC), defined below, does not hold). Moreover, in [FL90], several additional properties, which are almost never satisfied by the preference relations associated to the interesting circumscriptions, are required.

Proof of theorem 5.1: Let μ be a model of $(\mathbf{Circ_g}(\mathcal{T}) \cup \mathcal{T}')$, that is a model minimal for $(\mathcal{T}, \prec)$ which is also a model of $\mathcal{T}'$. Is μ minimal for $(\mathcal{T} \cup \mathcal{T}', \prec)$? Otherwise there would exist some ν, model of $\mathcal{T} \cup \mathcal{T}'$, such that $\nu \prec \mu$, and μ could not be minimal for $(\mathcal{T}, \prec)$. This proves $(\mathbf{Circ_g}(\mathcal{T}) \cup \mathcal{T}') \models \mathbf{Circ_g}(\mathcal{T} \cup \mathcal{T}')$, which is the *general (or infinitary) version* of RM.

Taking $\mathcal{T}' \equiv \mathcal{T} \cup \{\phi\}$ gives the *unitary version* (RM1) as a corollary. □

Note that (RM1) is verified by any preferential entailment (proof easy, it is a variant of the deduction principle, thanks to the full deduction theorem in classical logic). On the other hand, example 1.9-1 in [Sch92] shows that (RM) does not necessarily hold for a preferential entailment associated to a preference relation which is not definability preserving: Schlechta takes $\mathcal{T} = Th(\emptyset)$ (the set of all the tautologies), $\mathcal{T}' = Th(\mu_0) \cap Th(\mu_1)$ in example 3.1, which gives $\overline{\mathcal{T}} = \mathcal{T}, \overline{\mathcal{T}'} = Th(\mu_1)$, thus $\overline{\mathcal{T}} \cup \mathcal{T}' = \mathcal{T}'$, while $\overline{\mathcal{T} \cup \mathcal{T}'} = \overline{\mathcal{T}'} = Th(\mu_1)$. Thus $\overline{\mathcal{T}} \cup \mathcal{T}' \not\models \overline{\mathcal{T} \cup \mathcal{T}'}$.

Here is another form of (RM), obtained by renaming $\mathcal{T}$ as $\mathcal{T}_1$ and $\mathcal{T} \cup \mathcal{T}'$ as $\mathcal{T}_2$: Let $\mathcal{T}_1$ and $\mathcal{T}_2$ be two theories, we get :

(**RM′**) if $\mathcal{T}_2 \models \mathcal{T}_1$, then $\mathbf{Circ_g}(\mathcal{T}_1) \cup \mathcal{T}_2 \models \mathbf{Circ_g}(\mathcal{T}_2)$.

RM does not always hold for pre-circumscriptions, as a simple example in propositional logic shows:

Example 5.1 $\mathcal{L}$ is a propositional language with only one proposition symbol A. Thus, we have only four theories in $\mathcal{L}$, closed for entailment, which simplifies the definition of the operation $\overline{}$. We define the following pre-circumscription over $\mathcal{L}$: $\overline{Th(A \vee \neg A)} = Th(A)$, $\overline{Th(A)} = Th(A \wedge \neg A)$, $\overline{Th(\neg A)} = Th(\neg A)$, $\overline{Th(A \wedge \neg A)} = Th(A \wedge \neg A)$.

It is impossible to find any $\prec$ corresponding to this pre-circumscription. This is easy to verify as there are only here two propositionnal interpretations (one with A, one with $\neg A$), thus only $2^{(2^2)} = 16$ possible $\prec$'s. We may avoid this exhaustive search, because (RM1) is violated: $\overline{\overline{Th(\emptyset)} \cup \{A\}} \not\models \overline{Th(\emptyset \cup \{A\})}$. Indeed, $Th(\emptyset) = Th(A \vee \neg A)$, $Th(\emptyset \cup \{A\}) = Th(A)$, $\overline{Th(A)}$ is inconsistent while $\overline{Th(\emptyset)} \cup \{A\}$ is equivalent to $Th(A)$. Any preferential entailment satisfies (RM1), so this pre-circumscription is not associated to any preferential entailment (note that in the finite propositional case, the two notions of preferential entailment and of circumscription are equivalent).

(RM) is fundamental because a lot of the properties of circumscriptions may be considered as corollaries of (RM):

Theorem 5.2 If a pre-circumscription satisfies (RM), then it is idempotent, it satisfies the deduction principle, it is transitively cumulative and it allows to reason by cases.

Proof: We suppose that we have a pre-circumscription satisfying (RM).

- Deduction principle (theorem 4.9): We suppose $\overline{T \cup \{\phi\}} \models \psi$. Thus, using (RM1), we get: $(\overline{T} \cup \{\phi\}) \models \psi$, which is equivalent (thanks to the full deduction theorem for classical logic) to: $\overline{T} \models \phi \Rightarrow \psi$.
- Idempotence: We know that for any pre-circumscription we have: $\overline{T} \models T$, thus, using (RM') we get: $(\overline{T} \cup \overline{T}) \models \overline{\overline{T}}$, that is $\overline{T} \models \overline{\overline{T}}$. The other way (reflexivity) is verified by any pre-circumscription.
- Transitive cumulativity (theorem 4.5): We suppose $\overline{T} \models \phi$ and $\overline{T \cup \{\phi\}} \models \psi$. Then, using (RM1), we get $(\overline{T} \cup \{\phi\}) \models \psi$, which gives $\overline{T} \models \psi$ using the "cut rule" of classical logic.
- Reasoning by cases (theorem 4.8): Let $\{T_i\}_{i \in I}$ be a finite set of theories in $\mathcal{L}$, closed for entailment. $T_j \models \bigcap_{i \in I} T_i$, for any $j \in I$, thus, by (RM): $\overline{\overline{\bigcap_{i \in I} T_i} \cup T_j} \models \overline{T_j}$ for any $j \in I$, so we get: $\bigcap_{j \in I} Th(\overline{\bigcap_{i \in I} T_i} \cup T_j) \models \bigcap_{j \in I} \overline{T_j}$, i.e.: $(\overline{\bigcap_{i \in I} T_i}) \cup (\bigcap_{j \in I} T_j) \models \bigcap_{j \in I} \overline{T_j}$, now we have: $(\overline{\bigcap_{i \in I} T_i}) \models (\bigcap_{j \in I} T_j)$, thus we get the result:

(CR) $\overline{\bigcap_{i \in I} T_i} \models \bigcap_{j \in I} \overline{T_j}$ for any finite set of theories in $\mathcal{L}$, closed for entailment.

This is the *"finitary"* – or equivalently here *"binary"* – version of reasoning by cases, which implies the *"finitistic"*, or *"unary"* version given in theorem 4.8. □

Remarks 5.1 • About the relations between (RM) and (CR), it is interesting to note that any pre-circumscription satisfying (CR) satisfies also (RM1).

• Note that (RM1) is enough to imply the deduction principle (theorem 4.9), transitive cumulativity (theorem 4.5), and reasoning by cases (theorem 4.8).

Proof of the first point: For any theory T closed for entailment, and any formula ϕ in $\mathcal{L}$, we have: $T = Th(T \cup \{\phi\}) \cap Th(T \cup \neg\phi)$. Thus, from (CR) we get: $\overline{T} \models \overline{T \cup \{\phi\}} \cap \overline{T \cup \{\neg\phi\}}$, thus: $(\overline{T} \cup \{\phi\}) \models ((\overline{T \cup \{\phi\}} \cap \overline{T \cup \{\neg\phi\}}) \cup \{\phi\}$, i.e: $(\overline{T} \cup \{\phi\}) \models ((Th(\overline{T \cup \{\phi\}} \cup \{\phi\}) \cap (Th(\overline{T \cup \{\neg\phi\}} \cup \{\phi\}))$. Now, $\overline{T \cup \{\neg\phi\}} \models \neg\phi$ by reflexivity, thus $(Th(\overline{T \cup \{\neg\phi\}} \cup \{\phi\})) = Th(\bot)$ (the set of all formulas in $\mathcal{L}$), and $Th(\overline{T \cup \{\phi\}} \models \phi$, we have obtained: $(\overline{T} \cup \{\phi\}) \models \overline{T \cup \{\phi\}}$, i.e. (RM1).

In the finite propositional case, the two properties of (CR) and (RM) are thus equivalent for any pre-circumscription.

Now, the property of reflexivity, and the "and rule" are verified by any pre-circumscription. Thus, all the properties given in the preceding section, and which do not use some property explicitely involving the preference relation $\prec$ (such as well foundedness or well behaveness), are mere corollaries of (RM).

Another immediate consequence of (RM) is *relative consistency:*

Corollary 5.3 If $(\overline{\mathcal{T}} \cup \mathcal{T}')$ is consistent, then $\overline{\mathcal{T} \cup \mathcal{T}'}$ is consistent.
If ϕ is a formula , we get: If $\overline{\mathcal{T} \cup \{\phi\}}$ is inconsistent, then $\overline{\mathcal{T}} \models \neg\phi$.

Here, we introduce our last consequence of (RM), the "dual" of (CR):

Definition 5.1 A pre-circumscription has the property of *conjunctive coherence* (CC) if and only if, for any finite set $\{\mathcal{T}_i\}_{i\in I}$ of theories in $\mathcal{L}$, we have:

$$\text{(CC)} \qquad \bigcup_{i\in I} \overline{\mathcal{T}_i} \models \overline{\bigcup_{i\in I} \mathcal{T}_i}.$$

Theorem 5.4 Any pre-circumscription satisfying (RM) satisfies (CC), thus any circumscription satisfies (CC).

Proof: $\bigcup_{i\in I} \mathcal{T}_i \models \mathcal{T}_i$, thus, by (RM'), $\overline{\mathcal{T}_i} \cup (\bigcup_{i\in I} \mathcal{T}_i) \models \overline{\bigcup_{i\in I} \mathcal{T}_i}$, for any $i \in I$, thus: $\bigcup_{i\in I}(\overline{\mathcal{T}_i} \cup (\bigcup_{i\in I} \mathcal{T}_i)) \models \overline{\bigcup_{i\in I} \mathcal{T}_i}$, i.e: $(\bigcup_{i\in I} \overline{\mathcal{T}_i}) \cup (\bigcup_{i\in I} \mathcal{T}_i) \models \overline{\bigcup_{i\in I} \mathcal{T}_i}$, i.e, thanks to the reflexivity of any pre-circumscription: $(\bigcup_{i\in I} \overline{\mathcal{T}_i}) \models \overline{\bigcup_{i\in I} \mathcal{T}_i}$. □

Here is a last property of circumscriptions, which is not a consequence of (RM):

Theorem 5.5 Any circumscription has the property of *disjunctive coherence* (DC): for any finite set $\{\mathcal{T}_i\}_{i\in I}$ of theories in $\mathcal{L}$, closed for entailment, we have:

$$\text{(DC)} \qquad \bigcup_{i\in I} \mathbf{Circg}(\mathcal{T}_i) \models \mathbf{Circg}(\bigcap_{i\in I} \mathcal{T}_i).$$

Proof: Let μ be a model of $\mathbf{Circg}(\mathcal{T}_i)$ for any $i \in I$. μ is then a model of $\bigcap_{i\in I} \mathcal{T}_i$. Is μ a minimal model of $\bigcap_{i\in I} \mathcal{T}_i$? Otherwise, there exits ν model of $\bigcap_{i\in I} \mathcal{T}_i$ such that $\nu \prec \mu$. ν is a model of $\bigcap_{i\in I} \mathcal{T}_i$, thus it is a model of $\mathcal{T}_j$ for at least one $j \in I$. This contradicts the fact that μ is minimal for $\mathcal{T}_j$. □

Here is an example, completing example 5.1, showing that there exist some pre-circumscriptions satisfying (RM) which are not circumscriptions:

Example 5.2 $\mathcal{L}$ is a propositional language with two proposition symbols A, B. We have only sixteen theories in $\mathcal{L}$, closed for entailment, which simplifies the definition of the operation $\overline{}$. We take as definition:
$\overline{Th(A \wedge \neg A)} = Th(A \wedge \neg A)$; $\overline{Th(A \wedge B)} = Th(A \wedge B)$, $\overline{Th(A \wedge \neg B)} = Th(A \wedge \neg B)$, $\overline{Th(\neg A \wedge B)} = Th(\neg A \wedge B)$, $\overline{Th(\neg A \wedge \neg B)} = Th(\neg A \wedge \neg B)$, $\overline{Th(A)} = Th(A)$, $\overline{Th(B)} = Th(B)$, $\overline{Th(\neg(A \Leftrightarrow B))} = Th(\neg(A \Leftrightarrow B))$, $\overline{Th(A \Leftrightarrow B)} = Th(A \Leftrightarrow$

$\overline{B)}$, $\overline{Th(\neg B)} = Th(\neg B)$, $\overline{Th(\neg A)} = Th(\neg A)$, $\overline{Th(A \vee \neg B)} = Th(A \vee \neg B)$, $\overline{Th(\neg A \vee B)} = Th(\neg A \vee B)$, $\overline{Th(\neg A \vee \neg B)} = Th(\neg A \vee \neg B)$,

$\overline{Th(A \vee B)} = Th(\neg A \wedge B)$, $\overline{Th(A \vee \neg A)} = Th(\neg A \wedge B)$,

i.e. $\overline{T} = T$ except for the last two.

It is easy to verify that reverse monotony is respected here. However, it is not possible to find a relation $\prec$ associated with $\overline{}$. As there are only four models, the search for all the $2^{(4^2)} = 65536$ possible $\prec$'s is feasable. However, it is much shorter to prove that DC is violated:

We take $T_1 = Th(A)$, $T_2 = Th(B)$, with $I = \{1, 2\}$. $T_1 \cap T_2 = Th(A \vee B)$, $\overline{T_1 \cap T_2} = Th(\neg A \wedge B)$, $\overline{T_1} = Th(A)$, $\overline{T_2} = Th(B)$.

Thus, $\overline{T_1 \cap T_2} \models \neg A$, while $\overline{T_1} \cup \overline{T_2} \not\models \neg A$, which establishes the violation of (DC): $\overline{T_1} \cup \overline{T_2} \not\models \overline{T_1 \cap T_2}$.

This is why it does not exist any preference relation $\prec$ associated to this pre-circumscription (again, in this case, the notion of circumscription is equivalent to the notion of preferential entailment). □

This example proves that (RM) does not imply (DC).

Now, in order to precise the contour of the set of the properties enjoyed by any circumscription, as well as of the set of all the corollaries of (RM) for a pre-circumscription, we give an example of a circumscription falsifying the infinite version of (CR):

Definition 5.2 A pre-circumscrition satisfies the infinite version of (CR) iff, for any set of theories T_i in $\mathcal{L}$, closed for entailment, we have:

$$(\mathbf{CR\infty}) \qquad \overline{\bigcap_{i \in I} T_i} \models \bigcap_{j \in I} \overline{T_j}.$$

Example 5.3 We take the same language $\mathcal{L}$ as in example 3.1: it is a propositional language in which the only propositional symbols are the p_i's, for any $i \in \mathbb{N}$. Remind that the set of all the interpretations over $\mathcal{L}$ is here the set of all the subsets of $\mathcal{P} = \{p_i / i \in \mathbb{N}\}$. We define now $\prec$ by: $\mu \prec \nu$ iff $\mu = \nu$ and $\mu \neq \mu_0$, where μ_0 may be any chosen interpretation over $\mathcal{L}$, e.g. the μ_0 of example 3.1.

This preference relation is definability preserving: if μ_0 is not a model of T, then the set of the models minimal for T is the empty set, which is the set of all the models of $Th(\perp)$, if μ_0 is a model of T, the set of the models minimal for T is $\{\mu_0\}$, which is the set of all the models of $Th(\mu_0)$. Thus, we have a circumscription.

Now, we take the set C of all the complete theories in $\mathcal{L}$, less $T'_0 = Th(\mu_0)$. We have: $\bigcap_{T \in C} T = Th(\emptyset)$, $\mathbf{Circg}(Th(\emptyset)) = T'_0$ and $\mathbf{Circg}(T) = Th(\perp)$ for any $T \in C$, thus: $\bigcap_{T \in C} \mathbf{Circg}(T) = Th(\perp)$.

We do not have (CR∞): $\mathbf{Circg}(\bigcap_{T \in C} T) \not\models \bigcap_{T \in C} \mathbf{Circg}(T)$.

Thus, the infinitary version of (CR) is violated, and this proves that (RM) does not imply (CR∞). □

In the line of the preceding example, we examine the infinitary version of (DC). There exist some circumscriptions which violate this infinitary version, as the following example shows:

Example 5.4 We take the same language $\mathcal{L}$ as in examples 3.1 and 5.3. We choose two different interpretations over $\mathcal{L}$, μ_0 and μ_1. We define now $\prec$ by: $\mu \prec \nu$ iff $\mu = \mu_0$ and $\nu \neq \mu_0$. This preference relation is definability preserving: for any theory $\mathcal{T}$, either μ_0 is not a model, then the set of the models minimal for $\mathcal{T}$ is the set of all the models of $\mathcal{T}$, or μ_0 is a model of $\mathcal{T}$, and the set of the models minimal for $\mathcal{T}$ is $\{\mu_0\}$, which is the set of all the models of $Th(\mu_0)$. Thus, we have a circumscription again.

Now, we take $\mathcal{T}'_0 = Th(\mu_0), \mathcal{T}'_1 = Th(\mu_1)$, and we choose a formula ϕ, which is in $\mathcal{T}'_0$ and not in $\mathcal{T}'_1$ (if we choose the μ_0 and μ_1 as in example 3.1, we may choose the formula p_0 as our formula ϕ). V is the set of all the complete theories in $\mathcal{L}$ which contain ϕ. A is V less $\mathcal{T}'_0$. We consider $\mathcal{T}'_1 \cap \mathcal{T}'$, for any $\mathcal{T}' \in A$. Clearly, μ_1 is a model minimal for any of these theories. Now, we define $\mathcal{T} = \bigcap_{\mathcal{T}' \in A}(\mathcal{T}'_1 \cap \mathcal{T}') = \mathcal{T}'_1 \cap (\bigcap_{\mathcal{T}' \in A} \mathcal{T}')$. μ_0 is a model of $\bigcap_{\mathcal{T}' \in A} \mathcal{T}'$, thus of $\mathcal{T}$, which means that μ_1 is not a model minimal for $\mathcal{T}$.

Thus, the infinitary version of (DC) is violated for the set of all the $\mathcal{T}'_1 \cap \mathcal{T}'$, for any $\mathcal{T}' \in A$, we have: $\bigcup_{\mathcal{T}' \in A} \mathbf{Circg}(\mathcal{T}'_1 \cap \mathcal{T}') \not\models \mathbf{Circg}(\bigcap_{\mathcal{T}' \in A}(\mathcal{T}'_1 \cap \mathcal{T}'))$. □

Note that any pre-circumscription which satisfies (RM) and the infinite version of (DC) is a circumscription [Moi93].

We evoque now briefly the relations associated to two kinds of circumscriptions (see details in [MR94]), focusing on significative examples, which prove some (counter-)properties of preferential entailment or circumscription.

6 Some Examples with "Classical Circumscriptions"

6.1 Predicate circumscription

$\mathbf{P}=(P_1, \cdots, P_n)$ and $\mathbf{Q}=(Q_1, \cdots, Q_m)$ are two sequences of predicate symbols in $\mathcal{L}$. The *first order circumscription of* $\mathbf{P}$ *in* $\mathcal{T}$ *with the predicates of* $\mathbf{Q}$ *as variables*[McC80, PM86], noted $\mathbf{Circ}_1(\mathcal{T} : \mathbf{P}; \mathbf{Q})$, adds an infinite first order axiom schema to $\mathcal{T}$. The *second order circumscription* [McC86, Lif86] $\mathbf{Circ}_2(\mathcal{T} : \mathbf{P}; \mathbf{Q})$ is the set of all the first order formulas entailed (second order entailment) by $\mathcal{T}$ augmented with one second order formula.

Definitions 6.1 • A subset S of D^k_μ is *definable with parameters in* μ (e.g. [KK71, p.115–135]) if there is a formula Φ of arity k in $\mathcal{L}_\mu$, with $S = |\Phi|_\mu$.
• Let μ and ν be two interpretations over $\mathcal{L}$. We write $\mu =_{\mathbf{P};\, \mathbf{Q}} \nu$ when μ and ν are identical except for the extensions of the P_i's and of the Q_j's.
• If $\mu =_{\mathbf{P};\, \mathbf{Q}} \nu$ and moreover each $|P_i|_\mu$ and $|Q_j|_\mu$ is definable with parameters in ν, we write $\mu =^\delta_{\mathbf{P};\, \mathbf{Q}} \nu$. If $\mu =_{\mathbf{P};\, \mathbf{Q}} \nu$ (resp. $\mu =^\delta_{\mathbf{P};\, \mathbf{Q}} \nu$) and also $|P_i|_\mu \subseteq |P_i|_\nu$ for $1 \leq i \leq n$, with some $|P_i|_\mu \subset |P_i|_\nu$, we note $\mu <_{\mathbf{P};\, \mathbf{Q}} \nu$ (resp. $\mu <^\delta_{\mathbf{P};\, \mathbf{Q}} \nu$).

Theorem 6.1 [BHR88, Bes89] The models of $\mathbf{Circ}_1(\mathcal{T} : \mathbf{P}; \mathbf{Q})$ are the models minimal for $(\mathcal{T}, <^\delta_{\mathbf{P};\, \mathbf{Q}})$.
[Lif86] The models of $\mathbf{Circ}_2(\mathcal{T} : \mathbf{P}; \mathbf{Q})$ are the models minimal for $(\mathcal{T}, <_{\mathbf{P};\, \mathbf{Q}})$.

A model minimal for $<_{\mathbf{P};\,\mathbf{Q}}$ is minimal for $<^{\delta}_{\mathbf{P};\,\mathbf{Q}}$. If $\mathcal{T}$ is such that the converse is also true, we have the *completeness of first order circumscription* [PM86]. Universal (and thus Horn) theories are always well-founded for $<_{\mathbf{P};\,\mathbf{Q}}$ [BS85, EMR85, Lif86]. There is no simple relationship between well foundedness for $<_{\mathbf{P};\,\mathbf{Q}}$ and for $<^{\delta}_{\mathbf{P};\,\mathbf{Q}}$. For instance, universal theories are not guaranteed to be well-founded for $<^{\delta}_{\mathbf{P};\,\mathbf{Q}}$ (see [MR90, MR94]).

Concerning one corollary of RM (namely $\mathbf{Circ_g}(\emptyset) \cup \mathcal{T} \models \mathbf{Circ_g}(\mathcal{T})$), note that we have: $\mathbf{Circ}(\emptyset : P; \mathbf{Q}) \equiv \forall x\, \neg P(x)$ (for first order or second order version, and with any list $\mathbf{Q}$).

Here is an example showing the importance of well-foundedness in theorem 4.7.

Example 6.1 $\mathcal{T}$ is $\mathcal{T}'$, $\mathcal{S}$, where
$\mathcal{T}' \equiv \exists x \forall y\, (P(x) \wedge (P(y) \Rightarrow x \neq f(y)))$, $\forall x\, [P(x) \Rightarrow P(f(x))]$, 0=0; together with $\mathcal{S} \equiv \forall x\, (x \neq f(x))$, $\forall x \forall y\, [(f(x) = f(y)) \Rightarrow (x = y)]$ (Separation axioms).

$\mathbf{Circ}(\mathcal{T} : P) \models \bot$, thus e.g.: $\mathbf{Circ}(\mathcal{T} : P) \models \Psi$, where $\Psi \equiv P(0)$ (this is a well-known example of inconsistent circumscription, see e.g. [EMR85]). However $\mathbf{Circ}(\mathcal{T} \wedge \Psi : P)$ is consistent so $\mathbf{Circ}(\mathcal{T} : P)$ is not cumulative. Indeed, here is a model μ of $\mathcal{T} \cup \{\Psi\}$ minimal for $<_P$ (thus also for $<^{\delta}_P$): $D_\mu = |P|_\mu = \mathbb{N}$, $0_\mu = 0 \in \mathbb{N}$ and f_μ is the successor function in $\mathbb{N}$.

$\mathcal{T}$ is not well-founded, neither for $<_P$ nor for $<^{\delta}_P$.

Also, this is an example of a theory $\mathcal{S}$ which is well-founded for $<_P$ (being universal), while some $\mathcal{S} \wedge \mathcal{T}'$, i.e. $\mathcal{T}$, is not (cf theorem 2.1). □

Our next example of predicate circumscription shows that the relation naturally associated to $\mathbf{Circ}_1(\mathcal{T}$:P), $<^{\delta}_P$, is not necessarily semi-compatible with elementary equivalence (definition 3.3).

Example 6.2 $\mathcal{L}$ contains two predicates P (unary) and $\leq$ (binary, we use the notation $x \leq y$). $\mathcal{T}$ contains the axioms saying that $\leq$ is an order relation, and that P defines an initial segment: $\forall x\, \forall y\, ((P(x) \wedge x \leq y) \Rightarrow P(y))$.

We consider $\mathbf{Circ}_1(\mathcal{T}$:P), and its associated preference relation $<^{\delta}_P$.

We define the two following models of $\mathcal{T}$: $D_{\mu'} = E_0 \cup E_1 \cup E_2 \cup \cdots \cup E_n \cup \cdots$, $D_\mu = D_\nu = D_{\mu'} \cup E$, where each $E_n = \{0_n, 1_n, 2_n, \cdots, n_n\}$ is a copy of an initial segment of $\mathbb{N}$, and $E = \{0, 1, 2, \cdots, i, \cdots\}$ is a copy of $\mathbb{N}$. $\leq$ is interpreted in μ, μ' and ν by the natural order on each E_n and on E.

$|P|_\mu = D_\mu$, $|P|_\nu = D_\mu - E\ (= D_{\mu'})$ and $|P|_{\mu'} = D_{\mu'}$.

We have $\nu <_P \mu$. Also, $D_{\mu'}$ is definable with one parameter in μ: it suffices to take $0 \in E$ as parameter, an element $e \in D_\mu$ is in $D_{\mu'}$ iff $0 \leq_\mu e$. Thus we have $\nu <^{\delta}_P \mu$. Now, μ and μ' are elementarily equivalent. However, it does not exist a ν', elementarily equivalent to ν and such that $\nu' <_P \mu'$: indeed, $|P|_{\nu'}$ should be totally ordered, thus finite, but $|P|_\nu$ is not finite. □

We end this subsection by a word of caution about the second order circumscription. The preference relation naturally associated to this circumscription is definability preserving (second order version). But, if we stay in first order logic, which is generally the case in the literature about "second order versions" of circumscriptions, this is a different matter. Indeed, generally in the literature, we start from a first order theory $\mathcal{T}$, and we are interested only by the first order formulas which we may deduce from the "second order circumscription". This is clearly a preferential entailment notion, defined while staying in first order logic: $\overline{\mathcal{T}}$ is the theory of all the (first order) formulas which are true in any model minimal for $(\mathcal{T}, <_{\mathbf{P};\mathbf{Q}})$. But the relation naturally associated to this preferential entailment, i.e $<_{\mathbf{P};\mathbf{Q}}$, is not definability preserving (first order version), as the following example shows:

Example 6.3 (cf example 3.1 in [MR90])

$$\mathcal{T}: \forall x \exists y\, Q(x,y) \quad \text{(b1)}$$
$$\forall x \forall x' \forall y\, \{[(Q(x,y) \wedge Q(x',y)] \Rightarrow x = x'\} \quad \text{(b2)}$$
$$\forall x \forall y\, [(Q(x,y) \Rightarrow P(y)] \quad \text{(b3).}$$

$\mathcal{T}$ is consistent: a particular model μ has $\mathbb{N}$ (set of natural integers) for domain, $Q(e_1, e_2)$ is true if and only if e_2 is the successor of e_1, and $P(e)$ is true if and only if e is not 0. We study $\mathbf{Circ_2}(\mathcal{T}: P; Q)$. The circumscription axiom is:

(AC) $\equiv \forall p \forall q\, \{[\mathcal{T}[p,q] \wedge \forall z\, (p(z) \Rightarrow P(z))] \Rightarrow \forall z\, (P(z) \Rightarrow p(z))\}$.

Also, we define the following axiom: (AF) $\equiv \forall q\, \{\mathcal{H}(q) \Rightarrow \forall y \exists x\; q(x,y)\}$

(AF) is a *finiteness axiom*: it expresses that the domain is a finite set.

[MR90] establishes: $\mathbf{Circ_2}(\mathcal{T}: P; Q) \equiv \mathcal{T} \wedge \text{(AF)}$.

Here, the circumscription axiom is a finiteness axiom, which means that the models minimal for $(\mathcal{T}, <_{\mathbf{P};\mathbf{Q}})$ are finite models. On the other hand, there are minimal models of any finite cardinality (except 0), thus $\overline{\mathcal{T}}$ must have infinite models also (where $\overline{\mathcal{T}}$ is defined as explained above, it is thus a first order theory). □

6.2 Pointwise circumscription

Definition 6.2 [Lif88] P being a predicate, $P_{/\mathbf{y}}$ is defined by: $P_{/\mathbf{y}}(\mathbf{x}) \equiv (P(\mathbf{x}) \wedge \mathbf{x} \neq \mathbf{y})$. The *pointwise circumscription of P in $\mathcal{T}$*, noted $\mathbf{Cppp}(\mathcal{T}: P)$ adds to $\mathcal{T}$ the axiom $\mathcal{T}[P_{/\mathbf{y}}] \Rightarrow \neg P(\mathbf{y})$, which is one instance of the circumscription axiom schema of $\mathbf{Circ_1}(\mathcal{T}: P)$.

Pointwise circumscription is important for at least two reasons:

- It simulates and generalizes the notion of predicate completion [Moi88].
- We do not have to guess which instances of a possibly infinite axiom schema are useful, as with standard circumscription. And for some theories we have equivalence between $\mathbf{Circ_1}(\mathcal{T}: P)$ and $\mathbf{Cppp}(\mathcal{T}: P)$ [Lif88].

Definition 6.3 [Moi92] If $\mu =_P \nu$ and if $|P|_\mu$ is $|P|_\nu$ less exactly one element, we note $\mu <_{1:P} \nu$ (we may indifferently write $\mu =_P \nu$ or $\mu =^\delta_P \nu$ here, as $|P|_\mu$ is definable with one parameter – the element removed – in ν).

Theorem 6.2 A model of $\mathbf{Cppp}(\mathcal{T}: P)$ is a model of $\mathcal{T}$ minimal for $<_{1:P}$.

Example 6.4 $\mathcal{T}$: $P(a) \wedge \forall x\, (x = a \vee x = b \vee x = c)$.

Let μ, ν and ν' be three particular models of $\mathcal{T}$: $D_\mu = D_\nu = D_{\nu'} = |P|_\mu = \{a,b,c\}$; $|P|_\nu = \{a,b\}$; $|P|_{\nu'} = \{a\}$. ν' is minimal for $(\mathcal{T}, <_{1:P})$, $\nu' \overline{<_{1:P}}\, \mu$, but we do not have $\nu' <_{1:P} \mu$. The models of $\mathbf{Cppp}(\mathcal{T}: P)$ are the models of $\mathcal{T}$ in which $|P|_\mu = \{a_\mu\}$. $\mathbf{Cppp}(\mathcal{T}: P) \equiv \mathcal{T} \cup \{\forall x\, (P(x) \Rightarrow (x = a))\}$. $\mathcal{T}$ has ten models (up to isomorphism): it is easy to verify that $<_{1:P}$ is weakly well-founded (it is even "bounded" in Shoham's meaning), while it is neither well-founded nor mildly well-founded. **Cppp**($\mathcal{T} : P$) is not cumulative: it entails $\phi \equiv (a \neq b \wedge a \neq c) \Rightarrow (P(b) \Leftrightarrow P(c))$ and $\psi \equiv (a \neq b \wedge a \neq c) \Rightarrow \neg P(b)$ while **Cppp**($\mathcal{T} \wedge \phi : P$) does not entail ψ. □

In our next example, $\prec$ is mildly well-founded and not well-founded.

Example 6.5 (see [MR94] for more details) $\mathcal{L}$ contains five predicate symbols E, E_1, E_2, P (unary) and $<$ (binary, we note $x < y$), the symbol $=$ and one unary function f. We describe $\mathcal{T}$ informally, as a complete list of axioms would not be readable. Any x verifies one and only one formula among $E(x)$, $E_1(x)$ and $E_2(x)$. $<$ is a discrete and total strict order, defined only on E, without smallest or greatest element. f is a mapping from E_2 onto E_1, from E_1 onto E, and from E onto E_2 which satisfies $\forall x\ (f^3(x) = x)$. $P \cap E$ is a final interval of E (i.e: $\forall x\ (P(x) \wedge x < y) \Rightarrow P(y)$) which has a smallest element for $<$. We note e this element. We have also: $\forall x\ ((E_1(x) \wedge x \neq f^2(e)) \Rightarrow P(x))$, $\forall x\ (E_2(x) \wedge x \neq f(e)) \Rightarrow P(x))$, and $P(f(e)) \Rightarrow P(f^2(e))$. We have described all the axioms of $\mathcal{T}$, which is finitely axiomatizable.

$\mathcal{T}$ has only three kinds of models, which verify respectively $\phi_0 = \neg P(f^2(e))$, $\phi_1 = P(f^2(e)) \wedge \neg P(f(e))$ and $\phi_2 = P(f(e))$. $\mathcal{T}_0 = Th(\mathcal{T} \cup \{\phi_0\})$, $\mathcal{T}_1 = Th(\mathcal{T} \cup \{\phi_1\})$ and $\mathcal{T}_2 = Th(\mathcal{T} \cup \{\phi_2\})$ are three complete theories.

The models of $\mathcal{T}$ minimal for $<_{1:P}$ are the models of $\mathcal{T}_0$. Indeed, if μ is a model of $\mathcal{T}_0$, we cannot remove one single element in $|P|_\mu$ without contradicting some axioms in $\mathcal{T}$. If μ is a model of $\mathcal{T}_1$, we can remove $f_\mu^2(e)$ from $|P|_\mu$ (this gives a model μ' of $\mathcal{T}_0$). If μ is a model of $\mathcal{T}_2$, we can remove $f_\mu(e)$ from $|P|_\mu$ (this gives a model μ' of $\mathcal{T}_1$), and also we could remove e (this giving a model μ'' of $\mathcal{T}_2$). This describes all the possibilities, thus $\mathcal{T}$ is weakly well-founded for $<_{1:P}$ but it is not well-founded (note that $\mathcal{T}$ is not "bounded" in Shoham's meaning).

$\mathcal{T}$ is well-behaved. Indeed, let ϕ be a formula true in all the models minimal for $(\mathcal{T}, <_{1:P})$, and $\mathcal{T}'$ be $\mathcal{T} \cup \{\phi\}$. $\mathcal{T}_0 \models \mathcal{T}'$ and $\mathcal{T}' \models \mathcal{T}$. Thus, as $\mathcal{T}_0$, $\mathcal{T}_1$ and $\mathcal{T}_2$ are complete theories, $\mathcal{T}'$ is equivalent to $\mathcal{T}_0$, or to $\mathcal{T}_0 \cap \mathcal{T}_1$, or to $\mathcal{T}_0 \cap \mathcal{T}_2$, or to $\mathcal{T}_0 \cap \mathcal{T}_1 \cap \mathcal{T}_2$, i.e $\mathcal{T}$. For each of these four theories, the models minimal for $<_{1:P}$ are exactly the models of $\mathcal{T}_0$.

It can be shown that there are no examples of this kind with pointwise circumscription without an infinite sequence of models. □

Note that the preference relation $\overline{<_{1:P}}$ is antisymetric and antireflexive. However, pointwise circumscription is not an ordered preferential entailment notion, contrarily to predicate circumscriptions.

7 Conclusion

In order to study the main properties of the various existing circumscriptions, we have used a notion which in fact originated from the studies about circumscription: preferential entailment. We have precised exactly which kind of preferential entailment is needed: in fact, the simplest notion is the best one here, and all we need at the begining is a binary relation among models, we do not need the complicated versions using relations among sets of models. On the one hand, the preceding proposals were too constraint, e.g. requesting transitivity. On the other hand, they were not constrained enough: the nature of circumscription requires that preferential entailments are somehow coherent with respect to the syntax. We have precised this notion (called definability preservation by some authors), for the first order and for the second order circumscriptions.

We have studied carefully the unavoidable case of a non transitive preference relation. We have isolated a fundamental property of circumscriptions, reverse monotony, which implies the main general properties already known of circumscriptions: idempotence, the deduction principle, the ability to reason by cases and transitive cumulativity, and some other properties introduced here, such as conjunctive coherence. However, we have introduced a new property of circumscriptions (DC) which is not implied by reverse monotony. Finally, several examples illustrate the utility of an appropriate preferential entailment approach, and give some counter-examples, crucial when the objective is to precise all the relevant properties of circumscriptions.

All these results apply to a great variety of the existing circumscriptions. For instance this concerns the first order and the second order versions of domain and predicate (or formula) circumscriptions [McC80, McC86], all the varieties of pointwise circumscriptions of [Lif88], and also more exotic variants such as the closed or non recursive circumscriptions of [BMM89].

Acknowledgements

The two authors are glad to thank Philippe Besnard who initiated this work.

References

[Bes89] P. Besnard. *An Introduction to Default Logic.* Springer Verlag, Heidelberg, 1989.

[BHR88] Philippe Besnard, Jean Houdebine, and Raymond Rolland. A formula circumscriptively both valid and unprovable. In *ECAI*, pages 516–518, Munich, 1988.

[BMM89] P. Besnard, R. Mercer, and Y. Moinard. The importance of open and recursive circumscription. *Artificial Intelligence*, 39:251–262, 1989.

[BS85] Geneviève Bossu and Pierre Siegel. Saturation, nonmonotonic reasoning and the closed-world assumption. *Artificial Intelligence*, 25:13–63, 1985.

[BS88] Philippe Besnard and Pierre Siegel. The preferential-models approach to nonmonotonic logics. In *Non-Standard Logics for Automated Reasoning*, pages 137–161. Academic Press, 1988.

[CK73] C.C. Chang and H.J. Keisler. *Model Theory.* North-Holland, Amsterdam, 1973.

[EMR85] D.W. Etherington, R.E. Mercer, and R. Reiter. On the adequacy of predicate circumscription for closed-world reasoning. *Comput Intell.*, 1:11–15, 1985.

[FL90] Michael Freund and Daniel Lehmann. Deductive inference operations. In *JELIA-90, in LNCS 478*, pages 227–233, Amsterdam, Springer-Verlag, September 1990.

[Gab85] D.M. Gabbay. Theoretical foundations for nonmonotonic reasoning in expert systems. In *Proc. Logics and Models of Concurrent Systems, NATO ASI Series F*, volume 13. Springer Verlag, 1985.

[KK71] Georg Kreisel and Jean-Louis Krivine. *Elements of Mathematical Logic: Model Theory (2nd edition).* North Holland, Amsterdam, 1971.

[KLM90] S. Kraus, D. Lehmann, and M. Magidor. Nonmonotonic reasoning, preferential models and cumulative logics. *Artificial Intelligence*, 44:167–207, 1990.

[Lif86] Vladimir Lifschitz. On the satisfiability of circumscription. *Artificial Intelligence*, 28:17–27, 1986.

[Lif88] Vladimir Lifschitz. Pointwise circumscription. In Matthew L. Ginsberg, editor, *Readings in Nonmonotonic Reasoning*, pages 179–193. Morgan-Kaufmann, Los Altos CA, 1988.

[Mak88] David Makinson. General theory of cumulative inference. In *Non-Monotonic Reasoning, in LNAI-346*, pages 1–18. Springer-Verlag, June 1988.

[McC80] John McCarthy. Circumscription–a form of non-monotonic reasoning. *Artificial Intelligence*, 13:27–39, 1980.

[McC86] John McCarthy. Application of circumscription to formalizing common sense knowledge. *Artificial Intelligence*, 28:89–116, 1986.

[Moi88] Yves Moinard. Pointwise circumscription is equivalent to predicate completion (sometimes). In Robert A. Kowalski and Kenneth A. Bowen, editors, *Logic Programming*, pages 1097–1105, Seattle, MIT Press, 1988.

[Moi92] Y. Moinard. Circumscriptions as preferential entailments. In *ECAI*, pages 329–333, Wien, Wiley, August 1992.

[Moi93] Y. Moinard. Circumscriptions for preferential entailments. *submitted*, dec. 1993.

[MR90] Y. Moinard and R. Rolland. Unexpected and unwanted results of circumscription. In *Artificial Intelligence IV (methodology, systems, applications) (AIMSA)*, pages 61–70, Albena, Bulg., North-Holland, 1990.

[MR94] Y. Moinard and R. Rolland. Preferential entailments for circumscriptions. In *KR-94*, Bonn, Morgan Kaufmann, May 1994.

[PM86] Donald Perlis and Jack Minker. Completeness results for circumscription. *Artificial Intelligence*, 28:29–42, 1986.

[Ryc90] Piotr Rychlick. The Generalized Theory of Model Preference (Preliminary Report). In *AAAI*, pages 615–620, Boston, Morgan Kaufmann, August 1990.

[Sch92] Karl Schlechta. Some Results on Classical Preferential Models. *Journal of Logic Computation*, 2(6):675–686, December 1992.

[Sho88] Yoav Shoham. *Reasoning about change*. MIT Press, Cambridge, 1988.

[Som90] Léa Sombé. *Reasoning under Incomplete Information in Artificial Intelligence*. Wiley, New York, 1990.

The Computational value of Joint Consistency

Extended Abstract

Yannis Dimopoulos
Max-Planck-Institut für Informatik
Im Stadtwald, 66123 Saarbrücken, Germany
yannis@mpi-sb.mpg.de

Abstract. In this paper we investigate the complexity of some recent reconstructions of Reiter's Default Logic using graph-theoretical structures. It turns out that requiring joint consistency of the justification of the applied rules has serious effects on the computational features of default reasoning. Namely, many of the intractability problems of Reiter's original approach, in some cases disappear in the new frameworks. However, interesting problems remain intractable. We also present a propositional semantics for those approaches, stemming from the translation of the graph structures into propositional logic. Finally, the constraint stable model semantics is introduced, and proved to be related to the notion of joint consistency.

1 Introduction

Over the last years some variants of Reiter's Default Logic ([Rei80]) have been proposed ([Bre91], [DJ91], [Luk88], [Sch91]) in order to cope with some counterintuitive inferences that the original formalization sanction. These approaches differ in the way they define the applicability conditions of the default rules. Some of them require the *joint consistency* of the justifications of the applied defaults, a notion to be explained later.

The counterintuitive inferences is not the only critical issue for Default Logic. One of the most serious drawbacks of reasoning with Reiter's formalism is the intractability even in very simple cases of default theories ([KS91], [DM94], [DMP93], [PS92]). On the other hand, only a few things are known about the complexity of the new proposals for default reasoning. Only recently Gottlob and Mingyi, in [GM93], independently obtained algorithms and complexity results regarding Brewka's *Cumulative Default Logic* (CDL, [Bre91]). Those results indicate that reasoning with this logic is not harder than reasoning with Reiter's Default logic. However, while in [GM93] the general propositional case is considered, we restrict ourselves to the case of disjunction-free propositional theories.

In this paper we show that if we require joint consistency of the justifications of the defaults, in the case of propositional disjunction free Default Theories, reasoning becomes significantly easier. In particular, not only an extension can be computed in time polynomial in the size of the theory, but more interestingly, all the extensions can be effectively enumerated. On the negative side we show that, in general, goal directed and priority preserving reasoning remain intractable.

Roughly speaking, the critical issue turns out to be the number of the extensions that a default theory may have. If this number is bounded by a polynomial in the size of the theory then reasoning is tractable.

Our approach is based on a graph representation of the default theories. This allows us to employ some well known properties of graphs, and extends some recent results on the relation between Graph Theory and Nonmonotonic Reasoning ([DM94], [DMP93], [DT93], [PY92], [Tor94]). Slight modifications of the graph algorithms lead to derivation procedures for default theories under the joint consistency semantics.

We also obtain a propositional semantics for the default theories we consider, by translating the graph theoretic constructs into propositional logic. This semantics make possible the use of classical satisfiability algorithms in the case of default theories. Finally, we present a semantics for general logic programs, which is closely related to the J-extensions of an associated default theory. The problem whether this new semantics can be effectively computed is left open.

2 Preliminaries

A propositional Default Theory is a pair $\Delta = (D, W)$ where D is a set of rules of the form $a : Mb_1, Mb_2, \ldots, Mb_n/w$, where a, b_i, w are propositions and W is a set of propositions. Proposition a is called the *prerequisite* of the default, propositions $b_1, \ldots, b_n$ its *justifications* while w the *consequent* (denoted by *Prer*, *Just* and *Cons* respectively) . An agent holding such a default theory may have different "belief sets" (an empty belief set is possible) which are defined to be the extensions of the theory:

Definition 1. ([Rei80]) A set of propositions E is an *extension* of a default theory $\Delta = (D, W)$ iff it satisfies the equation $E = \cup_{i=0}^{\infty} E_i$ where $E_0 = W$ and for $i \geq 0, E_{i+1} = Th(E_i) \cup \{w \mid a : Mb_1, Mb_2, \ldots, Mb_n/w \in D, a \in E_i$ and $\neg b_j \notin E, 1 \leq j \leq n\}$. □

If E is an extension of a theory Δ, then the set of *generating defaults* of E, is the set $GD(E, \Delta) = \{a : Mb_1, Mb_2, \ldots, Mb_n/w \in D \mid a \in E, \neg b_i \notin E\}$ (similarly in the case of J-extensions and CDL extensions). In the above definition in order for E to be an extension there must be no justification in the applied defaults the negation of which occurs in E. However, in the definitions of the extensions in [Bre91] and [DJ91] joint consistency is required, that is, all the justifications of the applied rules must be jointly consistent with E. Obviously, in this case the justifications of the applied rules have to be recorded. The definition given in [DJ91] refers to seminormal theories, i.e. theories with rules of the form $a : Mb \wedge c/b$. In this paper we refer exclusively to seminormal theories.

Definition 2. ([DJ91]). Let $\Delta = (D, W)$ be a seminormal default theory. Define
$E_0 = (E_{J_0}, E_{T_0}) = (Th(W), Th(W))$
$E_{i+1} = (E_{J_{i+1}}, E_{T_{i+1}})$
$= (Th(E_{J_i} \cup \{b \wedge c\}), Th(E_{T_i} \cup \{b\}))$

where $i \geq 0$, $a : Mb \wedge c/b \in D$, $a \in E_{T_i}$ and $\neg(b \wedge c) \notin E_{J_i}$.
Then E is a *J-extension* for Δ iff $E = (E_J, E_T) = (\cup_{i=0}^{\infty} E_{J_i}, \cup_{i=0}^{\infty} E_{T_i})$. □

In this paper we investigate the case of *disjunction-free, seminormal, propositional* default theories. In a disjunction-free propositional theory, both W and the prerequisite, justifications and consequents of the defaults are conjunctions (sets) of literals. In the sequel, except if otherwise stated, the term default theory refers to a disjunction-free, seminormal, propositional theory.

Notice that in definition 2, the consistency check is performed wrt the set E_{J_i}. We show that, requiring consistency wrt to the set E_J, for the cases we consider, does not have any impact on the conclusions drawn from the default theories.

Definition 3. Let $\Delta = (D, W)$ be a seminormal default theory. Define
$E_0 = (E_{J_0}, E_{T_0}) = (Th(W), Th(W))$
$E_{i+1} = (E_{J_{i+1}}, E_{T_{i+1}})$
$= (Th(E_{J_i} \cup \{b \wedge c\}), Th(E_{T_i} \cup \{b\}))$
where $i \geq 0$, $a : Mb \wedge c/b \in D$, $a \in E_{T_i}$ and $\neg(b \wedge c) \notin E_J$.
Then E is a *JF-extension* for Δ iff $E = (E_J, E_T) = (\cup_{i=0}^{\infty} E_{J_i}, \cup_{i=0}^{\infty} E_{T_i})$. □

We prove that J-extensions and JF-extensions are identical.

Theorem 4. *A set E is a J-extension of a disjunction free propositional default theory Δ iff E is a JF-extension of Δ.*

Proof. (sketch) Let E be a JF-extension for Δ and $D' = (d_1, d_2, \ldots, d_k)$ the set of generating defaults of E, in the order they can be applied. We show that E is a J-extension. First see that for every default d_i, $\text{prer}(d_i) \subseteq \cup_{j<i}\text{cons}(d_j)$, holds. On the other hand since none of the literals in the justifications of the the defaults in D' occurs negated in E_J then it will not also occur in any of the sets E_{J_i} of the definition 2. Finally for every default $d \in D - D'$ either $\text{prer}(d) \not\subseteq \cup_{i=1}^{k}\text{cons}(d_i)$, or $b \in\text{just}(d)$ while $\neg b \in E_J$. The latter means that $\neg b \in E_{J_m}$, for some $m \leq k$. Hence E is a J-extension as well.
Let now E be a J-extension and $D' = (d_1, d_2, \ldots, d_k)$ the set of the generating defaults for E, in the order they can be applied. We prove, inductively on the defaults in D' that are applicable wrt the definition of the JF-extension. Assume that d_1 is not applicable wrt the JF-extension definition. This is because the negation of some literal p in the justifications of d_1, occurs in E_J. For this to happen it must be the case that there is a default $d_i \in D'$ such that $\neg p \in\text{just}(d_i)$, which is contradiction. Assume that every default d_i, $i < m$ is applicable wrt the JF-extension definition. Using arguments, similar to those for the default d_1, we can prove that d_{m+1} is also applicable. Hence every default in D' is applicable wrt the JF-extension definition.
On the other hand for every default $d \in D - D'$ either $\text{prer}(d) \not\subseteq \cup_{i=1}^{k}\text{cons}(d_i)$, or $b \in\text{just}(d)$ while $\neg b \in E_{J_k}$. The later means that $\neg b \in E_J$ as well, hence the default d is not applicable wrt the JF-extension definition as well. Hence E is also a JF-extension. □

Brewka's CDL ([Bre91]), even though equivalent to J-Default Logic, is briefly introduced here and used later in the paper. In CDL formulas are associated with supports, leading to *assertions* of the form $< p : \{q_1, \ldots, q_n\} >$, where $p, q_1, \ldots, q_n$ are formulas (in our case conjunctions of literals). Given a set of assertions W, $Form(W) = \{p| < p : \{q_1, \ldots, q_n\} > \in W\}$, while $Supp(W) = \{q_i| < p : \{q_1, \ldots, q_n\} > \in W\}$.

Definition 5. ([Bre91]) If A is a set of assertions then $Th_S(A)$, the set of *supported theorems* of A, is the smallest set such that,
(1) $A \subseteq Th_S(A)$,
(2) if $< p_1 : Q_1 >, \ldots < p_m : Q_m > \in Th_S(A)$ and $p_1, \ldots, p_m \vdash q$, then $< q : Q_1 \cup \ldots \cup Q_m > \in Th_S(A)$.

The symbol $\vdash$ in the above definition stands for classical derivability. An *assertion default theory* is again a pair (D, W), where D is a set of default rules and W is a set of assertions. The following is a quasi-inductive definition of the notion of the CDL extension.

Definition 6. ([Bre91]) E is a *CDL extension* of a assertion default theory $\Delta = (D, W)$ iff $E = \cup_{i=0}^{\infty} E_i$ where
$E_0 = W$ and for $i \geq 0$
$E_{i+1} = Th_S(E_i) \cup \{< C : \{q_1, \ldots, q_n, B, C\} > \mid A : B/C \in D, < A : q_1, \ldots, q_n\} > \in E_i$ and $\{B, C\} \cup Form(E) \cup Supp(E)$ is consistent $\}$. □

Reiter's Default Logic is in direct correspondence with the stable model semantics for (general) logic programs, introduced in [GL88]. Later in this paper, we introduce the *constraint stable models* semantics, which turn out to be closely related to the notion of joint consistency.

A logic program is a set of rules of the form

$$A \leftarrow B_1 \wedge \ldots \wedge B_n \wedge \neg C_1 \wedge \ldots \neg C_m$$

where $A, B_1, \ldots B_n, C_1, \ldots C_m$ are atoms. The lhs of the rule (i.e. the atom A) is called the head while the rhs the body.

Definition 7. ([GL88]) Let P be a logic program and M an interpretation. We define $G(M, P)$ to be the logic program obtained by

- Deleting every rule with a negative literal occurring in its body which does not belong to M
- Deleting all negative literals from the remaining rules

See that the resulting program $G(M, P)$ is definite and has a unique minimal model.

Definition 8. ([GL88]) Let P be a logic program and M an interpretation. Then M is a *stable model* of P iff the minimal model of $G(M, P)$ coincides with M.

3 Computational considerations

In [DM94] a method for constructing graphs from disjunction-free default theories has been proposed. We employ this method here and apply it to the case of J-extensions. We start with the simple class of theories consisting of rules with no prerequisites and $W = \emptyset$. We later extend this method to the case of theories with prerequisites.

3.1 Theories without prerequisites

Let $\Delta = (D, W)$ be a default theory, where $W = \emptyset$, and the defaults in D are prerequisite free. Given Δ we can construct a graph $G_R = (N, E)$ such that each node in N is associated with a default in D, while an edge $(n_i, n_j) \in E$ if $p \in$Cons(d_i) and $\neg p \in$Just(d_j). We call G_R the *rule graph of* Δ *under Reiter's semantics.* It can be proved that Reiter's extensions of Δ are in direct correspondence with the *kernels* of G_R. Kernel of a directed graph $G = (N, E)$ is a set of nodes $K \subseteq N$, such that the nodes in K are independent (meaning that there is no edge between any two of them) and for every node $n_j \in N - K$ there is a node $n_i \in K$ such that $n_i \in \Gamma^-(n_j)$.[1] Determining whether a graph has a kernel has been proved NP-complete by Chvatal ([GJ79]). Many problems regarding kernels are intractable, even in the case of graphs without odd loops, for which at least one kernel exists (see [DM94] and [DMP93]). However, we will show that the situation changes if we impose joint consistency on the justifications of the defaults. In this case more edges have to be added to the theory's graph. Note that the definition of J-extensions, in contrast to Reiter's extensions, is constructive, a fact which indicates that their computation is easier.

Definition 9. Let $\Delta = (D, \emptyset)$ be a default theory[2]. We define the *rule graph of* Δ *under J-extensions semantics* to be a graph $G_J(\Delta) = (N, E)$, such that for every rule in D there is a node N, while there is an edge $(n_i, n_j) \in E$ if either
(a) $p \in$Cons(d_i) and $\neg p \in$Just(d_j), or
(b) $p \in$Just(d_i) and $\neg p \in$Just(d_j). □

Since we only consider seminormal rules, case (a) in the above definition is redundant. Edges between the nodes depict negative interaction between the corresponding rules. That is, an edge (n_i, n_j) is present iff the application of the rule d_i can block the application of the rule d_j (and vise-versa). Like in the case of Reiter's extensions, J-extensions are in a direct correspondence with the kernels of the associated rule graph.

Theorem 10. *Let* $\Delta = (D, \emptyset)$ *be a default theory. Then* K *is a kernel for* $G_J(\Delta)$ *iff the set of rules which correspond to the nodes in* K *is a set of generating defaults* $GD(E, \Delta)$ *of a J-extension* E *of* Δ. □

[1] Let $G = (N, E)$ be a graph and $S \subseteq N$. Then $\Gamma^-(S) = \{n_i \mid (n_i, n_j) \in E, n_j \in S\}$, $\Gamma^+(S) = \{n_i \mid (n_j, n_i) \in E, n_j \in S\}$ and $\Gamma(S) = \Gamma^+(S) \cup \Gamma^-(S)$.

[2] In this section we refer exclusively to seminormal, prerequisite and disjunction free theories.

Example 1. Let $\Delta = (D, \emptyset)$ be a default theory where $D = \{: MB \wedge \neg E/B,: MC \wedge \neg B \wedge \neg E/C,: M\neg C \wedge D/D,: M\neg D \wedge E/E\}$. The two graphs of the theory, G_R and G_J, are depicted in figure 1 and correspond to Reiter's extensions semantics and the J-extensions semantics, respectively.

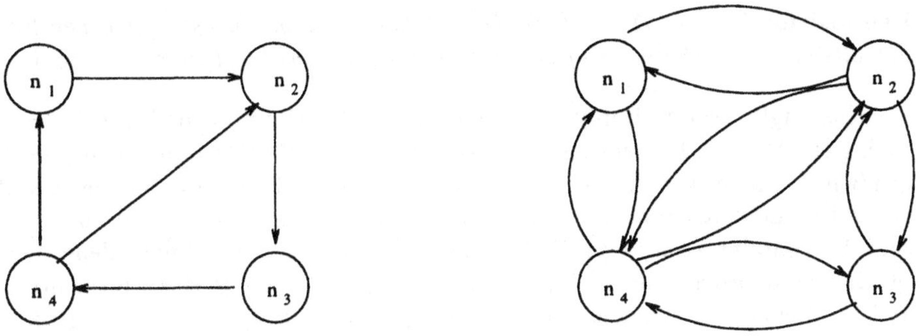

Figure 1

Graph G_R has exactly one kernel $K_1 = \{n_1, n_3\}$, while G_J has three kernels, namely $K_{21} = \{n_1, n_3\}, K_{22} = \{n_2\}, K_{23} = \{n_4\}$. Kernel K_1 corresponds to the Reiter's extension $E = \{B, D\}$ while the kernels of G_J to the J-extensions $E_1 = (\{B, D, \neg E, \neg C\}, \{B, D\})$, $E_2 = (\{\neg B, \neg E, C\}, \{C\})$, $E_3 = (\{\neg D, E\}, \{E\})$. □

It is easy to see that the rule graphs of theories under the J-extension semantics are symmetric, that is, if there is an edge (n_i, n_j) in the theory's graph, then (n_j, n_i) is also present. In the case of a symmetric graph G we can substitute the directed edges by undirected ones, obtaining in this way a new (undirected) graph G'. Then a set of nodes K is a *maximal independent set* (MIS) of G' iff K is a kernel of G. This property, together with theorem 10, lead to the following corollary.

Corollary 11. *Let $\Delta = (D, \emptyset)$ be a default theory. Then K is a MIS for the theory's undirected graph $G'(\Delta)$ iff the set of rules which correspond to the nodes in K is a set of generating defaults $GD(E, \Delta)$ of a J-extension E of Δ.* □

Given this correspondence between J-extensions and maximal independent sets, we can employ some well-known results from graph theory. Obviously computing a maximal independent set of a graph (which always exists) is trivial. Consequently, the same holds for the J-extensions of the default theories we consider here. Recall that the same problem (computing an extension) is intractable for Reiter's default logic (see [DM94]).

Moreover, even for an apparently harder problem, that of enumerating all of the MIS of graph, there is an effective algorithm. First we introduce the necessary notions. We say that a configurations generating algorithm is a *polynomial delay* ([JPY88]) one, if there is only a polynomial delay between any two configurations generated by the algorithm. Such algorithms may behave exponentially because of the number of the exponentially many different configurations, but this is obviously unavoidable. The time complexity of a polynomial delay algorithm is

$O(p(n)C)$ where n is the size of the input and C the number of configurations. In [JPY88] and [TIAS77] polynomial delay algorithms are presented for computing all maximal independent sets of an undirected graph. Hence, the next proposition follows:

Proposition 12. *Let $\Delta = (D, \emptyset)$ be a default theory. The sets of generating defaults of the J-extensions of Δ can be enumerated with polynomial delay.* □

This might seem to imply that the problem of performing goal-directed reasoning (is there an J-extension that contains the literal q?) is also of complexity $O(p(n)C)$, where n is the number of defaults and C the number of the sets of generating defaults. However, the problem is much easier since every node (except if it contains a self-loop) of a graph G belongs to some maximal independent set. Hence, in order to do goal-directed reasoning it suffices to check if the literal at hand belongs to the consequents of some default. Skeptical reasoning (does q belong to all J-extensions?) is trivial as well, since it suffices to check if the literal at hand occurs in the consequent of some default which is not connected to any other default. Recall again that all these problems are intractable for the same class of default theories under Reiter's semantics.

3.2 The general case

In the case of default theories with prerequisites computations become harder, but again the main computational burden comes from the number of solutions (J-extensions) that the problem (theory) may have. The method employed here is to divide the default theory into strata according to their prerequisites, solve the MIS problem within each stratum, and then propagate the solution to the lower strata.
Let *FindMIS(G)* be a polynomial delay algorithm which computes all the MIS of a graph G (see [JPY88], [TIAS77]). The algorithm will be of the form:

FindMIS(G)
While not finished iterate
    body1
    A MIS, S, is computed
    body2

Each iteration takes time polynomial in the size of G and computes a MIS, S. A simple modification of this procedure leads to an algorithm that computes all the J-extensions of a given default theory.

Findextensions($G, D, Cons$)
If $D = \emptyset$ then output Cons
else
while not finished iterate
    body1;
    A MIS, S is computed;

$D' = D - (S \cup \Gamma(S))$;
$Cons' = Cons \cup Cons(S)$;
Find all defaults in D', whose prerequisites is a subset of $Cons'$;
Compute the graph G' induced by these defaults, as described in the previous section;
Call Findextensions($G', D', Cons'$);
body2;
endwhile

When the procedure is called for the first time the arguments are instantiated as follows: The graph G is the rule graph of the subtheory consisting of rules all the prerequisites of which belong to W. The set of defaults D is the set of defaults of the theory at hand, while $Cons$ is first instantiated to W.
Since the depth of the search space is bounded by the size of the theory (number of the rules) the following proposition holds.

Proposition 13. *The above algorithm enumerates with polynomial delay all the sets of generating defaults of the J-extensions of a propositional disjunction free default theory.* □

Graph constructions are also possible for theories with prerequisites. In this case two kinds of edges are needed, depicting negative and positive interactions between rules respectively[3]. Negative edges are added between the rules like in the case of theories without prerequisites. For every literal p which occurs in the prerequisites of a rule d_i and the consequents of the rules $d_1, \ldots, d_k$, the "positive" edges $(n_j, n_i), 1 \leq j \leq k$ are added to the graph of the theory, marked with an *OR symbol*. Such structures are called *OR structures*. Since these constructions have two kind of edges, we call them *networks*, rather than graphs.

Example 2. Let $\Delta = (D, W)$ be a default theory, where $W = \{A\}$ and $D = \{A : MB/C, A : MB/D, A : MF/D, C \wedge D : ME/G\}$. The positive edges (OR structures) are depicted in figure 2A. □

Every network can be transformed to a graph using the technique shown in figure 2B. While all edges in figure 2B.1 are positive, all edges in 2B.2 are negative. The negative edges that may exist in the theory's network are copied into the theory's graph. The new nodes introduced during the transformation (denoted as "a" in Figure 2), are called *auxiliary* or *literal nodes*. Networks of default theories allow us a simple proof of the following result.

[3] The details of the method are presented in [DM94].

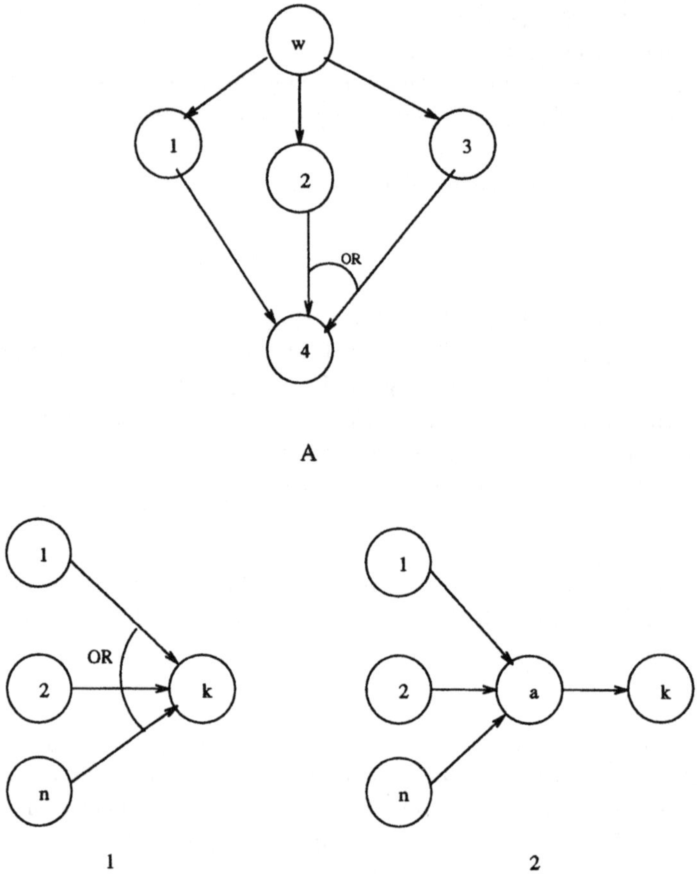

B

Figure 2

Theorem 14. *Let $\Delta = (D, W)$ be a disjunction free propositional default theory and q a literal. Deciding whether there is a J-extension of Δ that includes q is NP-complete.*

Proof. (sketch) Reduction from the kernel problem: Let $G = (N, E)$ be a directed graph. Given G we can construct a network $G' = (N_1 \cup N_2 \cup \{a\}, E')$ as follows: For every node in G there is a node in both N_1 and N_2 together with a distinguished node a. The edges between nodes in N_1 are the edges in G, without direction on them. These are the only negative edges of the network. The nodes in N_2 are connected, via positive edges, to the nodes in N_1 as follows: For every $n_i' \in N_2$ put an edge (n_j, n_i') in E', $n_j \in N_1$, iff $(n_j, n_i) \in E$, together with an edge (n_i, n_i') for every $n_i \in N_1$, $n_i' \in N_2$. Every such set of edges is an OR structure. Finally for every node $n_i' \in N_2$ add an edge (n_i', a) in E'. Figure 3 shows an example of such a transformation.

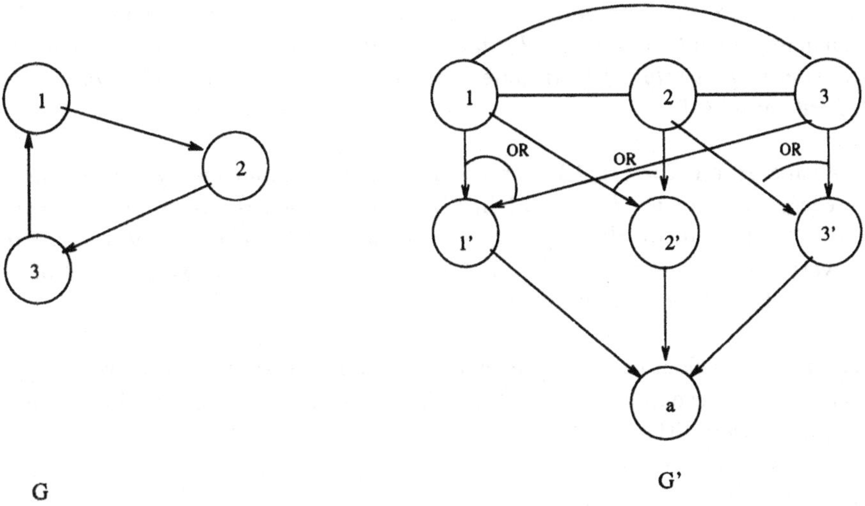

Figure 3

The network produced from G corresponds to a default theory Δ_G which can be constructed as follows. Let d_n denote the default in the theory associated with the node n. Then for every node $n'_i \in N_2$ construct the rule $d_{n_{i'}}$ with $\text{Prer}(d_{n_{i'}}) = \{P_i\}$, $\text{Cons}(d_{n_{i'}}) = \{C_i\} = \text{Just}(d_{n_{i'}})$. For every node $n_i \in N_1$, $\text{Prer}(d_{n_i}) = \emptyset$, $\text{Just}(d_{n_i}) = \{\neg J_i\} \cup \{J_m |$ for every node $n_m \in N_1$, $n_m \in \Gamma(n_i)\} \cup \text{Cons}(d_{n_i})$ and $\text{Cons}(d_{n_i}) = \{P_j | (n_i, n_{j'})$ belongs to network and $n_{j'} \in N_2\}$. Finally $\text{Prer}(d_a) = \bigcup \text{Cons}(d_{n_{i'}})$, and $\text{Cons}(d_a) = \{q\} = \text{Just}(d_a)$. Then every J-extension of Δ_G that contains q, induces a kernel for G, and vice-versa. Hence, if there is a polynomial algorithm answering the question whether there is a J-extension that contains q, then the kernel problem can be solved in polynomial time as well. This can not be true, unless NP=P. □

However, in practice the search space can be considerably reduced. Let q be a literal and R the set of rules which contain q in their consequents. Then the question "is there a J-extension including q", has time complexity $O(p(n)C)$, where n is the number of defaults of the subtheory of Δ which includes all rules on all possible paths through positive edges from W to some rule in R and C the number of the generating defaults of the J-extensions of this subtheory.

4 Semantical considerations

The graph model described in the previous sections allows a representation in propositional logic, of every propositional disjunction free default theory under Reiter extension or J-extensions semantics. We restrict ourselves to the case of theories without *cycles of positive edges* in the theory's network (see [DM94]). For these theories we can prove the following result.

Proposition 15. *Let $\Delta = (D, W)$ be a propositional seminormal disjunction free default theory without positive cycles in the theory's network. If $G_R(\Delta)$*

and $G_J(\Delta)$ are the rule graphs of the theory under Reiter's semantics and J-extensions semantics respectively, then E and E' is a Reiter extension and J-extension respectively, iff their generating defaults induce a kernel in the associated rule graphs. □

With a graph $G = (N, E)$ we associate a propositional theory G_P as follows: For every node $n_i \in N$, we add in G_P a set of clauses $\neg n_i \vee \neg n_j$, for every $n_j \in \Gamma^-(n_i)$. Additionally for every $n_i \in N$ we add the clause $n_i \vee n_{i1} \ldots \vee n_{ik}$ for every node $n_{ij} \in \Gamma^-(n_i), 1 \leq j \leq k$. We can prove that every model of G_P induces a kernel for G, and vice-versa.

Example 3. Consider the theory of example 1. The two graphs are shown in figure 1. The first corresponds to Reiter's extensions while the second to J-extensions. The propositional theory for G_R is:
$\neg n_1 \vee \neg n_4, \neg n_2 \vee \neg n_1,$
$\neg n_2 \vee \neg n_4, \neg n_3 \vee \neg n_2, \neg n_4 \vee \neg n_3,$
$n_1 \vee n_4, n_2 \vee n_1 \vee n_4,$
$n_3 \vee n_2, n_4 \vee n_3$
while for G_J
$\neg n_1 \vee \neg n_2, \neg n_1 \vee \neg n_4,$
$\neg n_2 \vee \neg n_1, \neg n_2 \vee \neg n_3, \neg n_2 \vee \neg n_4,$
$\neg n_3 \vee \neg n_4, \neg n_3 \vee \neg n_2,$
$\neg n_4 \vee \neg n_1, \neg n_4 \vee \neg n_2, \neg n_4 \vee \neg n_3$
$n_1 \vee n_4 \vee n_2, n_2 \vee n_1 \vee n_4 \vee n_3,$
$n_3 \vee n_2 \vee n_4$
The satisfying truth assignment for G_{R_P} is $\{n_1, n_3\}$, while for G_{J_P} are $\{n_1, n_3\}$, $\{n_2\}$ and $\{n_4\}$. These assignments correspond to the generating defaults of the extensions. □

Skeptical and credulous reasoning can be easily embedded in this framework. The question whether there is a J-extension including a literal p is equivalent to deciding the satisfiability of the set of clauses $G_{J_P} \cup \{\vee n_i | n_i$ is associated with a rule which contains p in its consequents $\}$. On the other hand proving that q skeptically follows form a theory amounts to proving the unsatisfiability of $G_{J_P} \cup \{\neg n_1\} \ldots \cup \{\neg n_k\}$, where $\{n_1, \ldots, n_k\}$ is the set of nodes associated with rules that include q in their consequents.

5 The case of Logic Programs

In this section we introduce a semantics for logic programs based on a modification of the Gelfond-Lifschitz transformation ([GL88]), that is closely related to the notion of the J-extensions.

Definition 16. Let P be a logic program and M an interpretation. We define $GLC(M, P)$ to be the program obtained after:

- Deleting every rule in P for which either $\neg p$ occurs in the body of the rule while p occurs in M or $\neg p$ occurs in M and p is the head of the rule.
- Deleting all negative literals from the body of the remaining rules.

Definition 17. Let P be a logic program and M an interpretation. We call M a *constraint model* of P iff M is identical to the minimal model of $GLC(M,P)$. If M is a maximal constraint model of P then M is a *constraint stable model* of P. □

We now show that the above semantics reflect the notion of joint consistency. Namely for every logic program P there exists a default theory Δ_P such that, the constraint stable models of P are related to the J-extensions of Δ_P. This relationship is demonstrated by the two propositions that follow the definition.

Definition 18. Let P be a logic program. Then the associated to P default theory Δ_P is defined to be the theory $\Delta_P = (D, W)$, $W = \emptyset$ and $D = \{a_1 \wedge \ldots \wedge a_k : M\neg b_1 \wedge \ldots \wedge \neg b_m \wedge c/c|$ where $c \leftarrow a_1 \wedge \ldots \wedge a_k \wedge \neg b_1 \wedge \ldots \wedge \neg b_m \in P\}$. □

Proposition 19. *Let P be a logic program and Δ_P its associated default theory. Then every constraint stable model of P, induces a J-extension for Δ_P.*

Proof. (sketch) Let M be a constraint stable model for P. Define $GR(M) \subseteq GLC(M,P)$ to be a set of rules which if applied forward lead to M (see that $M^+ = \text{head}(GR(M))$). The set of rules $GR(M)$ is associated with a set of rules $GD(M)$ in the default theory Δ_P. We will show that $E = (E_J, E_T)$, where $E_J = \{p|p \in \text{Just}(d_i) \text{ and } d_i \in GD(M)\}$[4] and $E_T = \{p|p \in \text{Cons}(d_i) \text{ and } d_i \in GD(M)\} = M^+$ is a J-extension. The defaults in $GD(M)$ can be applied in the order induced by M. It suffices to show that for any default $d_i \in GD(M)$, if $p \in \text{Just}(d_i)$ then $\neg p \notin E_T$. See that if this is the case then in $GD(M)$, and consequently in $GLC(M,P)$ there is a rule which contains $\neg p$ in its body, while $p \in M$, a contradiction.
Now assume that there is a rule, $d_i \in D$, $d_i \notin GD(M)$, $\text{Cons}(d_i) \notin E_T$, $\text{Prer}(d_i) \in E_T$, and for every $p \in \text{Just}(d_i)$, $\neg p \notin E_J$ holds. See that this can not be the case because then if $p \in \text{Cons}(d_i)$, $(M - \{\neg p\}) \cup \{p\}$ is a constraint model of P, hence M is not a maximal constraint model, a contradiction. Therefore, E is a J-extension for Δ_P. □

Proposition 20. *Let P be a logic program and Δ_P its associated default theory. Then every J-extension of Δ_P, induces a constraint model for P.*

Proof. (sketch) Let $E = (E_J, E_T)$ be a J-extension of Δ_P. Then E is generated by a set of defaults $GD(E)$. Denote by $GR(E)$ the rules of P, which correspond to the defaults in $GD(E)$. Define M to be an interpretation that assigns true to every atom in E_T and false to the rest. We will show that M is a constraint model of P.
The program $GLC(M,P)$ can not contain a rule the head of which does not

[4] See that E_J and E_J are deductively closed, but only the literals are considered here.

belong to M^+. On the other hand, every rule of $GR(E)$ will be contained in $GLC(M,P)$. This is because for every rule $r_i \in GR(E)$, head$(r_i) \in M$ and additionally if $\neg p \in$body(r_i) then $p \notin E_T$, hence $b_i \notin M$. Therefore, M is a model of $GLC(M,P)$. Furthermore, since $M^+ = E_T$, M will be a minimal model of $GLC(M,P)$, and consequently a constraint model of P. □

It is important to note that the proposition 20 is not the full reciprocal of proposition 19. The next example shows that the inverse of proposition 19 is not always true.

Example 4. Let $P = \{c \leftarrow \neg b, c \leftarrow, b \leftarrow\}$ and $\Delta_P = (D,W)$ its associated default theory, where $D = \{: M\neg b \wedge c/c, : Mc/c, : Mb/b\}$. While the program P has one constraint stable model, namely $M = \{b,c\}$, the theory Δ has two J-extensions, $E_1 = (\{\neg b, c\}), \{c\})$ and $E_2 \doteq (\{b,c\}), \{b,c\})$. See that the J-extensions are not maximal in the sense that $E_{1_T} \subset E_{2_T}$. □

Definition 21. Let Δ be a default theory and E a J-extension of Δ. Then E is a *maximal* J-extension of Δ, if there is no other J-extension E' of Δ, such that $E_T \subset E'_T$. □

Corollary 22. *A interpretation M is a constraint stable model for a logic program P iff M induces a maximal J-extension for the associated default theory Δ_P.* □

Example 5. Consider the program $P = \{p \leftarrow \neg q, r \leftarrow \neg p, q \leftarrow \neg r\}$. This program has three constraint stable models, namely $M_1 = \{p, \neg q, \neg r\}$, $M_2 = \{q, \neg p, \neg r\}$, and $M_3 = \{r, \neg q, \neg p\}$. See that P has no stable model. □

It is easy to see that every program has at least one constraint stable model. An important issue regarding constraint stable models is their computational features. It is open whether there is an polynomial algorithm which computes a constraint model for a given program P and whether the constraint models of P can be explicitly enumerated by an algorithm with polynomial delay.

6 Representing priorities

As pointed out by Brewka ([Bre91]), in the context of the above mentioned Default Logic formalisms much of the expressive power of the seminormal default is lost. In particular, in certain cases the priorities between the rules can not be represented (see [Bre91]). To overcome this difficulty he proposes to consider only *priority preserving* extensions, which are defined below.

Definition 23. ([Bre91]) An CDL extension E, of a assertion default theory (D,W) is called *priority preserving* if for no $A : B/C \in (D - GD(E))$, $A \in Form(E)$, $\{B,C\} \cup Form(E)$ is consistent, and $\{C\} \cup Form(E) \cup Supp(E)$ is inconsistent. □

While every assertion default theory has at least one CDL extension, this is not the case for priority preserving extensions. We prove now that the problem of computing priority preserving extensions is intractable, even in simple cases.

Theorem 24. *Let $\Delta = (D, \emptyset)$ be a propositional seminormal disjunction and prerequisite free assertion default theory. Deciding whether Δ has a priority preserving CDL extension is NP-complete.*

Proof. (sketch) The proof is by reduction from the kernel problem. Let $G = (N, E)$ be a directed graph. From G we can construct a propositional seminormal disjunction and prerequisite free theory $\Delta_G = \{D, \emptyset\}$, where D contains a rule d_i for every node $n_i \in N$, $d_i =: \neg n_1 \wedge \ldots \wedge \neg n_m \wedge n_i / n_i$, for every $n_j \in \Gamma^-(n_i)$, $1 \leq j \leq m$. We will show that E is a priority preserving CDL extension of Δ_G iff the rules in $GD(E)$ form a kernel for G. From the previous results of the paper follows that E is a CDL extension of Δ_G iff the rules in $GD(E, \Delta_G)$ form a MIS for G.
Let K be the nodes corresponding to the generating defaults of a priority preserving CDL extension E of Δ_G. Then K will be a MIS. We will show that K is dominating. Assume that K is not dominating. Therefore there is $n_i \notin K$ such that n_i is not receiving any edge from K. Hence for every negative literal $\neg n_k \in \mathrm{Just}(d_i)$, $n_k \notin E$, therefore for $d_i =: B/C$, $\{B, C\} \cup Form(E)$ is consistent. On the other hand, since K is a MIS, n_i must be adjacent to some node $n_j \in K$, $n_j \in \Gamma^+(n_i)$. This means that for the atom $n_i \in \mathrm{Cons}(d_i)$, $\neg n_i \in Supp(E)$ holds. Hence $\mathrm{Cons}(d_i) \cup Supp(E)$ is inconsistent, therefore E is not a priority preserving extension, which contradicts the assumption. Hence, K must be a kernel.
Let K be a kernel of G. The set of rules which correspond to nodes in K form a CDL extension for Δ_G. Since K is a kernel then for every node $n_i \notin K$, there will be a node $n_j \in K$ such that $n_j \in \Gamma^-(n_i)$. Hence, $n_j \in \mathrm{Cons}(d_j)$ and $\neg n_j \in \mathrm{Just}(d_i)$, which means that $\mathrm{Just}(d_i) \cup Form(E)$ is inconsistent, therefore the rules of K form a priority preserving extension.
We proved that the kernels of a graph G are in direct correspondence with the priority preserving CDL extensions of a propositional disjunction and prerequisite free theory Δ_G. Hence, deciding whether a propositional disjunction and prerequisite free assertion default theory has a priority preserving CDL extension can not be done in polynomial time, unless NP=P. □

7 Conclusions

In this paper we presented some complexity results for simple cases of default reasoning under semantics that require joint consistency of the justifications. This requirement makes default reasoning easier than in the original Reiter's formalism, but still, important fragments of reasoning remain intractable. However, it seems that for the cases we have examined here, the main source of the computational burden comes form the number of the different solutions that the problem may have. Theories with a sufficiently small number of J-extensions are expected to have a nice computational behavior.

The graph-theoretic formulation also provides a way to translate, in simple cases, default theories under different semantics, into classical propositional theories. In this way, a propositional semantics for J-extensions has been derived and methods for reasoning in classical logic can be used in the of default theories. Another translation from default theories under Reiter's semantics, into propositional logic has been presented in [BED91] and [BED92]. While their method is a direct translation into classical logic, the graph-theoretic approach offers the possibility to combine both graph and logic based techniques in tackling the reasoning problem.

The constraint stable models introduced here, is a preliminary attempt to define a semantics for general logic programs that reflects the notion of joint consistency. However the correspondence is not exact, since the constraint stable models of a program P correspond to the maximal J-extensions of the associated theory Δ_P. Many problems remain open. First, it seems possible that a 3-valued constraint stable model semantics is needed in order to fully capture the notion of joint consistency as implied by the J-extensions. The second open issue is the relation of the constraint stable models with other semantics for logic programs. Finally, the complexity of reasoning with the constraint stable models is an important question yet to be answered.

Acknowledgements The author thanks Torsten Schaub for many insightful comments and discussions.

References

[BED91] R. Ben-Eliyahu and R. Dechter. Default logic, propositional logic and constraints. In *Proc. of AAAI-91*, pages 379–385, 1991.

[BED92] R. Ben-Eliyahu and R. Dechter. Propositional semantics for default logic. In *Fourth International Workshop on Nonmonotonic Reasoning*, pages 13–27, Plymouth, VT, 1992.

[Bre91] G. Brewka. Cumulative default logic: in defence of nonmonotonic inference rules. *AI Journal*, 50:183–205, 1991.

[DJ91] J. Delgrande and W. Jackson. Default logic revisited. In J. Allen, R. Fikes, and E. Sandewall, editors, *Proc. of the Second Int. Conf. on Principles of Knowledge Representation and Reasoning*, pages 118–127. Morgan Kaufmann, 1991.

[DM94] Y. Dimopoulos and V. Magirou. A graph-theoretic approach to default logic. To appear in *Information and Computation*, 1994.

[DMP93] Y. Dimopoulos, V. Magirou, and C. Papadimitriou. On kernels, defaults and even graphs. Technical Report MPI-I-93-226, Max-Planck-Institut für Informatik, 1993.

[DT93] Y. Dimopoulos and A. Torres. Graph-theoretic structures in logic programs and default theories. Technical Report MPI-93-264, Max-Planck-Insitut für Informatik, 1993.

[GJ79] M. R. Garey and D. S. Johnson. *Computers and intractability: A guide to the theory of NP-completeness.* W. H. Freeman and Company, New York, 1979.

[GL88] M. Gelfond and V. Lifschitz. The stable model semantics for logic programming. In *Proc. Fifth International Conference and Symposium on Logic Programming*, pages 1070–1080, Cambridge, Mass., 1988. MIT Press.

[GM93] G. Gottlob and Z. Mingyi. Cumulative Default Logic: Finite Characterization, Algorithms, and Complexity. Draft paper, 1993.

[JPY88] D. Johnson, C. Papadimitriou, and M. Yannakakis. On generating all maximal independent sets. *Information Processing Letters*, 27(3):119–123, 1988.

[KS91] H. Kautz and B. Selman. Hard problems for simple default theories. *AI Journal*, 49, 1991.

[Luk88] W. Lukaszewicz. Considerations on default logic - an alternative approach. *Computational Intelligence*, 4:1–16, 1988.

[PS92] C. Papadimitriou and M. Sideri. On finding extensions of default theories. In J. Biskup and H. Hull, editors, *Proc. Fourth International Conference on Database Theory*, Berlin, Germany, 1992. Springer Verlag. LNCS, 646.

[PY92] C. Papadimitriou and M. Yannakakis. Tie-breaking semantics and structural totality. In *Proceedings Eleventh Symposium on Principles of Database Systems*, pages 16–22, 1992.

[Rei80] R. Reiter. A logic for default reasoning. *AI Journal*, 13:81–132, 1980.

[Sch91] T. Schaub. On commitment and cumulativity in default logics. In R. Kruse, editor, *European Conference on Symbolic and Quantitative Approaches to Uncertainty*, pages 304–309. Springer Verlag, 1991.

[TIAS77] S. Tsukiyama, M. Ide, H. Ariyoshi, and I. Shirakawa. A new algorithm for generating all maximal independent sets. *SIAM J. Comput.*, 6(3):505–517, 1977.

[Tor94] A. Torres. *The Hypothetical Semantics of Logic Programs*. PhD thesis, Stanford University, February 1994.

BELIEF DYNAMICS, ABDUCTION, AND DATABASES

Chandrabose ARAVINDAN and Phan Minh DUNG
Computer Science Program, School of Advanced Technologies
Asian Institute of Technology
P.O. Box 2754, Bangkok 10501. Thailand.
Internet: {arvind,dung}@cs.ait.ac.th

Abstract: In this paper, we introduce a new concept of generalized partial meet contraction for contracting a sentence from a belief base. We show that a special case of belief dynamics, referred to as knowledge base dynamics, where certain part of the belief base is declared to be immutable, has interesting connections with abduction, thus enabling us to use abductive procedures to realize contractions. Finally, an important application of knowledge base dynamics in providing an axiomatic characterization for deleting view atoms from databases is discussed in detail.

Keywords: epistemology, belief dynamics, rationality postulates, abduction, databases, view updates

§1. Introduction

We live in a constantly changing world, and consequently our beliefs have to be revised when there is new information. The central problem of epistemology is to study when we can be sure that we have revised our beliefs rationally. This has been studied extensively at a more general and abstract level by the researchers from the field of philosophy, leading to a new branch of study: the *belief dynamics*. Assuming a belief state of deductively closed set of sentences, Alchourrón, Gärdenfors, and Makinson, propose certain rationality postulates, popularly known as AGM-postulates, to be satisfied when the belief set is contracted or revised [e.g. 2,3,15,16,17,35]. The AGM-model for belief dynamics has gained popularity due to its attractive features like constructive models of contraction (partial meet contraction) and representation theorems. However, various researchers have pointed out the disadvantages of working with infinite belief sets and question the rationality of one of the AGM-postulates (recovery) [e.g. 21,22,23,25,36,37, 38]. Consequently Fuhrman, Hansson, Nebel, and others propose alternate belief models based on belief bases, set of sentences that need not be deductively closed. Other approaches to belief dynamics include: the model-theoretic approach of Katsuno and Mendelzon [29]; relational belief revision of Lindström and Rabinowicz [32]; and sphere model of Grove [18].

However, these philosophical works are mainly at an abstract level, paying little attention to computational aspects. It is important to study how the results of belief dynamics can be realized in real applications, such as view updates in databases, diagnosis, abductive reasoning, counterfactuals, non-monotonic reasoning etc. In his recent paper [24], Hansson has pointed out the gap between philosophical approaches and computational approaches, and reviews how the researchers from the philosophical side strive to move towards realism. The belief base approach of Hansson, mentioned above, is in fact a result of such an attempt. However, the results of belief dynamics are still far away from concrete applications, like view updates in databases, and computational tractability and efficiency of algorithms satisfying the rationality postulates are still unclear.

On the other hand, works on concrete problems like view updates, concentrate on operational aspects, without explicitly formalizing the intuition behind the update. A lot of algorithms have been proposed for view updates in relational databases [e.g. 10,31], and algorithms based on deduction trees have been proposed to carry out view updates on deductive databases [e.g. 9,11,19,20,28,39,40,41] ([1] provides a good survey of view updates in databases). Clearly, these algorithms for view updates are not at all arbitrary, and there are certain guiding factors behind them. For example, all the view definitions are considered as immutable and any algorithm for view update changes only the base facts. Moreover, when a view atom A is to be deleted, any algorithm has to make sure that A does not follow from the updated database. Are these two the only guidelines for view updates? Will any algorithm that satisfies these two policies be considered rational? Let us motivate the discussion with an example.

Example 1.1 Consider a deductive database, as follows:

student_dean(X,Y) ← student_school(X,Z), school_dean(Z,Y)
school_dean(cs,sharma) ←
student_school(dung,cs) ←
student_school(arvind,cs) ←

Suppose the view atom student_dean(dung,sharma), which currently follows from the database, is to be deleted. Assume that we have an algorithm that deletes both the base facts (student_school(dung,cs) ←) and (student_school(arvind,cs) ←) from the database. Clearly, this satisfies both the guidelines set above. But, no rational person will agree with such an algorithm, because the fact (student_school(arvind,cs) ←) is not "relevant" to the view atom to be deleted. ■

The above example highlights the need for some kind of "relevance policy" to be adopted when a view atom is to be deleted from a deductive database. How many such axioms and policies do we need to characterize a "good" view update? When are we sure that our algorithm for view update is "rational"? Clearly, there is a need for an axiomatic characterization of view updates. By axiomatic characterization, we mean explicitly listing all the rationality axioms that are to be satisfied by any algorithm for view update.

Thus we have abstract notions and results from the field of belief dynamics, whose acceptability in terms of computational efficiency has to be studied in concrete applications, in one hand; and algorithms for view updates in databases, whose rationality has to be studied against a scale, on the other hand. Our goal now is to bring these two fields closer to share the concepts and results across the fields.

As a first step towards this goal, in this paper we propose a restricted case of belief base dynamics, referred to as knowledge base dynamics, where certain part of belief is declared to be immutable, and bring out its relationship with abduction. We belief that knowledge base dynamics can provide the vital link between theory of belief dynamics and concrete applications, and to this effect we demonstrate how it provides an axiomatic characterization for deleting view atoms from definite deductive databases. We observe that the algorithms proposed in the literature, such as the ones proposed by Tomasic [41], Dayal & bernstein [10], do comply with our axioms, but are computationally inefficient. On the other hand, efficient algorithms, such as the ones proposed by Decker [11], Guessoum & Lloyd [19], Kakas & Mancarella [28], violate certain axioms, raising questions on rationality of those axioms on practical grounds. Thus, as a result of our axiomatization, we raise an important problem of balancing rationality with computational efficiency, which has to be addressed from both philosophical and computational perspectives.

The rest of the paper is organized as follows: In section 2, we review the concepts of belief dynamics that are relevant to this paper, along with our contributions to that field. Section 3 concentrates on knowledge base dynamics and shows how it is related to abduction. In section 4, we discuss an important application of knowledge base dynamics in providing an axiomatic characterization for deleting view atoms from databases. We also analyze how view update algorithms proposed in the literature comply with this axiomatic characterization, pointing out that computationally more efficient algorithms do not satisfy some rationality postulates, calling for further research in this direction. Section 5 concludes the paper with a discussion on the results of this research and directions for further research.

§2. Epistemology: Rationality postulates and constructions

Dynamics of belief has been studied at a more abstract level of philosophy resulting in certain *rationality postulates* that have to be satisfied while revising or contracting a state of belief. The model of belief dynamics proposed by Alchourrón, Gärdenfors, and Makinson, popularly known as the AGM-model, assume a belief state of deductively closed (according to a suitable logic) set of sentences from a language L [e.g. 2,3,15,16,17, and the references therein]. However, researchers like Fuhrmann, Hansson, Nebel, and others have studied revision and contraction of belief bases (set of sentences that need not be deductively closed) also [e.g. 14,22,23,25,37,38, and the references therein]. In this section, we briefly review the concepts from these works, that are relevant to this paper, along with our contribution to this field.

A state of belief can be *expanded* by simply adding a new information that is consistent with the present state of belief; *revised* with a new information that contradicts the current beleif state; or some belief from the current state can be *contracted*. AGM proposes the following rationality postulates that any contraction function $\dot{-}$ is expected to satisfy. The reader is referred to [e.g. 2,3, 15, 16, 17] for the motivation and intuition behind these postulates.

Definition 2.1 [17] (Rationality postulates for contraction of a sentence from a belief set)
For any sentence α and belief set[1] K:

(K$\dot{-}$1)	$K\dot{-}\alpha$ is a belief set.	(Closure)
(K$\dot{-}$2)	$K\dot{-}\alpha \subseteq K$	(Inclusion)
(K$\dot{-}$3)	If $\alpha \notin K$, then $K\dot{-}\alpha = K$	(Vacuity)

[1]A belief set K is a deductively closed (according to a suitable logic) set of sentences from L, i.e. K = Cn(K) where Cn is a consequence operator.

$(K\dot{-}4)$	If $\nvdash \alpha$, then $\alpha \notin K\dot{-}\alpha$	(Success)
$(K\dot{-}5)$	If $\vdash \alpha \leftrightarrow \beta$, then $K\dot{-}\alpha = K\dot{-}\beta$	(Preservation)
$(K\dot{-}6)$	If $\alpha \in K$, then $K \subseteq (K\dot{-}\alpha) + \alpha$	(Recovery)
$(K\dot{-}7)$	$K\dot{-}\alpha \cap K\dot{-}\beta \subseteq K\dot{-}(\alpha\wedge\beta)$	(Intersection)
$(K\dot{-}8)$	If $\alpha \notin K\dot{-}(\alpha\wedge\beta)$, then $K\dot{-}(\alpha\wedge\beta) \subseteq K\dot{-}\beta$	(Conjunction) ■

A function to contract α from K can be defined using the set of inclusion-maximal subsets of K that do not imply α, which is formally defined as: $K\downarrow\alpha = \{ K' \subseteq K \mid K' \nvdash \alpha$ and if $K' \subset K'' \subseteq K$, then $K'' \vdash \alpha \}$. Now, a *partial meet contraction* of α from K is defined as: $K \dot{-}_p \alpha = \bigcap \gamma(K\downarrow\alpha)$, where, γ is a selection function that selects a non-empty subset of $K\downarrow\alpha$, when the later is not empty. When $K\downarrow\alpha$ is empty, $\gamma(K\downarrow\alpha) = \{K\}$. It is not difficult to see that the set of all partial meet contractions of α from K, which we refer to as the *solution space*, forms a complete semi-lattice[2] wrt set-inclusion. Two interesting limiting cases of partial meet contractions are: *maxichoice contraction*, when γ selects a unique element of $K\downarrow\alpha$; and *full meet contraction*, when γ selects all the elements of $K\downarrow\alpha$. It is clear that all maxichoice contractions are the maximal elements in the semi-lattice, while the full meet contraction is the minimal element. A partial meet contraction is said to be *transitively relational*, iff there exists a transitive and reflexive ordering relation ≤ s.t. for all non-empty $K\downarrow\alpha$: $\gamma(K\downarrow\alpha) = \{ K' \in K\downarrow\alpha \mid \forall K'' \in K\downarrow\alpha\text{: } K'' \leq K' \}$.

Theorem 2.2 [17] For every belief set K, $\dot{-}_p$ is a partial meet contraction function iff it satisfies postulates $(K\dot{-}1)$ to $(K\dot{-}6)$. It is a transitively relational partial meet contraction function iff it satisfies postulates $(K\dot{-}1)$ to $(K\dot{-}8)$. ■

Working with deductively closed, infinite belief sets is not very attractive from computational point of view. Moreover, various researchers question the rationality of recovery postulate [e.g. 21,36, and the references therein], and consequently in [23] Hansson motivates the need for belief dynamics model based on belief bases that need not be deductively closed. The concepts of partial meet contraction can be easily carried over into belief bases, by simply dropping the condition that K is deductively closed. Partial meet and its limiting cases full meet and maxichoice contractions, and also transitively relational contractions, for belief bases are defined in the similar way as for belief sets. As in the case of belief set, the solution space of partial meet contractions forms a complete semi-lattice. Note that, given a belief base B, its corresponding belief set K_B can be obtained by simply deductively closing B, i.e. $K_B = Cn(B)$, where Cn is a consequence operator. Thus results of these two models (belief base and belief set) can be compared with each other.

The main result of Nebel [37] is that a maxichoice contraction for belief bases is highly rational, and is similar to a partial meet contraction for belief sets that satisfies all postulates $(K\dot{-}1)$ to $(K\dot{-}8)$. Hansson, on the other hand, proposes the following rationality postulates, replacing recovery postulate by relevance postulate. Apart from that, Hansson also discusses the need for "background" selection functions, i.e. when a belief base is contracted using a selection function, one should also know the selection function to be used

---

[2] a set S forms a complete semi-lattice wrt a partial order iff each non-empty subset of S has a greatest lower bound and each increasing sequence of S has a least upper bound.

for the contracted belief base. To this effect, he introduces the notion of *superselector*, which is defined as functions f that assign a selection function $f(B) = \gamma_B$ for each belief base B. A superselector is *unified*, iff $\forall B_1, B_2, \alpha_1, \alpha_2$: if $B_1 \downarrow \alpha_1 = B_2 \downarrow \alpha_2 \neq \phi$, then $(f(B_1))(B_1 \downarrow \alpha_1) = (f(B_2))(B_2 \downarrow \alpha_2)$. Any partial meet contraction function for belief bases, defined using a unified superselector is then referred to as *unified partial meet contraction* for belief bases.

Definition 2.3 [22] (Rationality postulates for contraction of a sentence from a belief base)
For any sentence α and belief base B:

(B$\dot{-}$1) $B \dot{-} \alpha \subseteq B$ (Inclusion)

(B$\dot{-}$2) If $\beta \in B \setminus (B \dot{-} \alpha)$, then $\exists B'$ with $B \dot{-} \alpha \subseteq B' \subset B$ s.t. $\alpha \notin Cn(B')$ and $\alpha \in Cn(B' \cup \{\beta\})$[3] (Relevance)

(B$\dot{-}$3) If $\nvdash \alpha$, then $\alpha \notin Cn(B \dot{-} \alpha)$ (Success)

(B$\dot{-}$4) If it holds for all subsets B' of B that $\alpha \notin Cn(B')$ iff $\beta \notin Cn(B')$, then $B \dot{-} \alpha = B \dot{-} \beta$ (Uniformity)

(B$\dot{-}$5) If $\nvdash \alpha$, and each element of Z implies α, then: $B \dot{-} \alpha = (B \cup Z) \dot{-} \alpha$ (Redundancy) ∎

Theorem 2.4 [22] For every belief base B, $\dot{-}_p$ is a partial meet contraction function iff it satisfies postulates (B$\dot{-}$1) to (B$\dot{-}$4). It is a unified partial meet contraction function iff it satisfies postulates (B$\dot{-}$1) to (B$\dot{-}$5). ∎

In [21], Hansson argues that there are certain rational cases where relevance need not be satisfied, and his example is reproduced below:

Example 2.5 [21] Consider the following real life conversation between two persons P1 and P2:

P1: Last summer I saw a three-toed woodpecker just outside my window. I could clearly see its red forehead and its red rump.

P2: You must be mistaken. The three-toed woodpecker does not have a red forehead or a red rump.

P1: You make me uncertain. Thinking about it, the only thing I am certain of is that the bird had a red forehead.

Here, P1's original beliefs were that the bird was a three-toed woodpecker (x), that it had a read forehead (y) and that it had a red rump (z), i.e. $B = \{x,y,z\}$. She contracted $x \wedge (y \vee z)$ that had been denied by P2, and retained only y, i.e. $B \dot{-} (x \wedge (y \vee z)) = \{y\}$. Clearly this result is not a partial meet contraction, since $B \downarrow (x \wedge (y \vee z)) = \{\{x\}, \{y,z\}\}$. ∎

To capture such cases, Hansson proposes a weak relevance policy referred to as core-retainment, and defines a preservative withdrawal of a sentence α from a belief base B, denoted as $B - \alpha$, as follows.

---

[3]Cn is a consequence operator.

Definition 2.6 [22] (Rationality postulates for preservative withdrawal of a sentence from a belief base)

For any sentence α and belief base B:

(B—1)	$B \cap Cn(B - \alpha) \subseteq B - \alpha$	(Extended Closure)
(B—2)	$B - \alpha \subseteq B$	(Inclusion)
(B—3)	If $\alpha \notin Cn(B)$, then $B - \alpha = B$	(Vacuity)
(B—4)	If $\nvdash \alpha$, then $\alpha \notin Cn(B - \alpha)$	(Success)
(B—5)	If $\vdash \alpha \leftrightarrow \beta$, then $B - \alpha = B - \beta$	(Preservation)
(B—6)	If $\beta \in B \backslash (B - \alpha)$, then $\exists B'$ s.t. $B' \subset B$; and $\alpha \notin Cn(B')$; and $\alpha \in Cn(B' \cup \{\beta\})$	(Core-retainment) ■

Characteristics of preservative withdrawal have not been studied further, and it is not clear what kind of solution space it provides. Unfortunately, there are no constructive model and representation theorem for the above postulates, and the limiting cases for the preservative withdrawals are not obvious. In our attempt to provide one, we generalize the partial meet contractions on belief bases as follows:

Definition 2.7 Let B be a belief base and α a belief to be contracted from it. Then, we define,

$$(B\downarrow\alpha)^* = \begin{cases} \{ B_i \mid \exists B_i' \in B\downarrow\alpha \text{ and } \bigcap(B\downarrow\alpha) \subseteq B_i \subseteq B_i' \} & \text{when } B\downarrow\alpha \text{ is not empty} \\ \{B\} & \text{otherwise} \end{cases}$$

Every element B_i of $(B\downarrow\alpha)^*$ is called as a *generalized partial meet* contraction of α from B, denoted as $B \dot{-}_g \alpha$. ■

It is clear that, the solution space of generalized partial meet contractions of α from B forms a complete semi-lattice, with the full meet contraction as the minimal bottom element, and all maxichoice contractions as maximal top elements. As it is evident from the following example, generalized partial meet contractions do not represent preservative withdrawals in general, as they do not satisfy the extended closure postulate.

Example 2.8 Consider a belief base $B = \{x, x\vee\neg y, z\}$ and a belief $\alpha \equiv (x\vee\neg y)\wedge z$ to be contracted from B. It is not difficult to see that there are two maxichoice contractions, i.e. $B\downarrow\alpha = \{\{x, x\vee\neg y\}, \{z\}\}$. The full meet contraction is the intersection of these two choices, i.e. ϕ. Clearly, $\phi \subseteq \{x\} \subseteq \{x, x\vee\neg y\}$, and so $\{x\}$ is a generalized partial meet contraction of α from B. But, $x\vee\neg y \in B$; $x\vee\neg y \in Cn(\{x\})$; and $x\vee\neg y \notin \{x\}$, and so the extended closure property is not satisfied. ■

However, generalized partial meet contractions do define the boundary of preservative withdrawal, and by adding an extra condition (that every B_i satisfies extended closure property) to the definition of the set $(B\downarrow\alpha)^*$, one can easily obtain a representation for preservative withdrawals. But, we believe that generalized partial meet contraction deserves independent treatment, because it has a constructive model and a representation theorem, to be presented next, and the violation of extended closure postulate is not really a handicap. We now show that any preservative withdrawal is a generalized partial meet contraction, and thus the solution space of generalized partial meet contractions contains that of preservative withdrawal.

Theorem 2.9 For every belief base B and belief α, any preservative withdrawal of α from B is a generalized partial meet contraction of α from B.

Proof

Let B—α be a preservative withdrawal of α from B. We want to show that B—α is a generalized partial meet contraction of α from B. When α is a valid statement, $B—\alpha = B \dot{-}_g \alpha = B$. When α is not valid, the required result follows from the following two observations:

1) $\exists B' \in B{\downarrow}\alpha$ s.t. $B—\alpha \subseteq B'$ (when α is not valid)

Note that B—α satisfies inclusion and thus contained in B. Also, success property is satisfied and hence $\alpha \notin Cn(B—\alpha)$. Every element of $B{\downarrow}\alpha$ is a inclusion maximal subset that does not derive α, and so any subset of B that does not derive α must be contained in a member of $B{\downarrow}\alpha$.

2) $\bigcap(B{\downarrow}\alpha) \subseteq B—\alpha$ (when α is not valid)

$\forall\beta \in \bigcap(B{\downarrow}\alpha)$, we show the following: if $\beta \notin B—\alpha$, then B—α does not satisfy core-retainment property. In other words, the full meet contraction is the "core" that a preservative withdrawal is aiming to "retain".

Note that β is a member of *every* maxichoice contraction. Hence, there dose *not* exist a set B' s.t. $B' \subseteq B$; $\alpha \notin Cn(B')$; and $\alpha \in Cn(B' \cup \{\beta\})$. This is because, any such B' is obviously contained in a maxichoice contraction and $B' \cup \{\beta\}$ is also contained in that maxichoice contraction, which does not derive α.

Thus it follows that B—α is a generalized partial meet contraction of α from B. ■

We next provide an axiomatic characterization of generalized partial meet contraction for belief bases, which serves as its representation theorem.

Theorem 2.10 For every belief base B, $\dot{-}_g$ is a generalized partial meet contraction iff it satisfies the postulates (B—2) to (B—6).

Proof

Let $B \dot{-}_g \alpha$ be a generalized partial meet contraction of α from B. The "if" part follows immediately from theorem 2.9. For the "only if" part, we have to show that $B \dot{-}_g \alpha$ satisfies the postulates (B—2) to (B—6).

(B—2) Obviously, $B \dot{-}_g \alpha \subseteq B$.

(B—3) If, $\alpha \notin Cn(B)$, then $B{\downarrow}\alpha = \{B\}$, and hence $B \dot{-}_g \alpha = B$. So, vacuity is satisfied.

(B—4) When α is not a valid statement, every member of $B{\downarrow}\alpha$ do not derive α. Thus $\alpha \notin Cn(B \dot{-}_g \alpha)$, and so success is satisfied.

(B—5) If, $\vdash \alpha \leftrightarrow \beta$, then $B{\downarrow}\alpha = B{\downarrow}\beta$. So, preservation is satisfied.

(B—6) Let $\beta \in B \backslash (B—\alpha)$. Then it is clear that $\beta \notin \bigcap(B{\downarrow}\alpha)$, and hence $\exists B' \in B{\downarrow}\alpha$ s.t. $\beta \notin B'$. From the definition of $B{\downarrow}\alpha$, it is clear that $B' \subset B$; $\alpha \notin Cn(B')$; and $\alpha \in Cn(B' \cup \{\beta\})$. So, core-retainment is satisfied. ■

Having identified that solution space of preservative withdrawals is contained in that of generalized partial meet contraction, we now show that the set of all preservative withdrawals of a sentence α from a belief base B forms a complete semi-lattice.

Lemma 2.11 Let B be a belief base and α a sentence. Then, the set of all preservative withdrawals of α from B forms a complete semi-lattice wrt set-inclusion.

Proof Let S be the set of all preservative withdrawals of α from B. To show that S forms a semi-lattice wrt set-inclusion, we need to show: (1) every non-empty subset S' of S has a greatest lower bound; and (2) every increasing sequence S'' of S has a least upper bound.

(1) To show that every non-empty subset S' of S has a greatest lower bound, we need to prove that $\cap_i B_i$ (intersection of all the elements B_i of S') is a preservative withdrawal of α from B, satisfying the postulates (B—1) to (B—6). Clearly $\cap_i B_i$ is a generalized partial meet and so satisfies (B—2) to (B—6). We just need to show that it satisfies (B—1) also. Since each B_i of S' is a preservative withdrawal satisfying (B—1), we have that $\cap_i(B \cap Cn(B_i)) \subseteq \cap_i B_i$. From this we get, $B \cap \cap_i(Cn(B_i)) \subseteq \cap_i B_i$. Since, $Cn(\cap_i B_i) \subseteq \cap_i(Cn(B_i))$, it now follows that, $B \cap Cn(\cap_i B_i)) \subseteq \cap_i B_i$. Thus, $\cap_i B_i$ satisfies (B—1).

(2) To show that every increasing sequence S'' of S has a least upper bound, we need to prove that $\cup_i B_i$ (union of all the elements B_i of S'') is a preservative withdrawal of α from B. Again, we need only to show that $\cup_i B_i$ satisfies (B—1). Since each B_i is a preservative withdrawal, all members of S'' satisfy (B—1), and so we have $\cup_i(B \cap Cn(B_i)) \subseteq \cup_i B_i$. From this it follows that $B \cap \cup_i(Cn(B_i)) \subseteq \cup_i B_i$. Since the consequnce operation is compact, we get, $B \cap Cn(\cup_i B_i) \subseteq \cup_i B_i$. Thus, $\cup_i B_i$ satisfies (B—1). ■

Finally, we show the direct relationship between generalized partial meet contractions on belief bases and partial meet contractions on belief sets.

Theorem 2.12 Let B be a belief base and α a belief. Then, $B \dot{-}_g \alpha$ is a generalized partial meet contraction of α from belief base B iff $Cn(B \dot{-}_g \alpha)$ is partial meet contraction of α from belief set Cn(B).

Proof Follows from theorem 2.10 and theorem T1 of [21]. ■

An alternate construction of contraction over belief bases, referred to as *kernel contraction*, satisfying the postulates (B—2) to (B—6), has been forwarded by Hansson in [26]. To delete a belief α from a belief base B, the idea of kernel contraction is to discard atleast one element from every inclusion-minimal subset of B that derives α.

Defintion 2.13 [26] Let B be a belief base and α a belief. Then, a *kernel set*, written as $B \perp \alpha$, is the set consisting of all inclusion-minimal subsets of B that derive α, i.e. $X \in B \perp \alpha$ iff: $X \subseteq B$ and $X \vdash \alpha$ and if $Y \subset X$, then $Y \nvdash \alpha$. Every element of $B \perp \alpha$ is referred to as an α*-kernel* of B.

Given a set of sets S, a *hitting set* H for S is defined as a set s.t. (1) $H \subseteq \cup S$ and (2) If $\phi \neq X \in S$, then $X \cap H \neq \phi$. A hitting set is said to be *maximal* when $H = \cup S$, and *minimal* if no proper subset of H is a hitting set for S.

An *incision function* σ for B is a function s.t. for all α, $\sigma(B \perp \alpha)$ is a hitting set for $B \perp \alpha$. An operator $\dot{-}_k$ for B is a kernel contraction iff there is some incision function σ for B s.t. $B \dot{-}_k \alpha = B \backslash \sigma(B \perp \alpha)$ for all beliefs α. ■

Theorem 2.14 [26] For every belief base B, $\dot{-}_k$ is a kernel contraction iff it satisfies the postulates (B–2) to (B–6). ■

From the theorems 2.10 and 2.14, it immediately follows that a contraction operation on a belief base is a kernel contration iff it is a generalized partial meet contraction. The following theorem formalizes this with additional insights into the relationship between kernel and generalized partial meet contractions.

Theorem 2.15

1. A contraction operation over a belief base B is a kernel contraction over B iff it is a generalized partial meet contraction over B.
2. When the incision function σ is *minimal*, i.e. the hitting set defined by σ is inclusion-minimal, then the kernel contraction defined by σ is a maxichoice contraction of α from B.
3. When the incision function σ is *maximal*, i.e. $\sigma(B \perp \alpha) = \cup(B \perp \alpha)$, then the kernel contraction defined by σ is the full meet contraction of α from B. [Observation 4 of [26]]
4. When the incision function σ is *relational*, i.e. based on a non-circular relation $<$ on sentences of B, then the kernel contraction defined by σ coincides with safe contraction and thus coincides with partial meet contraction of α from B. ■

Due to space restrictions, in this paper we restrict ourselves to contraction only, and in the full version we discuss belief revision also.

§3. Knowledge base dynamics and abduction

In this section, we consider contracting knowledge bases, a special kind of belief bases where certain portion of belief is declared to be immutable. Generally, we do not wish certain part of our knowledge or beliefs (the laws of science, for example) to be changed over period of time. This introduces a new *immutability* policy to be considered while contracting or revising a knowledge base. Formally, a *knowledge base* KB is defined as a finite set of sentences from language L, and divided into two parts: an *immutable theory* KB_I, which is fixed part of the knowledge; and *updatable theory* KB_U. We would like to emphasize here that this division of belief base into immutable and updatable parts, is only for practical purposes, and not to discuss belief dynamics at philosophical level. Certainly, this restricted case does not capture certain rare cases where the laws of science changes, but proves effective for most of the applications such as database updates. The additional imutability restriction reduces the solution space of generalized partial meet contraction of a sentence α from KB, to be those generalized partial meet contractions that retain KB_I in them. This idea is formalized below.

Definition 3.1 Let KB be a knowledge base with an immutable part $KB_I \subseteq KB$, and α a sentence. Then, the set of all *restricted generalized partial meet contractions* of α from KB, satisfying the immutability condition, is given as $\{KB' \mid KB' \in (KB{\downarrow}\alpha)^* \text{ and } KB_I \subseteq KB'\}$. ■

Clearly, every restricted generalized partial meet contraction satisfies the postulates (B—2) to (B—6), but the converse doesn't hold. In order to find an axiomatic characterization for restricted generalized partial meet contraction, we have to tighten the rationality postulates. The immutability policy obviously leads to an additional rationality postulate for contraction:

(B—7) $B_I \subseteq B \dot{-} \alpha$ (IT-inclusion)

Moreover, since immutable theory never changes, any knowledge that is implied by it can also never change. Hence, the success postulate has to be modified as follows:

(B—4.1) If $B_I \nvdash \alpha$, then $\alpha \notin Cn(B \dot{-} \alpha)$ (IT-success)

Theorem 3.2 For every knowledge base KB, $\dot{-}$ is a restricted generalized partial meet contraction function iff it satisfies the postulates (B—2), (B—3), (B—4.1), (B—5), (B—6), and (B—7). ■

In the sequel, we study the relationship between knowledge base dynamics and abduction, a different form of reasoning introduced by the philosopher Pierce [e.g. 27]. We show how abductive procedures could be used to realize contractions with immutability conditions. For abductive reasoning, a special subset of ground literals of language L, are designated as *abducibles* Ab. The predicates that appear in abducibles are referred to as *abducible predicates*. An *abductive framework* <P,A> stands for a theory P, which is a set of sentences from L, with possible hypotheses $A \subseteq Ab$. An abductive framework for a knowledge base $KB = KB_I \cup KB_U$ can be given as follows: Let $P = KB_I \cup \{\alpha \leftrightarrow \beta \mid \alpha$ is a sentence in KB_U and β is an abducible from Ab that does not appear in KB$\}$. Then the required abductive framework is <P,Ab>. In the sequel, we simply let the knowledge base KB itself to stand for its abductive framework. Further, without any loss of generality, we assume that KB_U is a set of abducibles.

Given a knowledge base KB and an observation (a sentence) α, the problem of abduction is to explain α in terms of abducibles, i.e. to generate a set of abducibles Δ s.t. $KB_I \cup \Delta \vdash \alpha$.

Definition 3.3 Let KB be a knowledge base and α an observation (sentence) to be explained. Then, a set of abducibles Δ is said to be an *abductive explanation* for α wrt KB_I iff $KB_I \cup \Delta \vdash \alpha$. Δ is said to be *minimal* wrt KB_I iff no proper subset of Δ is an abductive explanation for α, i.e. $\nexists \Delta' \subset \Delta$ s.t. $KB_I \cup \Delta' \vdash \alpha$. ■

The relationship between restricted generalized partial meet contraction and abduction can be captured by defining a special incision function, which we refer to as *abductive incision function*, for kernel contraction, that selects only the elements of updatable theory to be removed.

Definition 3.4 Let $KB = KB_I \cup KB_U$ be a knowledge base. An *abductive incision function* σ for KB is an incision function s.t. for all α: If $KB_I \nvdash \alpha$, then $\sigma(KB \perp \alpha) \subseteq Ab$. Else, if $KB_I \vdash \alpha$, then $\sigma(KB \perp \alpha) = \phi$ ■

Theorem 3.5 Let KB be a knowledge base and σ an abductive incision function. Then, the kernel contraction function $\dot{-}_k$ defined using σ for KB is a restricted generalized partial meet contraction function. ■

From the above discussion, it is clear that to compute a restricted generalized partial meet contraction of α from KB, we need to compute only the abducibles in every α-kernel of KB. So, it is now necessary to characterize precisely, the abducibles present in every α-kernel of KB. The notion of minimal abductive explanation is not enough to capture this, and we introduce kernel minimal and KB-closed abductive explanations.

Definition 3.6 Let KB be a knowledge base and α an observation (sentence) to be explained. Then, an abductive explanation Δ for α wrt KB_I is said to be *kernel minimal,* iff there exists a subset KB_I' of KB_I s.t. Δ is a minimal abductive explanation of α wrt KB_I' and Δ is not an abductive explanation for α wrt any proper subset of KB'. Δ is said to be *KB-Closed* iff $\Delta \subseteq KB_U$. ■

Example 3.7

Consider a knowledge base KB whose immutable part $KB_I = \{p \leftarrow q \wedge r, p \leftarrow r\}$, where q and r are abducibles. Clearly, $\Delta_1 = \{r\}$ is the only minimal abductive explanation for p wrt KB_I. $\Delta_2 = \{q,r\}$ is an abductive explanation for p wrt KB_I, but not minimal. However, Δ_2 is a kernel minimal abductive explanation for p wrt KB_I, since it is a minimal explanation for p wrt $\{p \leftarrow q \wedge r\}$ which is a subset of KB_I. The concept of kernel minimal abductive explanation is computationally attractive, since minimal abductive explanation is more expensive to compute. ■

The connection between kernel minimal abductive explanation for α wrt KB_I and an α-kernel of KB, which is formalized by the following lemma, follows immediately from their respective definitions.

Lemma 3.8 (1) Let KB be a knowledge base and α a sentence s.t. $\nvdash \alpha$. Let Δ be a KB-closed kernel minimal abductive explanation for α wrt KB_I. Then, there exists a α-kernel X of KB s.t. $X \cap KB_U = \Delta$.

(2) Let KB be a knowledge base and α a sentence s.t. $\nvdash \alpha$. Let X be a α-kernel of KB and $\Delta = X \cap KB_U$. Then, Δ is a KB-closed kernel minimal abductive explanation for α wrt KB_I.

Proof

(1) Since $\nvdash \alpha$, it is clear that there exists atleast one α-kernel of KB. Suppose Δ is empty (i.e. $KB_I \vdash \alpha$), then the required result follows immediately. If not, since Δ is a kernel minimal abductive explanation, there exists $KB_I' \subseteq KB_I$, s.t. Δ is a minimal abductive explanation of α wrt KB_I' and not an abductive explanation wrt any proper subset of KB'. Since, Δ is KB-closed, it is not difficult to see that KB_I' $\cup\, \Delta$ is a α-kernel of KB.

(2) Since X is a α-kernel of KB and Δ is the set of all abducibles in X, it follows that Δ is a minimal abductive explanation of α wrt $X \backslash \Delta$ and Δ is not an abductive explanation for A wrt any proper subset of X. It is obvious that Δ is KB-closed, and so Δ is a KB-closed kernel minimal abductive explanation for α wrt KB_I. ■

An immediate consequence of the above lemma is that, it is enough to compute all the KB-closed kernel minimal abductive explanations for α wrt KB_I to contract α from KB, because abductive incision functions select only abducibles from α-kernels of KB. Thus, well-known abductive procedures [e.g. 27] to compute an abductive explanation for α wrt KB_I could be useα to compute restricted kernel contraction.

KB_I could be used to compute restricted kernel contraction.

Algorithm 3.9 **(General Contraction Algorithm)**

Input: A knowledge base $KB = KB_I \cup KB_U$ and a sentence α to be contracted.

Output: A new knowledge base $KB' = KB_I \cup KB_U'$, s.t. KB' is a restricted kernel contraction of α from KB.

begin
1. Construct a set $S = \{X \mid X$ is a KB-closed kernel minimal abductive explanation for α wrt $KB_I\}$.
2. Determine a hitting set $\sigma(S)$.
3. Produce $KB' = KB_I \cup (KB_U \backslash \sigma(S))$ as a result.

end. ■

Theorem 3.10 Let KB be a knowledge base and α a sentence.

(1) If Algorithm 3.9 produces KB' as a result of contracting α from KB, then KB' is a restricted generalized partial meet contraction of α from KB.

(2) If KB' is a restricted generalized partial meet contraction of α from KB, then there exists an incision function σ s.t. KB' is produced by Algorithm 3.9 as a result of contracting α from KB, using σ at step 2. ■

§4. View updates in definite deductive and relational databases

An important application of knowledge base dynamics, discussed in the previous chapter, is in providing an axiomatic characterization of view updates in deductive and relational databases. A *definite deductive database* DDB consists of two parts: an *intensional database* IDB, a set of *definite* program clauses; and *extensional database* EDB, a set of *ground facts*. The intuitive meaning of DDB is provided by the *Least Herbrand model semantics* and all the inferences are carried out through SLD-derivation. The reader is referred to [33 and the references therein] for more information on definite programs and least Herbrand model semantics. All the predicates that are defined in IDB are referred to as *view predicates* and those defined in EDB are referred to as *base predicates*. Extending this notion, an atom with a view predicate is said to be a *view atom*, and similarly an atom with base predicate is a *base atom*. Further we assume that IDB does not contain any unit clauses and no predicate defined in a given DDB is both view and base.

Two kinds of view updates can be carried out on a DDB: An atom, that does not currently follow from DDB, can be *inserted*, denoted by (DDB + A); or an atom, that currently follows from DDB, can be *deleted*, denoted by (DDB - A). In this paper, we consider only deletion of an atom from a DDB. When an atom A is to be deleted, the view update problem is to delete only some relevant EDB facts, so that the modified EDB together with IDB will satisfy the deletion of A from DDB. View update problem, in the context of deductive databases, has been studied by various authors and algorithms based on SLD-trees have been proposed [e.g. 9,11,19,20,28,39,41, and the references therein]. As motivated in the introduction, our concern now is to discuss the rationality of view update, and provide an axiomatic characterization for it. This axiomatic characterization can be seen as a declarative semantics for view updates in deductive databases.

Note that a DDB can be considered as a knowledge base that is to be revised. The

IDB is the immutable part of the knowledge base, while the EDB forms the updatable part. Every base literal is an abducible, but since we deal only with definite databases, we require only positive abducibles. Thus, the rationality postulates (B—2), (B—3), (B—4.1), (B—5), (B—6), and (B—7) provide an axiomatic characterization for deleting a view atom A from a definite database DDB, and a restricted generalized partial meet contraction of A from DDB achieves deletion of A from DDB. Thus, a specific instance of Algorithm 3.9, which is produced below, can be used to compute deletion of a view atom from a definite deductive database.

Algorithm 4.1 **(Deletion Algorithm for Definite Deductive Databases)**

Input: a definite deductive database DDB = IDB ∪ EDB; and an atom A to be deleted

Output: DDB$\dot{-}$A (a contraction of A from DDB)

begin

1.0 Construct a *complete*[4] SLD-tree for ←A wrt DDB.

1.1 For every successful branch i: construct Δ_i = {C | C ∈ EDB and C is used as an input clause in branch i}. Let S be the set of all inclusion-minimal Δ_i's, i.e. ∀Δ∈S: ∄Δ'∈S s.t. Δ' ⊂ Δ.

2. Construct a hitting set D for S.

3. Produce DDB\D as the result.

end ■

It can be easily verified that algorithm 4.1 is a special instance of algorithm 3.9, and steps 1.0 and 1.1 compute all the DDB-closed abductive minimal explanations for A wrt IDB. This is formalized below.

Lemma 4.2

(1) Let DDB = IDB ∪ EDB be a definite deductive database and A an atom. Then, every Δ∈S constructed by algorithm 4.1 at step 1.1, given DDB and A as inputs, is a DDB-closed kernel minimal abductive explanation for A wrt IDB.

(2) Let DDB = IDB ∪ EDB be a definite deductive database and A an atom. Let Δ be a DDB-closed kernel minimal abductive explanation for A wrt IDB. Then, Δ is a member of set S constructed at step 1.1 of algorithm 4.1, given DDB and A as inputs.

Proof

Observe that any successful branch must have used atleast one EDB fact, because of our assumption that IDB does not contain any unit clauses.

(1) Consider a Δ∈S. Clearly it is DDB-closed. To show that Δ is a kernel minimal abductive explanation for A wrt IDB, alternatively we can show that there exists an A-kernel Z for DDB, s.t. Z ∩ EDB = Δ (see lemma 3.8). Let Δ be constructed from a successful branch i and X be the set of all input clauses used in that branch, i.e. X = {C | C is a input clause used in successful branch i}. Clearly, X⊑DDB and X ⊢ A. Suppose any proper subset Y of X does not derive A, then X is the A-kernel that we are looking for. Otherwise, let Y be a inclusion minimal proper

[4]A SLD-tree is said to be complete when all its leafs are either successful or failed. Note that some branches may be infinite leading to severe termination problems.

subset of X that derives A. Then, there must exist another succesful branch with only elements from Y used as input clauses. Let $\Delta' = Y \cap EDB$. Obviously, $\Delta' \subseteq \Delta$. Observe that Δ' can not be a proper subset of Δ, since in that case Δ can not be a member of S. Hence, $\Delta' = \Delta$ and Y is the A-kernel that we are looking for. Thus, Δ is a DDB-closed kernel minimal abductive explanation for A wrt IDB.

(2) Δ is a DDB-closed kernel minimal abductive explanation for A wrt IDB. To show that $\Delta \in S$, we have to show that Δ is indeed generated at step 1.1 from some successful branch i, and no proper subset of Δ is generated at step 1.1. From lemma 3.8, it is clear that there exists a A-kernel X of DDB s.t. $X \cap EDB = \Delta$. Since $X \vdash A$, there must exist a successful branch using only the elements of X as input clauses, and consequently Δ must have been generated. Since Δ is a kernel minimal abductive explanation, no proper subset of it can be generated, and so $\Delta \in S$. ■

Theorem 4.3 Let DDB be a definite deductive database and A an atom to be deleted. Let DDB' be a result of algorithm 4.1, given DDB and A as inputs. Then, DDB' is a restricted generalized partial meet contraction of A from DDB, satisfying the postulates (B—2), (B—3), (B—4.1), (B—5), (B—6), and (B—7).

Proof Follows from lemma 4.2 and theorem 3.10. ■

Algorithm 4.1 is inefficient, as it needs to build a complete SLD-tree. Unfortunately, any rational algorithm for deletion can *not* avoid constructing complete SLD-trees. If algorithm 4.1 is changed to extract input clauses from *incomplete* SLD-derivations, then the new algorithm should check the derivability of an atom from a deductive database, before any deletions are carried out (otherwise, vacuity can not be satisfied). Checking derivability is also computationally expensive and more than that, core-retainment (B—6) will *not* be satisfied in general. These points are illustrated by the following example.

Example 4.4

Consider the following definite database DDB:

p ← q,r
q ←

To delete p, suppose the algorithm does not construct a complete SLD-tree, then (q ←) can be deleted. Since p does not follow from the current database, deleting (q ←) violates the vacuity postulate. So, in order to avoid constructing a complete tree, any algorithm should check the derivability of p first. Consider another DDB:

p ← t
p ← q,r
q ← s
r ←
t ←

Again to delete p, without constructing a complete SLD-tree, the algorithm may delete both (r ←) and (t ←). Clearly, deletion of (r←) violates core-retainment, as it is deleted from a failing branch. So, any rational algorithm must construct a complete SLD-tree. ■

It is now interesting to ask, how rational are other algorithms that have been proposed in the literature to delete view atoms from deductive and relational databases. We first consider Tomasic's algorithm discussed in [41], and explore how rational it is.

Algorithm 4.5 (**Tomasic's Deletion Algorithm** [41])
Input: a definite deductive database DDB = IDB ∪ EDB, and an atom A to be deleted
Output: DDB$\dot{-}$A (a contraction of A from DDB)
begin
1.0 Construct a *complete* SLD-tree for ←A wrt DDB.
1.1 For every *successful* branch i: construct $\Delta_i = \{C \mid C \in$ EDB and C is used as input clauses in that branch$\}$. Let there be m such sets.
2.1 Let $D^* = \{ \{C_1,...C_m\} \mid C_i \in \Delta_i\}$
2.2 Let D be a inclusion-minimal set among D^*, i.e. $\nexists D' \in D^*$ s.t. $D' \subset D$.
3. Produce DDB\D as the result.
end ■

It is not difficult to see that Tomasic's algorithm produces a maxichoice contraction of A from DDB, thus fully in accordance with our axiomatic characterization for deleting a view atom. This result is formalized in the following theorem.

Theorem 4.6 Let DDB be a definite deductive database and A an atom to be deleted. Then, DDB' is a result of Tomasic's algorithm given DDB and A as inputs, iff DDB' is a restricted maxichoice contraction of A from DDB.

Proof From lemma 4.2, it is clear that step 1 generates all DDB-closed kernel minimal abductive explanations for A wrt IDB, and step 2 selects an inclusion-minimal hitting set. The required result now follows from theorem 3.10, and theorem 2.15. ■

Other algorithms, based on deduction tress (SLD or SLDNF), to delete a view atom from a deductive database have been forwarded by researchers like Decker [11], Guessoum & Lloyd [19,20], Kakas & Mancarella [28], and these algorithms do not necessarily construct complete trees. So, as discussed before, they do not satisfy vacuity and core-retainment as a rule. Obviously, these algorithms are more efficient, but it is not clear what kind of relevance policy is satisfied by them.

As observed by Kowalski in [30], logic can provide a conceptual level understanding of relational databases, and hence rationality postulates (B—2), (B—3), (B—4.1), (B—5), (B—6), and (B—7) can provide an axiomatic characterization for view deletion in relational databases also. A relational database and its view definitions can be represented by a definite deductive database (EDB representing tuples in the database and IDB representing the view definitions), and so algorithm 4.1 can be used to delete view extensions from relational databases also. Various algorithms have been proposed for relational databases [e.g. 10, 31] in this regard, and we consider the one by Dayal & Bernstein [10] here. In the full version of this paper, we discuss more about view updates and view maintenance in relational databases.

A *relational scheme* **R** can be thought of as a base predicate whose arguments define the attributes of the scheme. Its *relational extension* R, is a finite set of base atoms containing the predicate **R**. A *database schema* consists of finite collection of relational schemes $<\mathbf{R}_1,...,\mathbf{R}_n>$, and a *relational database* is a specific extension of database schema, denoted as $<R_1,...,R_n>$. A *view relation scheme* **V** (a view predicate) is defined using certain relational schemes $\mathbf{R}_{i1},...,\mathbf{R}_{im}$. Rels(V) stands for the set $\{\mathbf{R}_{i1},...,\mathbf{R}_{im}\}$. In our context, a relational database can be represented by a EDB = $\cup_{i=1,...,n} R_i$. Each view scheme V can be represented by a set of definite clauses of the form: $\mathbf{V(X)} \leftarrow \mathbf{R}_{i1}(\mathbf{Y}_1),....,\mathbf{R}_{im}(\mathbf{Y}_m)$, where $\mathbf{X},\mathbf{Y}_1,...,\mathbf{Y}_m$ stand for a vector of variables such that every variable in **X** appears in some $\mathbf{Y}_j$. Clearly, relational database, as discussed by Dayal & Bernstein, is a very special case of definite deductive databases. In the sequel, we present important definition and algorithm from [10], rewritten in deductive database terminology.

Definition 4.7 Let DDB= EDB $\cup$ IDB be a definite deductive database representing a relational database along with its views. Let A be a ground view atom (member of the extension of a view scheme V). Then, $\Delta = \{t_1,...,t_m \mid \forall i\ (1 \le i \le m): t_i \in R_i \in$ Rels(V)$\}$ is said to be a *generator* of A iff IDB $\cup\ \Delta \vdash$ A. Every member of Δ is referred to as a *source-tuple* for A. ■

Example 4.8

Consider a database scheme consisting of two relational schemes SD and SS with the atrributes "school" & "dean" and "student" & "school" respectively. Consider the following extension of this database scheme:

SD	SCHOOL	DEAN	SS	STUDENT	SCHOOL
	cs	sharma		arvind	cs
				dung	cs

Let us consider a view scheme STD defined using SD and SS, with attributes "student" of SS and "dean" of SD, with the condition that "school" attribute of SD coincides with "school" attribute of SS. Rels(STD) = {SD,SS}. This relational database, together with view scheme STD can be represented by a definite deductive database DDB as follows:

STD(Student,Dean) ← SS(Student,School), SD(School,Dean)
SD(cs,sharma) ←
SS(arvind,cs) ←
SS(dung,cs) ←

Consider a view atom STD(arvind,sharma). Its generator is given as {SS(arvind,cs), SD(cs,sharma)}. Both the base atoms SS(arvind,cs) and SD(cs,sharma) are source-tuples for STD(arvind,sharma). ■

Clearly, a generator Δ for a view atom is an abductive explanation for A wrt IDB. Further, because of the restrictions on IDB, it is a minimal abductive explanation for A wrt IDB and this is formalized by the following lemma.

Lemma 4.9 Let DDB= EDB $\cup$ IDB be a definite deductive database representing a relational database along with its views. Then, a set Δ of base atoms is a generator of a ground view atom A iff Δ is a DDB-closed kernel minimal abductive explanation for A wrt IDB. ■

Algorithm 4.10 **(Dayal & Bernstein's Deletion Algorithm** [10])

Input: A relational database (represented by a definite deductive database DDB) and a view atom A (over a view scheme V) to be deleted.

Output: A new relational database from which A has been deleted.

begin

0. Choose any base predicate $\mathbf{R}_i \in$ Rels(V).
1. Construct a set S = {X | X is a generator of some ground instantiation of A}
2. Let D = {t | t∈X∈ S and t contains the predicate $\mathbf{R}_i$}.
3. Produce DDB\D as a result.

end. ■

Clearly, Dayal & Bernstein's algorithm is a special case of algorithm 4.1, where a hitting set is constructed in a deterministic way. So, this algorithm produces a restricted generalized partial meet contraction of A from DDB, and thus satisfies the rationality postulates (B—2), (B—3), (B—4.1), (B—5), (B—6), and (B—7).

Theorem 4.11 Let DDB= EDB ∪ IDB be a definite deductive database representing a relational database along with its views. Then, the resulting database produced by Dayal & Berntein's algorithm to delete a view atom A from DDB, is a restricted generalized partial meet contraction of A from DDB.

Proof From lemma 4.9, it is clear that step 1 of Dayal&Bernstein's deletion algorithm generates all DDB-closed kernel minimal abductive explanations for A wrt IDB. Step 2 clearly determines a hitting set, and the required result follows from theorem 3.10. ■

§5. Concluding Remarks

The main contribution of this research is to provide a link between theory of belief dynamics and concrete applications such as view updates in databases. We have demonstrated how a restricted case of belief dynamics can provide an axiomatic characterization for deleting an atom from a definite deductive or a relational database. This bridging has helped us in identifying a useful notion of generalized partial meet contraction, in the belief dynamics context.

The central idea in bridging the gap between belief dynamics and view updates is to strike a balance between computational efficiency and rationality. While rationally attractive notions of partial meet contraction prove to be computationally inefficient, the rationality behind efficient algorithms based on incomplete trees is not clear at all. From the belief dynamics point of view, we may have to sacrifice some postulates, vacuity for example, to gain computational efficiency. Further, weakening of core-retainment has to be explored, to provide declarative semantics for algorithms based on incomplete trees.

On the other hand, from the database side, we should explore various ways of optimizing the algorithms that would comply with the proposed declarative semantics. Annotation technique, proposed by Tomasic in [41], is in line with this, and can effectively improve the algorithms of the previous section. However, the declarative semantics of annotation has to be studied from the belief dynamics angle. We belief that notions like epistemic entrenchment [5,15,32], epistemic relevance [38], order for safe contraction [3], will be useful in this regard.

We believe that partial deduction [e.g. 4,34, and the references therein] and loop detection techniques [e.g. 7 and the references therein,12] will play an important role in optimizing algorithms of the previous section. Note that, loop detection could be carried out during partial deduction [4,8], and complete SLD-trees can be effectively constructed wrt a partial deduction (with loop check) of a database, rather than wrt database itself. Moreover, we would anyway need a partial deduction for optimized query evaluation. This, however, opens up a new problem of dynamic partial deduction [6], to be scrutinized further.

For technical simplicity, we have considered definite deductive databases & relational databases, and deletion of atoms only. We are currently extending this framework for normal deductive databases, and consider insertion of atoms also. Moreover, we belief that knowledge base dynamics can also provide link between belief dynamics and other applications such as diagnosis, and we plan to explore it further in our future works. It would also be interesting to study how results from other models of belief dynamics, such as relational belief dynamics [32]; sphere models [18]; model-theoretic approach [29], can be carried into deductive database framework.

In [30], Kowalski proposes a related notion of *knowledge assimilation*, and argues that this new notion can provide an alternate account of the semantics of logic, with the implication that present conventional model theory is not necessary. It is important and useful to study how the results of this paper can be understood in Kowalski's knowledge assimilation framework.

Acknowledgements

We express our gratitude to H.N.Phien for his constant support and encouragements. It was S.O.Hansson who pointed out the connection between generalized partial meet and kernel contractions, and we are grateful to him for that. Encouraging and useful comments received from Decker, Hansson, Nebel, Rabinowicz, and the anonymous referees are also gratefully acknowledged. The first author would like to thank the Government of Japan for financially supporting his studies at the Asian Institute of Technology, Bangkok, Thailand.

References

1. **Abiteboul,S.**, Updates: A new Frontier, In: M.Gyssens, J.Paredaens, and D.Van Gucht (eds.), *Proc. of the second international conference on database theory*, LNCS 326, Springer-Verlag, 1988, pp. 1-18.
2. **Alchourrón,C.E., Gärdenfors,P., and Makinson,D.**, On the logic of theory change: Partial meet contraction and revision functions, *The Journal of Symbolic Logic* 50 Number 2 (1985) 510-530.
3. **Alchourrón,C.E., and Makinson,D.**, On the logic of theory change: Safe contraction, *Studia Logica* 44:405-422 (1985).
4. **Aravindan,C. and Dung,P.M.**, Partial deduction of logic programs wrt well-founded semantics, In: H.Kirchner and G.Levi (eds.), *Proc. of the 3rd International Conf. on Algebraic and Logic Programming*, LNCS 632, Springer-Verlag, 1992, pp. 384-402.

5. **Aravindan,C. and Dung,P.M.**, Relationship between Gärdenfors-Makinson's epistemic entrenchment and Katsuno-Mendelzon's faithful total pre-orders, To appear in: *Proc. of the International Conference on Expert Systems Development*, IEEE, March 1994.
6. **Benkerimi,K. and Shepherdson,J.C.**, Partial evaluation of dynamic logic programs, *Technical Report* TR-90-27, University of Bristol, U.K., November 1990.
7. **Bol,R.N., Apt,K.R., and Klop,J.W.**, An analysis of loop checking mechanisms for logic programs, *Theoretical Computer Science* 86 No. 1 (1991) 35-79.
8. **Bol,R.N.**, Loop checking in partial deduction, *The Journal of Logic Programming* 16 No. 1&2 (1993) 25-46.
9. **Bry,F.**, Intensional Updates: Abduction via deduction, In: *Proc. of the 7th International Conf. on Logic Programming*, 1990.
10. **Dayal,U. and Bernstein,P.A.**, On the correct translation of update operations on relational views, *ACM Transactions on Database Systems* 8 No. 3 (1982) 381-416.
11. **Decker,H.**, Drawing updates from derivations, *Technical Report* IR-KB-65, ECRC, Germany, September 1989.
12. **Demolombe,R.**, A strategy for the computation of conditional answers, In: *Proc. of ECAI '92*, 1992.
13. **Fagin,R., Kuper,G.M., Ullman,J.D., and Vardi,M.Y.**, Updating logical databases, In: *Advances in Computing Research*, Volume 3, Jai Press Inc., 1986, pp. 1-18.
14. **Fuhrmann,A.**, Theory contraction through base contraction, *Journal of Philosophical Logic* 20 (1991) 175-203.
15. **Gärdenfors,P. and Makinson,D.**, Revisions of knowledge systems using epistemic entrenchment, In: M.Y.Vardi (ed.), *Proc. of the second conf. on Theoretical aspects of Reasoning about Knowledge*, Morgan Kaufmann, 1988, pp. 83-95.
16. **Gärdenfors,P.**, Belief Revision: An Introduction, In: P.Gärdenfors (ed.), *Belief Revision*, Cambridge University Press, 1992, pp. 1-28.
17. **Gärdenfors,P. and Rott,H.**, Belief Revision, To appear in: Handbook of logic in AI and logic programming, Vol. IV: Epistemic and Temporal Reasoning.
18. **Grove,A.**, Two modellings for theory change, *Journal of Philosophical Logic* 17 (1988) 157-170.
19. **Guessoum,A. and Lloyd,J.W.**, Updating Knowledge Bases, *Technical Report* TR-89-05, Department of computer science, University of Bristol, U.K., December 1989.
20. **Guessom,A. and Lloyd,J.W.**, Updating Knowledge Bases II, *Technical Report* TR-90-13, Department of computer science, University of Bristol, U.K., May 1990.
21. **Hansson,S.O.**, Belief contraction without recovery, *Studia Logica* 50(2):251-260 (1991).
22. **Hansson,S.O.**, Belief base dynamics, *Doctoral dissertation*, Uppsala university, Sweden, 1991.
23. **Hansson,S.O.**, In defense of base contraction, *Synthese* 91:239-245 (1992).

24. **Hansson,S.O.**, Bridging a gap between AI research and philosophy, In: E.Sandewall, and C.G.Jansson (eds.), *Proc. of the Scandinavian conf. on Artificial Intelligence '93*, IOS Press, 1993, pp. 1-9.
25. **Hansson,S.O.**, Theory contraction and base contraction unified, *Journal of Symbolic Logic*, in press.
26. **Hansson,S.O.**, Kernel Contraction, *Journal of Symbolic Logic*, in press.
27. **Kakas,A.C., Kowalski,R.A., and Toni,F.**, Abductive logic programming, *Journal of Logic and Computation* 2 (1992) 719-770.
28. **Kakas,A.C., and Mancarella,P.**, Database updates through abduction, *Technical Report*, Department of Computing, Imperial College, London, U.K., 1990.
29. **Katsuno,H. and Mendelzon,A.O.**, Propositional knowledge base revision and minimal change, *Artificial Intelligence* 52 (1991) 263-294.
30. **Kowalski,R.**, Logic without model theory, *Technical Report*, Department of Computing, Imperial College, London, U.K., 1994. (available on Internet from LPNMR Archive <ftp.ms.uky.edu>)
31. **Langerak,R.**, View updates in relational databases with an independent scheme, *ACM Transactions on Database Systems* 15 No. 1 (1990) 40-66.
32. **Lindström,S. and Rabinowicz,W.**, Epistemic entrenchment with incomparabilities and relational belief revision, In: A.Fuhrmann and M.Morreau (eds.), *Proc. of the Workshop on The logic of theroy change*, LNAI 465, Springer-Verlag, 1991, pp. 93-126.
33. **Lloyd,J.W.**, Foundations of logic programming, Second extended edition, Springer-Verlag, 1987.
34. **Lloyd,J.W., and Shepherdson,J.C.**, Partial evaluation in logic programming, *Technical Report* No. CS-87-09, University of Bristol, U.K., 1987.
35. **Makinson,D.**, How to give it up: A survey of some formal aspects of the logic of theory change, *Synthese* 62 (1985) 347-363.
36. **Makinson,D.**, On the status of the postulate of recovery in the logic of theory change, *Journal of Philosophical Logic* 16 (1987) 383-394.
37. **Nebel,B.**, A knowledge level analysis of belief revision, In: R.J.Brachman, H.J.Levesque, and R.Reiter (eds.), *Proc. of the first international conference on principles of knowledge representation and reasoning*, Morgan Kaufmann, 1989, pp. 301-311.
38. **Nebel,B.**, Belief revision and default reasoning: Syntax-based approaches, In: J.A.Allen, R.Fikes, and E.Sandewall (eds.), *Proc. of the second international conference on Principles of knowledge representation and reasoning*, Morgan Kaufmann, 1991, pp. 417-428.
39. **Pereira,L.M., Calejo,M., and Aparício,J.N.**, Refining knowledge base updates, In: *Proc. of the 7th Brazilian Symposium on Artificial Intelligence*, November, 1990.
40. **Sadri,F. and Kowalski,R.**, A theorem proving approach to database integrity, In: J.Minker (ed.), *Foundations of deductive databases and logic programming*, Morgan Kaufmann, 1988, pp. 313-362.
41. **Tomasic,A.**, View update translation via deduction and annotation, In: M.Gyssens, J.Paredaens, and D.Van Gucht (eds.), *Proc. of ICDT '88*, LNCS 326, Springer-Verlag, 1988, pp. 338-352.

On the Logic of Theory Base Change

Mary-Anne Williams

Information Systems Group, Department of Management, University of Newcastle, NSW 2308 Australia

Abstract. Recently there has been considerable interest in change operators for theory bases, rather than entire theories. Especially since such operators could support computer-based implementations of revision systems. However, a perceived problem associated with theory base change operators is their sensitivity to the syntax of the theory base used. Although it has been argued that this sensitivity should reflect a higher level of commitment to formulae in the theory base than formulae derivable from the theory base. In this paper we develop a logic of theory base change, using constructions based on ensconcements. We show whenever two theory bases have equivalent ensconcements, the logical closure of their theory base revisions are identical. Moreover, we give explicit relationships associating theory base revision and theory base contraction, and provide explicit relationships between theory base change and theory change operations. We claim that these relationships show that our theory base revision and theory base contraction operators exhibit desirable behaviour.

1 Introduction

The AGM paradigm is a formal approach to informational change, in particular it provides a mechanism for the revision and the contraction of information. Within the AGM paradigm the family of revision operators, and the family of contraction operators are described by rationality postulates. The logical properties of a body of information are not strong enough to uniquely determine a revision or contraction operator, therefore the principal constructions for these operators rely on some form of underlying preference relation, such as a selection function [2], a system of spheres [13], or an epistemic entrenchment ordering [10].

In this paper we develop a logic of theory base change, using constructions based on *ensconcements* [28]. We show whenever two theory bases have equivalent ensconcements, the logical closure of their theory base revisions are identical. Moreover, we describe the relationships associating theory base revision and theory base contraction, as well as the relationships between theory base operations and theory operations. We argue that these relationships show our theory base revision and theory base contraction operators exhibit desirable behaviour.

In Section 2 we outline the AGM paradigm. In Section 3 we describe theory base change, and discuss the inherent problems associated with theory base change within the AGM paradigm. In Section 4 we propose a preference relation called an ensconcement ordering for a theory base, and show its relationship to an epistemic entrenchment ordering. In addition we describe a contruction for a finite ensconcement from an epistemic entrenchment, and vice versa. Then in Section 5 we show how to define theory base change operators using an ensconcement, and demonstrate that theory base revision is not dependent on the syntactical nature of the theory base, but rather is dependent on the nature of the ensconcement ordering. In Section 6 we describe the connection between theory base operators and theory operators. Related approaches to modeling theory base change are discussed in Section 7. Some proofs are outlined in the Appendix, the remainder can be found in [28].

2 The AGM Paradigm

Alchourron, Gärdenfors and Makinson [1,2,3,4,8,10,16,17] have identified sets of rationality postulates which appear to capture much of what is required of an ideal rational system of theory change. They have also explored several constructions for their theory change operators, we refer to their approach as the AGM paradigm. The rationality postulates embody the *principle of minimal change*, and circumscribe classes of theory change operators, however, they do not give a constructive method for defining contraction and revision operators. The construction of these operators requires extralogical information, for the purpose of the present discussion we focus on those based on epistemic entrenchment orderings.

We begin with some technical preliminaries. Let **L** denote a countable language which is closed under a complete set of Boolean connectives. We will denote formulae in **L** by lower case Greek letters. We assume **L** is governed by a logic that is identified with its consequence relation $\vdash$. The relation $\vdash$ is assumed to satisfy the following conditions [8]:

(a) If α is a truth-functional tautology, then $\vdash \alpha$.
(b) If $\alpha \vdash \beta$ and $\vdash \alpha$, then $\vdash \beta$ (*modus ponens*).
(c) $\vdash$ is consistent, that is, $\not\vdash \bot$, where $\bot$ denotes the inconsistent theory.
(d) $\vdash$ satisfies the deduction theorem.
(e) $\vdash$ is compact.

The set of all logical consequences of a set $T \subseteq \mathbf{L}$, that is $\{\alpha : T \vdash \alpha\}$, is denoted by $\mathrm{Cn}(T)$. The set of tautologies, $\{\alpha : \vdash \alpha\}$, is denoted by $\top$, and those formulae not in $\top$ are referred to as *nontautological.* A *theory* of **L** is any subset of **L**, closed under $\vdash$.The set of all theories of **L** is denoted by $\mathcal{T}_{\mathbf{L}}$. A *consistent* theory of **L** is any theory of **L** that does not contain both α and $\neg\alpha$, for any formula α of **L**. A theory is *finite* if the consequence relation $\vdash$ partitions the elements of T into a finite number of equivalence classes. Whenever for a set of formulae $\Gamma \subseteq \mathbf{L}$ we have that $T = \mathrm{Cn}(\Gamma)$, then Γ is referred to as a *theory base for* T, and if Γ is a finite set, then it is referred to as a *finite theory base.* Furthermore, if the elements of Γ are logically independent, then it is referred to as an *irredundant* theory base.

Finally, a *well-ranked* preorder on a set X is a preorder such that every nonempty subset of X has a minimal member, and similarly an *inversely well-ranked* preorder on a set X is a preorder such that every nonempty subset of X has a maximal member. We note that a total preorder on X is finite if and only if it is both well-ranked and inversely well-ranked.

2.1 Postulates for Theory Change

In the AGM paradigm information states are taken to be theories, and changes to the informational content of an information state are regarded as transformations on theories. There are three well known AGM transformations; expansion, contraction, and revision. These transformations allow us to model changes based on the *principle of minimal change*, indeed the rationality postulates for each of the AGM transformations attempt to encapsulate this principle [8].

The *expansion* of a theory T with respect to a formula α, is defined to be the logical closure of T and α, that is $T^+_\alpha = \mathrm{Cn}(T \cup \{\alpha\})$. Clearly, expansion is a monotonic operation, and if $\neg\alpha \in T$, then T^+_α is inconsistent.

In contradistinction to an expansion, it turns out that both contraction and revision, are nonunique operations and can not be represented logically and set theoretically [8].

Contraction of an information state involves the retraction of information, the difficulty is in determining those formulae that should be given up – a problem which in general presents a nonunique choice.

A *contraction* of T with respect to α, T^-_α, involves the removal of a set of formulae from T so that α is no longer implied, provided α is not a tautology.

Formally, a contraction operator $^-$ is any operator from $\mathcal{T}_\mathbf{L} \times \mathbf{L}$ to $\mathcal{T}_\mathbf{L}$, mapping (T, α) to T^-_α which satisfies the postulates $(^-1)$ – $(^-8)$, below, for any α, $\beta \in \mathbf{L}$ and any $T \in \mathcal{T}_\mathbf{L}$:

$(^-1)$ $T^-_\alpha \in \mathcal{T}_\mathbf{L}$
$(^-2)$ $T^-_\alpha \subseteq T$
$(^-3)$ If $\alpha \notin T$ then $T \subseteq T^-_\alpha$
$(^-4)$ If $\nvdash \alpha$ then $\alpha \notin T^-_\alpha$
$(^-5)$ If $\alpha \in T$, then $T \subseteq (T^-_\alpha)^+_\alpha$ *(recovery)*
$(^-6)$ If $\vdash \alpha \equiv \beta$ then $T^-_\alpha = T^-_\beta$
$(^-7)$ $T^-_\alpha \cap T^-_\beta \subseteq T^-_{\alpha\wedge\beta}$
$(^-8)$ If $\alpha \notin T^-_{\alpha\wedge\beta}$ then $T^-_{\alpha\wedge\beta} \subseteq T^-_\alpha$

We introduced the following postulates in [28] and use them to identify three special classes of contraction operators.

$(^-9)$ For every nonempty set Γ of nontautological formulae, there exists a formula $\alpha \in \Gamma$ such that $\alpha \notin T^-_{\alpha\wedge\beta}$ for every $\beta \in \Gamma$.

$(^-10)$ For every nonempty set Γ of nontautological formulae, there exists a formula $\alpha \in \Gamma$ such that $\beta \notin T^-_{\alpha\wedge\beta}$ for every $\beta \in \Gamma$.

Definition. A *well-behaved contraction* operator satisfies $(^-1)$ – $(^-9)$.

A *well-mannered contraction* operator satisfies ($^-1$) – ($^-8$) and ($^-10$).
A *very well-behaved contraction* operator satisfies ($^-1$) – ($^-10$).

A *revision* attempts to change an information state as 'little as possible' in order to incorporate a new formula, which may be inconsistent with this information state. In order to maintain consistency some old information may need to be retracted. Formally, a revision operator * is any operator from $\mathcal{T}_{\mathbf{L}} \times \mathbf{L}$ to $\mathcal{T}_{\mathbf{L}}$, mapping (T, α) to T^*_α which satisfies the postulates (^{*1}) – (^{*8}), below, for any α, $\beta \in \mathbf{L}$ and any $T \in \mathcal{T}_{\mathbf{L}}$:

(^{*1}) $T^*_\alpha \in \mathcal{T}_{\mathbf{L}}$
(^{*2}) $\alpha \in T^*_\alpha$
(^{*3}) $T^*_\alpha \subseteq T^+_\alpha$
(^{*4}) If $\neg\alpha \notin T$ then $T^+_\alpha \subseteq T^*_\alpha$
(^{*5}) $T^*_\alpha = \perp$ if and only if $\vdash \neg\alpha$
(^{*6}) If $\vdash \alpha \equiv \beta$ then $T^*_\alpha = T^*_\beta$
(^{*7}) $T^*_{\alpha\wedge\beta} \subseteq (T^*_\alpha)^+_\beta$
(^{*8}) If $\neg\beta \notin T^*_\alpha$ then $(T^*_\alpha)^+_\beta \subseteq T^*_{\alpha\wedge\beta}$

In the same way as we do for contraction, we introduce the following postulates [28] for revision, in fact they are the counterparts of ($^-9$) and ($^-10$).

(^{*9}) For every nonempty set Γ of nontautological formulae, there exists a formula $\alpha \in \Gamma$ such that $\alpha \notin T^*_{\neg\alpha\vee\neg\beta}$ for every $\beta \in \Gamma$.
(^{*10}) For every nonempty set Γ of nontautological formulae, there exists a formula $\alpha \in \Gamma$ such that $\beta \notin T^*_{\neg\alpha\vee\neg\beta}$ for every $\beta \in \Gamma$.

Definition. A *well-behaved revision* operator satisfies (^{*1}) – (^{*9}).
A *well-mannered revision* operator satisfies (^{*1}) – (^{*8}) and (^{*10}).
A *very well-behaved revision* operator satisfies (^{*1}) – (^{*10}).

Alchourrón et al. [2] have shown that contraction and revision operators are interdefinable, in particular they showed that $(T^-_{\neg\alpha})^+_\alpha$ defines a revision operator, and that $T \cap T^*_{\neg\alpha}$ defines a contraction operator. $T^*_\alpha = (T^-_{\neg\alpha})^+_\alpha$ is known as the *Levi Identity*, and $T^-_\alpha = T \cap T^*_{\neg\alpha}$ as the *Harper Identity*.

2.2 Epistemic Entrenchment Orderings

Epistemic entrenchment [8, 10] is based on an ordering of formulae in $\mathbf{L}$, and captures the importance of a formula in the face of change. In order to determine a unique revision or contraction operation a theory is endowed with an epistemic entrenchment ordering which can be used to determine the formulae to be retracted, retained, and acquired during contraction and revision.

Definition. Given a theory T of $\mathbf{L}$, an *epistemic entrenchment related to* T is any binary relation $\leq$ on $\mathbf{L}$ satisfying (EE1) – (EE5), below:
(EE1) $\leq$ is transitive.

(EE2) For all α, $\beta \in \mathbf{L}$, if $\alpha \vdash \beta$ then $\alpha \leq \beta$.
(EE3) For all α, $\beta \in \mathbf{L}$, $\alpha \leq \alpha \wedge \beta$ or $\beta \leq \alpha \wedge \beta$.
(EE4) When $T \neq \perp$, $\alpha \notin T$ if and only if $\alpha \leq \beta$ for all $\beta \in \mathbf{L}$.
(EE5) If $\beta \leq \alpha$ for all $\beta \in \mathbf{L}$, then $\vdash \alpha$.
(EE6) $\leq$ is well-ranked.
(EE7) $\leq$ is inversely well-ranked.

We define $\alpha < \beta$, as $\alpha \leq \beta$ and not $\beta \leq \alpha$. An epistemic entrenchment that satisfies (EE1) – (EE7), is well-ranked and inversely well-ranked, and therefore is a finite epistemic entrenchment.

It is well known that an epistemic entrenchment is a total preorder of the formulae in Lsuch that the following observation holds.

Observation 1. *If $\leq$ is an epistemic entrenchment, then for any $\alpha \in \mathbf{L}$, $\{\beta \in \mathbf{L} : \alpha \leq \beta\}$ is a theory.*

Definition. Let $\leq$ be an epistemic entrenchment.
For a formula $\alpha \in \mathbf{L}$ we define, $\text{cut}_{\leq}(\alpha) = \{\beta \in \mathbf{L} : \alpha \leq \beta\}$.
For a nontautological formula $\alpha \in \mathbf{L}$ we define, $\text{cut}_{<}(\alpha) = \{\beta \in \mathbf{L} : \alpha < \beta\}$.

Rott [24] uses a similar definition, in particular he defines a set of formulae S to be an EE-cut, if for any formula α in S all formulae β with $\alpha < \beta$ are also in S. Moreover, $S_\alpha = \text{cut}_{<}(\alpha)$.

From Observation 1 $\text{cut}_{\leq}(\alpha)$ is a theory for all $\alpha \in \mathbf{L}$. Since a subtheory of a finitely axiomatizable theory may not be finitely axiomatizable we make the following observation.

Observation 2. *An epistemic entrenchment ordering $\leq$ is finitely representable if and only if it has a finite number of natural partitions, and for all $\alpha \in \mathbf{L}$, $\text{cut}_{\leq}(\alpha)$ is finitely axiomatizable.*

Gärdenfors and Makinson [10] have shown that, for every contraction operator $^-$ there exists an epistemic entrenchment $\leq$ related to T such that the condition (E^-) below, is true for every $\alpha \in \mathbf{L}$, and conversely.

$$(E^-) \qquad T^-_\alpha = \begin{cases} \{\beta \in T : \alpha < \alpha \vee \beta\} & \text{if } \nvdash \alpha \\ T & \text{otherwise} \end{cases}$$

We extend this result in the following theorem to well-behaved, well-mannered, and very well-behaved contraction operators.

Theorem 3. *Let T be a theory of $\mathbf{L}$. For every well-behaved contraction (well-mannered contraction, very well-behaved contraction) operator $^-$ for T there exists a well-ranked epistemic entrenchment (inversely well-ranked epistemic entrenchment, finite epistemic entrenchment) $\leq$ related to T such that (E^-) is true for every $\alpha \in \mathbf{L}$, and conversely.*

From Theorem 3 and the Levi Identity it is straight forward to derive a similar result for revision using (E^*) below, see [28] for details.

$$(E^*) \qquad T^*_\alpha = \begin{cases} \{\beta \in \mathbf{L} : \neg\alpha < \neg\alpha \vee \beta\} & \text{if } \not\vdash \neg\alpha \\ \bot & \text{otherwise} \end{cases}$$

Peppas [22] obtains a similar result for well-behaved revision based on a well-ordered system of spheres, indeed he was the first to identify the class of well-behaved revision operators, and furthermore he shows that there exist revision operators that are not well-behaved. Hence, well-behaved revision operators are a strict subclass of revision operators. We note that (^{*9}) above is not identical to the well-behaved postulate given in [22], however it circumscribes the same class of operators.

3 Theory Base Change

Lakemeyer says, "it is widely accepted that agents, because of their limited resources, believe some but by no means all of the logical consequences of their beliefs" [15]. An agent who believes all the logical consequences of his beliefs is said to be *logically omniscient* [26], such an ability seems ponderous, and unrealistic. We, like others [6, 15, 19, 26], consider logical omniscience to be a fundamental problem with existing theoretical models which focus on modeling *ideal* reasoning agents.

Recently, there has been considerable interest in change operators for theory bases, rather than entire theories. Clearly, such operators would be useful in supporting computer-based implementations of revision systems. According to Makinson [17], theory change is "essentially a mathematical exercise on an idealized, ... elegant model". Objections to using theory change operators on theory bases have principally been that they are syntactically sensitive, that is, they are dependent on the syntax of the chosen theory base, and they do not satisfy all the AGM postulates, in particular the recovery postulate ($^-5$).

For the latter, we argue that recovery is inappropriate for a limited reasoning agent since it is inextricably tied to the consequence relation [17]. In the case of the former, we argue that theory bases can be augmented with some additional structure, namely an ensconcement ordering, so as to capture a epistemic entrenchment ordering. Although, it has been argued that this sensitivity should simply reflect a higher level of commitment to formulae in the theory base than formulae derivable from the theory base. According to Rott [24] "real life theories are generated from finite axiomatizations, and that the axioms may carry different epistemic weight. Both the syntactical encoding of a theory and the prioritization are held to be relevant for theory change".

Recovery, ($^-5$), is a nice property in the sense that if it is satisfied then no more information is lost then can be replaced in a straightforward expansion. Makinson [17] explores the use of contraction operators on theory bases, and shows that theory base contraction does not satisfy recovery, in general. However, Alchourrón and Makinson [1] show that if the theory base is irredundant and the contraction operator is maxichoice, then ($^-5$) is trivally satisfied. It is this observation that seems to endear irredundant bases to Alchourrón and Makinson. However we show in the next section that redundancy allows us to capture extralogical information, in particular an epistemic entrenchment.

In a convincing argument against recovery, Niederée [21] shows that a consequence of theory closure and recovery is that for all $\beta \in \mathbf{L}$, whenever a formulae α is contained in the theory T then $\beta \to \alpha \in T^-_{\alpha\vee\beta}$, and the consequences, $\alpha \in (T^-_{\alpha\vee\beta})^+_\beta$ and $\neg\beta \in (T^-_{\alpha\vee\beta})^+_{\neg\alpha}$, are not necessarily desirable for a limited reasoner.

Hansson [14] illustrates his objection to recovery using an example which highlights that recovery can be contrary to intuition.

The problem of syntax dependence stems from the inherent property observed by Makinson [17], that every maximal subset of $\mathrm{Cn}(\Gamma)$ that fails to imply α is a theory, however it need not have as a base a maximal subset of Γ that fails to imply α. As a consequence, the "result of a contraction or revision applied to the base of a theory in general depends upon the constitution of the base as much as on the theory itself" [1].

Finally, we note like others [7, 9, 14, 21] that operations on theory bases provide a mechanism for foundational belief change [9], Fuhrmann calls this *filtering* [7], where removing some information modifies an information state in such a way that all the consequences of that information are also retracted.

4 Ensconcement

In this section we define an ensconcement and describe its relationship to an epistemic entrenchment. It turns out that ensconcements are essentially identical to Rott's E-bases [25], which he developed independently as a mechanism for a canonical representation of an epistemic entrenchment. Rott uses E-bases to obtain results concerning conditionals, whilst we use ensconcements as a construction for theory base change operators. Although, Gärdenfors and Rott [12] discuss E-bases in relation to theory base change, indeed they discuss theory base change in a very broad context, and provide an enlightening analysis of many seemingly different approaches.

Definition. Define an *ensconcement* to be a set of formulae Γ together with a total preorder $\preceq$ of Γ satisfying the following conditions:

($\preceq$1) For all nontautological $\beta \in \Gamma$, $\{\alpha \in \Gamma : \beta \prec \alpha\} \not\vdash \beta$.
($\preceq$2) For all $\alpha, \beta \in \Gamma$, $\alpha \preceq \beta$ if and only if $\vdash \beta$.

Intuitively, ($\preceq$1) says that the formulae which are strictly more ensconced than an arbitrary formula α do not entail α. If there are any tautologies in Γ, then ($\preceq$2) says they are the most ensconced formulae.

Since the language $\mathbf{L}$ is countable, for an ensconcement $(\Gamma, \preceq)$, $\preceq$ will possess a countable number of natural partitions. If $\preceq$ is well-ranked, then we say that the ensconcement $(\Gamma, \preceq)$ is well-ranked, and similarly if $\preceq$ is inversely well-ranked, then we say that the ensconcement is inversely well-ranked. If $\preceq$ is both well-ranked and inversely well-ranked, then we refer to the ensconcement as finite. note however that Γ need not be finite for the ensconcement to be finite.

In his *plausibility theory* [23] Rescher defines something similar to an ensconcement for a finite propositional language. In particular each formula in a given set has a degree of plausibility that represents the reliability of its source. Plausibility indexes are ranked over the interval from 0 to 1, where the tautologies and possibly other consistent formulae are assigned 1, logical dependencies are respected in the sense that ($\preceq$1) holds, and therefore a plausibility index is very closely related to a finite ensconcement.

We now establish the relationship between ensconcements and epistemic entrenchment orderings.

Definition. Given an ensconcement, a *cut* for $\alpha \in \mathrm{Cn}(\Gamma)$ is defined below:

$$\mathrm{cut}_{\preceq}(\alpha) = \{\beta \in \Gamma : \{\gamma \in \Gamma : \beta \prec \gamma\} \not\vdash \alpha\}.$$

For a given ensconcement $(\Gamma, \preceq)$, if $\alpha \in \Gamma$ then the cut for α is a subset of Γ such that its members are equally or more ensconced than α. Clearly, if $\alpha \in \Gamma$, then α is in its cut, and if $\vdash \alpha$ then $\mathrm{cut}_{\preceq}(\alpha) = \emptyset$. Furthermore, if $(\Gamma, \preceq)$ is an ensconcement, $\leq$ is an epistemic entrenchment related to $\mathrm{Cn}(\Gamma)$, and $\leq = \preceq$, then $\mathrm{cut}_{\leq}(\alpha) = \mathrm{cut}_{\preceq}(\alpha)$.

Definition. Let $(\Gamma, \preceq)$ be an ensconcement. For α, $\beta \in \mathbf{L}$, define $\leq_{\preceq}$ to be given by: $\alpha \leq_{\preceq} \beta$ if and only if either

(i) $\alpha \notin \mathrm{Cn}(\Gamma)$, or

(ii) $\alpha, \beta \in \mathrm{Cn}(\Gamma)$ and $\mathrm{cut}_{\preceq}(\beta) \subseteq \mathrm{cut}_{\preceq}(\alpha)$.

From this definition it can be seen that, the tautologies are maximal, and formulae not in $\mathrm{Cn}(\Gamma)$ are minimal with respect to $\leq_{\preceq}$.

Theorem 4, below, follows from Observation 1, and the preceeding definitions, it demonstrates that $\leq_{\preceq}$ is an epistemic entrenchment, a proof is outlined in the Appendix.

Theorem 4. *If $(\Gamma, \preceq)$ is an ensconcement, then $\leq_{\preceq}$ is an epistemic entrenchment related to* $\mathrm{Cn}(\Gamma)$.

Corollary 5. *Given an ensconcement $(\Gamma, \preceq)$, $\preceq$ is well-ranked (inversely well-ranked, finite) if and only if $\leq_{\preceq}$ is well-ranked (inversely well-ranked, finite).*

Definition. Let $(\Gamma, \preceq)$ be an ensconcement. We refer to the epistemic entrenchment, $\leq_{\preceq}$, related to $\mathrm{Cn}(\Gamma)$ from Theorem 1 as *the epistemic entrenchment generated from $(\Gamma, \preceq)$*.

If $\leq$ is finitely representable, then there exists a finite Γ such that $(\Gamma, \preceq)$ generates $\leq$.

Definition. We define a *proper cut* for nontautological $\alpha \in \mathbf{L}$, below:

$$\mathrm{cut}_{\preceq}(\alpha) = \{\beta \in \Gamma : \{\gamma \in \Gamma : \beta \preceq \gamma\} \nvdash \alpha\}.$$

Equivalently, $\mathrm{cut}_{\preceq}(\alpha) = \{\beta \in \Gamma : \alpha <_{\preceq} \beta\}$. For a given ensconcement $(\Gamma, \preceq)$, if $\alpha \in \Gamma$ and nontautological, then the proper cut for α is a subset of Γ such that its members are strictly more ensconced than α. If $\alpha \notin \Gamma$, then $\mathrm{cut}_{\preceq}(\alpha)$ is a subset of Γ such that the members are strictly more epistemically entrenched than α, in accordance with $\leq_{\preceq}$, that is, the epistemic entrenchment ordering related to $\mathrm{Cn}(\Gamma)$ generated from $(\Gamma, \preceq)$. Clearly, a nontautological formula is not in its proper cut.

Intuitively, Theorem 6, below, says an ensconcement can be considered to be a (possibly minimal) specification of an epistemic entrenchment in the same way as Rott's E-Base. The outline of a proof can be found in the Appendix.

Theorem 6. *Let T be a theory in* $\mathbf{L}$, *and $\Gamma \subseteq T$. Let $\leq$ be an epistemic entrenchment related to T. For all $\alpha \in T$, if $\{\beta \in \Gamma : \alpha \leq \beta\} \vdash \alpha$, then $\leq|_\Gamma$ is an ensconcement ordering on Γ such that $\leq_{(\leq|_\Gamma)} = \leq$.*

In other words, the Γ in Theorem 6, while obviously not unique, must contain enough formulae of 'the right stuff' in each natural partition of the epistemic entrenchment ordering $\leq$ in order to regenerate it. In particular, it must satisfy the condition in the theorem.

Definition. Let $\leq$ be an epistemic entrenchment related to T. We denote $\leq|_\Gamma$ from Theorem 6 by $\preceq^{\leq}$ and refer to it as an *ensconcement ordering inherited from* $\leq$.

We can rephrase Theorem 6, using the definition above in the following way: the epistemic entrenchment generated from an ensconcement inherited from an epistemic entrenchment ordering $\leq$, is $\leq$ itself, that is, $\leq\, = \,\leq_{(\preceq^{\leq})}$.

Although the main focus for this paper is theory base change, we make the following observation concerning expectation orderings [11], which are orderings that satisfy (EE1) – (EE3) and used by Gärdenfors and Makinson to capture a form of nonmonotonic reasoning. It is not hard to see that if $\preceq$ satisfies ($\preceq$1) then $\leq_{\preceq}$ will be an expectation ordering. Conversely, if $\leq$ is an expectation ordering then $\preceq^{\leq}$ will satisfy ($\preceq$1). Consequently, Theorem 6 will hold, and we note that one of the conditions for theory base revision given in the next section is closely related to the expectation inference relation described in [11].

Corollary 7. *$\leq$ is well-ranked (inversely well-ranked, finite) if and only if $\preceq^{\leq}$ is well-ranked (inversely well-ranked, finite).*

The ensconcement inherited from an epistemic entrenchment ordering need not be unique since a theory does not have a unique theory base in general. Therefore, we will not obtain a corresponding result to Theorem 6 for ensconcements, that is, $\preceq$ need not equal $\preceq^{(\leq_{\preceq})}$. Although, for a fixed Γ we have $\preceq = \preceq^{(\leq_{\preceq})}$.

In the following definition we describe two ensconcements to be *equivalent* whenever they generate the same epistemic entrenchment ordering.

Definition. Let $(\Gamma_1, \preceq_1)$ and $(\Gamma_2, \preceq_2)$ be ensconcements.
Define $(\Gamma_1, \preceq_1)$ and $(\Gamma_2, \preceq_2)$ to be *equivalent*, denoted by $(\Gamma_1, \preceq_1) \equiv (\Gamma_2, \preceq_2)$, if and only if $\leq_{\preceq_1} = \leq_{\preceq_2}$.

It is obvious that, if $(\Gamma_1, \preceq_1)$ and $(\Gamma_2, \preceq_2)$ are equivalent, then $\mathrm{Cn}(\Gamma_1) = \mathrm{Cn}(\Gamma_2)$.

Finally we describe a stepwise construction for a finite ensconcement from a finite epistemic entrenchment, and vice versa. Similar constructions can be found for well-ranked and inversely well-ranked structures in [28]. In particular if the structure is inversely well-ranked then it is possible for the stepwise construction to proceed *outward*, for example from the most ensconced formulae to the least ensconced, so is this sense it is very similar to the construction for a finite ensconcement given below. Conversely for a well-ranked structure the construction proceeds *inward*.

Construction of an Epistemic Entrenchment from an Ensconcement. Let $(\Gamma, \preceq)$ be a finite ensconcement. Let $\Gamma_1, \Gamma_2, \ldots, \Gamma_n$ be the natural partitions of $\preceq$, indexed so that $\alpha \in \Gamma_i$, $\beta \in \Gamma_{i+1}$ implies $\alpha \prec \beta$ for $1 \leq i < n$, where Γ_n is a possibly empty a subset of the tautologies. We now proceed to construct $\leq_{\preceq}$ the epistemic entrenchment generated from the ensconcement $(\Gamma, \preceq)$.

Define,

$$
\begin{aligned}
T_n &= \mathrm{Cn}(\Gamma_n) = \mathrm{Cn}(\emptyset) \\
T_{n-1} &= \mathrm{Cn}(\Gamma_{n-1}) \\
&\ldots \\
T_i &= \mathrm{Cn}(\Gamma_{n-1} \cup \Gamma_{n-2} \cup \ldots \cup \Gamma_i) \\
&\ldots \\
T_1 &= \mathrm{Cn}(\Gamma_{n-1} \cup \Gamma_{n-2} \cup \ldots \cup \Gamma_i \ldots \Gamma_1) = T
\end{aligned}
$$

It is easily seen that $\preceq$ can be used to generate an epistemic entrenchment ordering $\leq_{\preceq}$ related to T in the following manner: $\alpha <_{\preceq} \beta$ if and only if either $\alpha \notin T$, or for some i, $1 \leq i \leq n$, $\alpha \notin T_i$, and $\beta \in T_i$. For each i, $1 \leq i < n$, the elements of $T_i - T_{i+1}$ are equally entrenched, and the elements of $\mathbf{L} - T$ are minimal.

Construction of an Ensconcement from an Epistemic Entrenchment. Let T be a theory. Let $\leq$ be a finite epistemic entrenchment ordering related to T. Let $P_0, P_1, P_2, P_3 \ldots, P_n$ be the natural partitions of $\leq$, indexed so that

$\alpha \notin P_i$, $\beta \in P_i$ implies $\alpha < \beta$ for $0 \leq i \leq n$. Let $T_i = \bigcup_{j=i}^{n} P_j$, in particular, then $\{\alpha : \vdash \alpha\} = T_n = P_n \subset T_{n-1} \subset \ldots \subset T_1 = T \subseteq \perp$.
Define,

$\Gamma_n = \emptyset$

Γ_{n-1} to be a possibly minimal subset of P_{n-1} such that $\mathrm{Cn}(\Gamma_{n-1}) = T_{n-1}$.

...

Γ_i to be a possibly minimal subset of P_i such that
$\mathrm{Cn}(\Gamma_{n-1} \cup \Gamma_{n-2} \cup \ldots \cup \Gamma_i) = T_i$

...

$\Gamma = \Gamma_{n-1} \cup \Gamma_{n-2} \cup \ldots \cup \Gamma_i \ldots \Gamma_1$

Γ together with the ordering inherited from $\leq$, namely $(\Gamma, \preceq^{\leq})$, is an ensconcement.

Definition. Let α be a formula in **L**. Let $(\Gamma, \preceq)$ be a finite ensconcement such that it has $n-1$ nontautological natural partitions indexed by Γ_i where $1 \leq i < n$, without loss of generality we assume $\Gamma_n = \emptyset$.
We define

$$\mathrm{rank}(\alpha) = \begin{cases} \text{largest } j \text{ such that } \bigcup_{i=j}^{n} \Gamma_i \vdash \alpha & \text{if } \Gamma \vdash \alpha \\ 0 & \text{otherwise.} \end{cases}$$

Clearly, if $\alpha \in \Gamma$, then $\mathrm{rank}(\alpha)$ is trivally determined, and if **L** is a finitary propositional language then it is decidable.

Observation 8. *Let $(\Gamma, \preceq)$ be a finite ensconcement.*
For a formula α in $\mathrm{Cn}(\Gamma)$, $\mathrm{cut}_{\preceq}(\alpha) = \bigcup_{i=\mathrm{rank}(\alpha)}^{n} \Gamma_i$, and
*for a nontautological formula α in **L**, $\mathrm{cut}_{\prec}(\alpha) = \bigcup_{i=\mathrm{rank}(\alpha)+1}^{n} \Gamma_i$.*

5 Constructing Theory Base Change Operators

In this section we describe how an ensconcement may be used to construct theory base change operators. Firstly, we define the *expansion* of a theory base Γ with respect to a formula α, to be $\Gamma_\alpha^{\otimes} = \Gamma \cup \{\alpha\}$. Clearly $(\mathrm{Cn}(\Gamma))_\alpha^{+} = \mathrm{Cn}(\Gamma_\alpha^{\oplus})$.

Definition. Let $(\Gamma, \preceq)$ be an ensconcement. Define a *theory base contraction operator* $^{\ominus}$ for Γ by the following condition, referred to as $(E^{\ominus})$:

$$\Gamma_\alpha^{\ominus} = \begin{cases} \{\beta \in \Gamma : \mathrm{cut}_{\prec}(\alpha) \cup \{\neg\alpha\} \vdash \beta\} & \text{if } \nvdash \alpha \\ \Gamma & \text{otherwise} \end{cases}$$

Intuitively, the condition $(E^{\ominus})$ defines a contraction for a theory base that retains as 'much as possible' of the theory base, this is demonstrated via theorems presented in the next section.

We now define a condition which can be used to construct another a theory base contraction, called a brutal contraction. In particular, a brutal theory base contraction retains as 'little as necessary' of the theory base.

Definition. Let $(\Gamma, \preceq)$ be an ensconcement. Define a *brutal theory base contraction operator* $^\ominus$ for Γ by the following condition, referred to as $(B^\ominus)$:

$$\Gamma_\alpha^\ominus = \begin{cases} \mathrm{cut}_\prec(\alpha) & \text{if } \nvdash \alpha \\ \Gamma & \text{otherwise} \end{cases}$$

Clearly, given an ensconcement $(\Gamma, \preceq)$, if $^\ominus$ is determined by $(E^\ominus)$, and $^{\ominus'}$ is determined by $(B^\ominus)$, then $\Gamma_\alpha^{\ominus'} \subseteq \Gamma_\alpha^\ominus$ for all $\alpha \in \mathbf{L}$.

In the same way as for contraction we describe two types of theory base revision operators.

Definition. Let $(\Gamma, \preceq)$ be an ensconcement. Define a *theory base revision operator* $^\otimes$ for Γ, by the following condition, referred to as $(E^\otimes)$:

$$\Gamma_\alpha^\otimes = \begin{cases} \{\beta \in \Gamma : \mathrm{cut}_\prec(\neg\alpha) \cup \{\alpha\} \vdash \beta\} \cup \{\alpha\} & \text{if } \nvdash \neg\alpha \\ \Gamma_\alpha^\oplus & \text{otherwise} \end{cases}$$

Intuitively, the condition $(E^\otimes)$ defines a revision for a theory base that retains as 'much as possible' of the theory base, and the brutal theory base revision operator, defined below, retains as 'little as necessary' of the theory base.

Definition. Let $(\Gamma, \preceq)$ be an ensconcement. Define a *brutal theory base revision operator* $^\otimes$ for Γ, by the following condition, referred to as $(B^\otimes)$:

$$\Gamma_\alpha^\otimes = \begin{cases} \mathrm{cut}_\prec(\neg\alpha) \cup \{\alpha\} & \text{if } \nvdash \neg\alpha \\ \Gamma_\alpha^\oplus & \text{otherwise} \end{cases}$$

Note, determination of the brutal operators is very simple and may not necessarily involve the consequence relation, for instance if $\alpha \in \Gamma$ for contraction and if $\neg\alpha \in \Gamma$ for revision. The brutal contraction operator is closely related to the withdrawal [17] function discussed in Rott [25], and brutal revision is closely related to Gärdenfors and Makinson's comparative expectation inference relation [11].

Theory base contractions, theory base revisions and their brutal counterparts adhere to the *principle of irrelevance of syntax* [5], to the extent that they are not dependent on the syntactical form of the formula to be removed or incorporated.

We show in Theorem 9, below, that the syntactic nature of two theory bases is irrelevant with respect to revision and brutal contraction, so long as the ensconcements are equivalent. Hence, what is important is not the syntactic nature of the theory base but the ensconcement ordering associated with it.

Theorem 9. *Let $(\Gamma, \preceq)$ and $(\Lambda, \preceq')$ be ensconcements. If $(\Gamma, \preceq)$ and $(\Lambda, \preceq')$ are equivalent, then the following conditions hold:*

(*i*) $\mathrm{Cn}(\Gamma_\alpha^\otimes) = \mathrm{Cn}(\Lambda_\alpha^\otimes)$, *where* $^\otimes$ *is determined by* $(E^\otimes)$.
(*ii*) $\mathrm{Cn}(\Gamma_\alpha^\otimes) = \mathrm{Cn}(\Lambda_\alpha^\otimes)$, *where* $^\otimes$ *is determined by* $(B^\otimes)$.
(*iii*) $\mathrm{Cn}(\Gamma_\alpha^\ominus) = \mathrm{Cn}(\Lambda_\alpha^\ominus)$, *where* $^\ominus$ *is determined by* $(B^\ominus)$.
(*iv*) $(\Gamma_\alpha^\ominus)_{\neg\alpha}^+ \cap \mathrm{Cn}(\Gamma) = (\Lambda_\alpha^\ominus)_{\neg\alpha}^+ \cap \mathrm{Cn}(\Lambda)$ *where* $^\ominus$ *is determined by* $(E^\ominus)$.

The following results follow immediately from the definitions, and show that we obtain something like a Levi Identity involving theory base operations, and conversely, something like a Harper Identity, so that a theory base revision may be constructed from a theory base contraction, and vice versa.

Theorem 10. *Given an ensconcement* $(\Gamma, \preceq)$. *Let* $^{\ominus}$ *be the theory base contraction operator for* Γ *uniquely determined by* $(E^{\ominus})$. *Let* $^{\otimes}$ *be the theory base revision operator for* Γ *uniquely determined by* $(E^{\otimes})$. *Then* $\Gamma^{\otimes}_{\alpha} = (\Gamma^{\ominus}_{\neg\alpha})^{\oplus}_{\alpha}$.

Theorem 11. *Given an ensconcement* $(\Gamma, \preceq)$. *Let* $^{\ominus}$ *be the theory base contraction operator for* Γ *uniquely determined by* $(E^{\ominus})$ *[or* $(B^{\ominus})$*]. Let* $^{\otimes}$ *be the theory base revision operator for* $\mathrm{Cn}(\Gamma)$ *uniquely determined by* $(B^{\otimes})$ *[or* $(E^{\otimes})$*]. Then* $\mathrm{Cn}(\Gamma^{\otimes}_{\alpha}) = \mathrm{Cn}((\Gamma^{\ominus}_{\neg\alpha})^{\oplus}_{\alpha})$.

Theorem 12. *Given an ensconcement* $(\Gamma, \preceq)$. *Let* $^{\otimes}$ *be the theory base revision operator for* Γ *uniquely determined by* $(E^{\otimes})$. *Let* $^{\ominus}$ *be the theory base contraction operator for* Γ *uniquely determined by* $(E^{\ominus})$. *Then* $\Gamma^{\ominus}_{\alpha} = \Gamma^{\otimes}_{\neg\alpha} \cap \Gamma$.

Theorem 13. *Given an ensconcement* $(\Gamma, \preceq)$. *Let* $^{\otimes}$ *be the theory base revision operator for* Γ *uniquely determined by* $(E^{\otimes})$ *[or* $(B^{\otimes})$*]. Let* $^{\ominus}$ *be the theory base contraction operator for* Γ *uniquely determined by* $(E^{\ominus})$. *Then* $\Gamma^{\ominus}_{\alpha} = \mathrm{Cn}(\Gamma^{\otimes}_{\neg\alpha}) \cap \Gamma$.

Theorems 12 and 13 above, capture the syntactical dependence of a theory base contraction operator $^{\ominus}$ on the theory base Γ. We also note that Theorems 10 and 12 hold for brutal base change operators as well.

6 Connections with Theory Change

So far we have seen that an ensconcement provides a representation for an epistemic entrenchment ordering, therefore theory operators can also be directly constructed from an ensconcement in a straightforward manner, for details see [28].

The theorems in this section establish the explicit relationship between theory base change operators and theory change operators.

Theorem 14. *Let* $(\Gamma, \preceq)$ *be an ensconcement. Let* $^{\ominus}$ *be the theory base contraction operator uniquely determined by* $(E^{\ominus})$. *Let* $^{-}$ *be the contraction operator for* $\mathrm{Cn}(\Gamma)$ *uniquely determined by* (E^{-}) *and* $\leq_{\preceq}$. *Then,* $\Gamma^{\ominus}_{\alpha} = (\mathrm{Cn}(\Gamma))^{-}_{\alpha} \cap \Gamma$.

Theorem 14 captures the dependence of a theory base contraction on the contents of the theory base, in particular, a formula is retained in the theory base contraction if and only if it is a member of the theory base and it would be retained in the corresponding theory contraction. This substantiates our claim that ($E^{\ominus}$) retains as 'much as possible' of the original theory base. Theorem 15 below establishes a similar result for revision.

Theorem 15. *Let* $(\Gamma, \preceq)$ *be an ensconcement. Let* $^{\otimes}$ *be the theory base revision operator uniquely determined by* $(E^{\otimes})$. *Let* * *be the revision operator for* $\text{Cn}(\Gamma)$ *uniquely determined by* (E^{*}) *and* $\leq_{\preceq}$. *Then,* $\Gamma^{\otimes}_{\alpha} = (\text{Cn}(\Gamma))^{*}_{\alpha} \cap \Gamma^{\oplus}_{\alpha}$.

It is now clear that an ensconcement with a finite theory base can be used to model belief change for a limited reasoner. If the theory base represents the reasoner's explicit beliefs then brutal theory base operations remove far too many explicit beliefs. However Theorems 14 and 15 clearly demonstrate that theory base operations determined by $(E^{\ominus})$ and $(E^{\otimes})$ contain as many of the explicit beliefs as would be retained in the corresponding theory change operations. In particular, we have established that each explicit belief retained in a theory change is also retained via the theory base change operators determined by $(E^{\ominus})$ and $(E^{\otimes})$.

Theorems 16 and 18, below, demonstrate that theory change operators can be formulated in terms of a theory base change or a brutal theory base change operators. Both these theorems are stated in their most general form, that is for equivalent ensconcements, the results for a particular ensconcement are given as corollaries which follow from Theorem 9. We note that Theorems 14 and 15 above can be generalized in such a way that the theory change operators $^{-}$ and * are determined by an equivalent ensconcement, rather than the same ensconcement, however the current reading is given for clarity.

Theorem 16. *Let* $(\Gamma, \preceq_1)$ *and* $(\Lambda, \preceq_2)$ *be equivalent ensconcements. Let* $^{\ominus}$ *be the theory base contraction operator uniquely determined by* $(\Gamma, \preceq_1)$ *and* $(E^{\ominus})$ *[or* $(B^{\ominus})$*]. Let* $^{-}$ *be the contraction operator for* $\text{Cn}(\Lambda)$ *uniquely determined by* (E^{-}) *and* $\leq_{\preceq_2}$. *Then,* $(\text{Cn}(\Lambda))^{-}_{\alpha} = (\Gamma^{\ominus})^{+}_{\alpha} \cap \text{Cn}(\Gamma)$.

Corollary 17. $(\text{Cn}(\Gamma))^{-}_{\alpha} = \text{Cn}(\Gamma^{\ominus}_{\alpha}) \cap \text{Cn}(\Gamma)$.

Theorem 18. *Let* $(\Gamma, \preceq_1)$ *and* $(\Lambda, \preceq_2)$ *be equivalent ensconcements. Let* $^{\otimes}$ *be the theory base revision operator uniquely determined by* $(\Gamma, \preceq_1)$ *and* $(E^{\otimes})$ *[or* $(B^{\otimes})$*]. Let* * *be the revision operator for* $\text{Cn}(\Lambda)$ *uniquely determined by* (E^{*}) *and* $\leq_{\preceq_2}$. *Then,* $(\text{Cn}(\Lambda))^{*}_{\alpha} = \text{Cn}(\Gamma^{\otimes}_{\alpha})$.

The corollary below, referred to as Foo's Lifting Lemma, follows from Theorem 9 (i) and (ii).

Corollary 19. $(\text{Cn}(\Gamma))^{*}_{\alpha} = \text{Cn}(\Gamma^{\otimes}_{\alpha})$.

By Corollary 7 and Theorem 3, if the ensconcement is well-ranked, inversely well-ranked or finite, then the corresponding theory change operators will be well-behaved, well-mannered, and very well-behaved, respectively.

Theorem 18 shows that the logical closure of theory base revision $\Gamma^{\otimes}_{\alpha}$ is equivalent to the corresponding theory revision operator.

From Theorems 10, 11 and subsequent remarks, it is obvious that for an ensconcement $(\Gamma, \preceq)$ if $^{\ominus}$ is a theory base contraction, or a brutal theory base contraction, then $\Gamma^{\ominus}_{\neg\alpha} \cup \alpha$ is a theory base for $\mathrm{Cn}(\Gamma^{\otimes}_{\alpha})$, and hence by Theorem 18, it is also a theory base for $(\mathrm{Cn}(\Gamma))^{*}_{\alpha}$ via (E^{*}) and $\leq_{\preceq}$. In other words, given an ensconcement a theory base contraction and a brutal theory base contraction are revision equivalent [17]. Furthermore, it is not hard to show that any proper subset of a brutal theory base contraction is not revision equivalent, and a proper subset of $\Gamma^{\otimes}_{\alpha}$ does not satisfy Theorem 18. Consequently, brutal operators retain as 'little as necessary' of the original theory base.

To obtain a similar lifting lemma for contraction we can modify $(E^{\ominus})$ so as to include the addition of certain formulae not in Γ so that recovery is satisfied, an observation made independently by Nebel [19] and Norman Foo. In particular, if the set $\{\alpha \rightarrow \beta : \beta \in \Gamma, \beta \notin \Gamma^{\ominus}_{\alpha}\}$ is added with each contraction, $\Gamma^{\ominus}_{\alpha}$, then we will obtain such a lifting lemma. In particular, $(\mathrm{Cn}(\Gamma))^{-}_{\alpha} = \mathrm{Cn}(\Gamma^{\ominus}_{\alpha} \cup \{\alpha \rightarrow \beta : \beta \in \Gamma, \beta \notin \Gamma^{\ominus}_{\alpha}\}$. However, it could be argued that, this procedure is somewhat ad hoc, since it is not compatible with the notion that Γ represents the explicit beliefs of a reasoning agent.

7 Related Work

In this section we outline the work of others, and contrast their results with those we obtained using an ensconcement. As noted earlier Gärdenfors and Rott [12] provide a comprehensive analysis of various prominent approaches to theory base change, and the interested reader should consult their work.

Nebel's approach [19, 20] is syntax-based, and his structures may be interpreted as assigning higher relevance to explicitly represented formulae, and this is of course quite different in principle to the structure of an ensconcement. Nebel proposes a *prioritized base revision* [20] and he defines an *epistemic relevance ordering* [19] which is a total preorder on all derivable formulae having at least one maximal element. He defines a revision using these structures which is guided by minimal loss of epistemically relevant propositions. Moreover, he obtains two representation theorems which state for each revision based on epistemic relevance ordering there exists a prioritized base revision which gives rise to the same resultant theory, and vice versa. He shows [20] that an epistemic relevance ordering may be used to construct a strict partial order on subsets of a theory base. This order turns out to be relational, and hence satisfies (*7) but is not transitive and hence does not satisfy (*8), unlike our theory base revision operator $^{\otimes}$ based on an ensconcement. An epistemic entrenchment ordering is a special epistemic relevance ordering both giving rise to the same revision operator, and when all the priority classes of a prioritized belief base are singletons the prioritized revision satisfies (*1) – (*8), in this case there exists an epistemic entrenchment that could be used to define such a revision operator. Given Nebel's results there is also an epistemic relevance ordering associated with this revision operator, however it need not be an epistemic entrenchment ordering itself, in particular it may not satisfy (EE2) and (EE3). Nebel also shows that there exists translations from an epistemic entrenchment ordering to an epistemic relevance ordering, via the priority classes.

Hansson [14] argues that derived formulae are dependent on formulae in the theory base, in particular derived formulae should disappear when the theory base is changed. Using a *superselector* which assigns a selection function to theory bases (sets of formulae), his approach also lends itself to iterated theory change. Furthermore, he provides for the change from one base to another for a single theory, this he refers to as a *reorganization*.

Nebel and Hansson use selection functions on maximally consistent sets of formulae that do not entail the undesired formula, whilst Fuhrmann uses minimal subsets of the theory base that entail the undesired formula. In other words, Fuhrmann's approach [7] is founded on an adaptation of safe theory contraction [3]. He uses degrees of *retractability*. Retractability is an ordering which is acyclic, transitive and possesses certain minimal existence properties, it is not total consequently some formulae are incomparable. Intuitively, if α is given up 'more easily' than β, then α has a higher degree of retractability than β. His theory base contraction satisfies ($^-1$) – ($^-4$) and ($^-6$), and in order to deal with local inconsistencies he considers using a paraconsistent logic. Fuhrmann argues like us that theory bases can provide extralogical meaning.

Nayak [18] modifies Fuhrmann's base contraction, so that it satisfies ($^-5$) and ($^-7$) for propositional languages. In particular, he changes the construction of the reject-set, and finds connections between this and Nebel's maxichoice base revision, and unambiguous partial meet revision [20].

For the purpose of a computer-based implementation, what is important is not just how theory base changes but how the underlying ensconcement ordering changes. For instance, does the *principle of minimal change* play a role in the transmutation of an ensconcement. Williams has explored the problem of iterated theory change in [28, 29] based on (E^-) and (E^*), and iterated theory base change operators in [27] based on ($E^{\ominus}$) and ($E^{\otimes}$). In [30] she has shown that iterated theory base change can be used to support Spohn's notion of *reason for*.

8 Discussion

Just as a theory does not uniquely determine a theory change operator, so too a theory base does not uniquely determine a theory base change operator. In the case of theory change additional structure in the form of an epistemic entrenchment ordering is used to construct unique operators, and for theory base change we used an ensconcement.

We described a mechanism for determining theory base contraction operators and theory base revision operators, based on an ensconcement, and showed they are related via, Levi and Harper identities.

We showed that well-behaved, well-mannered, and very well-behaved theory change operators can be constructed from well-ranked, inversely well-ranked, and finite ensconcements, respectively, and conversely.

Moreover, we gave explicit formulations of theory base change operators using theory change operators, and vice versa. We argued that the established relationships demonstrate that the nonbrutal theory base change operators retained as much explicit information as possible.

Appendix

Theorem 4. *If $(\Gamma, \preceq)$ is an ensconcement, then $\leq_\preceq$ is an epistemic entrenchment related to* $\mathrm{Cn}(\Gamma)$.

Proof: We show that $\leq_\preceq$ satisfies the epistemic entrenchment postulates.
Let α, β, γ be formulae in $\mathbf{L}$.
(EE4) trivally holds by definition.
For (EE5), if α is a tautology then $\mathrm{cut}_\preceq(\alpha) = \emptyset$ and (EE5) trivally holds by definition.
For (EE1), suppose $\alpha \leq_\preceq \beta$ and $\beta \leq_\preceq \gamma$. Assume $\alpha \notin \mathrm{Cn}(\Gamma)$ then $\alpha \leq_\preceq \gamma$ for all γ by definition. For the other case, assume $\alpha \in \mathrm{Cn}(\Gamma)$, by supposition $\alpha \leq_\preceq \beta$ and $\beta \leq_\preceq \gamma$, therefore both β and γ are in $\mathrm{Cn}(\Gamma)$ by (EE4) previously verified and since $\alpha \leq_\preceq \beta$ we have $\mathrm{cut}_\preceq(\alpha) \subseteq \mathrm{cut}_\preceq(\beta)$ by definition, and similarly $\mathrm{cut}_\preceq(\beta) \subseteq \mathrm{cut}_\preceq(\gamma)$. Hence we have that $\mathrm{cut}_\preceq(\alpha) \subseteq \mathrm{cut}_\preceq(\gamma)$, and therefore $\alpha \leq_\preceq \gamma$.
For (EE2), suppose $\alpha \vdash \beta$. If $\alpha \notin \mathrm{Cn}(\Gamma)$ then $\alpha \leq_\preceq \beta$ for all β by definition. Therefore, assume $\alpha \in \mathrm{Cn}(\Gamma)$. Since $\alpha \vdash \beta$ we have $\mathrm{cut}_\preceq(\beta) \subseteq \mathrm{cut}_\preceq(\alpha)$, and hence $\alpha \leq_\preceq \beta$ by definition.
For (EE3), if $\alpha \notin \mathrm{Cn}(\Gamma)$ then $\alpha \leq_\preceq \alpha \wedge \beta$ by definition and (EE3) holds. Hence we assume $\alpha \in \mathrm{Cn}(\Gamma)$. Since $\{\alpha\} \cup \{\beta\} \vdash \alpha \wedge \beta$, we have $\mathrm{cut}_\preceq(\alpha \wedge \beta) \subseteq \mathrm{cut}_\preceq(\beta) \cup \mathrm{cut}_\preceq(\alpha)$ by (EE2) and the definition of cut. Either $\mathrm{cut}_\preceq(\beta) \subseteq \mathrm{cut}_\preceq(\alpha)$, or $\mathrm{cut}_\preceq(\alpha) \subseteq \mathrm{cut}_\preceq(\beta)$, and hence, either $\mathrm{cut}_\preceq(\alpha \wedge \beta) \subseteq \mathrm{cut}_\preceq(\alpha)$ or $\mathrm{cut}_\preceq(\alpha \wedge \beta) \subseteq \mathrm{cut}_\preceq(\beta)$. Therefore, either $\alpha \leq_\preceq \alpha \wedge \beta$ or $\beta \leq_\preceq \alpha \wedge \beta$ by definition.

■

Theorem 6. *Let T be a theory in $\mathbf{L}$, and $\Gamma \subseteq T$. Let $\leq$ be an epistemic entrenchment related to T. For all $\alpha \in T$, if $\{\beta \in \Gamma : \ \alpha \leq \beta\} \vdash \alpha$, then $\leq |_\Gamma$ is an ensconcement ordering on Γ such that $\leq_{(\leq|_\Gamma)} = \leq$.*

Proof: Let Γ be a set of formulae such that for all $\alpha \in T$ $\{\beta \in \Gamma : \ \alpha \leq \beta\} \vdash \alpha$. We show that $(\Gamma, \leq |_\Gamma)$ is an ensconcement, if $\alpha \leq \beta$, then $\alpha \leq_{(\leq|_\Gamma)} \beta$ and if $\alpha < \beta$, then $\alpha <_{(\leq|_\Gamma)} \beta$.
If $\leq$ is an epistemic entrenchment related to T, then since any restriction of an epistemic entrenchment satisfies ($\preceq$1) and ($\preceq$2) we have that $(\Gamma, \leq |_\Gamma)$ is an ensconcement.

For all $\alpha \in T$, $\mathrm{Cn}(\mathrm{cut}_{\leq|_\Gamma}(\alpha)) = \mathrm{cut}_{\leq}(\alpha)$ by the definition of Γ. Therefore, if $\alpha, \beta \in T$ then $\alpha \leq_{(\leq|_\Gamma)} \beta$, and if $\alpha \notin T$ then $\alpha \leq_{(\leq|_\Gamma)} \beta$ for all β by definition of $\leq_{(\leq|_\Gamma)}$.

Establishing $\alpha <_{(\leq|_\Gamma)} \beta$ is similar to the argument above replacing cuts with proper cuts.

■

Theorem 14. *Let $(\Gamma, \preceq)$ be an ensconcement. Let $^\ominus$ be the theory base contraction operator uniquely determined by $(E^\ominus)$. Let $^-$ be the contraction operator for $\mathrm{Cn}(\Gamma)$ uniquely determined by (E^-) and $\leq_\preceq$. Then, $\Gamma^\ominus_\alpha = (\mathrm{Cn}(\Gamma))^-_\alpha \cap \Gamma$.*

Proof:

If $\vdash \alpha$, then the theorem immediately follows. Hence we assume $\not\vdash \alpha$. Firstly, we show $\Gamma^\ominus_\alpha \subseteq (\mathrm{Cn}(\Gamma))^-_\alpha \cap \Gamma$. Suppose $\beta \in \Gamma^\ominus_\alpha$. By $(E^\ominus)$ $\beta \in \Gamma$, and $\mathrm{cut}_\prec(\alpha) \cup \{\neg\alpha\} \vdash \beta$. By the deduction theorem, $\alpha \vee \beta \in \mathrm{Cn}(\mathrm{cut}_\prec(\alpha))$, therefore $\alpha <_\preceq \alpha \vee \beta$. Since $\beta \in \Gamma$ trivally $\beta \in \mathrm{Cn}(\Gamma)$, and by (E^-) we have $\beta \in (\mathrm{Cn}(\Gamma))^-_\alpha$, as desired.

Conversely, we show $(\mathrm{Cn}(\Gamma))^-_\alpha \cap \Gamma \subseteq \Gamma^\ominus_\alpha$. Suppose $\beta \in (\mathrm{Cn}(\Gamma))^-_\alpha \cap \Gamma$, then $\beta \in \Gamma$, and $\beta \in (\mathrm{Cn}(\Gamma))^-_\alpha$. Hence we have $\alpha <_\preceq \alpha \vee \beta$ by (E^-). Therefore, $\mathrm{cut}_\prec(\alpha) \vdash \alpha \vee \beta$, hence we have $\mathrm{cut}_\prec(\alpha) \cup \{\neg\alpha\} \vdash \beta$. Hence by $(E^\ominus)$, we have $\beta \in \Gamma^\ominus_\alpha$, since $\beta \in \Gamma$.

■

Theorem 18. *Let $(\Gamma, \preceq_1)$ and $(\Lambda, \preceq_2)$ be equivalent ensconcements. Let $^\otimes$ be the theory base revision operator uniquely determined by $(\Gamma, \preceq_1)$ and $(E^\otimes)$ [or $(B^\otimes)$]. Let * be the revision operator for $\mathrm{Cn}(\Lambda)$ uniquely determined by (E^*) and $\leq_{\preceq_2}$. Then, $(\mathrm{Cn}(\Lambda))^*_\alpha = \mathrm{Cn}(\Gamma^\otimes_\alpha)$.*

Proof:

If $\vdash \neg\alpha$ then the theorem immediately follows, hence we assume $\not\vdash \neg\alpha$.

$$
\begin{aligned}
(\mathrm{Cn}(\Lambda))^*_\alpha &= \{\beta \in \mathbf{L} : \neg\alpha <_{\preceq_2} \neg\alpha \vee \beta\} \qquad \text{by } (E^*). \\
&= \{\beta \in \mathbf{L} : \neg\alpha <_{\preceq_1} \neg\alpha \vee \beta\} \\
&\qquad \text{since } (\Gamma, \preceq_1) \text{ and } (\Lambda, \preceq_2) \text{ are equivalent ensconcements.} \\
&= \mathrm{Cn}(\mathrm{cut}_{<_{\preceq_1}}(\neg\alpha) \cup \{\alpha\}) \\
&= \mathrm{Cn}(\mathrm{cut}_{\prec_1}(\neg\alpha) \cup \{\alpha\}) \\
&= \mathrm{Cn}(\Gamma^\otimes_\alpha) \qquad \text{by } (\Gamma, \preceq_1) \text{ and } (E^\otimes) \ [\text{or } (B^\otimes)].
\end{aligned}
$$

■

Acknowledgements

This work has been partially supported by the Information Technology Division of the CSIRO. The author wishes to express her thanks and appreciation to David Makinson, whose comments and suggestions on an earlier version of this paper were exceedingly valuable, and to members of the Knowledge Systems Group, at the University of Sydney. The author is very grateful for comments received from Norman Foo, Peter Gärdenfors, Bernhard Nebel, Pavlos Peppas, Hans Rott and Brailey Sims.

References

1. Alchourrón, C., and Makinson, D., *On the logic of theory change: Contraction functions and their associated revision functions*, Theoria 48, 14 - 37, 1982.
2. Alchourrón, C., Gärdenfors, P., Makinson, D., *On the Logic of Theory Change: Partial Meet Functions for Contraction and Revision*, Journal of Symbolic Logic, 50: 510-530, 1985.
3. Alchourrón, C.E., and Makinson, D., *On the logic of theory change: Safe contraction*, Studia Logica 44, 405 - 422, 1985.
4. Alchourrón, C.E., and Makinson, D., *Maps between some different kinds of contraction function: The finite case*, Studia Logica, 45: 187- 198, 1986.
5. Dalal, M., *Investigations into a theory of knowledge base revision: preliminary report*, Proceedings of the seventh National Conference of the American Association for Artificial Intelligence, pp 475 - 479, 1988.
6. Fagin, R., Halpern, J.Y., *Belief awareness and limited reasoning*, Artificial Intelligence, 34: 39 - 76, 1988.
7. Fuhrmann, A., *Theory contraction through base contraction*, Journal of Philosophical Logic, 20: 175 - 203, 1991.
8. Gärdenfors, P., *Knowledge in Flux: Modeling the Dynamics of Epistemic States*, A Bradford Book, MIT Press, 1988.
9. Gärdenfors, P., *The dynamics of belief systems: Foundations vs. coherence*, Revue Internationale de Philosophie, 1989.
10. Gärdenfors, P. and Makinson, D., *Revisions of knowledge systems using epistemic entrenchment*, Proceedings of the Second Conference on Theoretical Aspects of Reasoning about Knowledge Conference, pp 83-95, Morgan Kaufmann, 1988.
11. Gärdenfors, P. and Makinson, D., *Nonmonotonic inference based on expectations*, Proceedings of the Third International Conference on Principles of Knowledge Representation and Reasoning, Morgan Kaufmann, 1991.
12. Gärdenfors, P. and Rott, H., *Belief Revision*, Chapter 4.2 in the Handbook of Logic in AI and Logic Programming, Volume IV: Epistemic and Temporal Reasoning, D. Gabbay (ed), Oxford University Press, Oxford, to appear.
13. Grove A., *Two modellings for theory change*, Journal of Philosophical Logic 17: 157-170, 1988.
14. Hansson, S.O., *New operators for theory change*, Theoria 55: 115 - 132, 1989.
15. Lakemeyer, *On the relation between explicit and implicit beliefs*, Proceedings of the Third International Conference on Principles of Knowledge Representation and Reasoning, Morgan Kaufmann, pp 368 - 375, 1991.

16. Makinson, D., *How to give it up: A survey of some formal aspects of the logic of theory change*, Synthese 62: 347 - 363, 1985
17. Makinson, D., *On the status of the postulate of recovery in the logic of theory change*, Journal of Philosophical Logic, 16: 383 - 394, 1987.
18. Nayak, A., *Foundational belief change*, Journal of Philosophical Logic, to appear.
19. Nebel, B., *A knowledge level analysis of belief revision*, Proceedings of the First International Conference on Principles of Knowledge Representation and Reasoning, pp 301–311, Morgan Kaufmann, 1989.
20. Nebel, B., *Belief revision and default reasoning: Syntax-based approaches*, Proceedings of the Third International Conference on Principles of Knowledge Representation and Reasoning, pp 301–311, Morgan Kaufmann, 1991.
21. Niederée, R., *Multiple contraction: A further case against Gärdenfors' principle of recovery*, The Logic of Theory Change, Proceedings for the Workshop at Konstanz, FRG, October 13-15, 1989, Springer-Verlag, Berlin, 1991.
22. Peppas, P., *Belief change and reasoning about action: An axiomatic approach to modelling inert dynamic worlds and the connection to the logic of theory change*, PhD Dissertation, University of Sydney, Australia, submitted 1993.
23. Rescher, N., *Plausible reasoning: An introduction to the theory and practice of Plausibilistic Inference*, Van Gorcum, Amsterdam, 1976.
24. Rott, H., *Two methods of constructing contractions and revisions of knowledge systems*, Journal of Philosophical Logic, 20: 149 - 173, 1991.
25. Rott, H., *A nonmonotonic conditional logic for belief revision I*, in A. Fuhrmann and M. Morreau (eds), *The logic of theory change*, Springer Verlag, LNAI 465, Berlin, 135 - 183.
26. Vardi, *On epistemic logic and logical omniscience*, Proceedings of the First Conference on Theoretical Aspects of Reasoning about Knowledge Conference, pp 293 - 305 Morgan Kaufmann, 1986
27. Williams, M.A., *Transmutations for theory base Change*, Information Systems Research Report, University of Newcastle, a revised and corrected version of a paper with the same title in the Proceedings of the Joint Australian Artificial Intelligence Conference, 1993.
28. Williams, M.A., *Transmutations of knowledge systems*, PhD dissertation, Department of Computer Science, University of Sydney, Australia, submitted 1993.
29. Williams, M.A., *Transmutations of knowledge systems*, in J. Doyle, E. Sandewall, and P. Torasso (eds), Principles of Knowledge Representation and Reasoning: Proceedings of the Fourth International Conference, Morgan Kaufmann, San Mateo, CA, 1994 (to appear).
30. Williams, M.A., *Explanation and theory base transmutations*, in the Proceedings of the European Conference on Artificial Intelligence, 1994 (to appear).

Belief, Provability, and Logic Programs

José Júlio Alferes and Luís Moniz Pereira

CRIA, Uninova and DCS, U. Nova de Lisboa*
2825 Monte da Caparica, Portugal
Phone: +351 1 295 31 56 Fax: +351 1 295 56 41
{jja|lmp}@fct.unl.pt

Abstract. The main goal of this paper is to establish a nonmonotonic epistemic logic $\mathcal{EB}$ with two modalities – provability and belief – capable of expressing and comparing a variety of known semantics for extended logic programs, and clarify their meaning. In particular we present here, for the first time, embeddings into epistemic logic of logic programs extended with a second kind of negation under the well–founded semantics, and contrast them to the recent embeddings into autoepistemic logics of such programs under stable models based semantics.
Furthermore, the language of the epistemic logic presented here being more general than that of extended programs, it offers a basic tool for further generalizations of the latter, for instance regarding disjunction and modal operators.

1 Introduction

The relationships between logic programming and several nonmonotonic reasoning formalisms bring them mutual benefits. Nonmonotonic formalisms provide semantics for logic programs, and help understand how these can express and compute solutions to AI problems. Conversely, the nonmonotonic formalisms benefit from the procedures and implementations of logic programming. Also, relations among nonmonotonic formalisms have been studied via logic programming shunting.

For normal logic programs the bridge to default theories [43] was first made in [6]. In [14] negation as failure of normal programs was first formalized as abduction, and in [12] was extended to capture both stable models [17] and the well founded semantics (WFS) of normal logic programs [15].

The view of logic programs as autoepistemic theories [27] first appeared in [16], which envisages every literal *not* L of logic programs as $\sim\mathcal{L}L$, i.e. *not* L has the epistemic reading: "there is no reason to believe in L"[2]. In [7] a variety of translations of negation as failure by belief literals are studied, in order to show how different logic programming semantics can be obtained from autoepistemic logics (AELs). In [41], Przymusinski assigns to *not* L the translation $\mathcal{B}\sim L$, with the reading "L is believed to be false".

* We thank Esprit BRA Compulog 2 (no. 6810), and JNICT - Portugal for their support. Thanks to Carlos Damásio and Teodor Przymusinski for helpful discussions.

[2] Referred to here as Gelfond's translation.

Several authors have stressed the importance of extending logic programming with a second kind of negation $\neg$, in addition to default negation, for use in deductive databases, knowledge representation, and nonmonotonic reasoning [18, 21, 34, 45]. Different semantics for extended logic programs with $\neg$-negation have appeared [13, 18, 21, 30, 38, 40, 45]. [3] contrasts some of these, where distinct meanings of $\neg$ negation are identified: classical, strong and explicit. It is also argued that explicit negation is preferable.

Some work exists comparing extended logic programs semantics and non-monotonic reasoning formalisms. In [18] the answer sets semantics for extended programs is introduced and compared to default theories. A comparison between the WFS with explicit negation (*WFSX*) [30] and default theories is given in [33]. *WFSX* is captured within an abductive framework in [2].

As noted by [9, 24, 25], Gelfond's translation cannot be generalized to extended programs.

Example 1. According to Gelfond's translation, P :

$$a \leftarrow b$$
$$\neg a$$

is rendered as the theory

$$T = \{b \Rightarrow a; \ \ \sim a\}.$$

This theory entails $\{\sim a, \sim b\}$, but the semantics of P under most of the approaches (e.g. under *WFSX* and answer sets) is $\{\neg a\}$.

A suitable translation between extended programs with answer sets semantics and reflexive AEL theories was proposed independently in [24] and [25]. Reflexive AEL, introduced in [44], views the operator $\mathcal{L}$ as "is known" instead of the "is believed" of Moore's AEL [27][3]. The translation renders an objective literal A (resp. $\neg A$) as $\mathcal{L}A$ (resp. $\mathcal{L} \sim A$, where $\sim$ denotes classical negation), i.e. "A is known to be true" (resp. "A is known to be false"), and renders *not* L as $\mathcal{L} \sim \mathcal{L}L$, i.e. "it is known that L is not known". In [24, 25] the authors prove that the answer sets of an extended program correspond to the reflexive expansions of its translation. Equivalently, the embedding of extended programs into reflexive AEL can also be defined for (non reflexive) AEL [24, 25], by translating any objective literal L into $L \wedge \mathcal{L}L$. This translation was proposed in [9] too.

In [29], the author surmises a translation also equivalent to the ones above. This translation is justified by first relating answer sets to constructive logics with strong negation [28], and then using the already known translation of the latter into nonmonotnic S4.

[3] Roughly, this is achieved by adding $F \equiv \mathcal{L}F$, instead of just $\mathcal{L}F$, when F holds.

The embedding of stable models semantics into AEL was generalized to WFS in [39], using Gelfond's translation, but where Generalized Closed World Assumption (GCWA) [26] replaces the Closed World Assumption (CWA) [42] in what regards the adoption of default literals. No study of embeddings of WFS with $\neg$ negation exists to date. One main purpose of this paper is to remedy this. Significantly, the embedding proposed in [9, 24, 25, 29] does not generalize to extended programs under WFS.

Example 2. Program $P = \{a \leftarrow not\ a\}$ translates into the non reflexive AEL theory $T = \{\sim\mathcal{L}a \Rightarrow \mathcal{L}a \wedge a\} = \{\mathcal{L}a\}$. It is easy to see that this theory has no expansion, even when GCWA is taken up instead of CWA. The same goes for the reflexive AEL translation.

Indeed, that translation is too specific, and can only be applied stable models based semantics (i.e. that are two valued).

In contradistinction, our stance is that, for greater generality, the second kind of negation introduced in logic programming represents and requires, for translation into some epistemic logic, an additional modality other than the one necessary for interpreting negation by default[4]. In our view, an objective literal $\neg A$ (resp. A) should be read "A is proven false", denoted by $\mathcal{E} \sim A$ (resp. "A is proven true"); and *not* L should be read "it is believed that L is not proven", denoted by $\mathcal{B} \sim\mathcal{E}L$. Thus, $\mathcal{E}$ refers to epistemic knowledge as defined by propositional provability, and relates to the consistency modality $\mathcal{M}$ by $\mathcal{E} \equiv\sim\mathcal{M} \sim$. The belief operator of this logic is $\mathcal{B}$, and is inspired by the one introduced in [41].

The main goal of this paper is to define, in section 2, an AEL augmented with the modality $\mathcal{E}$ which is capable of expressing and comparing various semantics of extended programs. The flexibility and generality of our approach are brought out in section 3, by establishing how different notions of provability and knowledge, and different semantics for extended programs are captured by it, and so providing for a better understanding of the different kinds of negation. The improved generality of our AEL language provides a tool for examining further generalizations of extended logic programming. This is discussed in section 4.

2 A logic of belief and provability

In this section we define an epistemic logic, $\mathcal{EB}$, with provability and belief modalities, and show how it captures the *WFSX* semantics for extended logic programming [30], which extends WFS with explicit negation [3], in addition to default negation.

We begin by considering definite extended logic programs only (i.e. extended programs without negation by default), and by defining a modal logic to interpret

[4] In [23] the author also proposes a bi–modal logic (MBNF) for interpreting extended logic programs. There is a MBNF rendering of answer–sets which, as shown in [9, 24], is equivalent to the AEL–unimodal translations, already discussed above, that express answer–sets too.

such programs. We then extend this logic to deal with belief propositions. Finally, we relate the $\mathcal{EB}$ logic to the full language of *WFSX*.

2.1 Provability in extended definite programs

To motivate and make clear the meaning of the provability modality, we begin with the simpler problem of how to capture the meaning of extended programs without negation by default, i.e. sets of rules of the form:

$$L_0 \leftarrow L_1, \ldots, L_n \quad n \geq 0 \tag{1}$$

where each L_i is an atom A or its explicit negation $\neg A$. Without loss of generality, as in [37], we assume that all rules are ground.

The semantics of these programs is desireably monotonic, and must be non-contrapositive, i.e. distinguish between $a \leftarrow b$ and $\neg b \leftarrow \neg a$, so that rules can be viewed as (unidirectional) "inference rules"; Gelfond's translation does not capture this distinction: both rules translate to $b \Rightarrow a$.

Example 3. In example 1, notice how $\sim b$ is derived in T via the contrapositive of the first rule.

The cause of the problem is that $\neg A$ translates into "A is false", and the rule connective $\leftarrow$ into material implication. In contrast, the semantics of extended logic programs wants to interpret $\neg A$ as "A is provenly false", in a grounded sense, and $\leftarrow$ as an inference rule. To capture this meaning we introduce the modal operator $\mathcal{E}$, referring to ***(propositional) "provability"***, or ***"epistemic knowledge"***, and accordingly translate rule (1) into:

$$\mathcal{E}L_1 \wedge \ldots \wedge \mathcal{E}L_n \Rightarrow \mathcal{E}L_0 \tag{2}$$

where any explicitly negated literal $\neg A$ is translated into $\mathcal{E} \sim A$ and reads "A is provenly false", and any atom A is translated into $\mathcal{E}A$ and reads "A is provenly true".

This translation directly captures the intuitive meaning of a rule "if all $L_1, \ldots, L_n$ are provable then L_0 is provable" and does not conflate contrapositives: $a \leftarrow b$ becomes $\mathcal{E}b \Rightarrow \mathcal{E}a$, whilst $\neg b \leftarrow \neg a$ is rendered as $\mathcal{E} \sim a \Rightarrow \mathcal{E} \sim b$.

Note the similarities to the translation defined in [24, 25] into reflexive AEL, where an atom A is translated into $\mathcal{L}A$, and $\neg A$ into $\mathcal{L} \sim A$, where $\mathcal{L}$ is the knowledge operator of modal logic **SW5**.

We need to assume little about $\mathcal{E}$, and this guarantees flexibility. $\mathcal{E}$ is defined as the necessity operator of the smallest normal modal system, modal logic **K**. This logic includes only modus ponens, necessitation, distribution over conjunctions, and the axiom[5]:

$$K: \ \mathcal{E}(F \Rightarrow G) \Rightarrow (\mathcal{E}F \Rightarrow \mathcal{E}G)$$

[5] For a precise definition of logic **K** and its properties see [8, 19].

In logic **K**, $\mathcal{E}$ is the dual of the modal consistency operator $\mathcal{M}$, i.e. $\mathcal{E} \equiv \sim\mathcal{M} \sim$. This weak modal logic, although sufficient for *WFSX* when combined with a belief modality and nonmonotonicity (as shown below), can also express other (stronger) meanings of $\mathcal{E}$ just by introducing more axioms for it. In section 3 in particular, we interpret $\mathcal{E}$ as knowledge by introducing, as usual, the additional axioms for the stronger logic **SW5**.

Since at this stage we are simply interested in the semantics of monotonic (definite) extended programs, we do not require yet a nonmonotonic version of this logic.

Above we said that translation (2) can capture the semantics of extended logic programs. The next theorem makes this statement precise for answer sets and *WFSX* semantics. It generalizes for almost every semantics of extended logic programs, the only exception being, to our knowledge, the "stationary semantics with classical negation" defined in [40], which is contrapositive.

Theorem 1. *Let P be an extended logic program, and T the theory obtained from P by means of translation (2). If $T \vdash_K \mathcal{E}A \wedge \mathcal{E} \sim A$, for no atom A, then:*

$$\begin{array}{lclcl} T \vdash_K \mathcal{E}A & \equiv & P \models_{AS} A & \equiv & P \models_{\mathrm{WFSX}} A \\ T \vdash_K \mathcal{E} \sim A & \equiv & P \models_{AS} \neg A & \equiv & P \models_{\mathrm{WFSX}} \neg A \end{array}$$

where $\vdash_S$ denotes, as usual, the consequence relation in modal logic S (in this case K), and $P \models_{AS} L$ (resp. $P \models_{\mathrm{WFSX}} L$) means that L belongs to all answer sets (resp. all WFSX *partial stable models) of P.*

Otherwise, the only answer set is the set of all objective literals, and P is contradictory wrt to WFSX.

2.2 Belief and provability

Besides explicit negation, extended logic programs also allow negation by default, which is nonmonotonic and usually understood as a belief proposition. Thus, we need to enlarge modal logic **K** with a nonmonotonic belief operator.

Before tackling the more general problem, we begin by defining what beliefs follow from definite extended programs. Such programs are readily translatable into sets of Horn clauses, thereby possessing a unique minimal model. So, as a first approach consider: "the agent believes in a formula if it belongs to the minimal model of the theory", i.e.

$$\text{if } T \models_{min} F \text{ then } \mathcal{B}F \ (\textit{introspection}).$$

Example 4. The program of example 1 translates into

$$T = \{\mathcal{E}b \Rightarrow \mathcal{E}a; \ \ \mathcal{E} \sim a\}$$

whose least model is $\{\mathcal{E} \sim a\}$. Thus an agent with knowledge T believes all of $\mathcal{B}\mathcal{E} \sim a$, $\mathcal{B} \sim \mathcal{E}a$, $\mathcal{B} \sim \mathcal{E}b$, and $\mathcal{B} \sim \mathcal{E} \sim b$.

Moreover we insist on the principle that, for rational agents, if $T \models \mathcal{E}L$ then $\mathcal{B} \sim\mathcal{E} \sim L$ (coherence). Coherence states that whenever L is provenly true then it is mandatory to believe that L is not provenly false[6]. The *coherence principle* introduced for extended logic programming in [30] is an instance of it. In the above example absence of coherencde does not interfere with the result. This is not in general the case:

Example 5. Consider $T = \{\mathcal{E}a;\ \ \mathcal{E} \sim a\}$ whose least model is $\{\mathcal{E}a, \mathcal{E} \sim a\}$. $\mathcal{B}\mathcal{E}a$ and $\mathcal{B}\mathcal{E} \sim a$ hold by introspection. Moreover, by coherence, an agent must sustain both $\mathcal{B} \sim\mathcal{E} \sim a$ and $\mathcal{B} \sim\mathcal{E}a$.

This kind of reasoning may seem strange since the agent must believe in complementary formulae (e.g. in $\mathcal{E}a$ and in $\sim\mathcal{E}a$.). But, as shown below, when the axioms for $\mathcal{B}$ are introduced we'll see these will detect inconsistency out from the intuitively inconsistent theory T, i.e. belief cannot be held of proven complements.

As for $\mathcal{E}$, little is assumed about $\mathcal{B}$, both for the sake of flexibility and because it is indeed enough for characterizing *WFSX*. More precisely, we assume the axioms introduced in [41] for the belief operator:

- For any tautologically false formula F: $\sim\mathcal{B}F$.
- For any formulae F and G: $\mathcal{B}(F \wedge G) \equiv \mathcal{B}F \wedge \mathcal{B}G$.

As proven in [41], from these axioms it follows for every formula F that $\mathcal{B}F \Rightarrow \sim\mathcal{B} \sim F$[7]. Consequently, from believing two complementary formulae, $\mathcal{B}F$ and $\mathcal{B} \sim F$, inconsistency follows because $\mathcal{B} \sim F \Rightarrow \sim\mathcal{B}F$.

In summary, for a theory T resulting from a definite extended program, the set of beliefs of an agent is the closure, under the above axioms, of:

$$\{\mathcal{B}F \mid T \models_{min} F\} \cup \{\mathcal{B} \sim\mathcal{E} \sim F \mid T \models \mathcal{E}F\}$$

as required by introspection and coherence, respectively.

In order to enlarge the logic **K** with a nonmonotonic belief operator we proceed as above, but now consider the case where formulae of the form $\mathcal{B}F$ or $\sim\mathcal{B}F$ (hereafter called belief formulae) occur in theories. In this case it is not adequate to obtain the belief closure as above. To deal with belief formulae in theories we must consider, as usual in AEL, the expansions of a theory.

An expansion T^* of a theory T is a fixpoint of the equation $T^* = T \cup Bel$, where Bel is a set of belief formulae depending on T^*. Intuitively, each expansion stands for a belief state of a rational agent. One issue arises: *which kind of nonmonotonicity to introduce in such theories?*

In this respect two main approaches have been followed in the literature: One, present in Moore's AEL and in reflexive AEL, is based on CWA an agent

[6] Note that $\mathcal{B} \sim\mathcal{E} \sim L \equiv \mathcal{B}\mathcal{M}L$.

[7] In fact, this implication is equivalent to $\sim(\mathcal{B}F \wedge \mathcal{B} \sim F)$, by the second axiom it is equivalent to $\sim\mathcal{B}(F \wedge \sim F)$, which is true because $F \wedge \sim F$ is tautologically false.

believes in F in an expansion T^* iff $T^* \models F$, and does not believe in F iff $T^* \not\models F$ and it captures two valued (or total) logic program semantics, i.e. those where whenever A does not belong to a model then *not* A belongs to it. The other approach is based on GCWA an agent believes in F in an expansion T^* iff $T^* \models_{min} F$, and does not believe in F iff $T^* \models_{min} \sim F$ and captures three valued (or partial) logic program semantics. The latter approach is followed in the AEL of closed beliefs [39], and in his static semantics [41][8].

Here we adopt the second approach too. The reasons for prefering a logic based on GCWA rather than on CWA are tantamount to those that prefer semantics based on WFS rather than on stable models, and are extensively discussed in the literature (e.g. in [3, 7, 15, 39, 41]). In this paper we do not go into the details for this preference, but summarize [39]: With CWA quite "reasonable" theories are often inconsistent; expansions are non cumulative, non rational, and non relevant even for theories resulting from logic programs[9]; expansions cannot be effectively computed (even for propositional logic programs); the insistance on total models often lacks expressivity. Non of this occurs with GCWA based expansions.

In the sequel we formally define our epistemic logic. We begin by extending the language of propositional logic with modal operators $\mathcal{E}$ and $\mathcal{B}$, standing for "provability" and "belief". Theories are recursively defined as usual. Moreover we assume every theory contains all axioms of logic **K** for $\mathcal{E}$, and the above two axioms for $\mathcal{B}$.

Definition 2. A *minimal model of a theory* T is a model M of T such that there is no smaller model N of T coinciding with M on belief propositions.

If F is true in all minimal models of T then we write $T \models_{min} F$.

An expansion T^* corresponds to a belief state where the agent believes in F if $T^* \models_{min} F$, and does not believe in F if $T^* \models_{min} \sim F$. With the axioms introduced for $\mathcal{B}$, the second statement is subsumed by the first. Indeed, by the first statement, if $T^* \models_{min} \sim F$ then $\mathcal{B} \sim F$, and from the axioms for $\mathcal{B}$ it follows, so we've seen, that $\sim \mathcal{B} F$.

Just as argued for definite extended programs, when considering theories with provability and belief one new form of belief obtention (coherence) is in place, namely if $T^* \models \mathcal{E} G$ then $\mathcal{B} \sim \mathcal{E} \sim G$. Thus, expansions should formalize the following notion of belief $\mathcal{B}$:

$$\mathcal{B}F \equiv F \text{ is minimally entailed, or } F = \sim \mathcal{E} \sim G \text{ and } \mathcal{E}G \text{ is entailed.}$$

[8] Note that the question of distinguishing between these two approaches is not relevant for definite programs, since in them nonderivability coincides with deriving the complement in the (single) minimal model.

[9] By cumulativity [10] we refer to the efficiency related ability of using lemmas. By rationality [10] we refer to the ability to add the negation of a non–provable conclusion without changing the semantics. By relevance [11] we mean that the top–down evaluation of a literal's truth–value requires only the call–graph below it

Definition 3. An *expansion of a theory* T is a consistent theory T^* satisfying the fixed point condition:

$$T^* = T \cup \{\mathcal{B}F \mid T^* \models_{min} F\} \cup \{\mathcal{B} \sim\mathcal{E} \sim G \mid T^* \models \mathcal{E}G\}$$

Example 6. [10] Consider an agent with the following knowledge:

- Peter is a bachelor;
- a man is not married if he is a bachelor;
- Susan is married to Peter, if we don't believe she's married to Tom;
- Susan is married to Tom, if we don't believe she's married to Peter;
- no one is married to oneself;

rendered by the autoepistemic theory T (with obvious abbreviations):

$$\begin{aligned} &\mathcal{E}b(p) \\ \mathcal{E}b(X) \Rightarrow\ &\mathcal{E} \sim m(X,Y) \\ \mathcal{B} \sim\mathcal{E}m(t,s) \Rightarrow\ &\mathcal{E}m(p,s) \\ \mathcal{B} \sim\mathcal{E}m(p,s) \Rightarrow\ &\mathcal{E}m(t,s) \\ &\mathcal{E} \sim m(X,X) \end{aligned}$$

The only expansion of T contains, among others, the belief propositions:

$$\{\mathcal{B}\mathcal{E}b(p), \mathcal{B}\mathcal{E} \sim m(p,s), \mathcal{B} \sim\mathcal{E}m(p,s), \mathcal{B}\mathcal{E}m(t,s)\}$$

In this example all of an agent's beliefs are completely decided, in the sense that for any proposition A the agent either believes or disbelieves A. This is not in general the case.

Example 7. Consider the statements:

- it is proven or it is believed that the car can be fixed;
- if it is not believed that one can fix the car then an expert is called for;
- an expert is not called for;

rendered by the autoepistemic theory T :

$$\begin{aligned} \mathcal{E}can\_fix\_car\ \vee\ &\mathcal{B}\mathcal{E}can\_fix\_car \\ \mathcal{B} \sim\mathcal{E}can\_fix\_car \Rightarrow\ &\mathcal{E}call\_expert \\ &\mathcal{E} \sim call\_expert \end{aligned}$$

The only expansion of T is:

$$T \cup \{\mathcal{B}\mathcal{E} \sim call\_expert, \mathcal{B} \sim\mathcal{E}call\_expert\}$$

stating that an agent believes that an expert is not called and that he disbelieves an expert is called for.

Note that about $\mathcal{E}can\_fix\_car$ the agent remains undefined. This is due to, on the one hand, believing it true is impossible since it is not a consequence in all minimal models; on the other hand, believing it false leads to an inconsistency.

[10] This example first appeared in [46], in the form of a logic program.

Like Moore's autoepistemic theories $\mathcal{EB}$ theories might have several expansions:

Example 8. Consider the theory T, describing the so called Nixon diamond situation:

$$\begin{array}{r}
\mathcal{E}republican(nixon)\\
\mathcal{E}quaker(nixon)\\
\mathcal{E}republican(X), \mathcal{B} \sim\mathcal{E}pacifist(X) \Rightarrow \mathcal{E} \sim pacifist(X)\\
\mathcal{E}quaker(X), \mathcal{B} \sim\mathcal{E} \sim pacifist(X) \Rightarrow \mathcal{E}pacifist(X)
\end{array}$$

T has three expansions, namely:

$$\begin{array}{l}
T \cup \{\mathcal{BE}r(n), \mathcal{BE}q(n), \mathcal{BE}p(n), \quad \mathcal{B} \sim\mathcal{E} \sim r(n), \mathcal{B} \sim\mathcal{E} \sim q(n), \mathcal{B} \sim\mathcal{E} \sim p(n)\}\\
T \cup \{\mathcal{BE}r(n), \mathcal{BE}q(n), \mathcal{BE} \sim p(n), \mathcal{B} \sim\mathcal{E} \sim r(n), \mathcal{B} \sim\mathcal{E} \sim q(n), \mathcal{B} \sim\mathcal{E}p(n)\}\\
T \cup \{\mathcal{BE}r(n), \mathcal{BE}q(n), \qquad\qquad \mathcal{B} \sim\mathcal{E} \sim r(n), \mathcal{B} \sim\mathcal{E} \sim q(n)\}
\end{array}$$

The first states that it is believed that Nixon is a pacisfist; the second that it is believed that Nixon is not a pacifist; and the third remains undefined in what concerns Nixon being a pacisfist or not.

When confronted with several expansions (i.e. several possible states of beliefs) a sceptical reasoner should only conclude what is common to all. Here that means the third expansion.

2.3 Relation to extended logic programs

An extended logic program is as set of rules of the form:

$$L_0 \leftarrow L_1, \ldots, L_m, not\ L_{m+1}, \ldots, not\ L_n \tag{3}$$

where each L_i is an objective literal, i.e. an atom A or its $\neg$ negation $\neg A$.

As argued above, an atom A is translated into $\mathcal{E}A$, and an explicitly negated atom $\neg A$ into $\mathcal{E} \sim A$. In [24, 25] literals of the form *not* L (default literals) are translated into $\mathcal{L} \sim\mathcal{L}L$ in reflexive AEL. [25] gives an intuitive reading of this formula: "it is known that L is not known". In our approach we translate *not* L into $\mathcal{B} \sim\mathcal{E}L$, i.e. "it is believed (or it is assumed) that L is not proven". So, each rule of the form (3) is translated into:

$$\mathcal{E}L_1, \ldots, \mathcal{E}L_m, \mathcal{B} \sim\mathcal{E}L_{m+1}, \ldots, \mathcal{B} \sim\mathcal{E}L_n \Rightarrow \mathcal{E}L_0 \tag{4}$$

Definition 4. A *WFSX* partial stable model M of an extended logic program P corresponds to an expansion T^* when:

- For an objective literal L : $L \in M$ iff $\mathcal{BE}L \in T^*$.
- For a default literal *not* L : *not* $L \in M$ iff $\mathcal{B} \sim\mathcal{E}L \in T^*$.

Theorem 5. *Let T be the theory obtained from an extended logic program P by means of translation (4). Then there is a one to one correspondence between the* WFSX *partial stable models of P and the expansions of T.*

This relationship brings mutual benefits to both *WFSX* and the $\mathcal{EB}$ logic. On the one hand, the logic allows for a more intuitive view of *WFSX*, specially in what concerns its understanding as modeling provability and belief in a rational agent. This allows for a clearer formulation within *WFSX* of some problems in knowledge representation and reasoning, and for a better understanding of *WFSX*'s results. In particular, it shows that explicit negation stands for proving falsity of a literal, default negation for believing that a literal is not provable, and undefinedness for believing neither the falsity nor the verity of a literal. The relationship also sheds light on several extensions of *WFSX* (cf. section 4).

On the other hand, for the class of theories resulting from some extended programs, the logic can be implemented using the top down procedures defined for *WFSX*[1]. Moreover, for this class, the logic enjoys the properties of cumulativity, rationality, relevance [11], and others proven for *WFSX* in [5]. In addition, the relationship also raises new issues in epistemic logics, and points towards their solution via the techniques in use in extended logic programming (cf. section 4).

3 Provability versus Knowledge

Above we claimed logic $\mathcal{EB}$ is flexible and general. Next we express with it different meanings for $\mathcal{E}$, and hence a variety of semantics for extended logic programs.

The logic **K** introduced for $\mathcal{E}$ is the simplest normal modal system, contained in any other. With additional axioms to our theories we can define other meanings for $\mathcal{E}$. In particular, with the axioms of logic **SW5**[11] $\mathcal{E}$ represents "knowledge", as in [44, 25]. Other formalizations of knowledge, such as that of logic **S4.2**[12], are similarly obtained.

Using the **SW5** meaning of $\mathcal{E}$, but keeping with the same translation, a different semantics for extended logic programs is obtained:

Theorem 6. *Let T be the theory obtained from an extended logic program P by means of translation (4), augmented with the* **SW5** *axioms for* $\mathcal{E}$. *Then there is a one to one correspondence between expansions of T and the partial stable models of P of WFS with strong negation, as defined in [3].*

Comparisons between WFS with strong negation and *WFSX*, found in [3, 5], prove the former is less suitable for implementation (as it does not enjoy relevance [11]), is more credulous, and assigns semantics to less programs.

Example 9. Program P:

$$\neg a$$
$$a \leftarrow not\ b$$
$$b \leftarrow not\ b$$

[11] I.e. axioms T: $\mathcal{E}F \Rightarrow F$, 4: $\mathcal{E}F \Rightarrow \mathcal{E}\mathcal{E}F$, and W5: $\sim\mathcal{E} \sim F \Rightarrow (F \Rightarrow \mathcal{E}F)$.

[12] [22] uses **S4.2** to formalize knowledge in a logic which also includes belief. We intend to compare this logic with ours when their final version becomes available to us.

translates into the theory:

$$T = \{\mathcal{E} \sim a;\ \ \mathcal{B} \sim \mathcal{E} b \Rightarrow \mathcal{E} a;\ \ \mathcal{B} \sim \mathcal{E} b \Rightarrow \mathcal{E} b\}.$$

Using logic **K**, there is one expansion:

$$T^* = T \cup \{\mathcal{B}\mathcal{E} \sim a, \mathcal{B} \sim \mathcal{E} a, \mathcal{B} \sim \mathcal{E} \sim b\}.$$

If logic **SW5** is used instead, there is no expansion. This happens because, by axiom T, $\mathcal{E} \sim a$ entails $\sim\mathcal{E} a$, and by the contrapositive of the second clause of T $\sim\mathcal{E} a$ entails $\sim\mathcal{B} \sim\mathcal{E} b$. Thus, by the third clause, every minimal model of every possible expansion has $\sim\mathcal{E} b$, and so $\mathcal{B} \sim\mathcal{E} b$ must be added. This is inconsistent with having $\sim\mathcal{B} \sim\mathcal{E} b$ in all models, and so no expansion exists. *WFSX* assigns a meaning to P, namely $\{\neg a, not\ a, not\ \neg b\}$, because axiom T is not assumed.

From theorem 6 and the results of [3] it follows that:

Theorem 7. *Let T be the theory obtained from an extended logic program P by means of translation (4), augmented with the* **SW5** *axioms for $\mathcal{E}$, and the axiom $\sim\mathcal{E} F \Rightarrow \mathcal{E} \sim F$. Then there is a one to one correspondence between the expansions of T and the partial stable models of P of the "stationary semantics with classical negation" of [40].*

Since answers sets are the total stable models of WFS with strong negation [3]:

Definition 8. An expansion T^* is total iff for every formula F :

$$T^* \not\models \mathcal{B} F \quad \Rightarrow \quad T^* \models \mathcal{B} \sim F$$

Theorem 9. *Let T be the theory obtained from an extended logic program P by means of translation (4), augmented with the* **SW5** *axioms for $\mathcal{E}$. Then there is a one to one correspondence between the total expansions of T and the answer sets of P.*

4 Further developments

Since the language of $\mathcal{E}\mathcal{B}$ is more general than that of extended logic programs, our logic is a tool for further generalizations of extended logic programming, for instance disjunction. All that is required is to define a translation of disjunctive extended logic programs into the logic. The study of possible translations, and the relationship between the resulting and extant semantics for disjunctive programs is the subject of ongoing investigations by us.

Another possible direct generalization of extended logic programming is with the modal operators of the logic, allowing for conjunction and disjunction within their scope. Examples of the use and usefulness of the belief operator for normal disjunctive programs can be found in [41].

With the relationship between $\mathcal{EB}$ logic and extended logic programming now established some issues already tackled in the latter can also be raised in the former. Furthermore, the former can profit from adapting techniques employed in the latter. One of the issues presented here in more detail is contradiction removal, or belief revision.

Recently, several authors have studied this issue in extended logic programming [13, 20, 31, 32]. The basic idea behind these approaches is that *not L* literals be viewed as assumptions, so that if an assumption partakes in a contradiction then its revision is in order. In epistemic logics this idea translates into: "If the results of introspection lead to the inexistence of expansions then revise your beliefs".

Example 10. The theory T:

$$\mathcal{B} \sim\mathcal{E}ab \Rightarrow \mathcal{E}fly$$
$$\mathcal{E} \sim fly$$

is consistent but has no expansion. This is so because $\sim\mathcal{E}ab$ is true in all minimal models and thus, by introspection, $\mathcal{B} \sim\mathcal{E}ab$ must be added causing a contradiction. In fact, this is a typical case where the result of introspection leads to contradiction[13].

In order to assign a meaning to consistent theories without expansions two approaches are possible: to define a more sceptical notion of expansion, introducing less belief propositions by introspection; or to minimally revise the theory in order to provide for expansions. [4] contrast these two approaches in the logic programming setting, dubbing the first "contradiction avoidance", and the second "contradiction removal", and showing them equivalent under certain conditions. The techniques used there to deal with this issue in logic programming can readily be employed in our epistemic logic:

Contradiction avoidance in the $\mathcal{EB}$ logic amounts to weakening the condition for introspection. This can be accomplished by allowing introduction of belief propositions solely for some chosen subset of the formulae minimally entailed by the theory. Of course, not all subsets are allowed. In particular, we are only interested in maximal subsets compatible with consistency. The study of additional preference conditions among these subsets is tantamount to the one in extended logic programming examined in [31].

Contradiction removal in the $\mathcal{EB}$ logic amounts to minimally adjoining, to some consistent theory without expansions, new clauses to inhibit the addition by introspection of belief propositions responsible for contradiction.

Example 11. The theory T' resulting from adjoining the "inhibiting clause"

$$\mathcal{B} \sim\mathcal{E}ab \Rightarrow \mathcal{E}ab$$

[13] Note this problem is not peculiar to our logic. The same occurs as well in e.g. AEL and reflexive AEL.

to the theory of example 10 has one expansion

$$T^* = T' \cup \{\mathcal{BE} \sim fly, \mathcal{B} \sim \mathcal{E} fly\}.$$

The extra clause states that believing $\sim\mathcal{E}ab$ is impossible, since it directly implies $\mathcal{E}ab$, and in this way inhibits the otherwise inevitable addition of $\mathcal{B} \sim\mathcal{E}ab$.

In extended logic programs the inhibition clause above translates to

$$ab \leftarrow not\ ab$$

and is called an inhibition rule for ab [32]. Similarly to what is done in [32] for extended logic programming, also for the $\mathcal{EB}$ logic we can define the revisions of a theory as those obtained by minimal additions of inhibiting clauses which achieve contradiction removal (i.e. allow the resulting theory to have an expansion). Thus, the revisions of a theory T are those theories

$$T' = T \cup InR$$

where InR is a minimal set of clauses of the form $\mathcal{B} \sim\mathcal{E} A \Rightarrow \mathcal{E} A$, such that T' has at least one expansion. Given the similarities of this approach and the contradiction removal techniques in extended logic programs, the procedures [4] and implementations of the latter, developed in order to avoid generating all possible revisions, are also applicable to the former.

5 Conclusions

To start, we've provided a first embedding, in a two-modality AEL, of well founded semantics based extensions of logic programs, either with explicit or strong negation [3], and shown how they differ in that setting.

Second, we've contrasted this embedding with recent ones for stable semantics based extensions of logic programs with a second kind of negation [9, 24, 25, 29].

Third, we've shown the usefulness of employing the combination of the two natural epistemic modalities of provability and belief for explicating the semantics of logic programs in AEL.

Fourth, we've introduced into epistemic logics the issue of belief revision in the face of inexistance of expansions, and how to tackle it inspired by similar questions and attending techniques previously developed, in the logic programming context, for contradiction removal [20, 32, 4].

Fifth, we've shown how (via the procedures and implementations of the extended logic programs we've embedded in our AEL) an important subset of this AEL can be employed to represent and solve a variety of nonmonotonic reasoning problems, already captured in and dealt with by extended logic programs: namely taxonomies, reasoning about actions, diagnosis, updates, and debugging [34, 35, 36].

Sixth, as the language of our epistemic logic is more general than that of extended programs, the former can now be used as a tool for further generalizations of the latter, for instance wrt disjunction and modalities.

References

1. J. J. Alferes, C. V. Damásio, and L. M. Pereira. Top-down query evaluation for well-founded semantics with explicit negation. In A. Cohn, editor, *European Conf. on AI*, pages 140–144. Morgan Kaufmann, 1994. To appear.
2. J. J. Alferes, P. M. Dung, and L. M. Pereira. Scenario semantics of extended logic programs. In L. M. Pereira and A. Nerode, editors, *2nd Int. Ws. on LP & NMR*, pages 334–348. MIT Press, 1993.
3. J. J. Alferes and L. M. Pereira. On logic program semantics with two kinds of negation. In K. Apt, editor, *Int. Joint Conf. and Symp. on LP*, pages 574–588. MIT Press, 1992.
4. J. J. Alferes and L. M. Pereira. Contradiction: when avoidance equal removal. In R. Dyckhoff, editor, *4th Int. Ws. on Extensions of LP*, volume 798 of *LNAI*. Springer–Verlag, 1994. To appear.
5. José Júlio Alferes. *Semantics of Logic Programs with Explicit Negation.* PhD thesis, Universidade Nova de Lisboa, October 1993.
6. N. Bidoit and C. Froidevaux. Minimalism subsumes default logic and circumscription in stratified logic programming. In *Symp. on Principles of Database Systems.* ACM SIGACT-SIGMOD, 1987.
7. P. Bonatti. Autoepistemic logics as a unifying framework for the semantics of logic programs. In K. Apt, editor, *Int. Joint Conf. and Symp. on LP*, pages 417–430. MIT Press, 1992.
8. B. Chellas. *Modal Logic: An introduction.* Cambridge Univ. Press, 1980.
9. J. Chen. Minimal knowledge + negation as failure = only knowing (sometimes). In L. M. Pereira and A. Nerode, editors, *2nd Int. Ws. on LP & NMR*, pages 132–150. MIT Press, 1993.
10. J. Dix. Classifying semantics of logic programs. In A. Nerode, W. Marek, and V. S. Subrahmanian, editors, *LP & NMR*, pages 166–180. MIT Press, 1991.
11. J. Dix. A framework for representing and characterizing semantics of logic programs. In B. Nebel, C. Rich, and W. Swartout, editors, *3rd Int. Conf. on Principles of Knowledge Representation and Reasoning.* Morgan Kaufmann, 1992.
12. P. M. Dung. Negation as hypotheses: An abductive framework for logic programming. In K. Furukawa, editor, *8th Int. Conf. on LP*, pages 3–17. MIT Press, 1991.
13. P. M. Dung and P. Ruamviboonsuk. Well founded reasoning with classical negation. In A. Nerode, W. Marek, and V. S. Subrahmanian, editors, *LP & NMR*, pages 120–132. MIT Press, 1991.
14. K. Eshghi and R. Kowalski. Abduction compared with negation by failure. In *6th Int. Conf. on LP.* MIT Press, 1989.
15. A. Van Gelder, K. A. Ross, and J. S. Schlipf. The well-founded semantics for general logic programs. *Journal of the ACM*, 38(3):620–650, 1991.
16. M. Gelfond. On stratified autoepistemic theories. In *AAAI'87*, pages 207–211. Morgan Kaufmann, 1987.
17. M. Gelfond and V. Lifschitz. The stable model semantics for logic programming. In R. Kowalski and K. A. Bowen, editors, *5th Int. Conf. on LP*, pages 1070–1080. MIT Press, 1988.
18. M. Gelfond and V. Lifschitz. Logic programs with classical negation. In Warren and Szeredi, editors, *7th Int. Conf. on LP*, pages 579–597. MIT Press, 1990.
19. G. Hughes and M. Cresswell. *A companion to modal logic.* Methuen, 1984.
20. K. Jonker. On the semantics of conflit resolution in truth maintenance systems. Technical report, Univ. of Utrecht, 1991.

21. R. Kowalski and F. Sadri. Logic programs with exceptions. In Warren and Szeredi, editors, *7th Int. Conf. on LP*. MIT Press, 1990.
22. P. Lamarre and Y. Shoham. On knowledge, certainty, and belief (draft). Personal communication of the second author, Stanford Univ., 1993.
23. V. Lifschitz. Minimal belief and negation as failure. Technical report, Dep. of Computer Science and Dep. of Philisophy, Univ. of Texas at Austin, 1992.
24. V. Lifschitz and G. Schwarz. Extended logic programs as autoepistemic theories. In L. M. Pereira and A. Nerode, editors, *2nd Int. Ws. on LP & NMR*, pages 101–114. MIT Press, 1993.
25. V. Marek and M. Truszczynski. Reflexive autoepistemic logic and logic programming. In L. M. Pereira and A. Nerode, editors, *2nd Int. Ws. on LP & NMR*, pages 115–131. MIT Press, 1993.
26. J. Minker. On indefinite databases and the closed world assumption. In M. Ginsberg, editor, *Readings in Nonmonotonic Reasoning*, pages 326–333. Morgan Kaufmann, 1987.
27. R. Moore. Semantics considerations on nonmonotonic logic. *Artificial Intelligence*, 25:75–94, 1985.
28. D. Nelson. Constructible falsity. *JSL*, 14:16–26, 1949.
29. D. Pearce. Answer sets and constructive logic, II: Extended logic programs and related nonmonotonic formalisms. In L. M. Pereira and A. Nerode, editors, *2nd Int. Ws. on LP & NMR*, pages 457–475. MIT Press, 1993.
30. L. M. Pereira and J. J. Alferes. Well founded semantics for logic programs with explicit negation. In B. Neumann, editor, *European Conf. on AI*, pages 102–106. John Wiley & Sons, 1992.
31. L. M. Pereira and J. J. Alferes. Optative reasoning with scenario semantics. In D. S. Warren, editor, *10th Int. Conf. on LP*, pages 601–615. MIT Press, 1993.
32. L. M. Pereira, J. J. Alferes, and J. N. Aparício. Contradiction Removal within Well Founded Semantics. In A. Nerode, W. Marek, and V. S. Subrahmanian, editors, *LP & NMR*, pages 105–119. MIT Press, 1991.
33. L. M. Pereira, J. J. Alferes, and J. N. Aparício. Default theory for well founded semantics with explicit negation. In D. Pearce and G. Wagner, editors, *Logics in AI. Proceedings of the European Ws. JELIA'92*, volume 633 of *LNAI*, pages 339–356. Springer–Verlag, 1992.
34. L. M. Pereira, J. N. Aparício, and J. J. Alferes. Nonmonotonic reasoning with well founded semantics. In Koichi Furukawa, editor, *8th Int. Conf. on LP*, pages 475–489. MIT Press, 1991.
35. L. M. Pereira, J. N. Aparício, and J. J. Alferes. Non–monotonic reasoning with logic programming. *Journal of Logic Programming. Special issue on Nonmonotonic reasoning*, 17(2, 3 & 4):227–263, November 1993.
36. L. M. Pereira, C. Damásio, and J. J. Alferes. Diagnosis and debugging as contradiction removal. In L. M. Pereira and A. Nerode, editors, *2nd Int. Ws. on LP & NMR*, pages 316–330. MIT Press, 1993.
37. H. Przymusinska and T. Przymusinski. Semantic issues in deductive databases and logic programs. In R. Banerji, editor, *Formal Techniques in AI, a Sourcebook*, pages 321–367. North Holland, 1990.
38. T. Przymusinski. Extended stable semantics for normal and disjunctive programs. In Warren and Szeredi, editors, *7th Int. Conf. on LP*, pages 459–477. MIT Press, 1990.
39. T. Przymusinski. Autoepistemic logic of closed beliefs and logic programming. In A. Nerode, W. Marek, and V. S. Subrahmanian, editors, *LP & NMR*, pages 3–20. MIT Press, 1991.

40. T. Przymusinski. A semantics for disjunctive logic programs. In Loveland, Lobo, and Rajasekar, editors, *ILPS'91 Ws. in Disjunctive Logic Programs*, 1991.
41. T. Przymusinski. Static semantics for normal and disjunctive programs. Technical report, Dep. of Computer Science, Univ. of California at Riverside, 1993.
42. R. Reiter. On closed-world data bases. In H. Gallaire and J. Minker, editors, *Logic and DataBases*, pages 55–76. Plenum Press, 1978.
43. R. Reiter. A logic for default reasoning. *Artificial Intelligence*, 13:68–93, 1980.
44. G. Schwarz. Autoepistemic logic of knowledge. In A. Nerode, W. Marek, and V. S. Subrahmanian, editors, *LP & NMR*, pages 260–274. MIT Press, 1991.
45. G. Wagner. A database needs two kinds of negation. In B. Thalheim, J. Demetrovics, and H-D. Gerhardt, editors, *Mathematical Foundations of Database Systems*, volume 495 of *LNCS*, pages 357–371. Springer–Verlag, 1991.
46. G. Wagner. Reasoning with inconsistency in extended deductive databases. In L. M. Pereira and A. Nerode, editors, *2nd Int. Ws. on LP & NMR*, pages 300–315. MIT Press, 1993.

Revision specifications by means of programs

Victor W. Marek and Mirosław Truszczyński

Department of Computer Science
University of Kentucky
Lexington, KY 40506-0027, USA
{marek,mirek}@ms.uky.edu

Abstract. We propose a formalism for specifying revisions in knowledge bases and belief sets. This formalism extends logic programming with stable model semantics. Main objects of our system are *revision programs* consisting of *revision rules*. A revision rule expresses a specification of change or a constraint on a knowledge base. There are two types of revision rules. *In-rules* require that an element be in a knowledge base whenever some other elements are in the knowledge base and yet other elements are absent from it. Similar conditions in an *out-rule* force the absence of an element from the knowledge base.
For a revision program P we introduce the notion of a *P-justified revision*, which we use to specify the meaning of the program. Main motivation for our formalism and for the semantics of P-justified revisions comes from default logic and logic programming with stable model semantics. In the paper, we show that if a knowledge base B is a model of a program P then B is the unique P-justified revision of B. We show that P-justified revisions are models of P. We also show that P-justified revisions of a given knowledge base satisfy some minimality criterion. We outline the proof theory for revision programs and show its adequacy for the proposed semantics. We generalize the notion of a revision program to the case of disjunctive revision programs. A simple example of an application is also discussed.

1 Introduction

In this paper we propose and study a new formalism for specifying necessary revisions in systems such as databases, knowledge bases and belief sets. This formalism properly extends logic programming with stable model semantics.

Imagine a knowledge base B. Say B is a collection of atomic facts from some universe U. We want to be able to specify that some elements must be in a knowledge base and some must be absent from it. We want to be able to require that at least one of a group of elements is in, that at least one is out, and that some two must not be in a knowledge base together. In principle, to make sure that a knowledge base B satisfies such specifications we will have to remove some facts from B and we will have to insert some new ones. If the change is specified explicitly in terms of the sets I (elements that must be inserted) and O (elements that must be deleted), then there is no problem with finding a revision.

It is simply given by $(B \cup I) \setminus O$ (assuming that $I \cap O = \emptyset$; otherwise, the revision is undefined).

Often, however, change is specified in terms of the initial knowledge base and in terms of **other changes**. For example, we may require to insert an element into a knowledge base **but only if another element is not in the knowledge base**. Similarly, we may want to delete an element **but only if several other elements are deleted, too.** There are even more complex specifications. For example, we may require that a knowledge base does not contain both a and b if it contains c. Such specifications can be regarded as a form of integrity constraints.

In this paper we propose a formalism which allows us to specify change and integrity constraints of the types mentioned above. Basic objects of our formalism are *revision programs* consisting of *revision rules*. Each revision rule provides a specification of change or a constraint on a knowledge base. Revision rules allow the programmer to specify a **preferred** way to satisfy constraints.

We assign to revision programs a natural and powerful semantics. This semantics is based on the concept of *necessary change* and a certain transformation of a program. It takes into account preferences expressed by the revision rules in the program.

Necessary change entailed by a revision program is a pair of sets (I, O). The elements of I must be **inserted**, and the elements of O must be **absent** after a knowledge base is revised. The main problem in defining such pairs is that our language of revision programs is quite expressive and its syntax goes much beyond rules explicitly stating unconditional insertions and deletions. An important insight is that, given an initial knowledge base B_I and a revision program P, the putative knowledge base B_R (a candidate for the revised version of B_I) also takes part in justifying the necessary insertions and deletions that convert B_I into B_R. This phenomenon is closely related to the way extensions are justified on the basis of a data set W and a default set D [Rei80, MT89] as well as to the way in which stable models of logic programs are defined [GL88].

Our approach can be summarized as follows. Given a revision program P and an initial knowledge base B_I, a knowledge base B_R is a *P-justified revision* of B_I if B_R is obtained from B_I by means of insertions and deletions specified by the *reduced* program (denoted below by $P_{B_R}|B_I$). Once the reduced program is formed, the necessary change is computed out of it. This change consists of two sets of elements I (to be included) and O (to be absent). If this change is *coherent* (that is, if $I \cap O = \emptyset$) and $B_R = (B_I \cup I) \setminus O$ then the passage from B_I to B_R is *P-justified*. Since $I \cap O = \emptyset$, it does not matter whether we first insert to B_I elements from I and then remove elements in O, or proceed in the opposite order. It should be stressed that the process of reduction is closely related to Gelfond-Lifschitz reduction [GL88].

Computing revisions is a "daily bread" while updating databases and it has been studied in this context by many researchers (see, for example, [AV90, AV91, Win90, Ull88]). The closest to our approach is the work of Manchanda and Warren [MW88]. In fact, the syntax of their programs is essentially identical with that of revision programs (in the non-disjunctive case). The difference is that we assign semantics to all revision programs, not only to stratified ones. In addition, revision programs we consider are **specifications** of change and thus, there may be several (or none) revisions satisfying them.

In this paper we shortly describe those aspects of revision programs and their semantics that pertain to knowledge representation. After providing motivating examples and basic definitions, we discuss properties of revision programs and their semantics. We show that if a knowledge base B satisfies a program P then B is the unique P-justified revision of B. We show that if a program possesses a model then it specifies a coherent change. We show that P-justified revisions are models of P. We also show that necessary changes leading to P-justified revisions are, in some sense, minimal. We prove a duality theorem showing a complete symmetry between deletion and insertion. We outline the proof theory for revision programs and show its adequacy to the proposed semantics. We generalize the notion of a revision program to the case of disjunctive revision programs. We briefly outline a simple application in belief revision. Finally, we mention predicate revision programs.

2 Definitions and examples

We will now introduce key concepts: *revision rule*, *revision program*, *necessary change* and *justified revision*. We will provide several examples to illustrate these notions.

The language we will consider is similar to the language of propositional logic programming. The restriction to the propositional case is not essential. Later in the paper, we briefly discuss the predicate case, too.

The key objects of the language, *revision programs*, are built of *revision rules* which, in turn, are built of atoms by means of special operators: $\leftarrow$, **in**, **out**, and ",".

Definition 1. Let U be a denumerable set. We call its elements *atoms*. A *revision in-rule* or, simply, an *in-rule*, is any expression of the form

$$\mathbf{in}(p) \leftarrow \mathbf{in}(q_1), \ldots, \mathbf{in}(q_m), \mathbf{out}(s_1), \ldots, \mathbf{out}(s_n), \tag{1}$$

where p, q_i, $1 \leq i \leq m$ and s_j, $1 \leq j \leq n$, are all in U. A *revision out-rule* or, simply, an *out-rule*, is any expression of the form

$$\mathbf{out}(p) \leftarrow \mathbf{in}(q_1), \ldots, \mathbf{in}(q_m), \mathbf{out}(s_1), \ldots, \mathbf{out}(s_n) \tag{2}$$

where p, q_i, $1 \leq i \leq m$ and s_j, $1 \leq j \leq n$, are all in U. Expressions $\mathbf{in}(a)$ and $\mathbf{out}(a)$ are called *literals*. All in- and out-rules are called *rules*. In (1 and 2 m, n or both may be equal 0. If both m and n are equal 0, we say that the rule has *empty* body. A collection of rules is called a *revision program* or, simply, a *program*.

Atoms of the language will represent beliefs or facts. Collections of atoms will be called *knowledge bases*.

Intuitively, revision programs can be regarded as operators which, given a current knowledge base, produce its revised version. An analogy with default logic has to be stressed here. In default logic, a default theory is a pair (D, W), where D is a set of defaults (can be regarded as a program) and W is the set of facts (input data). As the result of the set D of defaults operating on the set of data W, default extensions are produced. Here we will proceed similarly. Given a revision program P and a knowledge base B_I (I stands for initial) we will define the notion of a *P-justified revision* of B_I. It will turn out that, in an analogy with default logic, a knowledge base can have none, exactly one or several P-justified revisions. Thus, operators producing revisions according to the specification provided by a program P are multivalued.

We will now provide examples to motivate our definition of how a revision program specifies revisions of a knowledge base. First, revised knowledge bases must satisfy all constraints expressed by the rules in a program. There are several ways in which it is possible to achieve this. For instance, rule (1) will be satisfied in each of the following three cases:

1. if p is in the revised knowledge base,
2. if some q_i is not in the revised knowledge base,
3. if some s_i is in the revised knowledge base.

Similarly as in logic programming, by writing a specification (constraint) in the form of a **rule**, with clearly distinguished premises and consequent, the user expresses his/her preferences as to how to achieve that the rule is satisfied. In the case of rule (1) it is the first way that is preferred. In addition, each revision (insertion or deletion) that is performed in order to generate a new knowledge base and satisfy the program must be justified by at least one rule in the program. Consider an in-rule (1). If every q_i, $1 \leq i \leq m$, is **in** the knowledge base and if every s_j, $1 \leq j \leq n$, is **not in** the knowledge base, then the rule (1) provides a justification for having p in the revised knowledge base. Similarly, under the same assumptions about q_i and s_j, rule (2) provides a justification for absence of p from the knowledge base (deletion of p from the original knowledge base, if it is there).

We are a bit vague here. When we describe the way in which rules provide justifications for insertions or deletions (for believing or disbelieving), we refer to a knowledge base. But **which** knowledge base do we have in mind here? This turns out to be the key issue. Since the initial knowledge base needs to be revised it may contain atoms that must be deleted. Similarly, it is possible that the initial knowledge base does not contain atoms which, in view of a revision program, should be in the resulting one. Consequently, it is the **final** (new) knowledge base that needs to be used in the process. The circularity of reasoning evident from our discussion is reminiscent of the same phenomenon in the case of nonmonotonic logics such as default logic, autoepistemic logic or logic programming with negation as failure.

Example 1. Consider the program $P = \{\mathbf{in}(a) \leftarrow \mathbf{out}(b), \mathbf{in}(b) \leftarrow \mathbf{out}(a)\}$. Assume that the initial knowledge base is empty. This set of atoms needs to be revised by the program P. What is the revised knowledge base? One candidate is $\{a, b\}$. However, let us look for reasons to insert a. The rule $\mathbf{in}(a) \leftarrow \mathbf{out}(b)$ is the only potential justification. But, b is believed in (if we assume, as we do here, that $\{a, b\}$ is the new knowledge base). Hence, having a in a revised knowledge base is not justified. Consequently, $\{a, b\}$ should not be the result of revising the empty knowledge base by P. A similar reasoning shows that our program specifies two knowledge bases: $\{a\}$ and $\{b\}$.

Example 2. Let $P = \{\mathbf{out}(a) \leftarrow \mathbf{in}(a)\}$. Assume that the initial knowledge base is $\{a\}$. Then, no set of atoms can be regarded as the revision of $\{a\}$ by P. For example, suppose that $\emptyset$ is the revised knowledge base. If so, a must have been removed. The rule $\mathbf{out}(a) \leftarrow \mathbf{in}(a)$ is the only possible justification for the removal of $\{a\}$. But this rule "fires" precisely if a is in the knowledge base. However, according to our assumption, the revised (correct) knowledge base is empty. Thus, a is not there! In the same way, one can argue that $\{a\}$ is not a result of the revision by P.

Similarity to logic programming with stable model semantics [GL88] and to TMS [Doy79] is self-evident. Later, we will show that our formalism of revision programs generalizes logic programming with stable model semantics. However, revision programming is significantly more expressive. While logic programs do not allow one to state that an atom must be **absent** from a model, revision programs explicitly talk about deletions.

Let us discuss more examples. In all of them and throughout the paper B_I stands for the initial knowledge base (about to be revised) and B_R stands for the revised knowledge base.

Example 3. Let $B_I = \{a, b\}$ be the initial knowledge base. Let $P = \{\mathbf{in}(c) \leftarrow,$ $\mathbf{out}(b) \leftarrow\}$. Clearly, the meaning of this program is that c **must** be added and b **must** be removed. Hence, $B_R = \{a, c\}$.

Examples 1, 2 and 3 show that, as in nonmonotonic logics, given a set B of atoms, a revision program may define several, none or one possible revisions of B.

Programs such as the one discussed in Example 3 are easy to interpret. They consist of particularly simple rules that require unconditional insertion or deletion. The change in the knowledge base induced by these unconditional revisions will be referred to as the *necessary change*. Our next example shows that also more complicated types of rules contribute to the notion of necessary change.

Example 4. Consider the program $P = \{\mathbf{in}(c) \leftarrow, \mathbf{out}(b) \leftarrow \mathbf{in}(c)\}$. Assume that $B_I = \{a, b\}$. Since c must be inserted unconditionally, b — whose removal is conditioned only upon believing in c — must be removed. Hence, the necessary change is again determined by insertion of c and removal of b. Consequently, $B_R = \{a, c\}$.

Our next example shows that not all revision programs specify a *coherent* necessary change.

Example 5. Consider the program $P = \{\mathbf{out}(a) \leftarrow, \mathbf{out}(b) \leftarrow \mathbf{out}(a), \mathbf{in}(b) \leftarrow\}$. Notice that the necessary change consists of removing a, removing b and inserting b. These last two requirements cannot be reconciled. Hence, B_R is undefined (does not exist) in this case.

Programs such as the one discussed in Example 5, whose necessary change requires that for some atom a, a is both included into and deleted from the revised knowledge base, are called *incoherent*. In our setting these programs do not possess meaning.

Let us now define formally the notion of the *necessary change*. In the definition that follows, **we treat literals in(a) and out(a) occurring in P as separate propositional variables**. Under this interpretation, P becomes a Horn program. Consequently, it has a least model (least set of literals closed under the rules of P treated as definite Horn clauses). This observation provides a foundation for the next definition.

Definition 2. Let P be a revision program.

1. The *necessary change basis*, denoted $NCB(P)$, is the least model of P (assuming that all literals occurring in P are treated as separate propositional variables).
2. The *necessary change* for P is the pair (I, O), where $I = \{a : \mathbf{in}(a) \in NCB(P)\}$ and $O = \{a : \mathbf{out}(a) \in NCB(P)\}$.
3. If $NCB(P)$ does not contain any pair $\mathbf{in}(a)$, $\mathbf{out}(a)$, then P is called *coherent*.

Intuitively, the necessary change determined by a program P specifies those atoms that must be added and those atoms that must be deleted, independently of an initial knowledge base (Theorem 6).

Clearly, the necessary change has to be taken into account when computing B_R out of B_I. However, it is not always sufficient for generating B_R.

Example 6. Consider the program $P = \{\mathbf{out}(a) \leftarrow \mathbf{in}(b), \mathbf{out}(b) \leftarrow \mathbf{in}(a)\}$ and assume that $B_I = \{a, b\}$. Since the body of each rule is nonempty, the necessary change consists of no insertions and no deletions. When applied to B_I, it results in no change. But if $\{a, b\}$ were to be the revised knowledge base, then there would be a rule in P (the first rule) justifying the removal of a. Hence, $B_R \neq \{a, b\}$. In fact, one can see that there are two possibilities for B_R in this case: $\{a\}$ and $\{b\}$.

What we need is the concept of the necessary change but of a modified program. This modified program is obtained in the process of *a posteriori reducibility*. Given an initial knowledge base B_I and a hypothetical revised knowledge base B_R, we produce first the *reduct* of the program and then compute the necessary change determined by the reduct. If B_R can be obtained from B_I by executing insertions and deletions specified by the necessary change, then B_R is a viable revised knowledge base and it is called a *P-justified revision* of B_I.

Our concept of the reduct is similar, but not identical, to Gelfond-Lifschitz reduct in the case of stable models of logic programs [GL88]. We proceed in two stages. First, we eliminate rules which cannot be possibly used to justify any insertions or deletions. We use the hypothetical knowledge base B_R in this phase and eliminate a rule of type (1) or (2) if its body contains a literal that is not satisfied by B_R (B_R *satisfies* **in**(a) if $a \in B_R$ and B_R satisfies **out**(a) if $a \notin B_R$). Next, we subject the remaining clauses to the process of simplification. We use the initial knowledge base B_I here. We simplify each remaining rule by eliminating from its body literals that are satisfied by B_I.

Definition 3. Let B be a revision program and let B_I and B_R be two knowledge bases.

1. The reduct of P with respect to (B_I, B_R) is defined in two stages:
   Stage 1: Eliminate from P every rule of type (1) or (2) such that $q_i \notin B_R$, for some i, $1 \leq i \leq m$, or $s_j \in B_R$, for some j, $1 \leq j \leq n$.
   Stage 2: From the body of each rule that remains after Stage 1 eliminate each **in**(a) such that $a \in B_I$ and each **out**(a) such that $a \notin B_I$.
2. The program resulting from P after Stage 1 is denoted by P_{B_R}.
3. The program resulting from P after both stages is called *reduct of P with respect to* (B_I, B_R) and is denoted by $P_{B_R}|B_I$.
4. Let (I, O) be the necessary change determined by $P_{B_R}|B_I$. If $I \cap O = \emptyset$ (that is, if $P_{B_R}|B_I$ is coherent) and $B_R = (B_I \cup I) \setminus O$, then B_R is called a *P-justified revision of* B_I.

It is clear that in the process of reduction (both in Stage 1 and Stage 2) we treat all literals in the body of a rule in the same fashion. Rules are eliminated or simplified according to whether literals are true with respect to a set of atoms (B_R in Stage 1) and B_I in Stage 2. In the whole theory we have developed so far we obtain aesthetically satisfying duality between positive literals (**in**(a)) and negative literals (**out**(a)).

3 Basic properties of revision programs

In this section we will present several fundamental properties of the notions of necessary change and justified revisions. Our definition of justified revisions is closely patterned after the definition of stable models of logic programs. Our results in this section emphasize this connection.

First, we will introduce the notion of a model of a revision program. In the previous section we defined that a knowledge base B is a *model* of (*satisfies*) a literal **in**(p) (**out**(p), respectively) if $p \in B$ ($p \notin B$, respectively). We will now extend this definition.

Definition 4. A knowledge base B is a *model* of (*satisfies*) the body of a rule, if it satisfies each literal of the body. A knowledge base B is a *model* of (*satisfies*) a rule C if the following condition holds: whenever B satisfies the body of C, then B satisfies the head of C. A knowledge base B is a model of (*satisfies*) a revision program P if B satisfies each rule in P.

Intuitively, if a revision program P is satisfied by a knowledge base B then P should not force any new insertions or deletions. Indeed, our first result provides a formal version of this statement.

Theorem 5. *If a knowledge base B satisfies a revision program P then B is a unique P-justified revision of B.*

The notion of the necessary change determined by a revision program captures those insertions and deletions that do not depend on the initial state of the knowledge base but are forced, unconditionally, by the program. The next result shows that, indeed, every model of a revision program P "is consistent" with the necessary change, that is, contains its in-part and is disjoint from its out-part.

Theorem 6. *Let P be a revision program and let (I,O) be the necessary change determined by P. Then, for every model M of P, $I \subseteq M$ and $O \cap M = \emptyset$.*

This result immediately implies the following corollary for revision programs.

Corollary 7. *If a revision program P has a model then it is coherent.*

Coherence can be characterized in terms of three-valued interpretations.

Definition 8. 1. A three-valued interpretation is a pair of sets of atoms $\langle D_1, D_2 \rangle$ such that $D_1 \cap D_2 = \emptyset$.
2. Let $V = \langle D_1, D_2 \rangle$ be a three-valued interpretation. We say that $V \models_3 \mathbf{in}(a)$ if $a \in D_1$. Similarly, $V \models_3 \mathbf{out}(a)$ if $a \in D_2$. $V \models_3 c$ where c is a conjunction of literals if V satisfies ($\models_3$) every literal in the conjunction c.
3. Let V be a three-valued interpretation. Let r be a revision rule. We say that $V \models_3 r$ (V satisfies r) if from the fact that V satisfies the body of r it follows that V satisfies the head of r. V is a three-valued model of a revision program P ($V \models_3 P$) if V satisfies all the rules in P.

We have the following characterization of coherence.

Proposition 9. *Let P be a revision program. Then P is coherent if and only if there exists a three-valued interpretation V such that $V \models_3 P$.*

Justified revisions are meant to serve as possible revised versions of an initial knowledge base. Consequently, one would expect that justified revisions are models of the program that determines them.

Theorem 10. *Let P be a revision program and let B_I be a knowledge base. If a knowledge base B_R is a P-justified revision of B_I, then B_R is a model of P.*

Most major nonmonotonic reasoning systems as well as several revision theories are built on the principle of minimality (or parsimony). In the case of nonmonotonic systems, while describing them by semantic means we often impose on the class of models some minimality conditions. In the case of revision theories, we require that new knowledge bases, replacing the old ones, differ from them by "as little as possible". We will show now that the notion of necessary change also satisfies a certain minimality criterion.

Given two sets B_R and B_I, one can express the difference between them by means of their *symmetric difference*: $B_R \div B_I = (B_R \setminus B_I) \cup (B_I \setminus B_R)$. Minimality requirements for change leading from an initial knowledge base B_I to its revision B_R can be formulated as follows:

1. B_R satisfies the revision program P
2. subject to condition (1), B_R minimizes the symmetric difference $B \div B_I$.

We have already seen that justified revisions are models of revision programs (Theorem 10). We will now show that justified revisions satisfy the minimality requirement.

Theorem 11. *Let P be a revision program and let B_I be a knowledge base. If B_R is a P-justified revision of B_I, then $B_R \div B_I$ is minimal in the family $\{B \div B_I : B$ is a model of $P\}$.*

We will now introduce the notion of a dual revision program. Given a revision program P, let us define the *dual* of P (P^D in symbols) to be the revision program obtained form P by simultaneously replacing all occurrences of **in** by **out** and all occurrences of **out** by **in**. It turns out that whatever needs to be added to B_I according to revisions specified by P has to be removed from $\overline{B_I}$ according to P^D. Similarly, whatever has to be removed from B_I according to P has to be added to $\overline{B_I}$ according to P^D. Hence, in revision programming there is a full duality between **in** and **out** operators.

Theorem 12. *Let P be a revision program and let B_I be a knowledge base. Then, B_R is a P-justified revision of B_I if and only if $\overline{B_R}$ is a P^D-justified revision of $\overline{B_I}$, where $\overline{X}$ stands for the complement of X with respect to the set of all atoms in the language.*

Theorems 10 and 11 describe properties of justified revisions that are reminiscent of similar properties of stable models of logic programs: (1) stable models are, indeed models of a logic program, and (2) stable models are minimal models of a logic program. This should not be surprising. We have already stressed the similarity between the definitions of justified revisions and stable models. We will now describe an interpretation of logic programs as revision programs.

Given a logic program clause C

$$p \leftarrow q_1, \ldots, q_m, \mathbf{not}(s_1), \ldots, \mathbf{not}(s_n) \tag{3}$$

define the revision rule $r(C)$ as

$$\mathbf{in}(p) \leftarrow \mathbf{in}(q_1), \ldots, \mathbf{in}(q_m), \mathbf{out}(s_1), \ldots, \mathbf{out}(s_n). \tag{4}$$

In addition, for a logic program P, define the corresponding revision program $r(P)$ by

$$r(P) = \{r(C) : C \in P\} \tag{5}$$

We now have the following result.

Theorem 13. *Let P be a logic program. A set of atoms M is a model of P if and only if M is a model of $r(P)$. A set of atoms M is a stable model of P if and only if M is an $r(P)$-justified revision of $\emptyset$.*

Justified revisions have an elegant proof-theoretic characterization. It is similar to the proof-theoretic characterization of default extensions [MT89]. The notion of a proof it is based upon is closely related to the notions of an S-proof [MT89] and to the concept of a proof scheme [MNR90].

Let P be a revision program and let B and D be sets of atoms. A sequence of literals (that is, expressions of the form $\mathbf{in}(a)$ or $\mathbf{out}(a)$) $\langle \alpha_1, \ldots, \alpha_l \rangle$ is a *revision proof from B and P with respect to D* of a literal α if $\alpha = \alpha_l$, and if for every i, $1 \leq i \leq l$, at least one of the following conditions holds:

1. $\alpha_i = \mathbf{in}(a)$, for some $a \in B \cap D$
2. $\alpha_i = \mathbf{out}(a)$, for some $a \notin B \cup D$
3. there is a rule $\alpha_i \leftarrow \mathbf{in}(p_1), \ldots, \mathbf{in}(p_m), \mathbf{out}(q_1), \ldots, \mathbf{out}(q_n)$ in P such that
   (a) for every i, $1 \leq i \leq m$, $p_i \in B_R$ and $\mathbf{in}(p_i) = \alpha_j$, for some $j < i$
   (b) for every i, $1 \leq i \leq n$, $q_i \notin B_R$ and $\mathbf{out}(p_i) = \alpha_j$, for some $j < i$.

We denote the fact that a literal α has a revision proof from B with respect to D by writing: $B \vdash_{P,D} \alpha$.

Intuitively, proofs are context-dependent justifications for an atom to be added to or deleted from a knowledge base. Assume that B is an initial knowledge base. Assume also that a hypothetical revision of B, a knowledge base D, is given. Clearly, if an atom belongs to $B \cap D$, no justification is needed to have it in D as it was in the initial knowledge base to start with. Similarly, if an atom is not in $B \cup D$, no justification is needed for **not having** it in D since it was not in the initial knowledge base B. This motivates conditions (1) and (2). Other insertions and deletions have to be justified by rules in a revision program. But only those rules that have their bodies satisfied by the hypothetical knowledge base should be used in the process. In addition, all literals in their bodies must have been already justified. Hence, we get condition (3).

A knowledge base B is (P, D)-*consistent* if for every atom a we have $B \not\vdash_{P,D} \mathbf{in}(a)$ or $B \not\vdash_{P,D} \mathbf{out}(a)$.

Let us define now the revision operator. For a (P, D)-consistent knowledge base B it produces a new knowledge base that takes into account insertions and deletions that can be justified from B by means of revision proofs with respect to a program P and a knowledge base D. Namely, define

$$Rev^{P,D}(B) = (B \cup \{a: B \vdash_{P,D} \mathbf{in}(a)\}) \setminus \{a: B \vdash_{P,D} \mathbf{out}(a)\},$$

for every (P, D)-consistent knowledge base B. For (P, D)-inconsistent knowledge bases B, $Rev^{P,D}(B)$ is undefined.

Justified revision can be characterized by means of revision proofs (or, equivalently, by means of the revision operator Rev as follows.

Theorem 14. *Let P be a revision program. A knowledge base B_R is a P-justified revision of a knowledge base B_I if and only if B_I is (P, B_R)-consistent and $Rev^{P,B_R}(B_I) = B_R$.*

4 Disjunctive revision programs and an application

We will now extend the theory of revision programs to the case of programs with "nonstandard" disjunctions in heads of clauses. We will then present a simple disjunctive program with some applications in belief revision.

Definition 15. 1. A disjunctive revision rule is an expression of the form:

$$\mathbf{in}(p_1)|\ldots|\mathbf{in}(p_k)|\mathbf{out}(r_1)|\ldots|\mathbf{out}(r_l) \leftarrow \mathbf{in}(q_1),\ldots,\mathbf{in}(q_m),\mathbf{out}(s_1),\ldots,\mathbf{out}(s_n)$$

where all p_i, r_i, q_i and s_i are atoms from U, In addition we require that $k + l \geq 1$ (the head must be nonempty).
2. A disjunctive revision program is a set P of disjunctive revision rules.

The intuitive interpretation of a disjunctive revision rule is similar to that of an ordinary revision rule: if all q_i, $1 \leq i \leq m$, belong to a knowledge base B and all s_i, $1 \leq i \leq n$, do not belong to B then for some i, $1 \leq i \leq k$, $p_i \in D$, or for some j, $1 \leq j \leq n$, $r_j \notin D$.

Example 7. Let $P = \{\mathbf{in}(b)|\mathbf{out}(c) \leftarrow \mathbf{in}(a)\}$. Consider $B_I = \emptyset$. Then, since $a \notin B_I$ and since there is no way to justify the insertion of a, the resulting knowledge base B_R should be empty. Next, assume $B_I = \{a, c\}$. Now, we have a choice. There are two possible revisions of B_I: $B'_R = \{a\}$ and $B''_R = \{a, b, c\}$. Indeed, if a is a belief after revision, we have a choice of eliminating c or of inserting b.

The notions of a model and of satisfaction extend in a direct fashion to the case of disjunctive programs. We will now describe how to generalize the notions of necessary change and justified revision.

Let P be a disjunctive revision program. First, notice that the reduction process described in Section 3 depends only in the bodies of rules in P. Since disjunctive revision rules and ordinary revision rules have the same syntax of their bodies, the reduction process developed in Section 3 extends to the disjunctive case. From now on we assume that that the notion of the reduct $P_{B_R}|B_I$, where P is a disjunctive revision program, is well-defined.

Let us note that for a disjunctive program P there exist minimal sets of literals closed under rules of P, where $\mathbf{in}(a)$ and $\mathbf{out}(a)$ are treated as unrelated propositional atoms. The reader is referred to [GL91], where this fact is stated in the case of disjunctive programs without negation-as-failure symbol in the bodies). The family of all such minimal sets will be denoted (with some abuse of notation) by $NCB(P)$.

Definition 16. Let P be a disjunctive program. For every minimal set of literals $A \in NCB(P)$, define $I = \{a : \mathbf{in}(a) \in A\}$ and $O = \{a : \mathbf{out}(a) \in A\}$. Each such pair (I, O) is called a *necessary change* for P.

Notice that for a disjunctive revision program the necessary change is no longer unique. For example, if $P = \{\mathbf{in}(b)|\mathbf{out}(c) \leftarrow\}$, then $(\{b\}, \emptyset)$ and $(\emptyset, \{c\})$ are the two necessary changes determined by P.

Definition 17. Let P be a disjunctive program. A knowledge base B_R is a P-*justified revision* of a knowledge base B_I if for some necessary change (I, O) of $P_{B_R}|B_I$:

1. (I, O) is coherent (that is, $I \cap O = \emptyset$)
2. $B_R = (B_I \cup I) \setminus O$.

We will now state two results which show that our notion of a P-justified revision of a disjunctive program is a common generalization of both the notion of a P-justified revisions for ordinary revision programs, and of the notion of an answer sets of a disjunctive logic program.

Theorem 18. *Let P be an ordinary (no disjunctions) revision program. A knowledge base B_R is a P-justified revision of B_I (as defined in Section 3) if and only if B_R is a P-justified revision of B_I when P is treated as a disjunctive revision program.*

Now, let us extend the embedding (4) and (5) to the case of disjunctive revision programs. Given a disjunctive logic program clause C:

$$p_1|\ldots|p_k \leftarrow q_1, \ldots, q_m, \mathbf{not}(s_1), \ldots, \mathbf{not}(s_n),$$

define the corresponding revision rule $r(C)$ as

$$\mathbf{in}(p_1)|\ldots|\mathbf{in}(p_k) \leftarrow \mathbf{in}(q_1), \ldots, \mathbf{in}(q_m), \mathbf{out}(s_1), \ldots, \mathbf{out}(s_n)$$

In addition, define $r(P) = \{r(C) : C \in P\}$.

Theorem 19. *Let P be a disjunctive logic program. Then M is an answer set for P if and only if M is an $r(P)$-justified revision of $\emptyset$.*

We will look now at a simple application of disjunctive revision programs to belief revision. The problem of belief revision can be described as follows. Given a theory (knowledge base) T and a consistent formula φ, find a theory $T+\varphi$ which: is consistent, contains φ, and differs "as little as possible" from $T \cup \{\varphi\}$. We will show a simple disjunctive program encoding all maximal consistent theories contained in $Cn(T) \cup \{\varphi\}$ and containing φ. These theories are proposed as candidates for revision of T by φ in [AGM85].

Let $\mathcal{L}$ be a propositional language. We add a new atom c_ψ for every formula ψ of $\mathcal{L}$. This means that we treat the formulas as "entirely unrelated". The only connections will be provided by our program. The description of the program follows. Rules (1) "propagate" the results of deletions. Rules (2) and (3) ensure the results is closed under propositional provability. Rule (4) guarantees the result is consistent. Finally, rule (5) ensures φ belongs to a revised theory.

1. $\mathbf{out}(c_{\theta \supset \psi})|\mathbf{out}(c_\theta) \leftarrow \mathbf{out}(c_\psi)$ (for every pair of formulas θ, ψ)
2. $\mathbf{in}(c_\psi) \leftarrow \mathbf{in}(c_\theta), \mathbf{in}(c_{\theta \supset \psi})$ (for every pair of formulas θ, ψ)
3. $\mathbf{in}(c_\psi) \leftarrow$ (for every tautology ψ of $\mathcal{L}$)
4. $\mathbf{out}(c_\perp) \leftarrow$
5. $\mathbf{in}(c_\varphi) \leftarrow$

We will denote this program by R^φ. We have the following theorem.

Theorem 20. *Let T be a theory and ψ a formula. A theory S contains φ and is a maximal consistent theory included in $T \cup \{\varphi\}$ if and only if S^+ is an R^φ-justified revision of T^+, where $T^+ = \{c_\psi : \psi \in T\}$.*

An interesting feature of the program R^φ is that it explicitly describes the mechanism behind the (simplest) process of revision. On the other hand, R^φ is infinite — a rather undesirable feature. Below we outline a possible way in which the "infiniteness" of R^φ can be dealt with.

Our approach represents R^φ as a **finite predicate** revision program. (Disjunctive) predicate revision programs have the same structure as propositional revision programs, the only difference being that the arguments of **in** and **out** operators are atomic formulas of a certain language of predicate logic (rather than propositional variables).

Given a predicate program P, we define the *Herbrand base* H_P of the program P in the usual fashion [Apt90]. We assign to a revision program P its *ground* version P_{ground} by replacing each rule of P with its all ground substitutions. Thus the program P_{ground} can be regarded as a **propositional** revision program whose atoms come from H_P.

We say that a set $B_R \subseteq H_P$ is a P-justified revision of a set $B_I \subseteq H_P$, if B_R is a P_{ground}-justified revision of B_I. In this way, our theory of revision programs and justified revisions is lifted to the predicate case.

One of the benefits of considering the predicate case is that often a finite predicate program can be found to encode an infinite propositional program. In particular, the infinite program R^φ, given above, has such finite representation. We list its rules below. We use the "prefix" notation (with brackets) to describe a formula (for instance, we write $\supset (X, Y)$ instead of $X \supset Y$).

1. $\mathbf{out}(revised(\supset(X,Y)))|\mathbf{out}(revised(X)) \leftarrow \mathbf{out}(revised(Y))$
2. $\mathbf{in}(revised(Y)) \leftarrow \mathbf{in}(revised(X)), \mathbf{in}(revised(\supset(X,Y)))$
3. $\mathbf{in}(revised(A(X,Y,\ldots)) \leftarrow$, for every axiom schema $A(X,Y,\ldots)$ of some finite axiomatization of propositional calculus which assumes modus ponens as the only inference rule,
4. $\mathbf{out}(revised(\bot))$
5. $\mathbf{in}(revised(\psi))$

It should be clear, that the grounding of this program is equivalent to the program R^{φ} (in the sense that both define the same justified revisions).

5 Conclusions

We presented a formalism, called revision programming, to specify and compute minimal and justified change. The symmetric character of revision programming allows the user to describe both presence and absence of information in a uniform and explicit manner. Our system is derived from logic programming and generalizes stable model semantics for logic programs. It also extends the formalism of elementary updates in databases (insertions and deletions of sets of elements).

Revision programming has an elegant theory. The change is always minimal and justified by the appropriate use of the rules of the program. We outlined an extension of revision programming to the disjunctive case. We also briefly discussed the predicate case.

Future research will focus on applications of revision programming in reasoning about actions, combinatorial optimization, belief revision and abduction.

Acknowledgements

This work was partially supported by National Science Foundation under grant IRI-9012902.

References

[AGM85] C. E. Alchourrón, P. Gärdenfors, and D. Makinson. On the logic of theory change: Partial meet contraction and revision functions. *Journal of Symbolic Logic*, 50:510–530, 1985.

[Apt90] K. Apt. Logic programming. In J. van Leeuven, editor, *Handbook of theoretical computer science*, pages 493–574. MIT Press, Cambridge, MA, 1990.

[AV90] S. Abiteboul and V. Vianu. Procedural languages for database queries and updates. *Journal of Computer and System Sciences*, 41:181–229, 1990.

[AV91] S. Abiteboul and V. Vianu. Datalog extensions for database queries and updates. *Journal of Computer and System Sciences*, 43:62–124, 1991.

[Doy79] J. Doyle. A truth maintenance system. *Artificial Intelligence*, 12:231–272, 1979.

[GL88] M. Gelfond and V. Lifschitz. The stable semantics for logic programs. In R. Kowalski and K. Bowen, editors, *Proceedings of the 5th international symposium on logic programming*, pages 1070–1080, Cambridge, MA., 1988. MIT Press.

[GL91] M. Gelfond and V. Lifschitz. Classical negation in logic programs and disjunctive databases. *New Generation Computing*, 9:365–385, 1991.

[MNR90] W. Marek, A. Nerode, and J.B. Remmel. Nonmonotonic rule systems I. *Annals of Mathematics and Artificial Intelligence*, 1:241–273, 1990.

[MT89] W. Marek and M. Truszczyński. Stable semantics for logic programs and default theories. In E.Lusk and R. Overbeek, editors, *Proceedings of the North American conference on logic programming*, pages 243–256, Cambridge, MA., 1989. MIT Press.

[MW88] S. Manchanda and D.S. Warren. A logic-based language for database updates. In J. Minker, editor, *Foundations of Deductive Databases and Logic Programming*, pages 363–394, Los Altos, CA, 1988. Morgan Kaufmann.

[Rei80] R. Reiter. A logic for default reasoning. *Artificial Intelligence*, 13:81–132, 1980.

[Ull88] J.D. Ullman. *Principles of Database and Knowledge-Base Systems*. Computer Science Press, Rockville, MD, 1988.

[Win90] M. Winslett. *Updating Logical Databases*. Cambridge University Press, 1990.

Revision of Non-Monotonic Theories

Some postulates and an application to logic programming

Cees Witteveen[1] * and Wiebe van der Hoek[2] ** and Hans de Nivelle[1] ***

[1] Delft University of Technology, Dept of Mathematics and Computer Science, P.O.Box 356, 2600 AJ Delft, The Netherlands

[2] Utrecht University, Dept of Computer Science, Utrecht, The Netherlands

Abstract. We present some revision systems for non-monotonic theories and we concentrate on the revision of logic programs whenever they are classically consistent, but do not have an acceptable (non-monotonic) model. The revision method we propose is to expand an original theory (program) in order to obtain an acceptable model.
We distinguish between *weak* revision, *conservative* revision and *strong* revision systems. These systems differ to the extent the revision affects the set of classical models of the original theory.
We then show that there exist weak, conservative and strong expansion systems for normal logic programs using the stable model semantics.
In particular, we present a strong expansion method which makes it possible to construct for an arbitrary (incoherent) normal logic program P a -classically- equivalent expanded program P' such that P' always has a stable model.

Keywords: Revision, Non-monotonic Reasoning, Logic Programming.

1 Introduction

Most of the work done in belief revision concerns the question of how to revise one's beliefs when a theory is *classically* inconsistent. In such a setting, revision then amounts to *contracting* or *narrowing down* the (inconsistent) belief set B to a consistent subset B' (cf. [1, 2]).
The by now well-known Alchourrón-Gärdenfors-Makinson (AGM) framework ([1, 2]) approaches belief revision for belief sets that are closed under a (monotone and compact) consequence operator Cn. A main charm of the AGM-study is, that it provides us with a set of rationality postulates for belief revision, rather than proposing one unique rigid regime for undertaking such an enterprise. Since in the AGM-setup, the result of a revision is always a consistent set, it seems that consistency is adopted as a kind of meta-constraints for beliefs (cf. [11]).

* e-mail: witt@cs.tudelft.nl
** e-mail: wiebe@cs.ruu.nl
*** e-mail: nivelle@cs.tudelft.nl

Adopting this meta-constraint, and having an underlying monotonic consequence operator Cn, it is clear why revision of a belief set B which yields too many (unwanted) conclusions must at least *remove* some of its beliefs, since monotonicity of Cn simply imposes that any $B' \supseteq B$ ensures that $Cn(B') \supseteq Cn(B)$.

In this paper we argue that, even under the same meta-constraint of consistency of belief sets, the picture dramatically changes if one shifts to revising *non-monotonic* theories, i.e. belief sets of which the underlying consequence operator is not monotonic.

Firstly, in non-monotonic theories often one is forced to revise one's beliefs if the theory is *not* classically inconsistent. This situation can occur if, using a non-monotonic semantics, we prefer certain interpretations of the theory above others and such a preference gives rise to an interpretation containing mutually inconsistent beliefs. As a simple example, take the logic program P containing the rules $b \leftarrow\sim a$ and $\neg b \leftarrow$. If we prefer stable models, we are forced to consider both b and its negation $\neg b$ as true, while there exists a (non-preferred) classical model in which a is true and b is false.

Secondly, sometimes inconsistency does not even seem to play a role if revision of our beliefs is necessary. This latter situation arises if there exists no *acceptable* (non-monotonic) interpretation of the theory, although according to a classical reading of the theory, it is satisfiable. Again, take the program $a \leftarrow\sim a$. It has a classical model in which a is true, but now there is no stable model at all.

Following [9, 5] we will call the first kind of theories *contradictory* and the second kind *incoherent*. In both kinds we are faced with the problem that, although a classical reading of the theory might give no problems, i.e. there are classical models for a theory, a non-monotonic reading of the same theory would give us no *acceptable model*.

In order to find acceptable models for such theories we propose to revise our theory. In such a setting, however, *retraction* of some beliefs is not the only option we have, since the problem arises not because of a lack of *classical* models but a lack of *acceptable* models.

In such a situation then, a lack of acceptable models means that our preference criterion has to be changed. Therefore, we should perform belief revision by *changing our preference criteria*, i.e., we should select other models of the original theory as being acceptable.

We propose a special revision method to perform such a preference change. Instead of adapting the non-monotonic semantics, we adapt the theory and we will use preferred models of the adapted theory as the new acceptable models of the original theory. Syntactically, this revision method is a *theory-expansion* method: we expand the original theory by adding some extra statements in such a way that acceptable models of the expanded theory are always models of the original theory.

Remark. The idea of revision by expansion has –yet a short– history. In some specific non-monotonic formalisms, procedures have been suggested to perform revision by expansion and results (concerning existence and tractability) of some procedures were obtained. To be more precise, the work of Pereira et al. on contradiction removal semantics ([9, 10]) in '91 can be conceived as a first attempt to revise by expansion. Witteveen and Brewka defined an expansion procedure for truth maintenance systems with constraints ([12]). The idea of expanding was also mentioned in [13]. Although the main results in that paper were on truth maintenance, it also contained some first ideas for justifying more general revision-by-expansion strategies.

We need the following basic notions.
A *theory* T is just a set of sentences over some language $\mathcal{L}$. We assume that $\mathcal{L}$ contains $\perp$ and $\top$ denoting falsity and truth, respectively. The set of *classical* models for T is denoted by $Mod(T)$.
If T is intended as a non-monotonic theory, we distinguish the set $Acc(T) \subseteq Mod(T)$ of *acceptable models* of a theory. For example, if T is a logic program, the set of stable models might be such a set of acceptable models. We note that for several non-monotonic semantics it might occur that $Acc(T)$ is an empty set (the theory is contradictory) or even not defined (the theory is incoherent)[3] while $Mod(T) \neq \emptyset$.
This is considered to be a deficiency in most existing non-monotonic semantics, and, for us, a reason to introduce a more powerful operator Acc^*, which always returns a set of *revised acceptable models* of T, i.e. $Acc^*(T)$ will be defined for every theory T.

Remark. Note that we assume that somehow we have a way to classically interpret non-monotonic theories T; we think that, although in some cases it may take some effort to do so, in many other cases, like logic programming, such an interpretation is rather straightforward.

2 Postulates for Revision by Expansion

In this section we will discuss some (rationality) postulates for the set $Acc^*(T)$ of revised acceptable models of a theory T. First we will discuss some limiting cases, then we will introduce the expansion postulate.

The purpose of non-monotonic reasoning is to supplement standard reasoning methods. Therefore, we want to have a notion of non-monotonic inference that is at least as powerful as classical inference.
Therefore, we require non-monotonic inference to obey the principle of *supra-classicality*:

P1 For every $T \subseteq WFF(\mathcal{L})$, $Acc^*(T) \subseteq Mod(T)$. (Supra-classicality)

[3] In the literature, $Acc(T) = \emptyset$ is often used to indicate that T is incoherent *or* contradictory. To distinguish these notions, we will say that $Acc(T)$ is not defined if the theory is incoherent, while $Acc(T) = \emptyset$ iff the only acceptable 'models' that can be found are contradictory.

Non-monotonic inference should never result in deriving inconsistencies, whenever, classically speaking, the theory is consistent. Such a principle, for example, is obeyed in default theory, where an extension is inconsistent only if the classical part of the default theory is inconsistent. Since Acc^* is meant to extend existing non-monotonic semantics in a reasonable way, we should not allow $Acc^*(T)$ to be empty if T is classically satisfiable:

P2 $Mod(T) \neq \emptyset$ implies $Acc^*(T) \neq \emptyset$. (Consistency preservation)

To express that $Acc^*(T)$ should extend, rather than replace, an existing non-monotonic semantics we require it to agree with $Acc(T)$ whenever this is possible.

P3 Whenever $Acc(T)$ is defined and $Acc(T) \neq \emptyset$,
$Acc^*(T) = Acc(T)$. (Conservativity)

We would also like to relate Acc^* to Acc if T does ***not*** have acceptable models. The idea for the following expansion postulate can be motivated as follows. In a very general sense, non-monotonic reasoning is about reasoning with a theory and thereby ***assuming*** that some statements are true or false, without being able actually to ***prove*** so. If $Acc(T)$ is not defined or $Acc(T) = \emptyset$, it is not possible to reason with these assumptions in an acceptable way. Therefore, the assumptions themselves seem not be justifiable and therefore should be blocked in the reasoning process. One way to do so is to make explicit that some assumptions cannot be made, and therefore to ***add*** some statements to the theory. These statements should have the effect of blocking unwanted assumptions and make it possible to derive non-monotonic conclusions.
Hence, Acc and Acc^* should be related via ***expansions*** of the original theory. The following postulate expresses that it always should be the case that acceptable models of a theory T can be obtained by deriving them from a transformed theory T' containing T as a (strict) subset:

P4 For every $T \subseteq WFF(\mathcal{L})$, there exists some subset $\Phi_T \subseteq WFF(\mathcal{L})$, depending on T, such that $Acc^*(T) = Acc(T + \Phi_T)$. (Expansion)

From now on, we use 'Expansion' for both the postulate P4, as well as for the theory $T + \Phi_T$ that is a solution of the equivalence stated in P4.

2.1 Some Easy Consequences

We mention some easy consequences of these postulates:

First of all, observe that although $Acc(T)$ might be empty or undefined for a classically satisfiable theory T, $Acc^*(T)$, by Postulate P1 and P2, always contains at least one classical model which is acceptable:

Observation 2.1
For every satisfiable theory T, $\emptyset \neq Acc^*(T) \subseteq Mod(T)$. *(Existence)*

But obviously, although $Acc^*(T)$ always exists, it can never cure theories that are classically bad, since by $P1$, if $Mod(T) = \emptyset$, $Acc^*(T) = \emptyset$. Therefore Consistency preservation can be strengthened to:

Observation 2.2
$Acc^*(T) = \emptyset$ *iff* $Mod(T) = \emptyset$ *(Strong consistency preservation)*

The next observation implies that expansions of theories do not need to be expanded again:

Observation 2.3
For all T *and expansions* Φ_T: $Acc^*(T + \Phi_T) = Acc^*(T)$. *(Maximality)*

Moreover, expansions preserve satisfiability, as required:

Observation 2.4
T *is classically satisfiable iff* $T + \Phi_T$ *is.*

Proof. The if-part is obvious. The only-if part can be proven as follows:

$$Mod(T) \neq \emptyset \text{ iff } Acc^*(T) \neq \emptyset$$

by Strong consistency preservation. By Maximality and Supra-classicality we have

$$Acc^*(T) = Acc^*(T + \Phi_T) \subseteq Mod(T + \Phi_T)$$

Hence, $Mod(T) \neq \emptyset$ implies $Mod(T + \Phi_T) \neq \emptyset$.

2.2 Expansion Systems

We will call a system satisfying P1 to P4 a *weak* expansion system. The reason is that, although the expansion postulate P4 expresses the essence of our idea of defining an expansion operator, it is too weak to exclude expansion methods abusing this postulate by introducing almost trivial revision operators.
Such trivialization is likely to occur if we construct expansions T' which select small subsets of $Mod(T)$ as their set of classical models $Mod(T')$. Ideally, however, we would only change our preference for certain classical models of the original theory, *without* affecting this very set $Mod(T)$, since this class in a sense describes the set of all possible respectable interpretations of the theory. We only prefer certain models, but we do not want to exclude some of the alternatives. Therefore, we will introduce some stronger variants of P4.

The first one allows us to keep as much of $Mod(T)$ as possible in expanding the theory by allowing to extend our language while expanding our theory.
We say that a theory T' in a language $\mathcal{L}'$ is a *conservative extension of* a theory T in $\mathcal{L} \subseteq \mathcal{L}'$ if for every closed formula $\phi \in WFF(\mathcal{L})$, $Mod(T) = Mod_{\mathcal{L}}(T')$. Here, informally speaking, $Mod_{\mathcal{L}}(T')$ is the restriction of the models of T' to T. Then we say that a system is a *conservative expansion system* if it satisfies postulates P1 - P4 and the following postulate expressing that the classical models of T and the set of models of its expansion T', restricted to the language of T, are identical:

P4a $T + \Phi_T$ is a conservative extension of T. (Conservative expansion)

The strongest version of an expansion system can be obtained by requiring that the set of classical models is invariant under taking expansions.
We say that a system is a ***strong expansion system*** if it satisfies P1-P4 and the following postulate:

P4b $Mod(T + \Phi_T) = Mod(T)$. (Strong expansion)

Note that in a strong expansion system the information added to the revised theory in a certain sense is minimized in a global way: we do not add any classical knowledge, but only change our preference for certain interpretations.
In the next section we will discuss some particular expansion systems for normal logic programs, distinguishing weak, conservative and strong expansion systems.

3 Revisions of Logic Programs

To show a simple application of the expansion framework, we will show that there are techniques for revising ***normal logic programs*** satisfying the postulates we have discussed in the previous section.
For simplicity reasons, we have confined ourselves to this class of normal programs. Extensions to the class of normal programs with constraints or logic programs with explicit negation are straightforward.
All techniques we will discuss satisfy the postulates P1 - P4, but differ in satisfying the variants of the fourth postulate.
In particular we will show that

1. some existing techniques for belief revision in logic programs satisfy the postulates for weak expansion;
2. there is a simple technique satisfying the postulates for conservative expansion;
3. we show that there exists a revision strategy satisfying the strong expansion system. This technique extends some well-known dependency-directed backtracking techniques and is the first -as far as we know- proven to satisfy these postulates.

We assume the reader to be familiar with some basic terminology used in logic programming, as for example in Lloyd ([6]). A ***normal program rule*** is a directed propositional clause of the form $A \leftarrow B_1, \ldots, B_m, \sim D_1, \ldots, \sim D_n$, where $A, B_1, \ldots, B_m, D_1, \ldots, D_n$ are all atomic propositions. Often such a rule is denoted as $A \leftarrow \alpha$, where $\alpha = \alpha^+ \wedge \alpha^-$; α^+ denotes the conjunction of the B_i atoms and α^- denotes the conjunction of the negative atoms $\sim D_j$. A program is called ***positive*** if for every rule r, $\alpha^-(r) = \top$ (the empty conjunction). As usual, interpretations and models of a program are denoted by subsets of the Herbrand base B_P of P. T_P denotes the immediate consequence operator associated with the program P and $lfp(T_P) = T_P \uparrow \omega$ denotes the least fixpoint of this operator. The intended meaning of a normal logic program is given by the two-valued ***stable model*** semantics ([3]):

Definition 3.1 *A model M of a logic program P is called stable iff it equals the least fixpoint of the Gelfond-Lifschitz reduction $G(P,M)$ of P, where*

$$G(P,M) = \{c \leftarrow \alpha^+ \mid c \leftarrow \alpha \in P,\ M \models \alpha^-\}$$

Note that $G(P,M)$ is a positive program, i.e., does not contain any rule with a negative antecedent.
We will need the following lemma's pertaining to properties of stable models:

Lemma 1 (Marek & Truszczyński [8]). *Let M be a model of a program P. Then $T_{G(P,M)} \uparrow \omega \subseteq M$.*

Lemma 2. *If M is a stable model of P and $M \models r$ for some program rule r over B_P, then M is also a stable model of $P + r$.*

Proof. If P_1 and P_2 are positive programs, $P_1 \subseteq P_2$ implies $lfp(T_{P_1}) \subseteq lfp(T_{P_2})$. Hence, by the definition of stability and the previous lemma:

$$M = lfp(T_{G(P,M)}) \subseteq lfp(T_{G(P+r,M)}) \subseteq M.$$

So $M = lfp(T_{G(P+r,M)})$ and therefore, M is a stable model of $P + r$.

We will denote the set of stable models of a program P by $ST(P)$.
It is well-known that the stable model semantics is not supra-classical in the sense we discussed in the previous section: there are programs P, such as the program $a \leftarrow \sim a$, where $Mod(P) = \{\{a\}\}$ while $ST(P)$ is not defined. But if we add the rule $a \leftarrow$ to P, the resulting program P' has a unique stable model $M = \{a\}$. In the sequel we will consider such *expansions* P' such that $ST(P') \neq \emptyset$.

3.1 A Weak Expansion Strategy

The revision strategy proposed by Pereira et al. ([9, 10]) can be seen as an example of a weak revision by expansion system. Basically, their idea is that lack of stability is caused by the fact that some literals cannot become true in a stable model of the program and thereby prevent stability.
Therefore, they force such literals to become true by adding a minimal number of unary rules of the form $a \leftarrow$ to restore stability. Let us call such an expanded program a *unary expansion* of P.
For unary expansions, we have a property stating that every classical model of a program P can become an acceptable, i.e. stable, model of some unary expansion of P.

Proposition 3.2 *Let P be a program and $M \in Mod(P)$ a classical model of P. Then M is a stable model of the extended program $P' = P + \{a \leftarrow \ \mid a \in B_P, M \models a\}$*

Proof. By construction, M is a model of P'. Then by Lemma 1, $T_{G(P',M)} \uparrow \omega \subseteq M$ and by construction of P', $M \subseteq T_{G(P',M)} \uparrow 1$. Hence, $T_{G(P',M)} \uparrow \omega = M$ and by definition, $M \in ST(P')$.

This proposition shows that a unary expansion system can satisfy P4. It is easy to see that it also can satisfy P1 - P3. Hence, it can be used as a weak expansion system.
Note, however, that neither postulate P4a nor P4b can be satisfied, since the addition of true facts might change the set of classical models of the expanded program, as will be shown in the following example:

Example 3.3 Let P be the program

$$a \leftarrow \sim a, \sim b$$
$$b \leftarrow \sim a, \sim b$$
$$c \leftarrow a, b, \sim c$$

Adding the unary rule $a \leftarrow$ to P results in a program P' having a stable model $M' = \{a\}$. Note that $M = \{b\}$ is a model of P, but not a model of P'.

3.2 A Conservative Expansion Method

We introduce a simple alternative for weak expansion, based on providing a *conservative extension* P' of a normal logic program P.

Definition 3.4 *Let P be a logic program. Let P' be a program such that $P \subseteq P'$ and $B_P \subseteq B_{P'}$. We call P' a* conservative extension *of P iff $Mod(P) = \{M \cap B_P \mid M \in Mod(P')\}$*

We will now discuss a simple method to obtain a conservative extension P' of P which has the additional property that $ST(P') \neq \emptyset$. These extensions will be called *simple conservative expansions.*

Definition 3.5 *Let P be a normal program and $A \subseteq B_P$.*
Then P' is said to be a simple conservative expansion *of P* w.r.t. *A iff $P' = P + \{a \leftarrow \sim a^*,\ a^* \leftarrow \sim a \mid a \in A\}$. Here, for every $a \in A$, a^* is a unique new atom, not occurring in B_P.*

The following proposition shows that simple conservative expansions indeed are conservative:

Proposition 3.6 *Every simple conservative expansion P' of P w.r.t. some A, $A \subseteq B_P$, is a conservative extension of P.*

Proof. Note that classically $a \leftarrow \sim a^*$, $a^* \leftarrow \sim a$ both are equivalent to $a \vee a^*$. Hence, for every model M of P, $M' = M \cup \{a^* \mid a \in A,\ M \models \neg a\}$ is a model of P'. So $Mod(P) \subseteq \{M' \cap B_P \mid M' \in Mod(P')\}$.
Conversely, if M' is a model of P', $M = M' \cap B_P$ is a model of P, since $P \subseteq P'$. Therefore, $\{M' \cap B_P \mid M' \in Mod(P')\} \subseteq Mod(P)$.

Proposition 3.7 *Let P be a program such that $Mod(P) \neq \emptyset$. Let $A \subseteq B_P$ be the subset of atoms occurring negatively in some rule of P and let P' be the simple conservative expansion of P w.r.t. A. Then $ST(P') \neq \emptyset$.*

Proof. Let $P^+ \subseteq P$ be the largest positive program contained in P. Let M^+ be the least model of the positive program $P_A = P^+ + \{a \leftarrow \ \mid a \in A\}$. By definition of A, for every rule $r \leftarrow \alpha^+ \wedge \alpha^-$ in P such that α^- is nonempty, there exists at least one $a \in A$ such that $\sim a$ occurs in α^-. So none of these rules r can occur in $G(P', M^+)$. Therefore, since $a^* \notin M^+$, $G(P', M^+) = P^+ + \{a \leftarrow \ \mid a \in A\} = P_A$ and by definition of M^+, M^+ is a stable model of P'.

Example 3.8 Let P be the following program:

$$c \leftarrow \sim c, \sim b$$
$$a \leftarrow b$$

Note that P is incoherent, i.e., $ST(P)$ is not defined.
Let P' be the following simple conservative expansion of P w.r.t. $A = \{b, c\}$:

$$c \leftarrow \sim c, \sim b$$
$$a \leftarrow b$$
$$b \leftarrow \sim b^*$$
$$b^* \leftarrow \sim b$$
$$c \leftarrow \sim c^*$$
$$c^* \leftarrow \sim c$$

Then $M = \{a, b, c\}$ is a stable model of P'.

It is easy to see that such a conservative expansion strategy satisfies the postulates P1-P4 and postulate P4a, but not Postulate P4b. Therefore, it is an example of a conservative expansion system.

3.3 A Strong Expansion Method

In this section we will prove a rather strong result, namely that for every normal logic program P there exists a (classically) equivalent program P' such that $P \subseteq P'$ and $ST(P) \neq \emptyset$.
The idea we will develop is a strict generalization of the so-called *dependency-directed backtracking* approach used in truth-maintenance.
Like Giordano and Martelli ([4]) argued, dependency directed backtracking can be conceived as a procedure to find a stable model of a program by adding contrapositives of rules; our expansion system uses a modified version of adding contrapositives.
The idea essentially is the following:

1. We represent an incoherent program P by a classically equivalent set of propositional clauses Σ_P. Then Σ_P is reduced to a classically equivalent set of *minimally satisfiable* clauses Π .
2. We use the set Π to construct a *hierarchical* program P_Π *without constraints*, i.e., a normal logic program without circular dependencies. It is well-known that such a program has a unique stable model M_Π.

3. We show that M_Π is a model of P. Then, by Lemma 2, it follows that M_Π is also a *stable* model of the expansion $P' = P + P_\Pi$ of P. This shows that $ST(P') \neq \emptyset$.
4. Finally, we show that every model of P is also a model of P'. So $Mod(P) \subseteq Mod(P')$. Hence, since $P' \supseteq P$, implying that $Mod(P') \subseteq Mod(P)$, it follows that $Mod(P) = Mod(P')$.

To discuss this method, we need the following concepts and definitions.
A *literal* is an atomic proposition (atom) or its negation. A *clause* is a finite repetition-free disjunction of literals. A clause C is called *positive* if it contains only positive atomic propositions. $\sim C$ is the conjunction of the negations of literals occurring in C.
A clause C is said to be a *prime implicant* of a set of clauses Σ iff $\Sigma \models C$ and there is no proper subset $C' \subset C$ such that $\Sigma \models C'$.
Given a set of clauses Σ, let $\Pi(\Sigma)$ denote a (smallest) subset of prime implicants of Σ such that for every $C \in \Sigma$ there is a prime implicant $C' \subseteq C$ occurring in $\Pi(\Sigma)$.
The following proposition is immediate:

Proposition 3.9 $Mod(\Sigma) = Mod(\Pi(\Sigma))$.

Proof. Since $\Pi(\Sigma)$ is a set of prime implicants, $Mod(\Sigma) \subseteq Mod(\Pi(\Sigma))$. Since for every $C \in \Sigma$ there is a prime implicant C' such that $C' \models C$, we have $Mod(\Pi(\Sigma)) \subseteq Mod(\Sigma)$.

We will denote the set of clauses associated with a program P by Σ_P. It is obvious that Σ_P and $\Pi(\Sigma_P)$ are always satisfiable and $Mod(P) = Mod(\Pi(\Sigma_P))$.
The following observations are essential for the discussion of the expansion method:

Observation 3.10 *$M_e = \emptyset$ is a stable model of P iff $\Pi(\Sigma_P)$ contains no positive clause.*

Proof. Suppose that $M_e = \emptyset$ is a stable model of P and $\Pi(\Sigma_P)$ contains a positive clause C. Since $Mod(\Pi(\Sigma_P)) = Mod(P)$, it follows that every model M of P has to satisfy at least one positive literal in C; contradiction.
Conversely, suppose that $\Pi(\Sigma_P)$ contains no positive clause. Then, of course $M_e = \emptyset$ is a model of $\Pi(\Sigma_P)$ and hence, a model of P. Since $G(P, \emptyset)$ does not contain a unit rule $a \leftarrow$, it follows immediately that $M_e = \emptyset$ is a stable model of P.

Observation 3.11 *For every positive clause C in $\Pi(\Sigma)$ and for every atom c in C there exists at least one model M of Σ such that M minimally satisfies C and M makes c true.*

Proof. Obvious, since $\Pi(\Sigma)$ is a set of prime implicants.

We will now describe the method to extract a program P_Π having a stable model M_Π from the set of clauses $\Pi = \Pi(\Sigma_P)$.
First of all, we can assume that Π contains at least one positive clause $C = \{c_1, \ldots, c_m\}$. For else, according to Observation 3.10, $M_e = \emptyset$ would be a model of P and, since it is the stable model of the empty program, by Lemma 2, M_e would be an acceptable stable model of P.
According to Observation 3.11 then, there is at least one model M of Π, minimally satisfying C, for example by making true c_1 and making false the other literals c_j. Now consider the rule $r_C : c_1 \leftarrow \sim c_2, \ldots, \sim c_m$ and the model $\{c_1\}$. Note that this model is the unique stable model of r_C. So, let us extract r_C from Π and let P_Π contain r_C and M_Π contain c_1. Then this partial realization of M_Π now is a *partial* model[4] of Π and a stable model of r_C. To extend M_Π and P_Π, we have to consider the remaining part of Π after the extraction of r_C.
Since eventually, M_Π has to be a model of Π, we can *remove* every clause C' from Π containing the literal c_1 or the literals $\sim c_j$ for $j \geq 2$: such a clause is already satisfied by the partial realization of M_Π.
On the other hand, in the remaining clauses C', every literal of the form $\sim c_1$ or c_j, $j \geq 2$, has to be marked as *inactive*: such a literal is evaluated false by the current partial realization M_Π. If a literal is not marked as inactive, it is assumed to be active.
In general, we assume every clause C is partitioned into a set of active literals $Active(C)$ and inactive literals $Inactive(C)$: $C = Active(C) \cup Inactive(C)$.
Now let Π' be the set of remaining clauses and $Active(\Pi')$ the set of active parts of clauses in Π'. We would like to continue with this set Π' to extract new rules from it and to add them to P_Π. The set $Active(\Pi')$, however, in general is not a set of prime implicants. This implies that we cannot safely select positive clauses from $Active(\Pi')$ and assume that they can be minimally satisfied.
Therefore, for every $C' \in \Pi'$ we replace $Active(C')$ by a prime implicant $\pi(C') \subseteq Active(C')$ of $Active(\Pi')$ and we add the set of literals

$$\sim r_C = \{\sim c_1, c_2, \ldots, c_m\}$$

to $Inactive(C')$ to account for taking prime implicants *modulo* the realized part of the model M_Π.
Let $R(\Pi, r_C)$ denote the resulting set of clauses.
The addition to $Inactive(C')$ guarantees that every model of Π will also be a model of $R(\Pi, r_C)$. For, let $M \in Mod(\Pi)$ be an arbitrary model of the original set of clauses Π and let C' be an arbitrary clause of $R(\Pi, r_C)$. We show that M is a model of C'. If M is an extension of M_{r_C}, M will be a model of $Active(C')$; if, however, M is not an extension of M_{r_C}, $M \models \sim r_C$. Hence, M is a model of C', since $\sim r_C$ is included in $Inactive(C')$. So we have the following result:

Proposition 3.12 $\Pi \models R(\Pi, r_C) \cup \Sigma_{r_C}$

Now the extraction process can be continued as long as $Active(R(\Pi, r_C))$ contains a positive clause $C'' = \{d_1, \ldots d_k\}$.

[4] I.e., it can be extended to a model of Π

Then we add the rule $r'_C : d_1 \leftarrow \sim d_2, \ldots, \sim d_k, \sim Inactive(C'')$ to P_Π, we add d_1 to M_Π and we continue with $R(R(\Pi, r_C), r'_C)$. Note that, again, M_Π is the stable model of P_Π.

This process continues until either the empty set of clauses is returned or the result of extraction is a set of clauses without a positive clause.

The following procedure is a succinct description of our method to find a stable model of an expansion P' of P:

---

input: the set $\Pi = \Pi(\Sigma_P)$.
output: a program P_Π and a model M_Π.
begin
  Let every clause in Π consist of active literals;
  Let $P_\Pi := \emptyset$; let $M' := \emptyset$;
  while Π contains a clause C and $Active(C) = \{c_1, c_2, \ldots, c_m\}$ is positive **do**
    $r_C := c_1 \leftarrow \sim c_2, \ldots, \sim c_m, \sim Inactive(C)$;
    $P_\Pi := P_\Pi + r_C$;
    $M_\Pi := M_\Pi \cup \{c_1\}$;
    $\Pi := R(\Pi, r_C)$
  wend
  return (P_Π, M_Π);
end;

Clearly, since during the execution the total number of active literals is decreasing, this procedure halts for every normal program P.

The following observation is easily verified:

Observation 3.13 *P_Π is a hierarchical program and M_Π is the unique stable model of P_Π.*

Example 3.14 Consider the following incoherent program:

$$P: \quad b \leftarrow \sim a$$
$$c \leftarrow \sim b$$
$$a \leftarrow \sim c$$

Since $ST(P) = \emptyset$, we transform P into the following set of clauses $\Sigma_P = \{\{a, b\}, \{b, c\}, \{c, a\}\}$.

Now $\Pi = \Pi(\Sigma) = \Sigma$ and there is a positive clause $\{a, b\}$; so we add the rule $r : a \leftarrow \sim b$ to P_Π and we add a to M_Π. Now constructing $R(\Pi, r)$, the clauses $\{a, b\}$ and $\{c, a\}$ can be removed since a occurs in both. In the clause $\{b, c\}$, b is made inactive and $\sim a$ and b are added to the clause as inactive literals. So, $\Pi' = R(\Pi, r)$ contains one clause $C = \{b, c, \sim a\}$ where $Inactive(C) = \{\sim a, b\}$ and $Active(C) = \{c\}$ is positive. Therefore, $c \leftarrow a, \sim b$ is added to P_Π and

c to M_Π. Now $R(\Pi', c \leftarrow a, \sim b) = \emptyset$, so $P_\Pi = \{a \leftarrow \sim b, c \leftarrow a, \sim b\}$ and $M_\Pi = \{a, c\}$.
We see that M_Π is the unique stable model of the hierarchical program P_Π and that $M_\Pi \models C$ for every $C \in \Sigma_P$. Hence, M_Π is a stable model of $P + P_\Pi$.

Remark. This method can also be applied to normal programs with constraints (cf. [12, 13]):

Example 3.15 Consider the contradictory program P:

$$\bot \leftarrow c, \sim a$$
$$c \leftarrow \sim b, d$$
$$d \leftarrow$$
$$b \leftarrow a$$

Since $ST(P) = \{\{\bot, c, d\}\}$ and this stable model is contradictory, the set $Acc(P)$ is empty, while P is satisfiable. We transform P into an equivalent set of clauses $\Sigma_P = \{\{\sim c, a\}, \{c, b, \sim d\}, \{d\}, \{b, \sim a\}\}$.
Now $\Pi = \Pi(\Sigma) = \{\{\sim c, a\}, \{b\}, \{d\}, \{b\}\}$. Let us add the rule $r : b \leftarrow$ to P_Π and b to M_Π. Now constructing $R(\Pi, r)$, the clause $\{b\}$ can be removed and we add $\sim b$ to the inactive part of the remaining clauses. Next, we select $\{d, \sim b\}$, since its active part is positive. We add $d \leftarrow b$ to P_Π and d to M_Π. After the reduction, only the clause $\{\sim c, a, \sim b, \sim d\}$ remains. Its active part is $\{\sim c, a\}$. So this set of clauses does not contain an active positive clause. Therefore the procedure returns $P_\Pi = \{b \leftarrow, d \leftarrow b\}$ and $M_\Pi = \{b, d\}$.

To prove the correctness of the procedure, consider the expanded program $P' = P + P_\Pi$ and the set M_Π. We have to show that

1. M_Π is a stable model of P', hence $ST(P') \neq \emptyset$ and
2. $Mod(P') = Mod(P)$.

We will give a sketch of the proof of the second part.
By Proposition 3.12, we have

$$Mod(P) = Mod(\Pi) \subseteq Mod(R(\Pi, r_C) \cup \Sigma_{r_C})$$

Iterating these reductions, we see that

$$Mod(P) \subseteq Mod(\Sigma_{P_\Pi}) = Mod(P_\Pi)$$

Hence,

$$Mod(P') = Mod(P + P_\Pi) = Mod(P)$$

The following theorem summarizes the results we have obtained:

Theorem 3.16 *For every program P such that $Mod(P) \neq \emptyset$, there is an expansion P' of P such that $St(P') \neq \emptyset$ and $Mod(P) = Mod(P')$.*

4 Discussion and Conclusion

We have discussed a set of rationality postulates for revising non-monotonic theories which can be used as an extension to the Gärdenfors et al. postulates pertaining to revision of *monotonic* theories.

We have distinguished three expansion systems and we have illustrated them in the case of logic programming, showing that some existing revision methods satisfy the postulates for weak and conservative expansion systems. A new revision method for normal logic programs has been shown to satisfy the postulates of a strong expansion system.

We mention two possible extensions of the current approach. First of all, it should be possible to bring the expansion method more in line with the AGM-style approach to revision, that is, we should try to reformulate the expansion framework in terms of revising a non-monotonic theory with some new information instead of revising the theory itself.

Secondly, subsequent work has to be done on further strengthening of expansion systems. For example, it is clear that not every (weak, conservative or strong) expansion of a theory T is equally acceptable: it is clear that we would like to find an expansion T' *minimizing* the difference between a theory T and T'. One intuitive acceptable criterion could be to choose an expansion T' minimizing (i) the difference between $Mod(T)$ and $Mod(T')$ and (ii) minimizing the information added to T i.e. minimizing $T' - T$. For logic programming, this would mean a choice for minimal strong expansions of a program.

References

1. C. Alchourrón, P. Gärdenfors and D. Makinson, On the Logic of Theory Change: Partial Meet Contraction and Revision Functions, *Journal of Symbolic Logic*, **50**, 510–530, 1985
2. P. Gärdenfors, *Knowledge in Flux*, MIT Press, Cambridge, MA, 1988
3. M. Gelfond and V. Lifschitz, The Stable Model Semantics for Logic Programming. In: *Fifth International Conference Symposium on Logic Programming*, pp. 1070-1080, 1988.
4. L. Giordano and A. Martelli, Generalized Stable Models, Truth Maintenance and Conflict Resolution, in: D. Warren and P. Szeredi (eds) *Proceedings of the 7th International Conference on Logic Programming*, pp. 427-441, 1990.
5. K. Inoue, Hypothetical Reasoning in Logic Programs, *Journal of Logic Programming*, **18**, 3, 191–227, 1994
6. J. W. Lloyd, *Foundations of Logic Programming*, Springer Verlag, Heidelberg, 1987.
7. W. Marek, V.S. Subrahmanian, The relationship between stable, supported, default and auto-epistemic semantics for general logic programs, *Theoretical Computer Science* 103 (1992) 365–386.
8. V. W. Marek and M. Truszczyński, *Nonmonotonic Logic*, Springer Verlag, Heidelberg, 1993.
9. L. M. Pereira, J. J. Alferes and J. N. Aparicio, Contradiction Removal within well-founded semantics. In: A. Nerode, W. Marek and V. S. Subrahmanian, (eds.), *First*

International Workshop on Logic Programming and Non-monotonic Reasoning, MIT Press, 1991.

10. L. M. Pereira, J. J. Alferes and J. N. Aparicio, The Extended Stable Models of Contradiction Removal Semantics. In: P. Barahona, L.M. Pereira and A. Porto, (eds.), *Proceedings -EPIA 91*, Springer Verlag, Heidelberg, 1991.
11. H. Rott, Modellings for Belief Change: Base Contraction, Multiple Contraction, and Epistemic Entrenchment, in: D. Pearce and G. Wagner, *Logics in AI*, Springer Verlag Berlin, 1992.
12. C. Witteveen and G. Brewka, Skeptical Reason Maintenance and Belief Revision, *Artificial Intelligence*, **61** (1993) 1–36.
13. C. Witteveen and W. van der Hoek, Belief Revision by Expansion, in: M. Clarke et al. (eds), Symbolic and Quantitative Approaches to Reasoning and Uncertainty, LNCS 747, Springer, Heidelberg, 1993, pp. 380–388.

A Complete Connection Calculus with Rigid E-Unification*

Uwe Petermann

HTWK Leipsig, FB IMN
Postfach 66, D-04251 Leipsig (F.R.G.)
Net: uwe@imn.th-leipsig.de

Abstract. We present an approach to building-in equational reasoning into theorem provers which are based on the connection method. The approach is an instance of total theory reasoning. In order to achieve a completeness result we combine results concerning the simultaneous rigid E-unification problem with our general framework for building-in theories. We pose the problem whether for the construction of a complete goal-oriented prover with equality it is sufficient to be able to solve only a restricted version of the simultaneous rigid E-unification problem.

1 Introduction

One of the intriguing problems in automated theorem proving is the integration of equational reasoning into goal-oriented proof calculi. In the present paper we prove the completeness of a theory version of the connection method [5] with rigid E-unification. For this purpose we combine the following partial results:

- The completeness of the theory pool calculus as a theory version of the connection method [17].
- The decidability of the rigid E-unification problem. There are decision procedures due to Gallier et al. [8] and more recently to J. Goubault [9] as well as to G. Becher and the present author [2].
- The enumerability of complete sets of rigid E-unifiers. An enumeration procedure has been implemented by B. Beckert as part of a tableaux-based theorem prover [3].

The result carries over to other goal-oriented proof calculi like the matrix method [1], model elimination [12] and model elimination tableaux calculus [10].
Let us illustrate our approach by the following example.

Example 1.
We consider a set of three clauses which are written as rows of a matrix.

* This research has been supported by grants of the Deutsche Forschungsgemeinschaft and the Alexander von Humboldt-Stiftung.

$$\left[\begin{array}{cc} \multicolumn{2}{c}{q(b,a) \not\approx t} \\ q(U,V) \approx t, & U \approx V \\ \multicolumn{2}{c}{p(a,b,a,b) \approx t} \\ p(X,Y,Z,Z) \not\approx t, & X \not\approx Y \end{array}\right] \tag{1}$$

The set of clauses (1) is equationally unsatisfiable because one can find

(1) a set of copies of these clauses

$$\left[\begin{array}{cc} \multicolumn{2}{c}{q(b,a) \not\approx t} \\ q(U',V') \approx t, & U' \approx V' \\ \multicolumn{2}{c}{p(a,b,a,b) \approx t} \\ p(X',Y',Z',Z') \not\approx t, & X' \not\approx Y' \end{array}\right] \tag{2}$$

(2) three sets of literals (each written as a column)

$$\left\{\begin{array}{c} q(b,a) \not\approx t, \\ q(U',V') \approx t \end{array}\right\}, \left\{\begin{array}{c} U' \approx V', \\ X' \not\approx Y' \end{array}\right\}, \left\{\begin{array}{c} U' \approx V', \\ p(a,b,a,b) \approx t, \\ p(X',Y',Z',Z') \not\approx t \end{array}\right\} \tag{3}$$

and

(3) a substitution

$$\theta = \{V', X', Z' \mapsto a,\ U', Y' \mapsto b\} \tag{4}$$

(4) such that the following conditions are satisfied :
   (a) Every conjunct of the conjunctive normal form of the set of clauses (2) contains one the mentioned sets of literals as a subset.
   (b) For every set of literals C in (3) the instantiation $\theta(C)$ is unsatisfiable in any equational interpretation if the variables are treated as constants.

The literal sets in (3) are called *eq-connections*, a set of eq-connections satisfying condition (4a) is called a *spanning eq-mating* (following [5]). Sets of clauses will be called also *matrices*. The conditions (4a, 4b) imply that the instantiation of the clause set (2) by θ is unsatisfiable in any equational interpretation if

the variables are treated as constants. Therefore, the original set of clauses is equationally unsatisfiable. Now for $n = 3$ and $i = 1, \ldots, n$ we rewrite the eq-connection C_i as E_i, S_i where E_i is the set of equations in C_i and S_i consists of the single inequality in C_i. It is easy to see that the substitution θ is a rigid E_i-unifier of S_i for each i. It is called a ***simultaneous rigid E-unifier of S*** where $E = \{E_i\}_{i=1}^n$ and $S = \{S_i\}_{i=1}^n$.
The set of clause copies (2), the spanning eq-mating (3) and the simultaneous rigid E-unifier (4) represent a proof for the considered clause set. If one gives the clause set (1) as a proof task to a prover which relies on one of the calculi mentioned above then that prover will look for those components of the proof in a stepwise search. In order to describe this search a bit more precisely we need the notion of a path. A ***path*** in a clause set is a set which contains exactly one literal from each of the clauses. A ***partial path*** is a subset of a path. The search is driven by the aim to find for some partial path, called the ***actual path***, an appropriate eq-connection which is contained in a continuation of the actual path. For this purpose may be created new clause copies. It is characteristic for the mentioned calculi that for sake of completeness it may be necessary that elements of the actual path belong to that eq-connection. Our example illustrates that case. After finding the first two eq-connections

$$\left\{ \begin{matrix} q(b,a) \not\approx t, \\ q(U',V') \approx t \end{matrix} \right\} \quad \text{and} \quad \left\{ \begin{matrix} U' \approx V', \\ X' \not\approx Y' \end{matrix} \right\}$$

the literal $U' \approx V'$ belongs to the actual path. It is necessary in order to form the third eq-connection. One may observe that the equations which are taken from the actual path can not be instantiated arbitrarily. For each of them we can use only that instance which is already on the path. They must be treated rigid. This is an essential difference to general E-unification.
We need a carefully designed interface between a foreground reasoner which realizes the base calculus and the background reasoner which is dedicated for equational reasoning. Roughly speaking, the foreground reasoner has to keep track for the spanningness criterion. The background reasoner has to decide rigid E-unifiability of eq-connections and to supply rigid E-unifiers. More formally the interface is specified by the set of eq-connections and the algorithm solving the rigid E-unification problem. The set of eq-connections satisfies a certain completeness property. Those notions will be defined and studied in section 3. If both elements of the interface are given then the couple consisting of foreground and background reasoner forms a complete theory reasoner.

Related work

J. Gallier et al. claimed in [8] the completeness of a matrix calculus with rigid E-unification. There is a gap in their proof because they did not prove one of the essential preconditions of the procedure which has been intended for deciding the simultaneous rigid E-unifiability. W. Snyder and C. Lynch communicated in [19] that set-of-support-resolution with relaxed paramodulation is complete.

However, they could not supply a rigorous proof of that result. B. Beckert and R. Hähnle integrated rigid E-unification into a tableaux-prover [4]. Recently several groups have integrated D. Brand's modification method [7] into theorem provers which are close to model elimination [6, 13, 14]. This method offers an alternative approach to obtain complete goal-oriented provers with equality.

2 Preliminaries

We assume that the reader is familiar with the basic notions of first-order logic in clause form (cf. [11]).
A *clause* is a set of literals. A *matrix* is a set of clauses. A *(partial) path (in) through* a matrix M contains (at most) exactly one literal from each clause of M. We will distinguish paths and clauses with respect to their intended meaning. We consider a *clause* $L_1, \ldots, L_n$ to be the abbreviation of the universal closure $\bar{\forall}(L_1 \vee \ldots \vee L_n)$ of the disjunction of its elements. A *path* $L_1, \ldots, L_n$ abbreviates the existential closure $\bar{\exists}(L_1 \wedge \ldots \wedge L_n)$ of the conjunction of its elements. Clauses will be abbreviated also by Γ, Δ, Λ or Θ. Γ, Δ means the union $\Gamma \cup \Delta$ and Γ, L means $\Gamma \cup \{L\}$. Paths will be denoted also by p or q. A set of partial paths in a matrix M is called a *mating* in M. The meaning of a matrix $\{C_1, \ldots, C_n\}$ is the conjunction $\bar{\forall}C_1 \wedge \ldots \wedge \bar{\forall}C_n$. The meaning of a mating $\{p_1, \ldots, p_n\}$ is the disjunction $\bar{\exists}p_1 \vee \ldots \vee \bar{\exists}p_n$. A partial path u in matrix M is *spanning* a path p through M if $u \subseteq p$. A mating U in a matrix M is *spanning* if for every path p through M exists an element $u \in U$ which is spanning p. Clearly, if a mating U is unsatisfiable in a theory $\mathcal{T}$and U is spanning in a matrix M then M is $\mathcal{T}$-unsatisfiable too.
If L is a positive literal then $\bar{L}$ denotes the literal $\neg L$. If L has the form $\neg K$ then $\bar{L}$ denotes the literal K. If p is the path $L_1, \ldots, L_n$ then $\bar{p}$ denotes the clause $\overline{L_1}, \ldots, \overline{L_n}$. And, vice versa, if Γ is the clause $L_1, \ldots, L_n$ then $\overline{p}$ denotes the path $\overline{L_1}, \ldots, \overline{L_n}$.
The set of variables occurring in a term t, literal L, clause Γ or path p will be denoted by $Var(t)$, $Var(L)$, $Var(\Gamma)$ or $Var(p)$ respectively.
A *substitution* is a mapping from the set of variables into the set of terms which is almost everywhere equal to the identity. The *domain* of a substitution σ is the set $D(\sigma) = \{X \mid \sigma(X) \neq X\}$. The set of *variables introduced by* σ is the set $I(\sigma) = \bigcup_{x \in D(\sigma)} Var(\sigma(X))$. A substitution σ may be extended canonically to a mapping from the set of terms into the set of terms. This extension will be denoted by σ too. The *composition* $\sigma\theta$ of substitutions σ and θ is the substitution which assigns to every variable X the term $\theta(\sigma(X))$. A substitution σ is called *idempotent* if $\sigma = \sigma\sigma$. A substitution σ is idempotent iff $D(\sigma) \cap I(\sigma) = \emptyset$. Substitutions σ and λ are it identical with respect to a set of variables V iff for each $X \in V$ holds $\sigma(X) = \lambda(X)$. This will be denoted by $\sigma =_V \lambda$. We say that *σ is more general than* or *subsumes* λ if there exists a substitution δ such that $\sigma\delta =_{D(\lambda)} \lambda$.
A set of matrices which is closed w.r.t. the application of substitutions, forming sub-matrices and paths will be called a *query language*.

3 Complete sets of equational connections and their unification problem

In the present section we generalise the concepts of a connection and of a unifier of a connection towards theory reasoning. This generalisation follows the treatment of equality in [5]. We formalise what it means for a given theory to have "enough" theory connections in order to refute all theory unsatisfiable matrices which belong to a given query language. We formulate a Herbrand theorem by use of this notion. The general framework will be applied to the case of the theory of equality. A complete set of equational connections will be characterised. We discuss the decision complexity of this set. The notion of a complete set of unifiers for a theory connection generalises the notion of complete set of theory unifiers of a pair of terms. The notion of solvable unification problem describes that for each element of a given complete set of theory connections may be constructed a complete set of theory unifiers. We discuss the relation of the unification problem for equational connections to the rigid E-unification problem defined in [8].
Let $\mathcal{T}$ be a theory given as a set of clauses.

Definition 1. ($\mathcal{T}$-complementary)
A path u is called $\mathcal{T}$-complementary iff $\bar{\exists}\left(\bigwedge_{L\in u} L\right)$, i.e. the existential closure of the conjunction of the elements of u, is $\mathcal{T}$-unsatisfiable.

The notion of complementarity is defined via the $\mathcal{T}$-unsatisfiability according to the negative representation in the present paper. It is the advantage of the notion of complementarity that the definition of other notions of the calculus may rely on the $\mathcal{T}$-complementarity notion and become independent on the chosen (positive or negative) representation.

Definition 2. ($\mathcal{T}$-unifier, simultaneous $\mathcal{T}$-unifier, $\mathcal{T}$-connection)

(1) A substitution σ will be called a *$\mathcal{T}$-unifier* of a set of literals u iff $\sigma(u)$ is $\mathcal{T}$-complementary.
(2) A substitution σ will be called a *simultaneous $\mathcal{T}$-unifier* of a set U of sets of literals iff for every $u \in U$ $\sigma(u)$ is $\mathcal{T}$-complementary.
(3) A set of literals u is called a *$\mathcal{T}$-connection* if there exists a $\mathcal{T}$-unifier for u.
(4) A partial path p in a matrix M which is a $\mathcal{T}$-connection will be called a *$\mathcal{T}$-connection in M*.

Without restriction of generality we assume that the equality symbol is the unique predicate symbol. It is well known how to carry over the following results to the case with other predicate symbols. The reader should be aware that the treatment of non-equational atoms directly, i.e. without a standard translation into equations is important for efficiency.
Let $\mathcal{E}$ be the theory of equality, i.e. the set of clauses which express reflexivity, symmetry and transitivity of the equality symbol $\doteq$ and the instantiations of the axiom scheme for functional substitutivity. Let $\mathcal{Q}_{eq}$ be the query language

consisting of all those clause sets which have the equality sign as the unique predicate symbol. Sometimes we will write $s \approx t$ in order to denote an equation which may be either $s \doteq t$ or $t \doteq s$.

Definition 3. (eq-connections)
Let $\mathcal{U}_{eq}$ be the set of all $\mathcal{E}$-connections of the form $l_1 \doteq r_1, \ldots, l_m \doteq r_m, \neg s \doteq t$. Those $\mathcal{E}$-connections will be called eq-connections.

A decision procedure for the set of eq-connections has been first given by J. Gallier et al. [8]. Recently J. Goubault [9] and G. Becher and the present author [2] communicated further decision procedures. The complexity of the problem whether a literal set of the form $l_1 \doteq r_1, \ldots, l_m \doteq r_m, \neg s \doteq t$ is an eq-connection has been characterized first by J. Gallier et al. in [8].

Proposition 4.
The membership problem of the set of eq-connections is NP-complete.

Now we define the notion of a complete set of $\mathcal{T}$-connections $\mathcal{U}$. The completeness of $\mathcal{U}$ will be defined relative to a given query language $\mathcal{Q}$. We will show that the set of all eq-connections is $\mathcal{E}$-complete w.r.t. the query language consisting of all clause sets which have the equality sign as the unique predicate symbol.

Definition 5. ($\mathcal{U}$-connection in a matrix, complete set of $\mathcal{T}$-connections $\mathcal{U}$, spanning $\mathcal{U}$-mating)
Let $\mathcal{Q}$ be a query language. Let $\mathcal{U}$ be a decidable set of $\mathcal{T}$-connections such that $\mathcal{U} \subseteq \mathcal{Q}$.

(1) A *$\mathcal{U}$-connection* in a matrix $M \in \mathcal{Q}$ is a partial path u in M which is an element $u \in \mathcal{U}$.
(2) The set $\mathcal{U}$ will be called *complete w.r.t.* the query language $\mathcal{Q}$ if
  (a) for each $\mathcal{T}$-complementary ground path $p \in \mathcal{Q}$ exists $u \in \mathcal{U}$ such that $u \subseteq p$.
  (b) for each $\mathcal{T}$-complementary ground path of the form $\sigma(u) \in \mathcal{U}$ such that $u \in Q$ holds $u \in \mathcal{U}$.
(3) A set of $\mathcal{U}$-connections in a matrix M is called a *$\mathcal{U}$-mating* in M. $\mathcal{U}$-mating U is *spanning a set of partial paths* $\mathcal{P}$ in a matrix M if every path in $\mathcal{P}$ contains a connection being element of U. U is said to *span a matrix* M if U spans the set of all paths through M.

We are going to prove that the set of eq-connections $\mathcal{U}_{eq}$ is an $\mathcal{E}$-complete set of $\mathcal{E}$-connections. Because of proposition 4 it is sufficient to prove that $\mathcal{U}_{eq}$ satisfies the conditions (2a) and (2a) in definition 5. This property may be proved even for the more general assumption that the considered theory is an arbitrary set of conditional equations which contains $\mathcal{E}$. The more specific case with $\mathcal{T} = \mathcal{E}$ is then a simple corollary.

Proposition 6.
Let $\mathcal{T}$ be a set of conditional equations which contains $\mathcal{E}$. Moreover, let $\mathcal{U}$ be the set of all $\mathcal{T}$-connections which are of the form

$$l_1 \doteq r_1, \ldots, l_n \doteq r_n, \neg s \doteq t.$$

Then $\mathcal{U}$ is $\mathcal{T}$-complete with respect to $\mathcal{Q}_{eq}$ if it is decidable.

Proof. Suppose that $\mathcal{U}$ is decidable and satisfies the assumptions of the proposition. In order to prove that $\mathcal{U}$ is $\mathcal{T}$-complete let us first prove condition (2a) of definition 5. Suppose that there exists a ground path $p \in \mathcal{Q}_{eq}$ which does not contain any element of $\mathcal{U}$ as a sub-path. Let

$s_i \doteq s_i'$ for $i = 1, .., n$ be all equations occurring in p,
$t_j \doteq t_j'$ for $j = 1, .., m$ all equations occurring negated in p.

Then for all $j = 1, \ldots, m$ the conditional equation

$$\bigvee_{i=1}^{n} \neg(s_i \doteq s_i') \vee t_j \doteq t_j'$$

is not $\mathcal{T}$-derivable.
Now consider the $\mathcal{T}$-algebra **D** freely generated by the equations $s_i \doteq s_i'$ for $i = 1, \ldots, n$ (cf. definition 3.2.1., p.114 in [18]). It is a term algebra factored with respect to a minimal congruence relation $\equiv$ determined by $\mathcal{T}$ and the set of generating equations. In this algebra holds $[t_j]_\equiv \neq [t_j']_\equiv$ (cf. theorem 3.2.3., p.115 in [18]). Therefore **D** may be extended to a model of $\mathcal{T}$ such that this model together with a canonical valuation satisfies the formula represented by the path p. Therefore p is not $\mathcal{T}$-complementary.
It remains to prove condition (2b) of definition (5). Let us consider a $\mathcal{T}$-complementary ground path $\sigma(u) \in \mathcal{U}$ such that $u \in \mathcal{Q}_{eq}$. Then by point (3) of definition (2) u is a $\mathcal{T}$-connection and by definition $u \in \mathcal{U}$.

From propositions 6 and 4 we obtain the following corollary.

Corollary 7.
The set of eq-connections $\mathcal{U}_{eq}$ is $\mathcal{E}$-complete with respect to $\mathcal{Q}_{eq}$.

The notion of a $\mathcal{T}$-connection appeals to the semantical property to have a $\mathcal{T}$-unifier. For the purpose of interfacing the foreground and background reasoners we need a decidable set $\mathcal{U}$ of $\mathcal{T}$-connections. The decidability of $\mathcal{U}$ is desired because we expect the background reasoner to return a definite negative answer in case of lack of $\mathcal{T}$-unifiers for a given candidate for a $\mathcal{T}$-connection. The $\mathcal{T}$-completeness of a set of $\mathcal{T}$-connections $\mathcal{U}$ formalizes the idea that there are enough $\mathcal{T}$-connections in $\mathcal{U}$ in order to prove all theorems within a query language by the help of the "British Museum" procedure, i.e. by the exhaustive construction of all ground instantiations and checking them for the existence of a spanning $\mathcal{U}$-mating. The formal expression of this claim is the Herbrand theorem.

Theorem 8. *(Herbrand theorem)*
Let $\mathcal{Q}$ be a query language and $\mathcal{U}$ a set of $\mathcal{T}$-connections which is complete w.r.t. to $\mathcal{Q}$. Then a matrix $M \in \mathcal{Q}$ is $\mathcal{T}$-unsatisfiable if and only if there exists a set M' of copies of clauses from M, a $\mathcal{U}$-mating U which is spanning M' and a substitution σ such that $\sigma(u)$ is $\mathcal{T}$-complementary for each $u \in U$.

This theorem has been stated with a proof sketch in [17], a detailed proof may be found in [16]. Because $\mathcal{U}_{eq}$ is $\mathcal{E}$-complete w.r.t. the query language $\mathcal{Q}_{eq}$ with $\doteq$ as the unique predicate symbol we obtain the Herbrand theorem as a simple corollary.

Corollary 9. *(Herbrand theorem for $\mathcal{E}$)*
The Herbrand theorem holds for the theory of equality $\mathcal{E}$, the query language $\mathcal{Q}_{eq}$ and the set of eq-connections $\mathcal{U}_{eq}$.

The elements of the spanning $\mathcal{U}_{eq}$-mating and the substitution σ which exist according to theorem 9 should be found incrementally by use of a kind of unification. Now we introduce the notions for a formal treatment of this kind of unification.

Definition 10. (More general $\mathcal{T}$-unifier, $\mathcal{T}$-unification problem in $\mathcal{U}$)

(1) Let ϱ and σ be $\mathcal{T}$-unifiers of a path u such that $D(\varrho), D(\sigma) \subseteq Var(u)$. Then we say that ϱ is *more general than* or *subsumes* σ if there exists η such that $\varrho\eta =_{D(\sigma)} \sigma$. This will be denoted by $\varrho \leq \sigma$.
(2) A set S of $\mathcal{T}$-unifiers of a set $u \in \mathcal{U}$ will be called *complete* if for each $\mathcal{T}$-unifier σ of u exists a substitution $\varrho \in S$ such that $\varrho \leq \sigma$.
(3) Let $\mathcal{U}$ be a set of sets of literals. We say that the *$\mathcal{T}$-unification problem in $\mathcal{U}$ is solvable* if for every $u \in \mathcal{U}$ exists a enumerable complete set of $\mathcal{T}$-unifiers for u.

The following proposition is rather straightforward but important.

Proposition 11.
Suppose that the $\mathcal{T}$-unification problem is solvable for a set of $\mathcal{T}$-connections $\mathcal{U}$ which is subset of a query language $\mathcal{Q}$. Then for every simultaneous $\mathcal{T}$-unifier θ of a set of $\mathcal{T}$-connections $U \subseteq \mathcal{U}$ a simultaneous $\mathcal{T}$-unifier σ of U and a substitution η satisfying $\theta = \sigma\eta$ may be found incrementally.

Proof. Let S_u denote the complete set of $\mathcal{T}$-unifiers for each $u \in \mathcal{U}$. Moreover, let θ be the simultaneous $\mathcal{T}$-unifier of a finite set of $\mathcal{T}$-connections $U \subseteq \mathcal{U}$.
Claim: For each enumeration $u_1, \ldots, u_n$ of the elements of a non-empty set of $\mathcal{T}$-connections $U \subseteq \mathcal{U}$ may be constructed sequences $\{\sigma_i\}_{i=1}^n$, $\{\eta_i\}_{i=0}^n$, $\{\varrho_i\}_{i=0}^n$ such that

(1) $\eta_0 = \theta$ and $\varrho_0 = \{\ \}$ and
(2) for every $i, 1 \leq i \leq n$
  (a) $\sigma_i \in S_{\varrho_{i-1}(u_i)}$,

(b) $\sigma_i \eta_i = \eta_{i-1}$ and
(c) $\varrho_i = \varrho_{i-1}\sigma_i$.
(d) $\theta = \varrho_i \eta_i$.

This claim may be proven by induction on the cardinality n of U.

Unfortunately the case of equality is rather complicated. The known decision procedures don't allow to enumerate complete sets rigid E-unifiers. Unifiers which may be computed have a weaker subsumption property. We prepare the definition of this property by an auxiliary definition. Let $\vdash_0$ be the first-order inference relation without substitution rule.

Definition 12. (E-subsumes)
Let θ and σ be rigid E-unifiers of terms s and t such that $D(\theta), D(\sigma) \subseteq Var(E, s \not\approx t)$. Then σ *E-subsumes* θ if there exists η such that for all variables $X \in D(\theta)$

$$\mathcal{E} \cup \theta(E) \vdash_0 \sigma\eta(X) = \theta(X).$$

This property will be denoted by $\sigma \leq_E \theta$.

The rigid E-unifiability of an eq-connection $E, s \not\approx t$ may be decided by any of the decision procedures given in [8, 9, 2]. Those procedures cannot be used to enumerate complete sets of rigid E-unifiers which subsume a given rigid E-unifier of the eq-connection $E, s \not\approx t$. For this purpose a procedure given in [3] may be used. But this procedure is not a decision procedure. That procedure constructs in a first phase rigid E-unifiers which E-subsume a given rigid E-unifier. This first step is completion based. In a second step are constructed rigid E-unifiers which subsume the given rigid E-unifier. Therefore we can prove the proposition.

Proposition 13.
The $\mathcal{E}$-unification problem in the set $\mathcal{U}_{eq}$ is solvable.

With corollary 7 and proposition 13 we may deduce from a general result which has been proved in [17] that a connection calculus with rigid E-unification is complete. The calculus will be discussed in more detail in section 5. Before we will examine more precisely the simultaneous rigid E-unification problem.

4 The simultaneous rigid E-unification problem

Let us consider an eq-mating $U = \{E_i, s_i \not\approx t_i\}_{i=1}^n$. By E denote now the indexed family $\{E_i\}_{i=1}^n$. In order to compute a simultaneous rigid E-unifier it is necessary to compute rigid E_i-unifiers for each $s_i \not\approx t_i$. Then those unifiers may be "tuned" in order to obtain a simultaneous rigid E-unifier. This approach is quite expensive. According to a claim of J. Goubault simultaneous rigid E-unification is NEXPTIME-complete. Differently the single rigid E-unification problem is NP-complete. Perhaps for the purpose of theorem proving with equality it is not necessary to be able to decide the simultaneous rigid E-unification in full generality. We define a subclass of simultaneous rigid E-unification problems which should be NP.

Definition 14. (Reduced simultaneous rigid E-unifier, well behaved simultaneous rigid E-unification problem)
Let $\succ$ be a reduction ordering which is total on ground terms.
If U is the eq-mating $\{E_i, s_i \not\approx t_i\}_{i=1}^{n}$ then a simultaneous rigid E-unifier θ of U is called *reduced* if for each $j = 1, \ldots, n$ there are no $X \in Var(E_j, s_j \not\approx t_j)$ and $i = 1, \ldots, m_j$ such that $\theta(X)$ may be reduced by the equation $l_i^j \rightsquigarrow r_i^j$ which is oriented according to the ordering $\succ$. If U has a reduced simultaneous rigid E-unifier then U is called *well behaved.*

Proposition 15.
Suppose there exists a reduced simultaneous rigid E-unifier for an eq-mating $U = \{E_i, s_i \not\approx t_i\}_{i=1}^{n}$. Then it may be computed by use of the decision procedure for the single rigid E-unification problem.

Proof. Actually this is the claim of J. Gallier et al. in [8], theorems 10.9 and 10.10. Their algorithm may work nearly independently on each of the eq-connections $E_i, s_i \not\approx t_i$. The essential condition is that there exists a simultaneous rigid E-unifier which is reduced within each of the eq-connections. This condition makes sure that in each eq-connection of the considered mating the preconditions for the algorithm are satisfied.

The following example shows that, unfortunately, not all eq-matings have a reduced simultaneous rigid E-unifier. In the following set of clauses a spanning eq-mating and its simultaneous E-unifier have been annotated. The eq-connections are indicated by the arcs.

$$\left[\begin{array}{cc} & a \approx k(a) \\ a \not\approx X, & f(X) \approx g(k(a)) \\ \multicolumn{2}{c}{h(f(k(a))) \not\approx h(g(k(a)))} \end{array}\right] \{X \mapsto k(a)\} \qquad (5)$$

The matrix is equationally unsatisfiable because the eq-connections

$$\left\{\begin{array}{c} a \approx k(a), \\ a \not\approx X \end{array}\right\} \quad \text{and} \quad \left\{\begin{array}{c} f(X) \approx g(k(a)), \\ h(f(k(a))) \not\approx h(g(k(a))) \end{array}\right\}$$

have the substitution $\theta = \{X \mapsto k(a)\}$ as a simultaneous rigid E-unifier. Moreover it is the unique simultaneous rigid E-unifier of the considered eq-mating. One may observe that θ is not irreducible in variable X within the eq-connection

$$\left\{\begin{array}{c} a \approx k(a), \\ a \not\approx X \end{array}\right\}.$$

The simultaneous rigid E-unification problem determined by the eq-mating in (5) is not well behaved. Now, one may adjoin the equation $a \approx k(a)$ to the connection

$$\left\{ \begin{array}{c} f(X) \approx g(k(a)), \\ h(f(k(a))) \not\approx h(g(k(a))) \end{array} \right\}.$$

This yields the spanning eq-mating in matrix (6).

$$\left[\begin{array}{cc} & a \approx k(a) \\ a \not\approx X, & f(X) \approx g(k(a)) \\ & h(f(k(a))) \not\approx h(g(k(a))) \end{array} \right] \{X \mapsto a\} \qquad (6)$$

In opposite to the spanning eq-mating in matrix (5) the spanning eq-mating in matrix (6) is well behaved. Indeed, $\theta' = \{X \mapsto a\}$ is an irreducible simultaneous rigid E-unifier for the eq-mating consisting of

$$\left\{ \begin{array}{c} a \approx k(a), \\ a \not\approx X \end{array} \right\} \text{ and } \left\{ \begin{array}{c} a \approx k(a), \\ f(X) \approx g(k(a)), \\ h(f(k(a))) \not\approx h(g(k(a))) \end{array} \right\}.$$

Thus for the eq-mating (5) there exists another spanning eq-mating for the considered matrix which determines a well behaved simultaneous rigid E-unification problem. This example raises the following problem.
Problem: Does for every equationally unsatisfiable matrix M exist a set of clause copies M' and an eq-mating U spanning M' such that the rigid E-unification problem determined by U is well behaved?
If this would be the case it would be possible to use less complex decision procedures without lost of completeness.

5 A total equational connection calculus

In the present section we introduce a generalisation of the pool calculus [15] towards equational reasoning. The basic data structure of that calculus is a pool of so called hooks. Each hook is a pair (p, Γ) where p denotes a (partial) path through a matrix and Γ a (sub-)clause in that matrix. The hook will be denoted by $(p \perp \Gamma)$. It represents a number of subgoals which should be solved in order to complete a proof. Namely for the continuations of the path p via the elements of the clause Γ should be found eq-connections. A particular pool calculus is given by inference rules which describe how from a given hook (representing given subgoals) are produced new hooks (representing new subgoals).
We precede the formal definitions by an illustrative example. We consider the matrix

$$\left[\left[\begin{array}{c} a \approx b \\ \\ X \not\approx a, \quad f(a) \approx g(X) \\ \\ f(Y) \not\approx g(Y) \end{array}\right]\right]$$

and a derivation which in two steps finds the spanning eq-mating consisting of two eq-connections

$$\left\{\begin{array}{c} a \approx b, \\ X \not\approx a \end{array}\right\} \quad \text{and} \quad \left\{\begin{array}{c} a \approx b, \\ f(a) \approx g(X) \\ f(Y) \not\approx g(Y) \end{array}\right\}$$

and their simultaneous rigid E-unifier $\{X \mapsto b\}$. The actual goal in each matrix is pointed by an arrow. No actual literal is indicated in the third matrix because all goals have been proved. The elements of the actual path are surrounded by dashed boxes. The elements of an eq-connection are linked by (multiple) arcs.

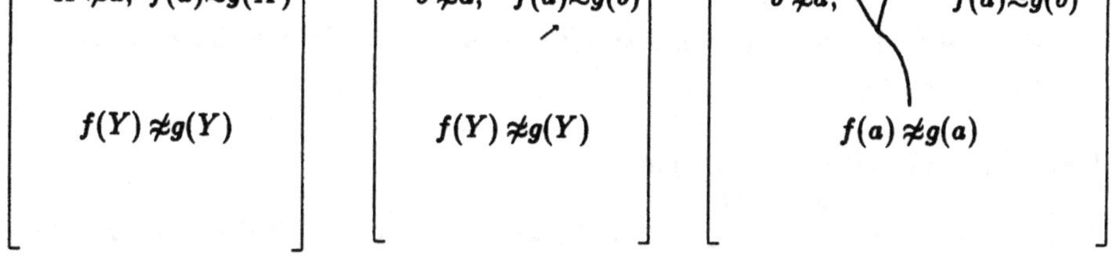

Now we show the same derivation written as a sequence of pools. A hook with empty path ($\perp$ Γ) represents an initial situation, a hook with empty clause (p $\perp$) may be considered as solved.

$$\{(\ \perp\ a\approx b),\}\vdash\left\{\begin{array}{c}(\ \perp\ ),\\ (a\approx b\ \perp\ f(a)\approx g(b))\end{array}\right\}\vdash\left\{\begin{array}{c}(\ \perp\ ),\\ (a\approx b\ \perp\ ),\\ (a\approx b, f(a)\approx g(b)\ \perp\ )\end{array}\right\} \tag{7}$$

The calculus is described by rules which determine how from a chosen hook may be constructed new hooks and which way this affects the pool. In the inference steps in derivation (7) also solved hooks have been adjoined to the resulting pool in order to illustrate the derivation. Below we prefer to avoid that kind of redundancy and adjoin only unsolved hooks to the new pool which has been obtained by an inference.

Definition 16. (eq-connection inference)
Let M be a matrix. An *eq-connection inference* is an inference rule of the form

$$\frac{(p\ \perp\ \Gamma_0, L_0)\qquad \Gamma_1\cup\{L_1\},\ \ldots\ ,\Gamma_n\cup\{L_n\}}{(p\ \perp\ \Gamma_0),\ (p, L_0\ \perp\ \Gamma_1),\ \ldots,\ (p, L_0,\ldots, L_{n-1}\ \perp\ \Gamma_n)}\ \sigma$$

where

(1) $(p\ \perp\ \Gamma_0, L_0)$ is a hook, called the *chosen hook*,
(2) if $0 < n$ then $\Gamma_1\cup\{L_1\},\ \ldots\ ,\Gamma_n\cup\{L_n\}$ are copies of clauses from M, called the *extension clauses*,
(3) σ is a idempotent substitution.
(4) $(p\ \perp\ \Gamma_0), (p, L_0\ \perp\ \Gamma_1), \ldots, (p, L_0, \ldots, L_{n-1}\ \perp\ \Gamma_n)$ are hooks, called *new hooks*, and
(5) there exists a sub-path q of p such that σ is a rigid E-unifier the eq-connection $u = q\cup\{L_0,\ldots, L_n\}$.

An eq-connection inference is called an *extension step* if $n \neq 0$ and a *reduction step* else.
An *initial hook* is a hook of the form $(\ \perp\ \Gamma)$ where Γ is a clause copy. A *solved hook* has the form $(p\ \perp\ )$.

The calculus is described by rules which determine how from a chosen hook may be constructed new hooks.

Definition 17. (Rule application)
A rule of the form $\dfrac{h\qquad \Gamma}{H}\ \sigma$ may be applied to a pool P if $h \in P$. The new pool P' is obtained from P by first removing h, then adjoining those hooks from H which are not solved and finally applying the substitution σ to the resulting pool. We will write $P \Rightarrow P'$.

Definition 18. (Derivation)
An *initial pool* for a matrix M consists of a single initial hook for that matrix. A *derivation* for a matrix M is a sequence of pools $\{P_i\}_{i=0}^{n}$ such that

(1) P_0 is an initial pool for that matrix and
(2) for every $i < n$ $P_i \Rightarrow P_{i+1}$ is an eq-connection inference for M and the variables occurring in clause copies which occur in the antecedent of that eq-connection inference must not occur in a clause copy which has been used in an earlier inference of that derivation.

A derivation is called *ground* if the unifier in every eq-connection step is empty. A derivation is *successful* if its last element is the empty pool.

Proposition 19. *(Soundness)*
Let M be a matrix and $\{P_i\}_{i<\lambda}$ be a successful eq-connection derivation for M. Then M is $\mathcal{E}$-complementary.

The completeness proof consists traditionally of the following steps: Herbrand theorem, proof of the ground completeness and lifting lemma. The Herbrand theorem (8) relies on the $\mathcal{E}$-completeness of the set of eq-connections $\mathcal{U}_{eq}$, the lifting lemma on the solvability of the rigid E-unification problem in $\mathcal{U}_{eq}$. The proof of the ground completeness relies on a number of interesting properties of minimal spanning matings.

Theorem 20. *(Completeness theorem)*
For every $\mathcal{E}$-unsatisfiable query from $\mathcal{Q}_{eq}$ exists a clause Γ and a successful derivation starting from the initial pool $\{(\ \perp\ \Gamma)\}$ such that in each inference according to definition 16 for the chosen connection u holds $u \in \mathcal{U}_{eq}$ and the chosen $\mathcal{E}$-unifier σ is an element of the complete set of $\mathcal{E}$-unifiers S_u which exists for u.

6 Towards an implementation

Together with G. Neugebauer and K. Hartmann we implemented the described calculus in a Prolog technology prover. For rigid E-unification we used a module of B. Beckert [3]. The experimental evaluation of the prover is still under work.

7 Conclusion

A complete calculus with equality based on the connection method has been presented. We formulated the problem whether it is sufficient to deal with well behaved simultaneous rigid E-unification problems.

Acknowledgements

The author wishes to thank to W. Bibel and his Intellectics Group, and to P. Enjalbert and the L.A.I.A.C. for fruitful discussions concerning the topic of this paper.

References

1. P. Andrews. Theorem Proving via General Matings. *J.ACM*, 28(2):193–214, 1981.
2. G. Becher and U. Petermann. Rigid E-Unification by Completion and Rigid Paramodulation (Extended version). Technical report, Dec. 1993. Les Cahiers du L.A.I.A.C. University of Caen. 1993.
3. B. Beckert. Ein vervollständigungsbasiertes Verfahren sur Behandlung von Gleichheit im Tableaukalkül mit freien Variablen. Master's thesis, Universität Karlsruhe (TH), 1993.
4. B. Beckert and R. Hähnle. An improved method for adding equality to free variable semantic tableaux. In D. Kapur, editor, *Proceedings, 11th International Conference on Automated Deduction (CADE), Saratoga Springs, NY*, LNCS 607, pages 507–521. Springer, 1992.
5. W. Bibel. *Automated Theorem Proving*. Vieweg, 1st edition, 1982.
6. W. Bibel, S. Brüning, U. Egly, and T. Rath. KoMeT. In W. Bibel, editor, *DFG-Kolloquium Schwerpunktprogramm "'Deduktion"'*, pages 19–20. TH Darmstadt, 1994.
7. D. Brand. Proving theorems with the modification method. *SIAM Journal of Computation*, 4:412–430, 1975.
8. J. Gallier, P. Narendran, D. Plaisted, and W. Snyder. Theorem Proving with Equational Matings and Rigid E-unification. *J. of ACM*, 1992.
9. J. Goubault. A Rule-Based Algorithm for Rigid E-Unification. In G. Gottlob, A. Leitsch, and D. Mundici, editors, *3rd Kurt Gödel Colloquium '93*, volume 713 of *Lecture Notes in Computer Science*. Springer-Verlag, 1993.
10. R. Letz, J. Schumann, S. Bayerl, and W. Bibel. SETHEO: A High-Performace Theorem Prover. *Journal of Automated Reasoning*, 8:183–212, 1992.
11. D. Loveland. *Automated Theorem Proving - A Logical Basis*. North Holland, 1978.
12. D. W. Loveland. Mechanical Theorem Proving by Model Elimination. *JACM*, 15(2), 1978.
13. M. Moser. Improving transformation systems for general e-unification. In *RTA '93: Rewriting Techniques and Applications*, number 690 in LNCS, pages 92–105. Springer, 1993.
14. G. Neugebauer and U. Petermann. CaPrI: Integrating theories into the prolog technology theorem prover ProCom. In *CADE '94, Workshop on Theory Reasoning*, 1994. to appear.
15. G. Neugebauer and T. Schaub. A pool-based connection calculus. Technical Report AIDA-91-02, FG Intellektik, Technische Hochschule Darmstadt, 1991.
16. U. Petermann. Completeness of the pool calculus with an open built in theory. NTZ-Report 24/93, Naturwissenschaftlich-Theoretisches Zentrum der Universität Leipzig, 1993.
17. U. Petermann. Completeness of the pool calculus with an open built in theory. In G. Gottlob, A. Leitsch, and D. Mundici, editors, *3rd Kurt Gödel Colloquium '93*, volume 713 of *Lecture Notes in Computer Science*. Springer-Verlag, 1993.
18. H. Reichel. *Initial Computability, Algebraic Specifications, and Partial Algebras*. Oxford University Press, Oxford, 1987.
19. W. Snyder and C. Lynch. Goal directed strategies for paramodulation. In R. Book, editor, *Rewriting Techniques and Applications*, Lecture Notes in Computer Science No. 488, pages 150 – 161, Berlin, 1991. Springer.

Equality and Constrained Resolution

Richard Scherl

Department of Computer Science
University of Toronto
Toronto, Ontario
Canada M5S 1A4
phone: 416-978-7390
email: scherl@cs.toronto.edu

Abstract. This paper looks at the addition of equality reasoning to systems that perform constrained resolution as defined within the substitutional framework. These systems reason with a constraint logic in which the constraints are interpreted relative to a constraint theory. First, a special case is considered when equality can be treated as a constraint. Then the general case is dealt with by developing and proving correct the rule of constrained paramodulation, which along with the rule of constrained resolution (and factoring) yields a refutationally complete set of inference rules for languages with equality. It is shown that if certain conditions are met, paramodulation into variables is not necessary and the functionally reflexive axioms need not be present. The modal case satisfies these conditions. If other weaker conditions are met, paramodulation into variables is necessary, but the functionally reflexive axioms are not needed. Some sorted logics satisfy these conditions. The analysis provides a means of extending restrictions on resolution and paramodulation (e.g. ordering restrictions) to constrained deduction, a relatively clean and simple mechanism for adding paramodulation to sorted logics and a proof of the conjectures of Walther and Schmidt-Schauss on the need for paramodulation into variables and the functionally reflexive axioms in the case of sorted logics.

1 Introduction

The substitutional framework [Frisch, 1991] is a general methodology for developing systems that perform hybrid deduction. These systems reason with a constraint logic in which the constraints are interpreted relative to a constraint theory. The sentences in the constraint theory, are not utilized directly by the main deductive method, but are only accessed when performing unification. The resulting architecture is hybrid since a subsystem, distinct from the main deductive process, is used to access the constraint theory. Hybrid architectures are ubiquitous in implemented A.I. systems because they provide a method for obtaining significant gains in efficiency. One important instance of the framework has been the development and analysis of theorem proving methods for sorted logics [Frisch, 1991]. Another is a framework for developing automated deduction methods for modal logics [Frisch and Scherl, 1991; Scherl, 1992].

This paper develops and proves correct the rule of constrained paramodulation, which along with the rule of constrained resolution (and factoring) yields a sound and complete set of inference rules for languages with equality. It is shown that if certain conditions on the constraint theory are met, paramodulation into variables is not necessary and the functionally reflexive axioms need not be present. If other weaker conditions are met, paramodulation into variables is necessary, but the functionally reflexive axioms are not needed. Otherwise, some functionally reflexive axioms are needed. Additionally, the special case when equality can be treated as a constraint is discussed.

Section 2 outlines the substitutional framework and Section 3 provides the needed background on equality. The special case of equality as a constraint is covered in section 4. Section 5 develops the rule of constrained paramodulation. Some examples are given in Section 6.

2 The Substitutional Framework

This section presents the method of performing deduction with constrained sentences as an instance of a general class of hybrid deduction systems, called substitutional systems. This substitutional framework [Frisch, 1989; Frisch, 1991] provides a general method for systematically transforming an automated deduction system for first-order logic into one for constraint logic[1], and for transforming the completeness proof of the original deduction system into one for the resulting deduction system. These transformations can be applied to any first-order deduction system that treats variables schematically through the use of unification.

A substitutional system for constraint logic treats quantified sentences as sentence schemas in much the same way that an ordinary deduction system does. The only difference is that a sentence with constraints does not stand for the set of all its ground instances, but only for those obtained by replacing variables with terms that respect the constraints attached to the variables. Whether or not the constraints are respected depends upon the constraint theory Σ. The ground instances of a constrained formula are therefore referred to as ***Σ-ground instances*** and defined as such:

Definition 1. Let Σ be a constraint theory, e/C be a constrained expression, and θ be a substitution such that $e\theta$ is ground. Then $e\theta$ is a ***Σ-ground instance*** of e/C if $\Sigma \models \overline{\exists}\, C\theta$.[2] The set of all Σ-ground instances of e is denoted by $e_{\Sigma gr}$.

Observe that by this definition a ground instance of a constrained expression has no variables and ***no constraints***. Also observe that some constrained expres-

[1] Note that the constraints here are conjunctions of predicates and so the work reported in this paper differs from that of [Kirchner *et al.*, 1990] where the constraints are in the form of equations and inequations.

[2] Here, and in general, an expression of the form $\overline{\exists}\phi$ denotes the existential closure of ϕ—that is, the formula $\exists x_1 \cdots x_n\, \phi$ where $x_1, \ldots, x_n$ are the freely-occurring variables of ϕ.

sions, such as $P(x)/Q(x) \wedge \neg Q(x)$, have no Σ-ground instances. We say that a constraint C is *Σ-solvable* if $\Sigma \models \bar{\exists} C$. Hence, a constrained expression, e/C, has a Σ-ground instance if, and only if, C is Σ-solvable[3].

These results enable a proof of the correctness of a constrained deductive system (e.g. constrained resolution) based on the proof of correctness of the original deductive system. If the rules of the original deductive system specified that $D\theta$ can be deduced from A and B, the rules of the constrained deductive system will specify that from A/C and B/C' the formula $(D\theta/E)$ can be deduced where $E \equiv (C \wedge C')\theta$ and E is Σ-solvable ($\Sigma \models \exists E$).

This treatment of restricted quantifiers is justified by the *Constraint Herbrand Theorem*. This theorem states that, under certain circumstances, a constraint theory Σ and a set α of constrained sentences are jointly satisfiable if, and only if, the set of all Σ-ground instances of α is satisfiable. In saying that α and Σ can be replaced by the Σ-ground instances of α, the Constraint Herbrand Theorem justifies using Σ only in instantiating formulas. Notice that Σ must be used in generating the correct instances of α, but once they are obtained Σ is irrelevant because ground instances have no constraints.

Theorem 2 Constraint Herbrand Theorem. *Let α be a set of constrained Skolem normal form sentences with only positive constraints[4] and let Σ be a constraint theory that has a least Herbrand model. Then $\alpha \cup \Sigma$ is unsatisfiable if, and only if $\alpha_{\Sigma gr}$ is unsatisfiable[5].*

As the rule of resolution operates on clauses, it is necessary to develop a clausal form for the prenex-normal form constrained sentences. Now, since the convention is to drop quantifiers when converting to clause form, a clause like $\forall w_1{:}c(w_1) \ldots \forall w_n{:}c(w_n)\ \varphi$, where φ is a disjunction of literals can be written as $\varphi/c(w_1) \wedge \ldots \wedge c(w_n)$. With this convention, after all existential quantifiers have been eliminated, universal quantifiers may be dropped, the remaining formula can be put in conjunctive normal form and separated into clause form. Each separate clause will have a (possibly empty) constraint expression which is a conjunction of literals expressing the constraints on the variables in that clause.

Here a version of the resolution rule that combines resolution with factoring will be used. The rule states that from two clauses $P \cup L$ and $N \cup L'$, where the literals in P and N are of opposite polarity, the clause $(L \cup L')\theta$ can be deduced where θ is the most general unifier of the atoms in P and N. The resulting constrained rule of resolution takes two constrained clauses $P \cup L/C_1$ and $N \cup L'/C_2$ where C_1 and C_2 are the constraints on the two clauses and yields $(L \cup L')\theta/(C_1 \wedge C_2)\theta$ provided that θ is the most general unifier of $P \cup N$ and $\Sigma \models \bar{\exists}(C_1 \wedge C_2)\theta$.

The Constraint Herbrand theorem is then used to obtain a proof of the completeness of constrained resolution. Consider the completeness theorem for

[3] This assumes that Σ has a least Herbrand model.

[4] See [Frisch, 1991] for the full definition of positive constraints. Here it suffices to view the constraints, less generally, as conjunctions of positive literals.

[5] See [Frisch, 1991] and [Scherl, 1992] for the proof of similar theorems.

constrained (ΣRES) resolution. The proof of this theorem, parallels the completeness proof for ordinary resolution.

Theorem 3 Completeness Theorem for Constrained Resolution. *Let α be a set of constrained clauses with only positive constraints and Σ be a constraint theory with a least Herbrand model. If $\Sigma \cup \alpha$ is unsatisfiable then $\alpha \vdash_{\overline{\Sigma\text{RES}}} \Box$.*

So, a completeness proof of the form used for ordinary resolution can be transformed systematically to a completeness proof for the corresponding constrained deduction system. All references to the Herbrand Theorem are replaced with references to the Constraint Herbrand Theorem. The ground completeness theorem remains unchanged. All one needs to do is prove a lifting theorem for the constrained deduction system. This too can be done in a systematic manner by transforming the Lifting theorem for ordinary resolution.

The Lifting theorem for ordinary resolution states that for any set S of clauses and any clause C, there is a resolution derivation of C from S if there is a resolution derivation of some ground instance of C from S_{gr}. The Constrained Lifting Theorem is as follows:

Theorem 4 Lifting Theorem for Constrained Resolution. *Let Σ be a constraint theory with a least Herbrand model and S be a set of constrained clauses (with only positive constraints) with Σ-solvable constraints and let C be a constrained clause with a Σ-solvable constraint. Then, $S \vdash_{\overline{\Sigma\text{RES}}} C$ if $S_{\Sigma gr} \vdash_{\overline{\Sigma\text{RES}}} C'$ for some $C' \in C_{\Sigma gr}$.*

The proof of the constrained Lifting Theorem was produced by systematically transforming the proof of the Lifting Theorem for ordinary resolution [Frisch, 1989]. The insight behind this transformation is that for any expressions E and E' which have a most general unifier θ and constraints C and C', the intersection of the Σ-ground instances of E/C and E'/C' is $(E/C \wedge C')\theta$.

Note that the soundness of the constrained resolution rule of inference follows immediately from the soundness of the ordinary rule of resolution. The constraints can first be translated out. Clearly if $D\theta$ is a resolvent of A and B, then $((C \wedge C') \rightarrow D)\theta$ is a resolvent of $C \rightarrow A$ and $C' \rightarrow B$. The resolvent can then be translated into constrained form as $(D/(C \wedge C')\theta$.

In [Frisch, 1991] the framework is used to develop a deduction method for sorted logic. Previously existing sorted logics can also be reconstructed within this framework. The sorted logic case is illustrated below with a simple example. Consider a constraint theory (Σ) consisting of $elephant(x) \rightarrow animal(x)$, $cat(x) \rightarrow animal(x)$, $elephant(jumbo)$. The main clause set contains the clause $grey(jumbo)$. It is desired to prove $\exists x : animal(x)\ grey(x)$. The negation in clause form is $\neg grey(x) \,/\, animal(x)$. This does resolve with $grey(jumbo)$ to yield the empty clause since $\Sigma \models animal(jumbo)$. In other words, $\neg grey(jumbo)$ is a Σ-ground instance of $\neg grey(x) \,/\, animal(x)$. Note that if $elephant(x) \rightarrow animal(x)$ is removed from Σ, $grey(jumbo)$ is no longer a Σ-ground instance of $\neg grey(x) \,/\, animal(x)$ since it is not the case that $\Sigma \models animal(jumbo)$ and the resolution step cannot occur.

An illustrative modal example is to show that positive-introspection is a feature of S_4 by proving the unsatisfiability of

$$\neg(Knows(P) \rightarrow Knows(Knows(P))) \tag{1}$$

The basic idea is to translate modal logic into a reified modal language, i.e., a fist-order language with terms denoting possible worlds and a predicate symbol (K) denoting the accessibility relation. Sentence 1 is translated (following [Frisch and Scherl, 1991; Scherl, 1992]) into the constrained sentence 2

$$\forall w_1{:}K(0, w_1)\ P(w_1) \wedge \neg P(sk') \tag{2}$$

and constraint theory 3.

$$\begin{array}{rrl} & K(w_1, w_1) & \text{a.} \\ K(w_1, w_2) \wedge K(w_2, w_3) \rightarrow & K(w_1, w_3) & \text{b.} \\ K(0, sk) & K(sk, sk') & \text{c.} \end{array} \tag{3}$$

Sentence 2 forms two constrained clauses $P(sk')$ and $P(w_1)/K(0, w_1)$. These clauses resolve to yield the empty clause as long as the new constraint $K(0, sk')$ is entailed by the constraint theory. This test can be understood as a proof that the world sk' is accessible from world 0. It is (as the axiom for transitivity is included) and a refutation has been obtained.

3 Equality

When dealing with languages with equality, it is necessary to restrict our attention to models in which equality is given the usual interpretation. "Identity is a binary relation in a structure or 'real world environment' which is true precisely when the two objects being related are indeed the same object [Loveland, 1978, page 263]." These models are called E-interpretations [Chang and Lee, 1973]. The definition is as follows:

Definition 5. An E-interpretation I of a set S of clauses is an interpretation of S satisfying the following four conditions. Let α, β and γ be any terms in the Herbrand universe of S and let L be a literal in I. Then

1. $(\alpha = \alpha) \in I$;
2. if $(\alpha = \beta) \in I$, then $(\beta = \alpha) \in I$;
3. if $(\alpha = \beta) \in I$ and $(\beta = \gamma) \in I$ then $(\alpha = \gamma) \in I$;
4. if $(\alpha = \beta) \in I$ and L' is the result of replacing some one occurrence of α in L by β, then $L' \in I$.

Of course, equality may be treated as an ordinary predicate. Then it is necessary to add additional axioms to specify the meaning of equality. These are the well known set of equality axioms.

Definition 6. Let S be a set of clauses. Then the *set of equality axioms for S* is the set of the following clauses.

1. $x = x$.
2. $x \neq y \vee y = x$.

3. $x \neq y \lor y \neq z \lor x = z$.
4. $x_j \neq x_0 \lor \neg P(x_1, \ldots x_j, \ldots x_n) \lor P(x_1, \ldots x_0, \ldots x_n)$ for $j = 1, \ldots n$, for every n-place predicate symbol P occurring in S.
5. $x_j \neq x_0, \lor f(x_1 \ldots x_j \ldots x_n) = f(x_1, \ldots x_0, \ldots x_n)$ for $j = 1, \ldots n$, for every n-place functions symbol f occurring in S.

Clause 1 establishes the reflexivity of equality. Clauses 2 and 3 state the symmetric and transitive properties of equality. Clause schemas 4 and 5 are the substitutivity axioms for equality.

The following theorem shows that the set of equality axioms is correct.

Theorem 7 Chang and Lee. *Let S be a set of clauses and K be the set of equality axioms for S. Then S is E-unsatisfiable if and only if $(S \cup K)$ is unsatisfiable.*

4 Equality as a Constraint: A Special Case

Consider the special case where the clause set can be divided into two groups such that the only predicate symbol that occurs in one group is the equality symbol and the equality symbol does not occur in the other group of clauses. For example, the following clause set satisfies this requirement.

$$P(f(a,b)), \;\; \neg P(f(b,a)), \;\; f(x,y) = f(y,x) \tag{4}$$

Now I define a transformation[6] on this set of clauses called $\mathcal{E}$. Wherever a non-variable term t_i occurs, it needs to be replaced with a new variable x_i and the literal $t_i \neq x_i$ is added to the clause. This process is continued until all arguments to predicate symbols in the clause set are variables, constants or function terms with only variable arguments. Furthermore the equality axioms for reflexivity, symmetry and transitivity are added to the clause set.

The set $\mathcal{E}(\ 4)$ is as follows:

$$\begin{array}{l} x \neq y \;\lor\; x \neq f(z_1, z_2) \;\lor\; a \neq z_1 \;\lor\; b \neq z_2 \;\lor\; P(y), \\ x \neq y \;\lor\; x \neq f(z_1, z_2) \;\lor\; b \neq z_1 \;\lor\; a \neq z_2 \;\lor\; \neg P(y), \\ f(x,y) = f(y,x), \\ x = x, \;\; x \neq y \lor y = x, \;\; x \neq y \lor y \neq z \lor x = z \end{array} \tag{5}$$

The following theorem shows that the $\mathcal{E}$ transformation is correct.

Theorem 8. *Given a set of clauses S that is divisible into groups S_1 and S_2 such that S_1 does not contain the equality symbol and S_2 does not contain any predicate symbol other than the equality symbol, S is E-unsatisfiable if and only if $\mathcal{E}(S)$ is unsatisfiable.*

[6] This transformation is similar to one utilized by Brand[1975].

Proof. $\Rightarrow$ Suppose S is E-unsatisfiable, but $\mathcal{E}(S)$ is satisfiable. Then there exists an interpretation I that satisfies $\mathcal{E}(S)$. Since $\mathcal{E}(S)$ contains the axioms for reflexivity, symmetry and transitivity, I satisfies the first three conditions of an E-interpretation.

Consider the sentences in $\mathcal{E}(S)$ other than the three equality axioms. By the construction of the transformation $\mathcal{E}$, for any predicate symbol P that occurs in S, if $\langle\alpha_1,\ldots\alpha_n\rangle \in I(P)$, then if $\alpha_i = \beta \in I(=)$, then $\langle\alpha_1,\ldots,\alpha_{i-1},\beta,\alpha_{i+1}\ldots\alpha_n\rangle$ must also be in $I(P)$. Therefore I (being in effect an E-interpretation with regard to the predicate symbols in S) can be extended to an E-interpretation that satisfies S. This contradicts the assumption that S is E-unsatisfiable.

$\Leftarrow$ Suppose $\mathcal{E}(S)$ is unsatisfiable, but S is E-satisfiable. Then, there exists an E-interpretation I_E such that I_E satisfies S. Clearly I_E satisfies the reflexivity, symmetry, and transitivity axioms in $\mathcal{E}(S)$. Since I_E is an E-interpretation, it must satisfy property 4 and therefore since each sentence in S is true in I_E, by the construction of the $\mathcal{E}$ transform, each of the transformed sentences must be true in I_E and therefore I_E must also satisfy $\mathcal{E}(S)$. □

Now the important thing is that the resulting set of clauses is in a form suitable for treating the equality symbol as a constraint, as long as the set S_2 has a least Herbrand model. So the result of transforming 5 into constraint logic form is a set of constrained clauses consisting of 6 and 7

$$P(y)/x = y \wedge x = f(z_1, z_2) \wedge a = z_1 \wedge b = z_2 \tag{6}$$

$$\neg P(y')/x' = y' \wedge x' = f(z_1', z_2') \wedge b = z_1' \wedge a = z_2' \tag{7}$$

along with the following constraint theory:

$$\begin{array}{l} f(x,y) = f(y,x), \\ x = x, \;\; x \neq y \vee y = x, \;\; x \neq y \vee y \neq z \vee x = z \end{array} \tag{8}$$

Clauses 6 and 7 resolve to yield the empty clause since the resulting constraint (9) is solvable.

$$\begin{array}{c} x = y \wedge x = f(z_1, z_2) \wedge a = z_1 \wedge b = z_2 \wedge x' = y \wedge x' = f(z_1', z_2') \wedge \\ b = z_1' \wedge a = z_2' \end{array} \tag{9}$$

The results here show that E-resolution [Plotkin, 1972] is a special case of the substitutional framework. Similar results arrived at by a different route have been obtained in [Frisch, 1992].

5 Constrained Paramodulation

For classical first-order logic with equality, the rules of resolution and paramodulation together form a refutationally complete set of inference rules. In this section, it is shown that the rules of constrained resolution and constrained paramodulation form a complete set of inference rules for sentences in constraint logic. The ordinary rule of paramodulation was first proposed by Robinson and Wos [1969; 1973], but although they conjectured that the functionally reflexive axioms are not needed, their proof of completeness required that the axioms be added to the clause set. Later, it was shown by Brand [1975] and Peterson [1983] that the axioms are not needed. Further refinements have been made by Hsiang and Rusinowitch[1991].

The rule of paramodulation can be transformed into a rule of constrained paramodulation as follows:

Definition 9. Given two clauses φ_1 and φ_2 with no variables in common

$$\varphi_1\colon (L[t] \cup \varphi_1')/C_1 \qquad \varphi_2\colon (r = s \cup \varphi_2')/C_2$$

we can deduce the paramodulant

$$L\sigma[s\sigma] \cup (\varphi_1')\sigma \cup (\varphi_2')\sigma/(C_1 \wedge C_2)\sigma$$

where σ is the most general unifier of t and r.

A version of the constraint Herbrand theorem needs to be developed for languages with equality.

Theorem 10 Constraint Herbrand Theorem with Equality. *Let α be a set of constrained Skolem normal form sentences, constructed in a language that contains a special equality symbol, with only positive constraints and let Σ be a constraint theory that has a least Herbrand model. Then $\alpha \cup \Sigma$ is E-unsatisfiable if, and only if $\alpha_{\Sigma gr}$ is E-unsatisfiable.*

Proof. $\alpha \cup \Sigma$ is unsatisfiable in all E-models
iff (by Theorem 7)
$\alpha \cup \Sigma \cup E$ is unsatisfiable in all uninterpreted[7] models
iff (by Constrained Herbrand Theorem)
$\alpha_{\Sigma gr} \cup E_{gr}$ is unsatisfiable in all uninterpreted models
iff (by Theorem 7 and the ordinary Herbrand Theorem)
$\alpha_{\Sigma gr}$ is unsatisfiable in all E-models

□

So given a proof of the ground completeness of resolution and paramodulation, all we need is a constrained lifting lemma for paramodulation to obtain the proof of completeness for constrained resolution and paramodulation.

Theorem 11 Ground Completeness of Resolution and Paramodulation. *Let α be a set of ground clauses. If α is $E-unsatisfiable$ the combination of resolution and paramodulation will yield the empty clause.*

The proof of this theorem can be taken from [Wos and Robinson, 1973] or [Peterson, 1983]. Wos and Robinson did not clearly distinguish the ground proof from the lifting theorem, but this can easily be done.

The term that is being paramodulated into on the ground level may or may not be visible on the level of clauses with variables. If the outermost function

[7] The predicate representing equality is an ordinary predicate symbol, i.e., it is not interpreted in a special fashion.

symbol of the term is present on the level of clauses with variables, then the term is visible. Otherwise the paramodulation has been made into what is on the general level a variable position. Thus we need two lifting lemmas—one for each case

The first lemma is for the case when the term is visible at the level of clauses with variables. The proof of the following lifting lemma can be found in Peterson [1983].

Theorem 12 Lifting Lemma 1 for Paramodulation. *If φ' is a (visible) paramodulant of $\varphi_1\theta$ into node*[8] *n of $\varphi_2\theta$, then there is a paramodulant φ of φ_1 into node n of φ_2 such that φ' is an instance of φ.*

The second lemma is for the case when the paramodulation has been made into what is on the level of clauses with variables a variable position. The proof can be found in [Wos and Robinson, 1973].

Theorem 13 Lifting Lemma 2 for Paramodulation. *Given a set of (functionally reflexive) clauses, assume that paramodulation on the ground level is performed into position n of a literal B', a ground instance of a literal B that occurs in some clause in S, where a variable x occurs at position i of B where $i < n$. Let $B\theta = B'$. Assume that θ maps x to $f_1(\ldots f_2(\ldots(\ldots f_p(\ldots u' \ldots))))$ where $p = n - i$. If the set S is closed under paramodulation, there will be a clause in S in which there is a literal B^* such that there is a substitution γ that maps x to $f_1(\ldots f_2(\ldots(\ldots f_p(\ldots u \ldots))))$ and $B\gamma = B^*$ and there is some substitution δ such that $B^*\delta = B'$.*

Since the set of clauses S is functionally reflexive let $G_1, G_2, \ldots G_p$ be the functionally reflexive axioms corresponding to $f_1, f_2, \ldots f_p$. If the set S is closed under paramodulation, then if there is a literal B with a variable x in position n, there will be a clause in S in which there is a literal B with the term $f_1(\ldots f_2(\ldots(\ldots f_p(\ldots y \ldots))))$ in position n. This is obtainable by p applications of paramodulation with each of the functionally reflexive axioms used in turn.

Each of the lifting lemmas is needed to prove completeness on the general level. Of course, Peterson's proof does not need the second lifting lemma as paramodulation does not operate into variables and the functionally reflexive axioms are not needed.

The first lifting lemma can be transformed into a lifting lemma for constrained paramodulation as follows:

Theorem 14 Lifting Lemma 1 for Constrained Paramodulation. . *Let Σ be a constraint theory with a least Herbrand model and let φ_1/C_1, φ_2/C_2 and φ/C be constrained clauses with Σ – solvable (positive) constraints. If φ' is a paramodulant of $\varphi_1\theta$ into node n of $\varphi_2\theta$ (where θ is a ground substitution, and $\varphi_1\theta$ contains a literal of the form $t = s$), then there is a paramodulant φ of φ_1 into node n of φ_2 such that φ' is an instance of φ and $C = (C_1 \wedge C_2)\delta$ and $\Sigma \models \exists(C_1 \wedge C_2)\delta$ where δ is the most general unifier of s and node n of φ_2.*

[8] Terms are viewed as labelled trees. A node is a subtree of the term beginning at a particular position in the tree.

Proof. Let φ_1/C_1 be equal to $(s = t) \vee \varphi_1^*/C_1$; then $(\varphi_1/C_1)\theta$ is $(s\theta = t\theta) \vee \varphi_1^*\theta$ and $\Sigma \models C_1\theta$. Then

$$\varphi' = (\varphi_2\theta[n\theta \leftarrow t\theta] \vee \varphi_1^*\theta)$$

Since $s\theta$ and node n of $\varphi_2\theta$ are unifiable, it follows that node n of φ_2/C_2 and s are unifiable. Let σ be their mgu. Then there is a substitution ψ such that $\theta = \sigma\psi$ and there is a paramodulant φ/C of φ_1/C_1 into node n of φ_2 given by

$$\varphi = (\varphi_2[n \leftarrow t] \vee \varphi_1^*)\sigma/(C_1 \wedge C_2)\sigma$$

Additionally, $\Sigma \models (C_1 \wedge C_2)\sigma$. Furthermore,

$$\begin{aligned} \varphi\psi &= (\varphi_2[n \leftarrow t] \vee \varphi_1^*)\sigma\psi \\ &= (\varphi_2[n \leftarrow t] \vee \varphi_1^*)\theta \\ &= (\varphi_2\theta[n\theta \leftarrow t\theta] \vee \varphi_1^*\theta) \\ &= \varphi' \end{aligned} \tag{10}$$

It must be the case that $\Sigma \models (C_1 \wedge C_2)\sigma\psi$ since $\theta = \sigma\psi$ and both $\varphi_1\theta$ and $\varphi_2\theta$ are Σ-ground instances of φ_1/C_1 and φ_2/C_2. □

The second lifting lemma can easily be transformed in a similar fashion. The proof for both of these lemmas can easily be produced by transforming the proof of the non-constrained version of the lifting lemma. Note that the variables that occur in the functionally reflexive axioms are not constrained.

The two lemmas combined, along with the lifting lemma for constrained resolution, yield a lifting theorem for resolution and paramodulation. This theorem along with the ground completeness of resolution and paramodulation are used to construct the completeness theorem for resolution and paramodulation. This discussion has glossed over the circumstances under which paramodulation into variables is needed and when the functionally reflexive axioms are needed. This is the topic that will now be covered.

The major difficulty with the paramodulation lifting lemma that does not occur with the resolution lifting lemma occurs because paramodulation operates within terms. The problem with the lifting lemma here is that when paramodulating from $C_1\phi$ into $C_2\theta$, it is necessary to ensure that the replaced part of $C_2\theta$ consists of a subtree of $C_2\theta$ that begins in C_2. In other words, the lifting lemma for paramodulation only works if the paramodulation that occurs on the ground level was done to a term that can be "seen" at the general level.

The problem is solved by Robinson and Wos by allowing paramodulation into variables and requiring the presence of the functionally reflexive axioms. Any term can then be constructed at a variable position through repeated applications of paramodulation with one of the functionally reflexive axioms. Although this solves the problem of proving completeness, the size of the search space is greatly enlarged.

The solution used by Peterson is to show that one can use only completely reduced substitutions in considering ground clauses. This means that only a subset of the ground instances need to be considered. Those are the ground instances produced by reduced substitutions. The notion of reduced substitution is defined in detail in [Peterson, 1983]. The important characteristic of the reduced substitutions is that paramodulation cannot apply to any of the terms in the substitution. Thus it is not necessary to paramodulate into variables and the functionally reflexive axioms are not needed.

But this solution does not hold for constraint logics. The problem has been noted in sorted logics where the terms that are arguments to an equality literal are not of identical sorts and are not subsorts. For example, consider a constraint theory containing $D(a), C(b)$ and a main set of clauses as follows: $P(x)/D(x)$, $\neg P(y)/C(y)$, and $a = b$. These clauses are clearly unsatisfiable. It is clear that without paramodulation into variables, it is not possible to derive the empty clause.

But note that if there are no equalities between constrained terms[9], then the issue does not arise.

Fact 5.1 (Need For Paramodulation into Variables) *If constrained terms do not occur as arguments to any equality literal, paramodulation into variables is not needed.*

Clearly, the reified modal logic satisfies this condition. The translation procedure introduces the constrained terms and does not introduce any equality literals. In other words, all equality literals in the reified modal logic are the result of equality literals in the initial modal logic, and therefore the arguments of the equality literals range over elements of the domain of the original modal logic and not over worlds introduced by the translation procedure. Thus, in the case of reified modal logic, paramodulation does not need to operate into variables and the functionally reflexive axioms are not needed.

Now, consider again the sorted logic example given above. The Σ-ground instances of the clause set $P(x)/D(x)$, $\neg P(y)/C(y)$, and $a = b$ are $P(a)$, $\neg P(b)$, $a = b$. On the ground level, a refutation is obtainable. But without paramodulating into variables, a refutation is not obtainable on the general level. Because $P(a)$ and $P(b)$ are now hidden in the constraint theory, they are no longer available as positions for paramodulation.

Schmidt-Schauss[1989] conjectures that the functionally reflexive axioms are not needed if the signature is simple. Sorted logics with simple signatures are those without polymorphism. Walther's[1987] sorted logic is an example. Walther demonstrates that paramodulation into variables is needed for his sorted logic and also conjectures that the functionally reflexive axioms are not needed. The definition of a simple signature when translated into the constraint logic formulation is as follows:

---

[9] A constrained term is a term that occurs both in the body of a constrained clause and in the constraint, or in the body of a constrained clause and in the constraint theory. If the term is a variable, its ground instances are then constrained by the constraint theory.

Definition 15. A simple constraint theory consists of only definite clauses. Furthermore, the heads of clauses do not contain function symbols with variable arguments that occur in the literals in the body of the clause. Additionally, in a simple constraint theory every function symbol can only occur once.

So, if a constraint theory has a clause like $D(x) \rightarrow G(f(x))$, it is not simple. But clauses like $D(a)$, $G(f(x))$, and $D(x) \wedge E(x) \rightarrow G(x)$ are acceptable.

Property 5.1 *If the constraint theory is simple, then constrained variables can only be instantiated by constants or function symbols with unconstrained arguments.*

Theorem 16 Need For Functionally Reflexive Axioms. *The functionally reflexive axioms are not needed if the constraint theory Σ is simple.*

In this case, paramodulation into variables is needed because the Σ-ground terms needed for paramodulation to operate cannot be seen from the variable level. But since constrained terms only occur in the constraint theory as arguments to predicate symbols—not as arguments to function symbols, the operation of paramodulation into variables using the equality literals that exist in the clause set suffices. Peterson's results on reduced substitutions apply to the non-constrained variables that occur as arguments to both predicates and function symbols.

Consider a more general case as follows: The clause set is $\neg P(x)/D(x)$, $P(y)/C(y)$, and $a = b$ and the constraint theory is $A(a)$, $B(b)$, $A(x) \rightarrow D(f(x))$, $B(x) \rightarrow C(f(x))$. Note that once again, resolution and paramodulation cannot be used to deduce the empty clause even if paramodulation is allowed to operate into variables. The Σ-ground instances of the clause set are $\neg P(f(a))$, $P(f(b))$, $a = b$. On the ground level, the rules of resolution and paramodulation can be used to obtain a refutation. The problem at the variable level is that the outer term needs to be constructed before paramodulation can operate. Thus, in the general case, when the constraint theory is not simple, at least some functionally reflexive axioms will be needed.

6 Examples

Before concluding, the rule of constrained paramodulation is illustrated with two examples. A sorted logic example comes from Walther[1987].

> (i) Reptiles and birds are animals, which breathe by lungs and lay eggs. (ii) There are reptiles as well as birds. (iii) There is a bird which hatches out the eggs it lays. (iv) There is a reptile or a bird which does not hatch out its eggs. (v) There are at least two animals which breathe by lungs.

The literal $H(x)$ is used to represent the statement "animal x hatches out its eggs." The predicates R, B, L and E represent reptiles, birds, animals with lungs, and animals that lay eggs.

The set of constrained clauses are as follows:

$$\{H(b')\} \tag{11}$$

$$\{\neg H(b), \neg H(r)\} \tag{12}$$

$$\{x_1 = x_2\}/L(x_1) \wedge L(x_2) \tag{13}$$

Note that 13 results from the negation of (iv) $\exists x{:}L(x)\ \exists y{:}L(y)\ \neg x = y$. The constraint theory contains the following clauses:

$$\begin{array}{l} R(x) \rightarrow L(x),\ R(x) \rightarrow E(x),\ B(x) \rightarrow L(x),\ B(x) \rightarrow E(x), \\ B(b),\ B(b'),\ R(r) \end{array} \tag{14}$$

Paramodulating 13 into 11 yields 15

$$\{H(x_2)\}/L(b') \wedge L(x_2) \tag{15}$$

Clause 15 resolves with 12 to yield a resolvent that resolves with 15 to yield 16.

$$\Box/L(b') \wedge L(r) \wedge L(b) \tag{16}$$

As the constraint is solvable, the hypothesis is proven.

This treatment of sorted logic with equality is particularly clean and simple in comparison with the very complex Walther [1987]. The difference results from the fact that Walther attaches sorts to argument positions of predicate symbols, while here it is the terms that are sorted (constrained).

For a modal example, consider the following based on an example in [Moore, 1980]:

$$\Box num(bill) = num(mary) \tag{17}$$

$$\exists x \Box x = num(bill) \tag{18}$$

The hypothesis to be proven is 19

$$\exists x \Box x = num(mary) \tag{19}$$

The translation into the reified modal logic (along with negating the hypothesis results in the following clauses:

$$num(bill, w_1) = num(mary, w_1)/K(0, w_1) \tag{20}$$

$$f = num(bill, w_1)/K(0, w_1) \tag{21}$$

$$\neg(x = num(mary, sk(x)))/\emptyset \tag{22}$$

The constraint theory for this problem contains the clause $K(0, sk(x))$ along with the accessibility axioms for the modal logic being used.

A proof that 20 — 22 are unsatisfiable is obtained by first paramodulating clause 20 into clause 21 to obtain 23.

$$f = num(mary, w_1)/K(0, w_1) \tag{23}$$

Next, the empty clause 24 is obtained by resolving clauses 23 and 22.

$$\Box/K(0, sk(x)) \tag{24}$$

Note that $\Sigma \models K(0, sk(x))$ since Σ contains $K(0, sk(x))$.

A number of closely related methods for automated modal deduction deduction [Ohlbach, 1988; Auffray and Enjalbert, 1989; Frisch and Scherl, 1991] have recently been proposed. The analysis given here provides all of these methods with a rule of constrained paramodulation. Such a rule was suggested in [Ohlbach, 1988], but here a proof is given of the refutational completeness of the rule along with the rule of constrained resolution.

7 Summary and Future Research

This paper has developed and proven correct the rule of constrained paramodulation, which along with the rule of constrained resolution (and factoring) yields a refutationally complete set of inference rules for languages with equality. For the modal logic instances of the framework, the result is a proof that paramodulation can be restricted to not operate into variables and the functionally reflexive axioms are not needed. The results provide a delimitation of the circumstances under which sorted logics do not require paramodulation into variables and the functionally reflexive axioms. It can be seen as a proof of the conjectures on this topic of Walther[1987] and Schmidt-Schauss[1989]. Additionally, as the methods for modal deduction satisfy all of these conditions, the result is an automated deduction method for modal logics with equality in which the functionally reflexive axioms are not needed and paramodulation may be restricted so as to not operate into variables.

The results presented here should easily extend the use of ordering restrictions on resolution and paramodulation (as discussed in [Hsiang and Rusinowitch, 1991]) for both the sorted and modal deduction methods. Additionally, it seems likely that the results can be extended to non-substitutional methods as well when these are viewed as constraint deduction with a constraint theory that is not required to have a least Herbrand model (by drawing upon [Bürckert, 1990]).

Acknowledgments

Helpful discussions on the topic of this paper have been held with Alan Frisch and Thomas Uribe. I would also like to thank various anonymous reviewers for their suggestions.

References

[Auffray and Enjalbert, 1989] Auffray, Yves and Enjalbert, Patrice 1989. Modal theorem proving: An equational viewpoint. In *Proceedings of the Eleventh International Joint Conference on Artificial Intelligence*, Detroit. 441–445.

[Brand, 1975] Brand, D. 1975. Proving theorems with the modification method. *Siam Journal of Computation* 4(4):412–430.

[Bürckert, 1990] Bürckert, Hans Jürgen 1990. A resolution principle for clauses with constraints. In Stickel, M. E., editor 1990, *10th International Conference on Automated Deduction*, volume 449 of *Lecture Notes in Artificial Intelligence*. Springer-Verlag, Kaiserslautern, FRG. 178–192.

[Chang and Lee, 1973] Chang, Chin-Liang and Lee, Richard Char-Tung 1973. *Symbolic Logic and Mechanical Theorem Proving*. Academic Press.

[Frisch and Scherl, 1991] Frisch, Alan and Scherl, Richard 1991. A general framework for modal deduction. In Allen, J.A.; Fikes, R.; and Sandewall, E., editors 1991, *Principles of Knowledge Representation and Reasoning: Proceedings of the Second International Conference*, San Mateo,CA : Morgan Kaufmann. 196–207.

[Frisch, 1989] Frisch, Alan M. 1989. A general framework for sorted deduction: Fundamental results on hybrid reasoning. In Brachman, Ronald J.; Levesque, Hector J.; and Reiter, Raymond, editors 1989, *Proceedings of the First International Conference on Principles of Knowledge Representation and Reasoning,* Toronto. 126–136.

[Frisch, 1991] Frisch, Alan M. 1991. The substitutional framework for sorted deduction: Fundamental results on hybrid reasoning. *Artificial Intelligence* 49(1–3):161–198.

[Frisch, 1992] Frisch, Alan M. 1992. Deduction with constraints: The substitutional framework for hybrid reasoning.

[Hsiang and Rusinowitch, 1991] Hsiang, Jieh and Rusinowitch, Michael 1991. Proving refutational completeness of theorem-proving strategies: The transfinite semantic tree method. *Journal of the Association for Computing Machinery* 38(3):559–587.

[Kirchner *et al.*, 1990] Kirchner, Claude; Kirchner, Hélène; and Rusinowitch, M. 1990. Deduction with symbolic constraints. *Revue d'Intelligence Artificielle* 4(3):9–52. Special issue on Automatic Deduction.

[Loveland, 1978] Loveland, Donald W. 1978. *Automated Theorem Proving: A Logical Basis,* volume 6 of *Fundamental Studies in Computer Science.* North-Holland Publishing Company, Amsterdam.

[Moore, 1980] Moore, R.C. 1980. Reasoning about knowledge and action. Technical Note 191, SRI International.

[Ohlbach, 1988] Ohlbach, Hans Jürgen 1988. A resolution calculus for modal logic. In Lusk, E. and Overbeek, R., editors 1988, *9th International Conference on Automated Deduction,* volume 310 of *Lecture Notes in Computer Science.* Springer-Verlag, Berlin. 500–516.

[Peterson, 1983] Peterson, Gerald E. 1983. A technique for establishing completeness results in theorem proving with equality. *SIAM Journal of Computing* 12(1):82–100.

[Plotkin, 1972] Plotkin, Gordon 1972. Building in equational theories. *Machine Intelligence* 7:73–90.

[Robinson and Wos, 1969] Robinson, G. and Wos, L. 1969. Paramodulation and theorem proving in first-order theories with equality. In Meltzer, B. and Michie, D., editors 1969, *Machine Intelligence 4.* Edinburgh University Press, Edinburgh. 135–150. Reprinted in Automation of Reasoning 2 edited by Siekmann and Wrightson.

[Scherl, 1992] Scherl, Richard 1992. *A Constraint Logic Approach To Automated Modal Deduction.* Ph.D. Dissertation, University of Illinois.

[Schmidt-Schauss, 1989] Schmidt-Schauss, M. 1989. *Computational Aspects of an Order-Sorted Logic with Term Declarations,* volume 395 of *Lecture Notes in Artificial Intelligence.* Springer-Verlag, Berlin.

[Walther, 1987] Walther, Christoph 1987. *A Many-Sorted Calculus Based on Resolution and Paramodulation.* Morgan Kaufman, Los Altos, CA.

[Wos and Robinson, 1973] Wos, L.T. and Robinson, G.A. 1973. Maximal models and refutation completeness: Semidecision procedures in automatic theorem proving. In Boone, W.; Cannonito, F.; and Lyndon, R., editors 1973, *Word Problems: Decision Problems and the Burnside Problem in Group Theory.* North Holland, New York, N.Y. 609–639. Reprinted in Automation of Reasoning 2, edited by Siekmann and Wrightson.

Efficient Strategies for Automated Reasoning in Modal Logics

Stéphane Demri*

LIFIA-IMAG
46, Avenue Félix Viallet, 38031 Grenoble Cedex, France

Abstract. In Automated Deduction for non classical logics and specially for modal logics, efficient (and not only complete) strategies are needed. Ordering strategies are presented for the Fitting's tableaux calculi. Besides orderings of tableaux rules, different variants for backtracking are used. The strategies apply to most usual propositional modal logics: K, T, K4, S4, D, D4, C, CT, C4, CS4, CD, CD4, G. More precisely, they apply to logics for which there exists a tableaux calculus such that the number of sets of formulas introduced in a tree is finite -the analytic tableaux systems satisfy this requirement. The strategies are proved to be complete for most of these logics. The results are presented for S4. The strategies have been implemented and extensively experimented in the tableaux theorem prover running on our Inference Laboratory **ATINF**. Experiments have shown the efficiency of some of the proposed strategies.

1 Introduction

In the fields of Artificial Intelligence and Computer Science, modal logics represent powerful tools for formalizing numerous problems. For instance, as it is well-known, there exist logics for knowledge and belief based on particular modal logics (e.g., [Hin62, HM85, Kon86b, Orl91]). To mechanize modal logics, two main approaches can be broadly distinguished: the *direct* approach and the *translation* approach. The *direct* approach consists in defining specific proof systems for these logics. For instance, tableaux, resolution, and connection methods have modal versions (e.g., [Fit83, FdC83, Wal90]). The *translation* approach (e.g., [Orl79, Ohl93, Her89, FB90, CM93]) roughly consists in defining translations between modal logics and classical first-order logic. In automated deduction for logics of knowledge, as Konolige pointed out in [Kon86a], two major issues are practical proof methods and theoretical computational issues. In this paper we deal with efficient proof methods for mechanizing modal logics with tableaux calculi.

It is worth mentioning that to define specific proof systems for modal logics does not always work. For example, normal forms required by resolution do

* This work has been supported by a grant of the French Ministry of Research and Space.

not always exist for first-order modal logics. In the case of modal tableaux, a backtracking mechanism is usually added, which is mainly related to π-rule (i.e., possibility rule) inferences. As such, this drawback, as well as others, affects the efficiency of theorem provers. In this paper we give experimental evidence that "good" strategies can be proposed in order to overcome criticisms usually addressed to tableaux method.

Three families of ordering strategies for tableaux are defined. As backtracking and periodicity tests (required for the sake of completeness and decidability) are expensive, it is important to reduce their application. The strategies are partly based on the orderings of tableaux rules (or formulas), and on the use of a restricted form of backtracking. Of course the idea of introducing strategies for modal calculi is not new. In [Fit88] *ideas* about rules ordering are *informally* discussed and the possibility of including strategies is clearly mentioned but the prover described in [Fit88] does not use the strategies as a parameter of the theorem prover unlike the present work. For modal resolution, strategies have been proved to be complete in [AEH90].

The defined strategies are complete[2] for the modal propositional logics K, T, K4, S4, D, D4, C, CT, C4, CS4, CD, CD4, G (terminology used in [Fit83]) but in the presentation of our work, the attention is focused on propositional S4 [HC68]. Logic S4 is widely considered as an adequate logic for reasoning about knowledge when the agents are ideal reasoners. Our work applies to the multi-agent case though in the present paper a single agent is considered. Tableaux methods have also been implemented in [Cat91] for different classes of non classical logics but all the logics Catach considers are *propositional* and *no strategy* is mentioned. Nevertheless the theorem prover described in [Cat91] seems to be very efficient. From a practical point of view, very few *running* implementations exist for the modal tableaux (e.g., [Fit88, TMM86, DG88]). We have implemented and experimented the proposed strategies with the tableau prover for modal logics within the Inference Laboratory **ATINF** (see e.g., [CH93]). Our implementation is mainly based on Fitting's work [Fit83]. Experiments in Section 4 show the efficiency of some strategies. The inefficiency of some other strategies is also detected with the experiments.

There are other ways to consider strategies for modal logics. For example, in the translation approach, strategies for classical logic are used and therefore there is no need to define strategies for modal proof systems. The approach by backward translation defined in [CD93, CDH93] can be seen as a way of keeping trace of the strategy used in the *target* logic and thus this is a way of translating strategies from classical into modal logics.

Some of the ideas developed in this paper have been presented in [Dem93].

---

[2] For each logic, the adequate tableaux rules defined in [Fit83] are considered. The strategies for the logics D, D4, CD and CD4 provide to the ν-rule a treatment similar to the one for the π-rule (see in [Dem] the definition of the strategies for the logics previously mentioned).

2 Definitions

We present a set of tableau rules that can be adapted to different modal logics. The tableau method for classical logic is fully explained in [Smu68] and the modal counterpart of the classical case can be found in [Fit83].

2.1 Tableaux Rules

We use in the sequel the standard language for modal propositional logics, i.e., the classical propositional language with the addition of the unary operators $\Box$ and $\Diamond$. The letters P, Q,... denote propositional variables unless otherwise stated. A possibility (respectively necessity) formula, noted π (respectively ν), is a modal formula of the form $\Diamond F$ or $\neg\Box G$ (respectively $\Box F$ or $\neg\Diamond G$). For a π formula $\Diamond F$ (respectively $\neg\Box F$), we note π_0 the formula F (respectively $\neg F$). Similarly, for a ν formula $\Box F$ (respectively $\neg\Diamond F$), we note ν_0 the formula F (respectively $\neg F$). We use the different types of formulas defined in [Fit83]. Throughout the paper, in a standard way, if R denotes a binary relation on the set U then we note R^* the reflexive and transitive closure of R. Furthermore, if s denotes a sequence of objects $(o_1, \ldots, o_n)$, we note $l(s) = n$ the length of s and for $i \in [1 \ldots l(s)]$ $s[i]$ the ith object of s.

Definition 1. A *branch* is a set of modal formulas. A *tree* is a sequence of branches. A branch is said to be *closed* iff it contains two contradictory formulas. A tree is said to be *closed* iff every branch is closed.

For example $(\{P, \Box Q, P, \neg R\}, \{(R \vee S), \Diamond Q\})$ is a tree. In the following definitions the operator $\sharp$ maps a branch to another branch. It is a generic operator: the nature of the operator depends on the logic and especially on the properties of the accessibility relation (see Kripke semantics in [Kri63]). The following are all the rules for the transformation of trees. The rules Neg-rule, α-rule and β-rule are the same as those in classical logic. The different types of rules for modal tableaux can be found in [Fit83]. The present definition is only another formulation.

Definition 2. The tableaux rules for modal logics are (derivation of trees):

Neg-rule: $(B_1, \ldots, B_{i-1}, B \cup \{\neg\neg F\}, B_{i+1}, \ldots, B_N) \rightarrow (B_1, \ldots, B_{i-1}, B \cup \{F\}, B_{i+1}, \ldots, B_N)$

α-rule: $(B_1, \ldots, B_{i-1}, B \cup \{\alpha\}, B_{i+1}, \ldots, B_N) \rightarrow (B_1, \ldots, B_{i-1}, B \cup \{\alpha_1, \alpha_2\}, B_{i+1}, \ldots, B_N)$

β-rule: $(B_1, \ldots, B_{i-1}, B \cup \{\beta\}, B_{i+1}, \ldots, B_N) \rightarrow (B_1, \ldots, B_{i-1}, B \cup \{\beta_1\}, B \cup \{\beta_2\}, B_{i+1}, \ldots, B_N)$

ν-rule: $(B_1, \ldots, B_{i-1}, B \cup \{\nu\}, B_{i+1}, \ldots, B_N) \rightarrow (B_1, \ldots, B_{i-1}, B \cup \{\nu, \nu_0\}, B_{i+1}, \ldots, B_N)$

π-rule: $(B_1, \ldots, B_{i-1}, B \cup \{\pi\}, B_{i+1}, \ldots, B_N) \rightarrow (B_1, \ldots, B_{i-1}, \sharp B \cup \{\pi_0\}, B_{i+1}, \ldots, B_N)$

We note $\sharp_{S4} B = \{\nu \mid \nu \in B\}$ (B denotes a branch) and $R_{S4} = \{$Neg-rule, α-rule, β-rule, π-rule, ν-rule$\}$ the set of tableaux rules for propositional modal logic S4. The usual formulation of the π-rule, as it is described in [Fit83] is the following $\frac{S, \pi}{\sharp S, \pi_0}$. Similarly the usual formulation of the ν-rule, as it is described in [Fit83] is the following $\frac{\nu}{\nu_0}$. Other rules should be presented for the previously mentioned logics. Since the attention of the paper is focused on S4, we omit them.

From Definition 2, branch derivations can be straightforwardly defined.

Definition 3. $B \rightarrow_{r,f} B'$ denotes that rule r has been applied to formula f in branch B. A branch is said to be *complete* if no rule can be applied to it.

A tree derivation is characterized by occurrences of formulas, rules and branches. We shall therefore note $t \rightarrow_{r,b,f} t'$, the transformation of t in t' by applying the rule r on the bth branch of t with the formula f. The notion of path introduced in the following definition corresponds to the construction of a tableau.

Definition 4. A sequence of trees $(t_1, ..., t_n)$ is said to be a *path* iff for $1 \leq i \leq (n-1)$, there exists r_i, k_i and f_i such that $t_i \rightarrow_{r_i,k_i,f_i} t_{i+1}$. Each path $(t_1, \ldots, t_j)$ with $j \in [1 \ldots n]$ is said to be a *subpath* of the path $(t_1, ..., t_n)$ -also noted $(t_1, \ldots, t_j) \subseteq (t_1, \ldots, t_n)$. A path is said to be *closed* iff its last tree is closed.

We now state a known result about the soundness and the completeness of the S4 tableau system.

Fact 5. *[Fit83] A modal formula f is S4-valid iff there exists a closed tree derived from $(\{\neg f\})$ by using the set of rules R_{S4} (see Definition 2).*

In the sequel we shall only deal with special forms of paths, namely left-right depth-first paths.

Definition 6. Let p be a path such that for $1 \leq i \leq (l(p)-1)$, there exists r_i, k_i and f_i such that $p[i] \rightarrow_{r_i,k_i,f_i} p[i+1]$. The path p is said to be a *LRDF-path* (left-right depth-first path) iff for $1 \leq j \leq (l(p)-1), 1 \leq i \leq (k_j - 1), p[j][i]$ is a closed branch.

The notion of *current branch* naturally arises from the previous one.

Definition 7. Let t be a tree. The index of the *current branch* of t, noted $ib(t)$, is the following:$Max(\{i \mid \text{for } 1 \leq j \leq (i-1), t[j] \text{ is closed}\} \cup \{0\})$. The *current branch* of t is $t[ib(t)]$ if $ib(t) \in [1 \ldots l(t)]$, otherwise t has no current branch. Similarly, the current branch of the path p, if it exists, is the current branch of the tree $p[l(p)]$.

In the sequel, for any entity e for which the notion of current branch has a sense, we note $b(e)$ the current branch of e if it exists. We note $t \Rightarrow_{r,f} t'$ for $t \rightarrow_{r,ib(t),f} t'$. In a LRDF-path, the branches of the tree have to be closed from left to right with a depth-first strategy on the branches. Completeness of the rules is preserved with such a strategy.

Proposition 8. *If $(\{\neg f\}) \rightarrow^*_{R_{S4}} t$ with t a closed tree then there exists a closed tree t' such that a $(\{\neg f\}) \Rightarrow^*_{R_{S4}} t'$.*

In the sequel we assume that all the paths are LRDF-paths. Before going any further, we specify what is formally the detection of cycles.

Definition 9. Let p be a LRDF-path. The path p *contains a cycle* iff the following conditions hold:

1. The path p has a current branch
2. There exist $k, k' \in [1 \ldots l(p)]$ with $k' > k$ such that the following conditions hold
   (a) $b(p[k]) = b(p[k'])$ (equality between two current branches)
   (b) There is a set of indices $\{u_0, \ldots, u_s\} \subseteq [k \ldots k']$ such that
      i. $k = u_0$, $k' = u_s$ and for $i < j$, $u_i < u_j$
      ii. for $i \in [0 \ldots (s-1)]$, if $p[u_i] \Rightarrow_{r_i, g_i} p[u_{i+1}]$ then $b(p[u_i]) \rightarrow_{r_i, g_i} b(p[u_{i+1}])$ (the branch $b(p[k'])$ is derived from $b(p[k])$)
      iii. There is some $j \in [0 \ldots (s-1)]$ such that $r_j = \pi - rule$
   (c) The set $\{p[k'][\mathcal{K}] \mid \mathcal{K} > ib(p[k'])\}$ contains the set $\{p[k][\mathcal{K}] \mid \mathcal{K} > ib(p[k])\}$. These two sets contain the branches to be explored in the respective trees.

By the term periodicity test, we mean a procedure detecting a cycle in a path.

2.2 Strategy

As usual a "strategy" is informally a set of rules to guide the construction of proofs and to reorder the development of deductions -see for example [Lov78]. In the case of tableaux proof systems, a deduction can be considered as a sequence of tree derivations, namely a path. A strategy guides the choices in the tree derivations. The choices for the tree derivations engender choices for the path derivations. Indeed, every path derivation rests on tree derivations. So, to specify what are the paths that can be derived from a given path necessarily takes into account the used strategy. Sequences of paths are therefore adequate objects to describe strategies. In the sequel, sequences of paths shall be defined as *metapaths* and we define a strategy as a binary relation on metapaths. A strategy can therefore be identified with metapath derivations. The specification of strategies in terms of metapaths (instead of paths) is justified by the following. A metapath mp contains all the paths that have been generated before generating its last element. That is why, the derivation of a new metapath mp' from mp (with a given strategy) takes into account the history of the paths already generated. *If a strategy were defined as a binary relation on paths then the derivation of a new path will not take into account the set of paths already generated.* The necessity of the knowledge of such a set is justified by the backtrackings introduced by the π-rule inferences.

Definition 10. A *metapath* mp is a finite sequence of paths. The *current tree* of mp is the tree $mp[l(mp)][l(mp[l(mp)])]$. The current branch of mp is the current branch of the current tree if it exists. A metapath is said to be closed iff its current tree is closed.

Roughly speaking, a strategy is a tool to guide the derivation of paths in a privileged direction. That is why, in the following definition, a strategy is characterized by pairs of metapaths. The conditions for deriving metapaths represent (meta)rules that control the application of the tableaux rules.

Definition 11. A *strategy* is a binary relation on metapaths, noted $\rhd$, such that

1. The relation $\rhd$ holds only for pair of metapaths of the form $((p_1, \ldots, p_n), (p_1, \ldots, p_n, p_{n+1}))$
2. It is decidable whether $(mp_1, mp_2) \in \rhd$

Definition 11 can be related to the one of ***search strategies for abstract theorem proving problems*** elegantly defined in [Kow69] (see in [Dem] a detailed discussion). A metapath mp represents the different nodes explored in the search graph before reaching the current tree of mp. The **ATINF** modal prover does not explicitly use a structure for coding the metapaths. However such a notion is used in order to describe the exploration of nodes in the search graph.

A strategy is said to be *complete* iff for any valid formula, there exists a closed metapath mp such that $(((\{\neg f\}))) \rhd^* mp$. This definition is rather standard. We shall also use a stronger version for completeness. A strategy $\rhd$ is *strongly complete* iff for any valid formula f, if $(((\{\neg f\}))) \rhd^* mp$ then there is a closed metapath mp' such that $mp \rhd^* mp'$. Complete and strongly complete strategies will deserve special attention.

3 Families of Complete Ordering Strategies

In Section 2 tableaux rules are presented but no indication is given about the way to apply them. Let $p_1 = (t_1, \ldots, t_n)$ and $p_2 = (t_1, \ldots, t_n, t_{n+1})$ be two paths. We note $p_1 \hookrightarrow p_2$ iff $t_n \Rightarrow t_{n+1}$. We may use indices to specify which rule (or formula) is applied.

We propose a first example of strategy. To introduce the backtracking in the definition of a strategy is a necessary condition to ensure completeness. We start defining a strategy $\rightsquigarrow_{back}$ with backtracking for S4. Let mp be a metapath such that $mp = (p_1, \ldots, p_n)$. For $1 \leq i \leq n$, p_i is equal to $(t_i^1, \ldots, t_i^{h(i)})$. Let mp' be another metapath such that $mp' = (p_1, \ldots, p_n, p_{n+1})$. The derivation $mp \rightsquigarrow_{back} mp'$ holds iff one of the conditions below holds:

1. If $b(t_n^{h(n)})$ exists, is complete and if Condition 2. below does not hold then there exists a maximal index $I \in [1 \ldots (n-1)]$ satisfying the conditions below:
   (a) $p_I \subseteq p_n$
   (b) For $i \in [1 \ldots n], p_{n+1} \not\subseteq p_i$
   (c) Either (i) or (ii) below holds:
      i. If $t_I^{h(I)} \hookrightarrow_{\pi-rule} t_n^{h(I)+1}$ and if for all the path p such that $p_I \hookrightarrow_{\pi-rule} p$, there exists $d \in [1 \ldots n]$ such that p is a subpath of p_d then $p_I \hookrightarrow_r p_{n+1}$ with $r \neq \pi$-rule

ii. $p_I \hookrightarrow_{\pi\text{-}rule} p_{n+1}$
2. $p_n \hookrightarrow_r p_{n+1}$, $r \in R_{S4}$ (see Section 2)

The derivation of mp' rests on the derivation of p_{n+1}. The path p_{n+1} can be built from p_n by applying a tableaux rule to the last tree of p_n (see 2.). In that case, the structure of mp is not useful to compute p_{n+1}. However if the last tree of p_n cannot derive other trees, a backtracking is performed. The search for a path p_I (see 1.(a)) that is able to engender other paths is made. In this example, the choices for backtracking are *only* based on π-rule inferences. However the strategy $\rightsquigarrow_{back}$ is not strongly complete. Indeed no detection of cycles is made. For instance, the branch $\{\Box\Diamond P, \Diamond P, \Diamond(P \wedge \neg P)\}$ may generate an infinite number of branches.

3.1 Backtracking and Periodicity Test

In comparison with the classical propositional tableau method, the modal case has a higher[3] complexity [Lad77]. In the construction of a modal tableau, control features are necessary to ensure completeness unlike tableaux for classical propositional logic. For the latter case, it is well known that if there is a closed tableau for a set of formulas then the rules can be applied in any order. However it is enough to apply a rule only once on a formula of the branch. This last property does not hold for the modal counterpart (see for example [Fit88]) since cycles must be detected so that infinite branches are pruned. When the accessibility relation is transitive (see Kripke-style semantics in [Kri63]), a mechanism to detect cycles must be implemented: the size of the formulas does not strictly decrease for each inference.

A π-rule inference on a branch cancels the other possible applications of this rule later on -at least with the logics we shall work on this paper. When the application of a π-rule cannot close the tree then a backtracking is necessary. At a certain point of the construction of a branch the theorem prover must choose how to apply the π-rule (π-choices).

3.2 The Elementary Ordering Strategy

Let mp be a metapath such that $mp = (p_1, \ldots, p_n)$. For $1 \le i \le n$, p_i is equal to $(t_i^1, \ldots, t_i^{h(i)})$. Let mp' be another metapath such that $mp' = (p_1, \ldots, p_n, p_{n+1})$. The derivation mp $\rightsquigarrow$ mp' holds iff one of the three conditions below is satisfied:

1. If $b(mp)$ exists and $mp[l(mp)]$ contains a cycle (Definition 9) or $b(mp)$ is complete then there exists a maximal index $I \in [1 \ldots (n-1)]$ satisfying the conditions below (backtracking):
   (a) $p_I \subseteq p_n$
   (b) Either (i) or (ii) below holds.

[3] However, for propositional S5, the problem of deciding S5-satisfiability is NP-complete.

i. If $t_I^{h(I)} \hookrightarrow_{\pi\text{-rule}} t_n^{h(I)+1}$ and if for all the path p such that $p_I \hookrightarrow_{\pi\text{-rule}} p$, there exists $d \in [1 \ldots n]$ such that p is a subpath of p_d then $p_I \hookrightarrow_r p_{n+1}$ with $r \neq \pi$-rule (all the π-choices must be tried)

ii. $p_I \hookrightarrow_{\pi\text{-rule}} p_{n+1}$ (new π-choice)

(c) For $i \in [1 \ldots n], p_{n+1} \not\subseteq p_i$ (no redundancy)

2. $p_n \hookrightarrow_r p_{n+1}$ with $r \in R_{S4} \setminus \{\pi\text{-rule}\}$
3. $p_n \hookrightarrow_{\pi\text{-rule}} p_{n+1}$

We assume that a rule is applied only once on a branch with a given formula and Condition 1. is tested in priority. The reason for backtracking is either the completeness of the current branch (see Definition 3) or the detection of cycles (see Definition 9). The relation $\leadsto$ is called the *elementary ordering strategy*. Since the presentation of the strategies is for S4, periodicity tests are performed. However, for numerous logics, such as K, T, D or C the periodicity tests do not need to be part of the definition of the strategies.

When a rule r different from the π-rule is applied on the tree, if the current branch cannot be closed later on, then no backtracking is performed to the choice point where r has been applied. On the contrary, if the π-rule is applied and if the current branch cannot be closed later on then new choices are possible:(1) to apply the π-rule with a new possibility formula, (2) to apply an other tableaux rule. This can be represented in the figures (1)-(3). Each node is a node in the proof search tree. The label of each arrow is the name of the rule that has been applied. These restrictions limit the number of backtrackings. In the figures, a 'failure' means that the current branch cannot be closed.

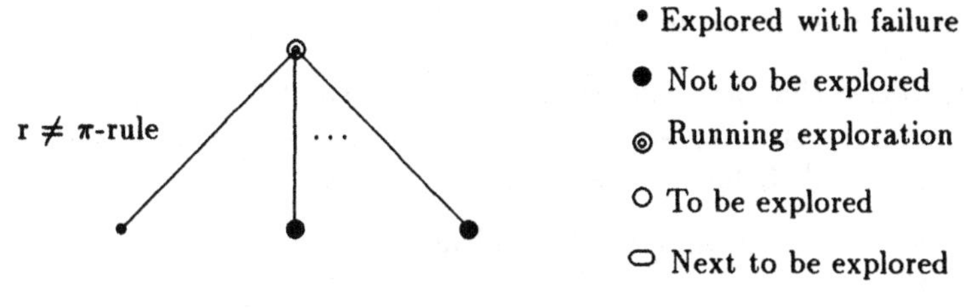

Fig. 1. Elimination of Nodes

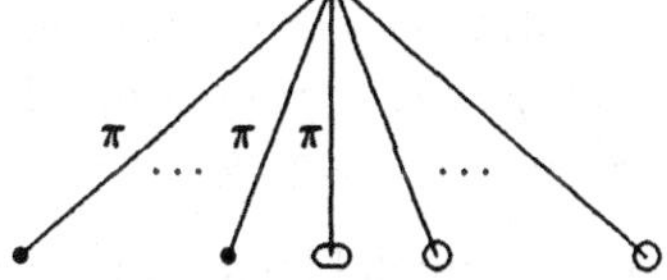

Fig. 2. Successive Exploration of the π-choices

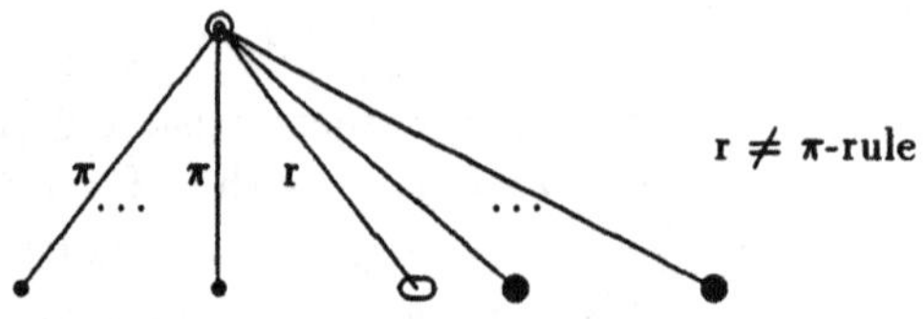

Fig. 3. Elimination of Nodes After the Exploration of all the π-choices

Proposition 12. *The binary relation* $\rightsquigarrow$ *is strongly complete for S4.*

Proof(sketch). The key idea of the proof is to relate the metapath derivations and the searches in the corresponding abstract theorem-proving graphs -see in [Kow69] the definition of such graphs. Let f be a modal formula. The different steps of the proof are the following:

1. to build a *finite* abstract theorem proving graph depending on f -the finiteness is guaranteed by the periodicity tests. Indeed the set of branches that can be derived from the branch $\{\neg f\}$ is finite. This can be related to the property (*) defined in [Rau83] p.408.
2. to prove that if f is valid then a given search in the graph always reaches a closed node -corresponding to the construction of a closed tree. The completeness of the original tableaux proof system for S4 [Fit83] is used.
3. to show that this search corresponds to a metapath derivation with $\rightsquigarrow$.

A detailed proof can be found in [Dem].

All the strategies that are defined in the sequel shall use the proof search tree of the elementary ordering strategy but the branches of exploration will be ordered or even pruned. Different criteria are introduced:

- to order the search according to the tableaux rules with a uniform priority (Section 3.4),
- to order the search according to the formulas to be decomposed with a uniform priority (Section 3.5),
- to prune additional branches (Section 3.3).

3.3 Strict Ordering Rules Strategies

The first hint to define strategies is to build paths with some structural properties. The *strict ordering rules strategies* are defined by ordering the tableaux rules in the way stated in Definition 14. First the following definition is needed.

Definition 13. Let L be a modal logic and R_L be the set of tableaux rules for L (defined for example in [Fit83]). A *rules sequence* rs is a sequence $(r_1, \ldots, r_k)$ such that each rule of R_L appears at most once in rs. For $i, j \in [1 \ldots k]$, we note $r_i \leq_{rs} r_j$ iff $i \leq j$ and for $r \in R_L \setminus \{rs\}$, $i \in [1 \ldots k]$, $r_i <_{rs} r$.

Name	Rules sequence
RS1	(Neg-rule, α-rule, ν-rule, β-rule, π-rule)
RS2	(Neg-rule, α-rule, ν-rule, π-rule, β-rule)
RS3	(Neg-rule, α-rule, β-rule, π-rule, ν-rule)
RS4	(π-rule, Neg-rule, α-rule, β-rule, ν-rule)

Fig. 4. Definition of rules sequences

We use in the sequel the set of rules sequences defined in Figure 4. Let $rs = (r_1, \ldots, r_k)$ be a rules sequence and p be a path. With the strict ordering rules strategies, when a rule must be applied to the current branch b of p, one first checks whether r_1 is applicable otherwise we check whether r_2 is applicable and so on. The paths are almost in normal form but a non-determinism still exists: the formula for which the rule is applied, is not specified.

Definition 14. Let rs be a rules sequence. The *strict ordering rules strategy* rs^- is obtained from the definition of $\rightsquigarrow$ by deleting the condition 1.(b).i and, by replacing the conditions 2. and 3. by the condition 2′.
Condition 2′: $p_n \hookrightarrow_r p_{n+1}$ with $r \in R_{S4}$ and there is no path p and no rule $u <_{rs}$ r such that $p_n \hookrightarrow_u p$.

Fewer paths are built with that class of strategies. So for some rules sequences, completeness is missed.

Proposition 15. *The strategies $RS2^-$, $RS3^-$ and $RS4^-$ are not strongly complete for the logic S4. The strategy $RS1^-$ is strongly complete for the logic S4.*

Proof(sketch). For example, the formula $f = (\Box P \vee \Box Q) \Rightarrow \Box(P \vee Q)$ can be used to prove the incompleteness of $RS2^-$. The only branches derived from $\{\neg f\}$ are $\{(\Box P \vee \Box Q), \neg\Box(P \vee Q)\}, \{\neg(P \vee Q)\}$ and $\{\neg P, \neg Q\}$. Since f is S4-valid and no closed branch can be derived then $RS2^-$ is not complete. The completeness of the strategy $RS1^-$ can be roughly proved with the following arguments. The strict ordering rules strategies cancel the application of the rule r in the case described in Figure 3. That is why numerous strict ordering strategies are not complete. However, with the strategy $RS1^-$, this case cannot happen since the π-rule is the last rule to be applied on a branch.

3.4 Flexible Ordering Rules Strategies

In Section 3.3 it has been shown that a simple structural criterion on paths may cause completeness to be missed. Though the strict ordering rules strategies limit the search space, for some specific rules sequences the completeness can be preserved. We now propose another class of strategies. A systematic normal form is not required on paths so that completeness is preserved.

Definition 16. Let rs be a rules sequence. The *flexible ordering rules strategy* rs^+ is obtained from the definition of $\leadsto$ by replacing the conditions 2. and 3. by the condition 2′ -see Definition 14. Furthermore the condition 1.(b).i is completed with: There is no path p and no rule $u \neq \pi$-rule such that $u <_{rs} r$ and $p_n \hookrightarrow_u p$.

This class of strategies can be seen as an intermediate class between the elementary ordering strategy and the class of strict ordering rules strategies. Indeed, the tableaux rules are ordered, a restricted form of backtracking is performed but (and it is the main difference with the strict ordering rules strategies) a backtracking is possible when all the π-choices have been attempted -see Figure 3. We note $\mathcal{S}_c$ the union set of complete flexible ordering rules strategies and complete strict ordering rules strategies.

Proposition 17. *Every flexible ordering rules strategy is strongly complete for S4.*

Unlike the previous family of strategies, every flexible ordering rules strategy is strongly complete. The proof of Proposition 17 is a consequence of the proof of Proposition 12. If a valid formula f cannot derive a closed tree with rs^- then the application of certain rules must be forced against the order $\leq_{rs}$, which corresponds to generate a closed metapath with rs^+ from $(((\{\neg f\})))$. Such a family of complete strategies is very useful for experiments - see Section 4.

3.5 π-Ordering Strategies

Most often, a π-rule can be applied to different possibility formulas (or π-formulas) in a branch. Each π-formula corresponds to a unique choice of a π-rule inference -π-choice. We propose to define orderings on π-formulas to establish priorities for π-rule inferences. A *good π-ordering strategy* should avoid as much as possible backtracking. To do so, we define a *formula valuation* as an application mapping a modal formula to an integer.

Definition 18. Let Val be a formula valuation. The *π-ordering strategy* Val^π is obtained by replacing condition 3. by condition 3′. and, 1.(b).ii. by 1.(*b*).*ii*.′ in the definition of $\leadsto$.
Condition 3.′: Let $\{\pi_1, \ldots, \pi_s\}$ the set of π-formulas in $b(p_n)$ such that for $1 \leq i \leq (s-1)$, $Val(\pi_i) \leq Val(\pi_{i+1})$. Then $p_n \hookrightarrow_{\pi-rule,\pi_1} p_{n+1}$.
Condition 1.(*b*).*ii*.′: Let $\{\pi_1, \ldots, \pi_s\}$ the set of π-formulas in $b(p_I)$ such that for $1 \leq i \leq (s-1)$, $Val(\pi_i) \leq Val(\pi_{i+1})$. There exists $K \in [1 \ldots (s-1)]$ such that, for $1 \leq i \leq K, p_I \hookrightarrow_{\pi-rule,\pi_i} p'_i$ and p'_i occurs in mp. Then $p_I \hookrightarrow_{\pi-rule,\pi_{K+1}} p_{n+1}$.

Proposition 19. *Every π-ordering strategy is strongly complete for S4.*

In the **ATINF** modal prover, the following valuations have been implemented: age on the branch, random valuation, modal degree, modal weight, possible modal weight, weight.

3.6 An Alternative Periodicity Test

An unexpected improvement can be stated about the implementation of the ordering rules strategies. The proofs will be sketched for the logics S4 but they can be extended to other logics, as it has been done in the **ATINF** modal prover. One of the main factors for the cost of the tree constructions remains the periodicity tests. As a matter of fact, for logics such as D4 or S4, this detection is necessary to preserve the decidability of these calculi. The cycles may appear when the application of the rules does not decrease systematically the size of formulas in the branch. Proposition 21 states that the periodicity tests can be executed only in some special occasions while insuring the detection of any cycle on the branch.

However the reduction of the number of periodicity tests may cause the introduction of useless formulas on the branch.

Definition 20. Let $\triangleright$ be a member of $\mathcal{S}_c$ (see Section 3.4). The strategy $\triangleright^{\downarrow}$ is defined as $\triangleright$ except that before running the periodicity test the following condition must hold:

$$b(mp[l(mp)-1]) \rightarrow_{\pi-rule} b(mp[l(mp)])$$

For the derivation of metapaths with the strategy $\triangleright^{\downarrow}$, the periodicity tests are performed only under a certain condition. It restricts the number of periodicity tests. The periodicity test is performed when a π-rule has been applied without closing the current branch.

Proposition 21. *Let $\triangleright$ be in $\mathcal{S}_c$. The strategy $\triangleright^{\downarrow}$ is complete.*

Proof. Let mp and mp' be two metapaths. We show that if $mp \triangleright mp'$ then there exists a metapath mp'' such that $mp \triangleright^{\downarrow *} mp''$ and $m''[l(m'')] = m'[l(m')]$. Since $\triangleright$ is complete, we get the completeness for $\triangleright^{\downarrow}$.

If $mp[l(mp)]$ does not contain a cycle or $b(mp[l(mp)-1]) \rightarrow_{\pi-rule} b(mp[l(mp)])$ then we naturally get $mp \triangleright^{\downarrow} mp'$. Now assume that none of these conditions holds. We note $(p_1, \ldots, p_n)$ the metapath mp and $(p_1, \ldots, p_n, p_I)$ the metapath mp' with $I < n$. We note $(T_1, \ldots, T_{h(n)})$ the path p_n. Since p_n contains a cycle and since the periodicity test has priority in $\triangleright$ then there exists $k < h(n)$ such that $b(T_k) = b(T_{h(n)})$ and there exists a set of indices $\{u_0, \ldots, u_s\} \subseteq [k \ldots h(n)]$ such that $u_0 = k$, $u_s = h(n)$ and for $i \in [0 \ldots (s-1)]$, $b(T_{u_i}) \rightarrow_{r_i, g_i} b(T_{u_{i+1}})$. Let J be the smallest element of the set $\{j \mid j \in [0 \ldots (s-1)], r_j = \pi - rule\}$. This set is not empty from Condition 2.b.iii in Definition 9.

Let $(T^i_{h(n)})_{i \in [0 \ldots J+1]}$ be a family of trees such that $T^0_{h(n)} = T_{h(n)}$ and for $i \in [0 \ldots J]$, $T^i_{h(n)} \Rightarrow_{r_i, g_i} T^{i+1}_{h(n)}$. Such a family exists since a cycle has been detected. At the path level, we note $(p^i_n)_{i \in [0 \ldots J+1]}$ the family of paths such that $p^i_n = p_n$ and for $i \in [0 \ldots J]$, $p^i_n \hookrightarrow_{r_i, g_i} p^{i+1}_n$. It can be checked that $mp \triangleright^{\downarrow *} (p_1, p_2, \ldots, p_n, p^1_n, \ldots p^{J+1}_n)$. We are now in a position to backtrack. The backtracking cannot stop on p^i_n $(i \in [0 \ldots J])$ since the π-rule has not been applied along $(p^1_n, \ldots p^{J+1}_n))$. The maximal index for backtracking is therefore I which leads to $mp'' \triangleright^{\downarrow} (p_1, p_2, \ldots, p_n, p^1_n, \ldots p^{J+1}_n, p_I)$.

4 Experimental Results

We have incorporated the strategies defined in the previous sections to a tableau theorem prover for standard modal logics. The prover has been implemented within the Inference Laboratory **ATINF** (see e.g., [CH93]) as an extension of the one developed for classical logic -see [KZ90]. It uses Fitting's method for different propositional and first-order modal logics but it also incorporates different translations that allow us to deal with particular logics *without explicitly implementing tableau systems for them.* This section is devoted to present some of the significant tests performed with our modal theorem prover. The prover admits different parameters such as the choice for a strategy. These parameters enable to prune the search space, or to accelerate the closure of branches. For example, an ordering rules strategy can be defined during the tree constructions at any moment. Furthermore, when a proof is produced, the user can access to some statistical information about it. The theorem prover has been written in Common Lisp. The tests have been run on a SUN4/52Mb machine.

Different ordering strategies are compared on the following S4-valid formulas.

1. $\Box(\Box P \vee \Box Q) \Leftrightarrow (\Box P \vee \Box Q)$
2. $\Box(\Box(P \Leftrightarrow Q) \Rightarrow R) \Rightarrow \Box(\Box(P \Leftrightarrow Q) \Rightarrow \Box R)$
3. $\Box P \Leftrightarrow (\Box(Q \Rightarrow P) \wedge \Box(\neg Q \Rightarrow P))$
4. $\Box_C(\neg PC \Rightarrow \Box_B \neg PC) \wedge \Box_C\Box_B\Box_A(PA \vee PB \vee PC) \wedge \Box_C\Box_B(\neg PB \Rightarrow \Box_A \neg PB) \wedge$ $\Box_C\Box_B(\neg PC \Rightarrow \Box_A \neg PC) \wedge \Box_C \neg \Box_B PB \wedge \Box_C\Box_B \neg \Box_A PA \Rightarrow \Box_C PC$[4]
5. $\Diamond\Box(\Box(P \vee \Box Q) \Leftrightarrow (\Box P \vee \Box Q))$
6. $\Diamond\Box((P \Rightarrow Q) \Leftrightarrow F(Q, F(P, Q)))$ with $F(P, Q) = \neg P \vee (Q \wedge \Diamond(P \wedge \neg Q)) \vee \neg\Diamond(P \wedge Q)$[5]
7. $\Box(\Box(\Box P \Rightarrow \Box(\Box Q \Rightarrow \Box R)) \Rightarrow \Box(\Box(\Box P \Rightarrow \Box Q) \Rightarrow \Box(\Box P \Rightarrow \Box R)))$
8. $\Box(\Box(\Box P \Rightarrow \Box(\Box Q \Rightarrow \Box R)) \Rightarrow \Box(\Box Q \Rightarrow \Box(\Box P \Rightarrow \Box R)))$
9. $\Box(\Box(\Box P \Rightarrow \Box(\Box Q \Rightarrow \Box R)) \Rightarrow \Box(\Box(\Box P \Rightarrow \Box Q) \Rightarrow \Box R))$[6]

For each strategy in $\{RS1^-, RS2^+, RS3^+, RS4^+\}$ and for each formula the number of π-rule inferences is given in Figure 5. Furthermore, the π-ordering strategy based on the modal degree[7] of formulas is used to build the proofs. For example the construction of the proof for the formula number 8. with the strategy $RS4^+$ needs 8 π-rule inferences. The strategies $RS1^-$ and $RS2^+$ are the most efficient for this test set especially when the number of π-rule inferences

[4] This formula represents a way to formalize the 'Wise Man Puzzle' -see for example [McC78]. In particular, the number of hypothesis has been minimized.

[5] The formulas (5) and (6) are originally valid formulas for S5. In [Mat55], it has been shown that f is S5-satisfiable iff $\Diamond\Box f$ is S4-satisfiable. Moreover the formula (6) is one of the conditions to prove that all the operators in S5 can be defined in terms of the operator F [DdM81].

[6] The formulas (7)-(9) are originally valid formulas of the intuitionistic logic. A translation from the intuitionistic logic to S4 has been used.

[7] If $md(f)$ denote the modal degree of f then md can be easily defined as follows: $md(\neg F) = md(F)$, $md(F \wedge G) = max(md(F), md(G))$, $md(\Box F) = md(\Diamond F) = 1 + md(F)$ and $md(P) = 0$ for P propositional variable.

	1	2	3	4	5	6	7	8	9
$RS1^-$	6	45	5	45	32	19	285	52	19
$RS2^+$	9	18	3	51	62	1096	37	6	5
$RS3^+$	7	18	3	138	255	841	53	8	72
$RS4^+$	10	32	4	393	371	4812	61	8	80

Fig. 5. Number of π-rule inferences for different ordering strategies

increases. The number of π-rule inferences is experimentally proportional to the CPU time to obtain the proof. These tests remain relevant to underline the difference of performances between the ordering rules strategies as well as the relativity of each example.

5 Conclusion

Three families of ordering strategies have been defined for the Fitting's modal tableaux calculi for the logic S4. Similar strategies can be defined for propositional logics K, T, K4, D, D4, C, CT, C4, CS4, CD, CD4, G (see [Dem]). They can also be naturally extended to the multiagent case and very likely to the first-order case though new difficulties are expected. The strategies are based on orderings of tableaux rules with different variants for backtracking. Most of the defined strategies have been proved complete. It is worth noting that performing all possible backtracking is not required for completeness.

Experiments with our modal tableau prover have underlined the relevance of some strategies, such as $RS1^-$ and $RS2^+$. A common point shared by these two efficient strategies is the late application of the π-rule on a branch. The proposed formal framework has been shown to be a suitable one to define strategies and to experiment them. A formal analysis of the efficiency of the strategies would be of great help for theorem provers but we were not able to find such results.

We are aware of the scope of the present work. For instance, only tableau systems have been discussed although the defined strategies can be adapted to other proof systems. Furthermore the first-order case and the multi-modal case are not treated in this paper but ordering of rules can be applied to them.

As mentioned earlier there exist also translation methods for mechanizing non-classical logics. A comparison between the present work and these methods is worth being made since it would be of special interest to improve the efficiency of the mechanization of non-classical logics. Future investigations will be also oriented towards the extension of the existing strategies to first-order logics (with their experimentation in the `ATINF` modal prover for first-order modal logics) and to multi-modal logics.

Acknowledgments: The author wishes to thank Ricardo Caferra and the anonymous referees for their relevant and precise comments on an earlier version of this paper.

References

[AEH90] Y. Auffray, P. Enjalbert, and J-J. Herbrard. Strategies for modal resolution: results and problems. *Journal of Automated Reasoning*, 6:1–38, 1990.

[Cat91] L. Catach. TABLEAUX: A General Theorem Prover for Modal Logics. *Journal of Automated Reasoning*, 7:489–510, 1991.

[CD93] R. Caferra and S. Demri. Cooperation between direct method and translation method in non classical logics: some results in propositional S5. In *IJCAI-13*, pages 74–79, 1993.

[CDH93] R. Caferra, S. Demri, and M. Herment. A framework for the transfer of proofs, lemmas and strategies from classical to non classical logics. *Studia Logica*, 52(2):197–232, 1993.

[CH93] R. Caferra and M. Herment. $\mathtt{GLEF}_{\mathtt{ATINF}}$: a graphic framework for combining theorem provers and editing proofs for different logics. In A. Miola, editor, *DISCO'93*, pages 229–240. Springer-Verlag, LNCS 722, 1993.

[CM93] M. Cerioli and J. Meseguer. May I borrow your logic? In A. Borzyskowski and S. Sokolowski, editors, *Mathematical Foundations of Computer Science*, pages 342–351. Springer Verlag, LNCS 711, August 1993.

[DdM81] L. Dubikajtis and L. de Moraes. On single operator for Lewis S5 modal logic. *Reports on Mathematical Logic*, 11:57–61, 1981.

[Dem] S. Demri. PhD thesis. Forthcoming.

[Dem93] S. Demri. Ordering strategies for tableau-based modal theorem provers. In *IJCAI'93 Workshop on Executable Modal and Temporal Logics*, 1993.

[DG88] J. Delgrande and C. Groeneboer. Tableau-based theorem proving in normal conditional logics. In *AAAI-7*, pages 171–176, 1988.

[FB90] A. M. Frisch and Scherl R. B. A constraint logic approach to modal deduction. In *JELIA' 90*, pages 234–250. Springer-Verlag, 1990.

[FdC83] L. Fariñas del Cerro. Un principe de résolution en logique modale. *RAIRO*, 18:161–170, 1983.

[Fit83] M. C. Fitting. *Proof methods for modal and intuitionistic logics*. D. Reidel Publishing Co., 1983.

[Fit88] M. C. Fitting. First-order modal tableaux. *Journal of Automated Reasoning*, 4:191–213, 1988.

[HC68] G. E. Hughes and M. J. Cresswell. *An introduction to modal logic*. Methuen and Co., 1968.

[Her89] A. Herzig. *Raisonnement automatique en logique modale et algorithmes d'unification*. PhD thesis, Université P. Sabatier, Toulouse, 1989.

[Hin62] J. Hintikka. *Knowledge and Belief*. Cornell University Press, 1962.

[HM85] J. Y. Halpern and Y. Moses. A guide to the modal logics of knowledge and belief: preliminary draft. In *IJCAI-9*, pages 480–490, 1985.

[Kon86a] K. Konolige. *A deduction model of belief*. Pitman, 1986.

[Kon86b] K. Konolige. Resolution and quantified epistemic logics. In J. H. Siekmann, editor, *CADE-8*. Springer-Verlag, LNCS 230, 1986.

[Kow69] R. Kowalski. Search strategies for theorem-proving. *Machine Intelligence*, 5:181–201, 1969.

[Kri63] S. Kripke. Semantical considerations on modal logics. *Modal and Many-valued logics, Acta Philosophica Fennica*, 1963.

[KZ90] T. Kaufl and N. Zabel. Cooperation of decision procedures in a tableaux-based theorem prover. *Revue d'Intelligence Artificielle, Special Issue on Automated Deduction*, 4(3):99–125, 1990.

[Lad77] R. E. Ladner. The computational complexity of provability in systems of modal propositional logic. *SIAM J. Comp.*, 6(3):467–480, September 1977.

[Lov78] P. W. Loveland. *Automated Theorem Proving: A Logical Basis*. North-Holland, 1978.

[Mat55] K. Matsumoto. Reduction theorem in Lewis's sentential calculi. *Mathematica Japonicae*, 3:133–135, 1955.

[McC78] J. McCarthy. Formalization of two puzzles involving knowledge. Stanford University, 1978.

[Ohl93] H. Ohlbach. Optimized translation of multi modal logic into predicate logic. In A. Voronkov, editor, *LPAR'93*, pages 253–264. Springer-Verlag, LNAI 698, 1993.

[Orl79] E. Orlowska. Resolution systems and their applications I. *Fundamenta Informaticae*, 3:253–268, 1979.

[Orl91] E. Orlowska. Relational proof systems for some AI logics. In Ph. Jorrand and J. Kelemen, editors, *FAIR'91*, pages 33–47. Springer-Verlag, LNAI 535, 1991.

[Rau83] W. Rautenberg. Modal tableau calculi and interpolation. *The Journal of Philosophical Logic*, 12:403–423, 1983.

[Smu68] R. M. Smullyan. *First-Order Logic*. Springer-Verlag, 1968.

[TMM86] P. B. Thistlewaite, M. A. McRobbie, and R. K. Meyer. The KRIPKE Automated Theorem Proving System. In *CADE-8*, pages 705–706. Springer-Verlag, LNCS 230, 1986.

[Wal90] L. A. Wallen. *Automated Deduction in Nonclassical Logics*. MIT Press, 1990.

TAS-D^{++}: Syntactic Trees Transformations for Automated Theorem Proving*

Gabriel Aguilera, Inma P. de Guzmán, Manuel Ojeda

Dept. Matemática Aplicada. Universidad de Málaga.
Plz. El Ejido, s/n. 29071 Málaga, Spain

Abstract. In this work a new Automated Theorem Prover (ATP) via refutation for classical logic and which does not require the conversion to clausal form, named TAS-D^{++}, is introduced. The main objective in the design of this ATP was to obtain a parallel and computationally efficient method naturally extensible to non-standard logics (concretely, to temporal logics, see [8]).
TAS-D^{++} works by using transformations of the syntactic trees of the formulae and, as tableaux and matrix style provers [3, 11, 12], it is Gentzen-based. Its power is mainly based in the efficient extraction of implicit information in the syntactic trees to detect valid, unsatisfiable, equivalent or equal subformulae (in difference with the standard ATPs via refutation which have a general algorithm for all formulae). TAS-D^{++} is sound and complete and, moreover, it is a method that generates countermodels in a natural way. This method is implemented in [1].

1 Introduction

The use of the resolution method during the last 20 years has given rise to sophisticated implementation techniques which has allowed to develop theorem provers to solve high complexity problems. However, this method has several inconvenients, not solved yet, in spite of the many attempts made in research with this goal. The following can be highlighted:

It is difficult to extend the resolution method to non-standard logics. This is because no adequate normal forms have been found for such logics. This prevents the *nice* applications of the techniques based on resolution.

In this work a new ATP for classical logic, by refutation and that does not require the normal form conversion, named TAS-D^{++} is presented. It has been a goal during its design, to get a parallel and computationally efficient method which is extensible in a natural way to non-standard logics (in particular to

* This work is based on the results of work carried out for his Doctoral Thesis by Francisco Sanz, colleague of the authors in the GIMAC investigation group. This Thesis could not be defended because of the sudden and untimely death of Francisco Sanz. Our gratitude to Francisco permeates every section of this work.

temporal logics, see [8]). TAS-D^{++} is strongly based on the structure of the formula, concretely, in the structure of its syntactic tree. It works by using transformations on these syntactic trees (TAS stands for *Transformaciones de Árboles Sintácticos*, Spanish translation for Syntactic Tree Transformations.).

Its power is based not only on the intrinsically parallel design of the involved transformations, but also on the fact that the exponential complexity in the worst case of TAS-D^{++} is due to only one of the transformations (named $(\wedge$-$\vee)$-*par* and commissioned on the distributions of $\wedge$ over $\vee$). On the other hand, these transformations are not just applied one after the other; through the efficient determination and manipulation of sets of unitary models of a formula and its negation, the method *investigates exhaustively* the formula (or formulae), to detect valid, unsatisfiable, equivalent or equal subformulae. With this tool, it can be concluded whether the structure of the syntactic tree has or has not direct information about the validity of the formula (or the inference). This way, either the method stops giving this information or, otherwise, it decreases the size of the problem before applying the next transformation. So, it is possible to decrease the number of distributions or, even, the application of $(\wedge$-$\vee)$-*par*.

TAS-D^{++} is *sound and complete* and, as Beth's method of *semantic tableauxs*, is a model building method [2]. It is necessary to underline that Automated Demonstration has not paid much attention to build countermodels; in fact, the number of works on this subject is not significant compared with the number of works based on refutation or deduction methods. However, the construction of models is considered as one of the most remarkable facts in Automated Demonstration [4, 6] as shown in outstanding papers as [13] and [14]. This work pretends to give an advance on this subject. It is notable the natural way of generating countermodels during the execution of TAS-D^{++}.

To introduce this new method, this work has necessarily to be restricted to the basic case; anyway, in the last section the extension to first-order is briefly sketched (due to space restrictions).

2 Concepts and Previous Results

In the rest of the paper L will denote the language of the standard propositional logic, Q is the set of propositional symbols, T_A is the standard concept of binary syntactic tree for a well formed formula (*wff*), and $\widehat{T}_A$ is the generalized n-ary syntactic tree of a negation normal formula (*nnf*).

The following definition gives an elemental transformation that converts a *wff* A in an equivalent *nnf*:

Definition 1. Given a *wff* A and $\sigma \in \{+,-\}$, $A(\sigma)$ is recursively defined as follows:

$$\begin{array}{ll}
p(+) = p & p(-) = \neg p \\
(\neg A)(+) = A(-) & (\neg A)(-) = A(+) \\
(A \vee B)(+) = A(+) \vee B(+) & (A \vee B)(-) = A(-) \wedge B(-) \\
(A \wedge B)(+) = A(+) \wedge B(+) & (A \wedge B)(-) = A(-) \vee B(-) \\
(A \rightarrow B)(+) = A(-) \vee B(+) & (A \rightarrow B)(-) = A(+) \wedge B(-)
\end{array}$$

Given a *wff* A, to *sign* a tree T_A will mean "to obtain $\widehat{T_{A(+)}}$". It is obvious that if A is a wff then $A(+)$ is a nnf equivalent to A.

The idea to use the information given by partial interpretations, exhaustively used in Quine's method [10], is taken in TAS-D^{++} just for unitary partial interpretations but, as it will be seen, in a surprisingly powerful manner.

Definition 2. Let Φ be a set of valuations and A and B *wffs*. It will be said that A and B are *Φ-equivalent* if $V(A) = V(B)$ for all $V \in \Phi$.

Theorem 3. *Let A be a nnf, p a propositional symbol in A, $p1 = \{V \in 2^Q \mid V(p) = 1\}$ and $p0 = \{V \in 2^Q \mid V(p) = 0\}$, then:*

1. *If A and $\top$ are $p1$-equivalent then $A \equiv p \vee A[p/\bot]$*
2. *If A and $\top$ are $p0$-equivalent then $A \equiv \neg p \vee A[p/\top]$*
3. *If A and $\bot$ are $p1$-equivalent then $A \equiv \neg p \wedge A[p/\bot]$*
4. *If A and $\bot$ are $p0$-equivalent then $A \equiv p \wedge A[p/\top]$*

As an immediate consequence of the previous theorem, the following corollary can be stated:

Corollary 4. *Let A be an nnf, p a propositional symbol in A and $b \in \{0, 1\}$, then:*

1. *If A and $\top$ are pb-equivalent then A is satisfiable.*
2. *If A and $\bot$ are $p1$-equivalent then*

   A unsatisfiable if and only if $A[p/\bot]$ is unsatisfiable.

3. *If A and $\bot$ are $p0$-equivalent then*

   A unsatisfiable if and only if $A[p/\top]$ is unsatisfiable.

The previous theorem and its corollary will be used all over this paper by means of the sets Δ_0 and Δ_1, the key tools of TAS-D^{++}, defined below:

Definition 5. Given a *nnf* A, the sets $\Delta_0(A)$ and $\Delta_1(A)$ are recursively defined as follows:

$$\Delta_0(p) = \{p\,0\}; \qquad \Delta_1(p) = \{p\,1\}; \qquad \Delta_0(\neg p) = \{p\,1\}; \qquad \Delta_1(\neg p) = \{p\,0\}$$

$$\Delta_0\left(\bigvee_{i=1}^{n} A_i\right) = \bigcap_{i=1}^{n} \Delta_0(A_i); \qquad \Delta_1\left(\bigvee_{i=1}^{n} A_i\right) = \bigcup_{i=1}^{n} \Delta_1(A_i)$$

$$\Delta_0\left(\bigwedge_{i=1}^{n} A_i\right) = \bigcup_{i=1}^{n} \Delta_0(A_i); \qquad \Delta_1\left(\bigwedge_{i=1}^{n} A_i\right) = \bigcap_{i=1}^{n} \Delta_1(A_i)$$

$$\Delta_0(\bot) = \Delta_1(\top) = Q\,0 \cup Q\,1; \qquad \Delta_1(\bot) = \Delta_0(\top) = \emptyset$$

where $Q\,b = \{p\,b \mid p \in Q\}$ and $b \in \{0,1\}$. In the following Q_0^1 will mean $Q\,0 \cup Q\,1$.

It is obvious that the elements of $\Delta_0(A)$ can be seen as unitary models of $\neg A$ and that the elements of $\Delta_1(A)$ can be seen as unitary models of A.

Definition 6. If A is a *nnf*, to *label* $\widehat{T_A}$ will mean to label each node N in $\widehat{T_A}$ with the ordered pair $(\Delta_0(B), \Delta_1(B))$ where B is the subformula of A such that N is the root of $\widehat{T_B}$.

For technical details concerning efficiency matters, a subset Δ_0^* of Δ_0 is defined below. The usefulness of this subset will be explained later.

Definition 7. Given a *nnf* $A = \bigwedge_{i \in I} A_i$ and $J = \{i \in I \mid A_i \text{ is a literal } l_i\}$, the set $\Delta_0^*(A)$ denotes the subset of $\Delta_0(A)$ defined as

$$\Delta_0^*(A) = \{p_i\, b_i \mid i \in J\}$$

where $b_i = 0$ if $l_i = p_i$, and $b_i = 1$ if $l_i = \neg p_i$.

Example 1. For the formula

$$A = \Big(\big(p \vee (r \rightarrow t)\big) \wedge \big(q \vee (t \rightarrow s)\big)\Big) \rightarrow \Big(\big(p \wedge \neg(q \rightarrow \neg t)\big) \vee \big(r \rightarrow ((q \rightarrow (s \vee r)) \wedge s)\big)\Big)$$

the syntactic tree $T_{\neg A}$ appears in Fig. 1. After *signing* and *labeling* $\widehat{T_{\neg A(+)}}$ the tree $\widehat{T_B}$ in Fig. 2 is obtained. In the figure the labels (Δ_0, Δ_1) in the literal subtrees have been omitted.

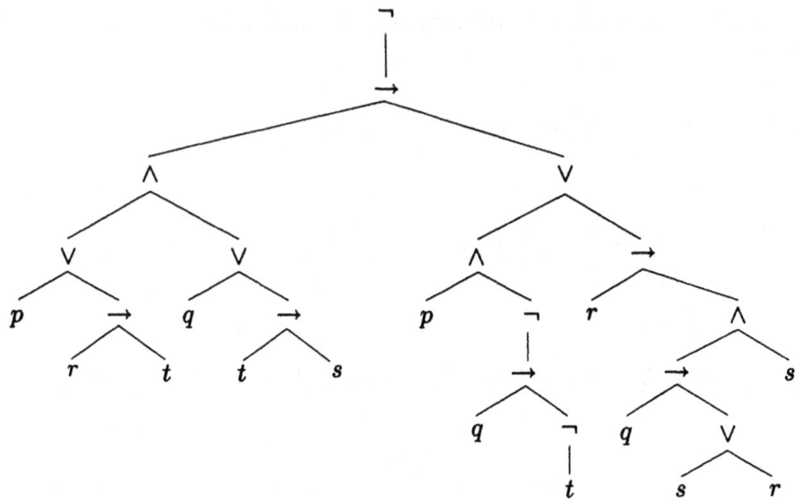

Fig. 1. The tree $T_{\neg A}$.

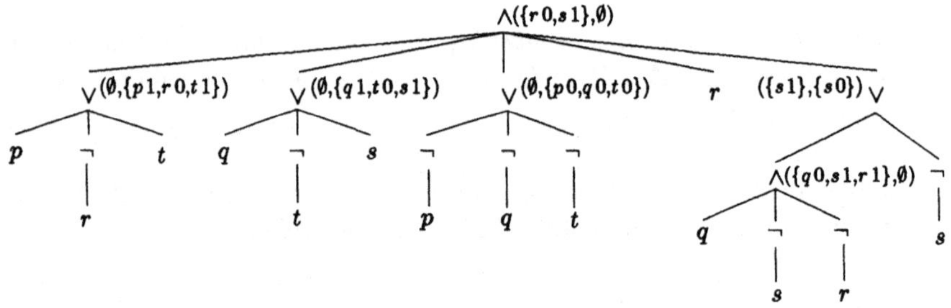

Fig. 2. The tree $\widehat{T_B}$.

3 The Method TAS-D^{++} for a Formula

Given a *wff* A with no occurrences of constants $\top$ or $\bot$, the input of TAS-D^{++} is the syntactic tree of $\neg A$, $T_{\neg A}$. After *signing* and *labeling*, the tree-transformations denoted by $\mathcal{F}$ (described in Sect. 3.1), $\mathcal{S}$ (described in Sect. 3.2), $\mathcal{E}$ (described in Sect. 3.3) and ($\wedge$-$\vee$)-*par* (described in Sect. 3.4) are applied following the flow diagram in Fig. 3, where the outputs are either VALID or NON-VALID and a model for $\neg A$ (countermodel).

For an infinitude of formulae A, if TAS-D^{++} does not finish after executing $\mathcal{F}$, $\mathcal{S}$ or $\mathcal{E}$, the size of the input tree of ($\wedge$-$\vee$)-*par* has been drastically decreased.

The interest of this strategy is based on the fact that every *good* method must take an acceptable time for most of its inputs and, only rarely, will take an excessive time. So, strategies which can reduce the size of the problem before applying the transformation which holds most of the complexity (($\wedge$-$\vee$)-*par* in our case) will be useful.

Remark. The philosophy of TAS-D^{++} does not require the process of *signing*, however concerning propositional classical logic, the conversion of the formula to *nnf* eases the description of the proper algorithms of TAS-D^{++}.

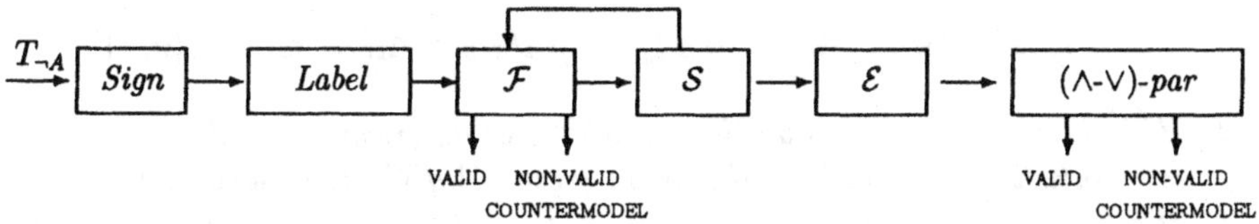

Fig. 3. The method TAS-D^{++}.

3.1 The Tree-Transformation $\mathcal{F}$

The input of this tree-transformation is a (Δ_0, Δ_1)-labeled generalized syntactic tree, $\widehat{T_C}$. This stage tries to detect, by means of the labels (Δ_0, Δ_1), whether the structure of $\widehat{T_C}$ provides either complete information about the unsatisfiability of C or useful information to decrease the size of C before distributing.

Concretely, the labels (Δ_0, Δ_1) will allow:

- to conclude that a subformula B, in particular the whole formula, is equivalent to $\top$, $\bot$ or a literal (using the process *simplify* to be defined later).
- to decrease the size of the formula, substituting A by a simultaneously unsatisfiable formula in which the symbols in $\Delta_0(A)$ or $\Delta_1(A)$ does not occur more than once (using the processes *reduce* and *synthesize* to be defined later).

So, to describe the tree-transformation $\mathcal{F}$ will require following concepts:

Definition 8. Let A be an *nnf*, $b \in \{0, 1\}$, $\bar{0} = 1$ and $\bar{1} = 0$, it is said that $\widehat{T_A}$ is:

- *finalizable* if either A is $\bot$ or $\Delta_1(A) \neq \emptyset$,
- *b-conclusive* if there exists a propositional symbol p such that the labels $p0, p1 \in \Delta_b(A)$,
- *p-simple* if A is not a literal and $pb \in \Delta_0(A)$ and $p\bar{b} \in \Delta_1(A)$,

- *simplifiable* if there exists a subtree b-conclusive or p-simple,
- If A is neither a clause nor a cube then: $\widehat{T_A}$ is said to be 0-*reducible* if $\Delta_0^\star(A) \neq \emptyset$, 1-*reducible* if $\Delta_1(A) \neq \emptyset$, and *completely 0-reducible* if $\Delta_0(A) \neq \emptyset$.
- *reducible* if either it is completely 0-reducible or there exists a b-reducible proper subtree.

The following theorem is the semantic foundation of the tree-transformation $\mathcal{F}$.

Theorem 9. *Let A be a wff then:*

1. *If $\widehat{T_A}$ is finalizable and $\Delta_1(A) \neq \emptyset$ then A is satisfiable.*
2. *If $\widehat{T_A}$ is 1-conclusive then $A \equiv \top$, and if $\widehat{T_A}$ is 0-conclusive then $A \equiv \bot$.*
3. *If $\widehat{T_A}$ is p-simple then $A \equiv l_p$ where*

$$l_p = p \text{ if } p\,b \in \Delta_b(A) \text{ and } l_p = \neg p \text{ if } p\,b \in \Delta_{\bar{b}}(A) \text{ for } b \in \{0, 1\}$$

4. *If $\widehat{T_A}$ is completely 0-reducible and $p\,1 \in \Delta_0(A)$ (resp. $p\,0 \in \Delta_0(A)$), then A is unsatisfiable if and only if $A[p/\bot]$ (resp. $A[p/\top]$) is unsatisfiable.*
5. *If $\widehat{T_A}$ is reducible, B a proper subformula of A and $p\,1 \in \Delta_1(B)$ (resp. $p\,0 \in \Delta_1(B)$), then $B \equiv p \vee B[p/\bot]$ (resp. $B \equiv \neg p \vee B[p/\top]$).*
6. *If $\widehat{T_A}$ is reducible, B a proper subformula of A and $p\,1 \in \Delta_0^\star(B)$ (resp. $p\,0 \in \Delta_0^\star(B)$), then $B \equiv \neg p \wedge B[p/\bot]$ (resp. $B \equiv p \wedge B[p/\top]$).*

The previous theorem is used below to define *simplify, reduce, synthesize* and *update*, the basic actions involved in $\mathcal{F}$.

Definition 10. If $\widehat{T_A}$ is simplifiable, to *simplify* $\widehat{T_A}$ will mean to traverse depth-first $\widehat{T_A}$, making in the first b-conclusive or p-simple subtree $\widehat{T_B}$, the substitutions determined by items 2 and 3 in Theorem 9 (i.e. substitutions of subformulae for their equivalent: $\top$, $\bot$ or p).

Definition 11. If $\widehat{T_A}$ is reducible, to *reduce* $\widehat{T_A}$ will mean:

a) if $\widehat{T_A}$ is completely 0-reducible, to make in $\widehat{T_A}$ the substitutions determined by item 4 in Theorem 9 (i.e. substitute A for a simultaneously unsatisfiable formula in which the propositional symbols in $\Delta_0(A)$ do not occur),
b) otherwise, to traverse $\widehat{T_A}$ depth-first making, in the first 0-reducible or 1-reducible proper subtree, the substitutions determined by items 5 and 6 in Theorem 9 (i.e. substitution of subformulae B for their smaller-sized equivalent in which the propositional symbols in $\Delta_0^\star(B)$ or $\Delta_1(B)$ occur just once).

Now we introduce the process *synthesize*; this is a partial reduction to treat the cases in which $\Delta_1(\bigwedge_{i\in I} A_i) = \emptyset$ but there exist a subset $J \subset I$ such that $\Delta_1(\bigwedge_{i\in J} A_i) \neq \emptyset$. The definition of this partial reduction needs some previous concepts:

Consider a formula $A = \bigodot_{i\in I} A_i$, where $\bigodot$ is $\bigvee$ or $\bigwedge$, such that $\widehat{T_A}$ is neither simplifiable nor reducible, and let $\bigcup_{i\in I} \Delta_1(A_i) = \{p_1\,b_1, \ldots, p_n\,b_n\}$. The integers denoted by $m(p_j\,b)$, and defined below, are associated to the generalized syntactic tree $\widehat{T_A}$:

$$m(p_j\,b) = \left|\{i \in I \mid p_j\,b \in \Delta_1(A_i)\}\right|$$

where $|X|$ denotes the cardinality of the finite set X.

Consider $p\,b \in \bigcup_{i\in I} \Delta_1(A_i)$, it is said that $\widehat{T_A}$ is *$p\,b$-remarkable* if $m(p\,b) > 1$ and

$$m(p\,b) = \max\{m(p_j\,b) \text{ associated to } \widehat{T_A}\,\}$$

Consider $A = \bigodot_{i\in I} A_i$, if $\widehat{T_A}$ is *$p\,1$-remarkable* and $J = \{i \in I \mid p\,1 \in \Delta_1(A_i)\}$ then $\widehat{T_{A,p\,1}}$ will denote the syntactic tree of the formula

$$(p \vee \bigodot_{i\in J} A_i[p/\bot]) \odot (\bigodot_{i\in I-J} A_i)\ .$$

Similarly, if $\widehat{T_A}$ is *$p\,0$-remarkable* and $J = \{i \in I \mid p\,0 \in \Delta_1(A_i)\}$ then $\widehat{T_{A,p\,0}}$ will denote the syntactic tree of the formula

$$(\neg p \vee \bigodot_{i\in J} A_i[p/\top]) \odot (\bigodot_{i\in I-J} A_i)\,.$$

Definition 12. $\widehat{T_A}$ is *synthesizable* if $\widehat{T_A}$ has some $p\,b$-remarkable subtree. To *synthesize* $\widehat{T_A}$ will mean to traverse depth-first the tree $\widehat{T_A}$ in order to find the first $p\,b$-remarkable subtree, $\widehat{T_B}$, and substitute it for $\widehat{T_{B,p\,b}}$; i.e. to *synthesize* $\widehat{T_A}$ is, essentially, to consider A as the formula $(\bigodot_{i\in I} A_i) \odot (\bigodot_{i\in I-J} A_i)$ and then to *reduce* $B = \bigodot_{i\in I} A_i$ with respect to $p\,b \in \Delta_1(B)$.

After *simplifying, reducing* or *synthesizing* $\widehat{T_A}$, the labels (Δ_0, Δ_1) of some nodes have to be recalculated (these nodes are just the ascendents of the reduced nodes), and the constants $\top$ and $\bot$ have to be eliminated. These eliminations and the recalculations are made by the process *update* defined below:

Definition 13. Let $\mathcal{P}$ be any of the processes *simplify, reduce* or *synthesize* and consider that $\mathcal{P}$ has been applied to $\widehat{T_A}$, then if a subtree $\widehat{T_B}$ has been substituted for $\widehat{T_C}$ to *update* $\mathcal{P}(\widehat{T_A})$ consists in labeling $\widehat{T_C}$ and traversing the ascendents of the root $N(\Delta_0, \Delta_1)$ of $\widehat{T_C}$ recalculating (Δ_0, Δ_1) for any node and eliminating the leaves $\top$ and $\bot$ just using the elemental equivalence laws.

The sequence of checking the *finalizability, simplifying, reducing* and *synthesizing* with their corresponding *updates*, as many times as possible, is the core of the tree-transformation $\mathcal{F}$. This sequence is shown in Fig. 4.

Remark.

1. *Simplify* can force the method to end or decrease the size of the involved formula.

2. *Reduce* uses again the information contained in (Δ_0, Δ_1) in order to decrease the size of the tree before distributing. Note that after *reducing* a tree $\widehat{T_B}$, a simplifiable tree $\widehat{T_C}$ can be obtained and thus, a new *reduction* may avoid distributions (or end TAS-D^{++}).
   It is noticeable that the definition of 0-reducible does not use the set Δ_0 but $\Delta_0^\star$. To justify this restriction just note that the *reduction* (over a proper subtree) of $p\,b \in \Delta_0 - \Delta_0^\star$ will lead to substitute disjunctions by conjunctions, and such a substitution may increase the number of distributions.
3. *Synthesize*, a partial *reduction*, has the same goal that *reducing* proper subtrees. The justification to use only Δ_1 in the process *synthesize* is, once again, to avoid undesirable distributions.

The following theorem, a direct consequence of the previous definitions and Theorem 9, states the correction of the tree-transformation $\mathcal{F}$.

Theorem 14. *If $\mathcal{P}$ is any of the processes of simplifying, reducing, synthesizing or updating, and $\mathcal{P}(\widehat{T_B}) = \widehat{T_C}$, then B is unsatisfiable if and only if C is unsatisfiable.*

Example 2. For the formula A in example 1, the tree $\widehat{T_B}$ obtained after *signing* and *labeling* $T_{\neg A}$, shown in Fig. 2, is not *finalizable* but it is *simplifiable* since it has an s-simple subtree.

After *simplifying* $\widehat{T_B}$, the tree $\widehat{T_C}$ of Fig. 5 is obtained. This tree is neither *finalizable* nor *simplifiable* but it is *reducible*, more concretely, it is *completely* 0-*reducible* since $\Delta_0(C) = \{r\,0, s\,1\}$.

After *reducing* $\widehat{T_C}$, a tree $\widehat{T_D}$ where $D = (p \vee t) \wedge (q \vee \neg t) \wedge (\neg p \vee \neg q \vee \neg t)$. This tree is neither *finalizable*, *simplifiable* nor *reducible* but it is *synthesizable*, more concretely, it is $t\,0$-*remarkable*.

After *synthesizing* $\widehat{T_D}$, the tree $\widehat{T_E}$ in Fig. 6 is obtained.

Now, $\widehat{T_E}$ is neither *finalizable* nor *simplifiable*, but *reducible*. Concretely, the subtree $\widehat{T}_{q\wedge(\neg p\vee\neg q)}$ (of $\widehat{T_E}$) is 0-reducible and, after *reducing*, the tree $\widehat{T_F}$ of Fig. 7 is obtained. This tree is neither *finalizable*, *simplifiable*, *reducible* nor *synthesizable*, therefore it is the output of $\mathcal{F}$.

3.2 The Tree-Transformation $\mathcal{S}$

This tree-transformation tries once again to decrease the size of the problem before distributing $\wedge$'s over $\vee$'s. To begin with, the notions of positive and negative occurrences of a propositional symbol in a *nnf* are introduced.

Definition 15. If A is a *nnf* and p a propositional symbol, it is said that *p is positive* (resp. *negative*) in A if all the occurrences of p in A are not (resp. are) preceded by the conective $\neg$.

It is easy to prove the following theorem:

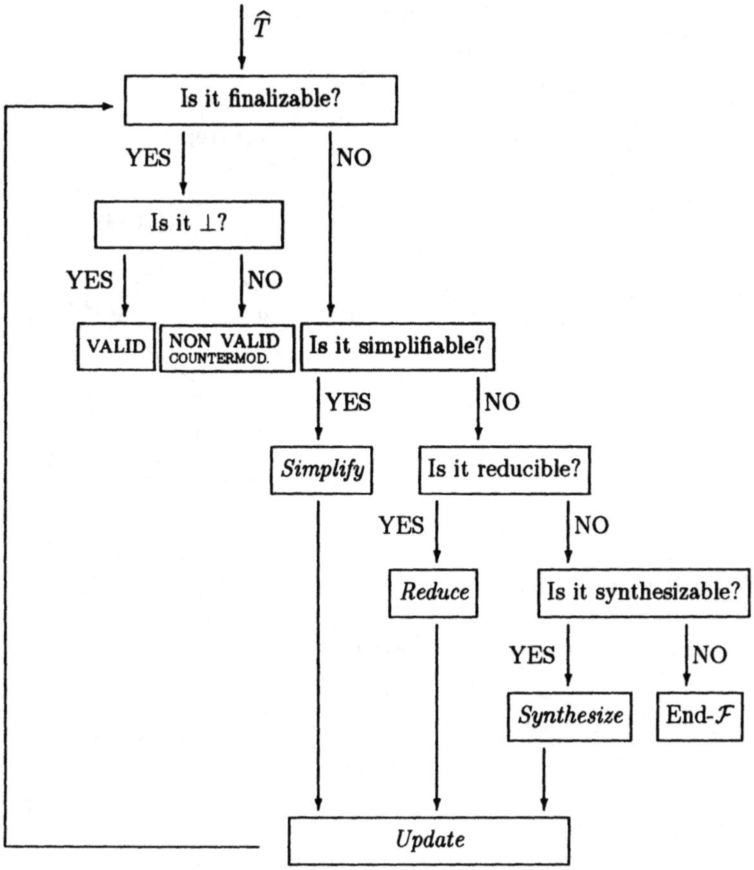

Fig. 4. The tree-transformation $\mathcal{F}$

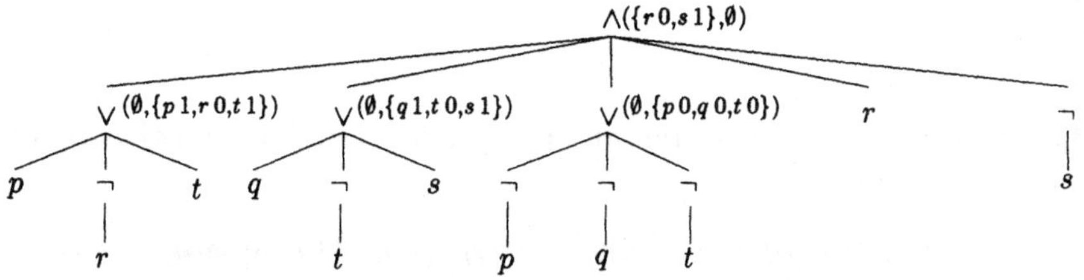

Fig. 5. The tree $\widehat{T_C}$.

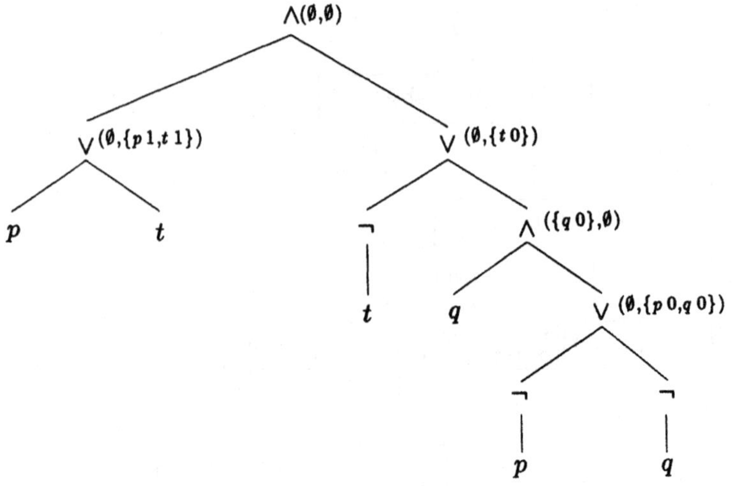

Fig. 6. The tree $\widehat{T_E}$.

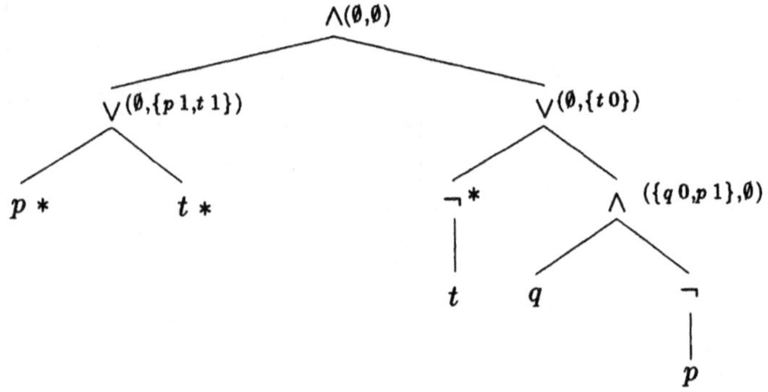

Fig. 7. The tree $\widehat{T_F}$. The asterisk (∗) beside a literal denotes that such literal have been *reduced*.

Theorem 16. *Let A be a nnf and p a positive (resp. negative) propositional symbol in A, then*

$$A \text{ is unsatisfiable iff } A[p/\top] \text{ (resp. } A[p/\bot]) \text{ is unsatisfiable.}$$

The tree-transformation $\mathcal{S}$, while positive or negative propositional symbols exist, makes the substitutions determined by the previous theorem together with the corresponding *updates*. If $\widehat{T}$ is the input of $\mathcal{S}$ and $\mathcal{S}(\widehat{T}) \neq \widehat{T}$ then $\mathcal{S}(\widehat{T})$ will be the input of $\mathcal{F}$ (once again), otherwise $\mathcal{S}(\widehat{T})$ will be the input of the tree-transformation $\mathcal{E}$ (to be defined in the section below).

Example 3. For the tree $\widehat{T_F}$ of Fig. 7 the symbol q is positive, therefore, substituting q by $\top$ and *updating*, the tree $\widehat{T_G}$ in Fig. 8 is obtained. On this tree neither $\mathcal{F}$ nor $\mathcal{S}$ can make any transformation.

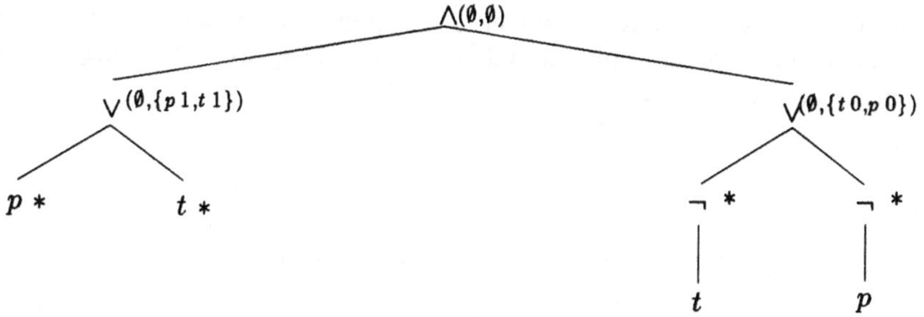

Fig. 8. The tree $\widehat{T_G}$.

3.3 The Tree Transformation $\mathcal{E}$

The transformation $\mathcal{E}$ is just the translation in terms of tree-transformations of the idempotence laws $A \wedge A \equiv A$ and $A \vee A \equiv A$. Therefore, if $\mathcal{E}(\widehat{T_A}) = \widehat{T_B}$ then $A \equiv B$.

To execute $\mathcal{E}$ one has to check whether a node has two equal children and, in that case, remove the duplications. In this process, the labels (Δ_0, Δ_1) are powerful tools to avoid comparations since for the equality of two subtrees, it is necessary the equality of the labels (Δ_0, Δ_1) of their roots.

In the example described, $\mathcal{E}$ does not act on $\widehat{T_G}$ and the input of the next transformation, ($\wedge$-$\vee$)-*par*, will be the tree $\widehat{T_G}$ in Fig. 8.

3.4 The Tree-Transformation ($\wedge$-$\vee$)-*par*

As indicated in the introduction, TAS-D^{++} confirms the suspects derived from Cook's theorem [5] that any algorithm to treat SAT has exponential complexity in the worst case. On the other hand, the problems solved by $\mathcal{F}$, $\mathcal{S}$ and $\mathcal{E}$ are "tractable" problems. Consequently, the exponential complexity of TAS-D^{++} is due to its last stage, thus TAS-D^{++} will be a good method provided this last stage has an acceptable behaviour for most inputs. In our opinion, this will only be possible if the following aims are reached:

1. Make feasible the parallel execution of non-avoidable distributions.

2. Avoid as many distributions of $\wedge$ over $\vee$ as the structure of the formula admits. For this, the minimum non-avoidable distribution is executed (using ($\wedge$-$\vee$) below), and then the tree-transformations $\mathcal{F}$, $\mathcal{S}$ and $\mathcal{E}$ are used again on each generated subtask.

These two goals have led the design of the tree-transformation ($\wedge$-$\vee$)-*par*, which is based on the *tree-transformation* ($\wedge$-$\vee$) defined below:

The tree-transformation ($\wedge$-$\vee$) is what we mean as the minimum unavoidable distribution, it associates to every tree, $\widehat{T}$, shaped as in the left of Fig. 9 the tree to the right, and *updates* only the labels of the nodes $\wedge(i)$ (these are the roots of each generated subtask).

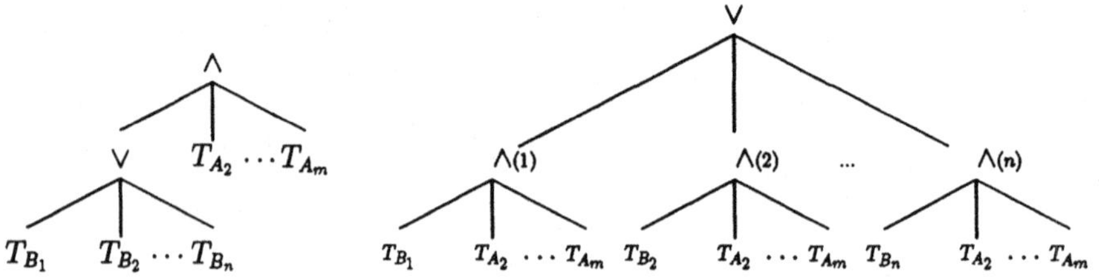

Fig. 9. The tree-transformation ($\wedge$-$\vee$).

The following theorem, which proof is immediate by the distributive laws of $\wedge$ with regard to $\vee$, justifies the soundness of the previous transformation.

Theorem 17. *If* $(\wedge\text{-}\vee)(\widehat{T_B}) = \widehat{T_C}$ *then* $B \equiv C$.

Now, ($\wedge$-$\vee$)-*par* is described by the flow diagram in Fig. 10.

The output of ($\wedge$-$\vee$)-*par* is:

- NON-VALID if some of the generated subtasks has the output NON-VALID.
- VALID if all of the generated subtasks have the output VALID.

Remark. The trace of the execution of ($\wedge$-$\vee$)-*par* can be represented by an n-ary tree, denoted $\mathrm{Tr}_{(\wedge\text{-}\vee)\text{-}par}$, which nodes are labeled with syntactic trees (this is a tree of trees). The root of the *execution trace tree* is the input of ($\wedge$-$\vee$)-*par*; any node has just one child if it is the output of $\mathcal{F}$, $\mathcal{S}$ or $\mathcal{E}$ and, otherwise, it has as many children as generated subtasks. Obviously, every leaf of the *execution trace tree* is $\perp$ or $\top$ (all of them are $\perp$ if the output is VALID or at least one leaf is $\top$ if the output is NON-VALID).

Example 4. For the formula in the example 1 the input to ($\wedge$-$\vee$)-*par* is the tree $\widehat{T_G} = (\mathcal{E} \circ \mathcal{S} \circ \mathcal{F})(\widehat{T_{\neg A(+)}})$ in Fig. 8 (see examples 1 to 3). ($\wedge$-$\vee$)-*par* begins with an application of $\wedge$-$\vee$ producing the tree $\widehat{T_H}$ in Fig. 11 in which there are two generated subtasks, marked in the figure by (1) and (2).

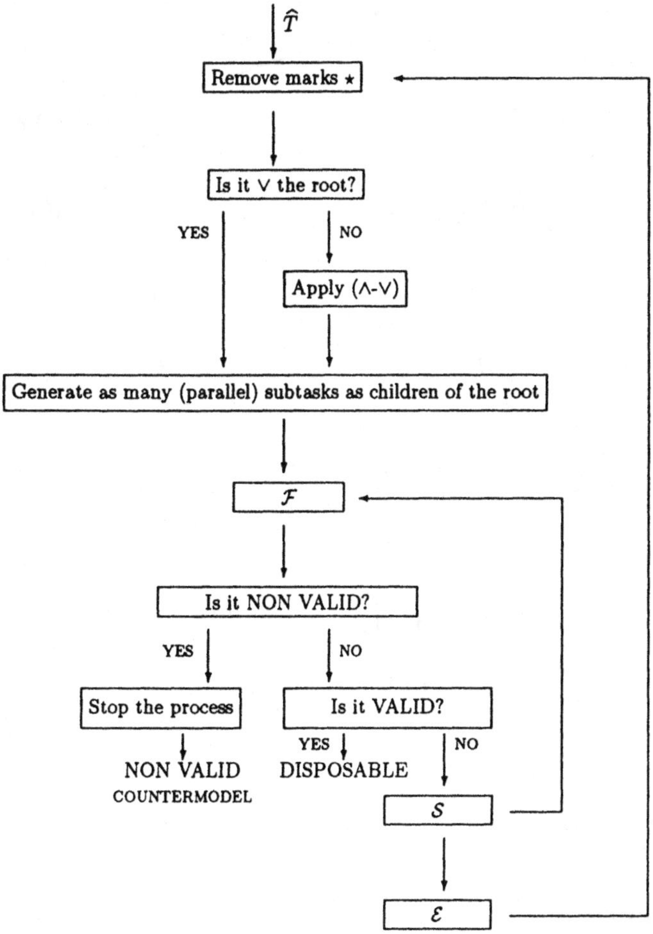

Fig. 10. The tree-transformation (∧-∨)-*par*.

As these two subtasks are *reducible* (completely 0-reducible), after applying *reduce* and *update* in parallel processes the tree $\widehat{T_I}$ in Fig. 12 is obtained.

Now, (1) and (2) are *finalizable* trees and the execution of $\mathcal{F}$ over any of them will produce the output NON-VALID. Therefore TAS-D^{++} ends with the output NON-VALID and the countermodel described in Sect. 4.

Now, when all of the processes of TAS-D^{++} have been introduced and the distinct stages have been defined, the following theorem, consequence of theorems 14, 16 and 17, states the soundness and completeness of TAS-D^{++}.

Theorem 18 Soundness and completeness of TAS-D^{++}.
Let A be a wff then A is valid if and only if TAS-D^{++}($T_{\neg A}$) = "VALID".

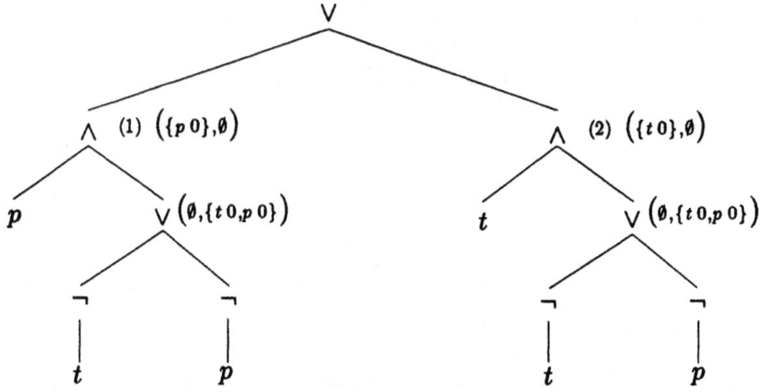

Fig. 11. The tree $\widehat{T_H}$.

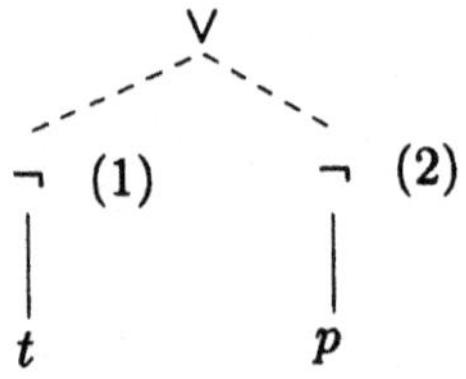

Fig. 12. The tree $\widehat{T_I}$.

3.5 Examples

1. For the formula $A = (p \to (q \to r)) \to ((p \to q) \to (p \to r))$ TAS-D^{++} takes $\neg A$ and executes this way:

   $\neg A \Longrightarrow$ *(sign)* $\Longrightarrow (\neg p \vee \neg q \vee r) \wedge (\neg p \vee q) \wedge p \wedge \neg r \Longrightarrow$ *(reduce p and $\neg r$)* $\Longrightarrow$
   $\Longrightarrow \neg q \wedge q \Longrightarrow$ *(simplify)* $\Longrightarrow \bot$ and then A is VALID.
2. For the formula A defined by

$$\Big(\neg\big((t \to (q \wedge \neg r)) \to ((t \vee p) \wedge \neg(r \to s))\big) \to \big((p \to s) \vee ((q \vee t) \to r)\big)\Big)$$

$$\to \neg\Big(\big(((p \vee q) \to (q \to s)) \to s\big) \vee \neg\big((\neg r \wedge t) \to (q \vee (t \wedge \neg p))\big)\Big)$$

TAS-D^{++} stops after executing $\mathcal{F}$, since $\widehat{T_{\neg A}}$ is *finalizable* with $\Delta_1(\neg A) = \{s\,1\} \neq \emptyset$, thus, just labeling $\widehat{T}_{\neg A}$ we obtain the information A is NON-VALID.

3. For $A = \Big(\big((p \rightarrow \neg s) \rightarrow (q \wedge \neg r)\big) \wedge (\neg q \vee r)\Big) \rightarrow$
$$\Big((p \rightarrow \neg s) \rightarrow \Big((p \rightarrow q) \rightarrow \big((\neg p \vee \neg s) \wedge (\neg q \vee r)\big)\Big)\Big)$$
after applying $\mathcal{F}$, (since $\widehat{T_{\neg A(+)}}$ is *synthesizable*) the tree $\widehat{T_B} = \widehat{T_{\neg A,p0}}$ is obtained, where
$$B = ((p \wedge s) \vee (q \wedge \neg r)) \wedge (\neg q \vee r) \wedge (\neg p \vee (\neg s \wedge q)) \wedge ((p \wedge s) \vee (q \wedge \neg r))$$
After executing $\mathcal{E}$ (eliminating a subtree $T_{(p \wedge s) \vee (q \wedge \neg r)}$) the tree $\widehat{T_C}$ is obtained, where
$$C = ((p \wedge s) \vee (q \wedge \neg r)) \wedge (\neg q \vee r) \wedge (\neg p \vee (\neg s \wedge q))$$
the tree $\widehat{T_C}$ is the input of ($\wedge$-$\vee$)-*par* which, after the application of ($\wedge$-$\vee$), generates two subtasks with trees the $\widehat{T_D}$ and $\widehat{T_E}$ corresponding to the formulae
$$D = p \wedge s \wedge (\neg q \vee r) \wedge (\neg p \vee (\neg s \wedge q)) \quad \text{and} \quad E = q \wedge \neg r \wedge (\neg q \vee r) \wedge (\neg p \vee (\neg s \wedge q))$$
Both tasks are completely 0-reducible ($\{p\,0, s\,0\} = \Delta_0(D)$ and $\{q\,0, r\,1\} = \Delta_0(E)$), after *reducing* the *updating* produces a $\perp$ in both tasks; therefore, both of them are *finalizable* with output VALID and, thus, the output of TAS-D^{++} is VALID.

4 TAS-D^{++} is a Model Building Method

As the method of "semantic tableauxs" [11], TAS-D^{++} gives a model for $\neg A$ if A is non-valid. This model is obtained just collecting the information of the unitary partial interpretations that allowed either *completely 0-reduce* (when applying $\mathcal{F}$), or the elimination of positive and/or negative symbols when applying $\mathcal{S}$) or, possibly, *finalize* because $\Delta_1 \neq \emptyset$.

Concretely, suppose that

- $\widehat{T}_{A_1}, \ldots, \widehat{T}_{A_n}$ are the trees in the trace of TAS-D^{++} before applying ($\wedge$-$\vee$)-*par*.
- if the transformation ($\wedge$-$\vee$)-*par* has been used, $\widehat{T}(n)$ is the node in $\mathrm{Tr}_{(\wedge\text{-}\vee)\text{-}par}$ that ends the execution of ($\wedge$-$\vee$)-*par*, and $\rho(n)$ is the branch in $\mathrm{Tr}_{(\wedge\text{-}\vee)\text{-}par}$ determined by $\widehat{T}(n)$, let $\widehat{T}_{B_1}, \ldots, \widehat{T}_{B_m}$ the trees in $\rho(n)$.

The following will be denoted

1. $C_0 = \bigcup_{i=1}^{n} \Delta_0(\widehat{T}_{A_i});\qquad C_{\rho(n),0} = \bigcup_{i=1}^{m} \Delta_0(\widehat{T}_{B_i});$
2. $C_{\mathcal{S}} = \bigcup_{i=1}^{n} \Delta_+(\widehat{T}_{A_i}) \cup \bigcup_{i=1}^{n} \Delta_-(\widehat{T}_{A_i}) \qquad C_{\rho(n),\mathcal{S}} = \bigcup_{i=1}^{m} \Delta_+(\widehat{T}_{B_i}) \cup \bigcup_{i=1}^{m} \Delta_-(\widehat{T}_{B_i})$ where
   $\Delta_+(\widehat{T}) = \{p_i\,1 \mid p_i$ is a positive prop. symbol removed to obtain $\mathcal{S}(\widehat{T})\}$ and
   $\Delta_-(\widehat{T}) = \{p_i\,0 \mid p_i$ is a negative prop. symbol removed to obtain $\mathcal{S}(\widehat{T})\}$

With the previous definitions, the following theorem shows how a countermodel can be constructed for every non-valid formula.

Theorem 19. *If TAS-D*$^{++}(T_{\neg A})$ = *"NON-VALID" then every valuation* V *such that* $V(p_i) = \bar{b}$ *if* $p_i b \in C_0 \cup C_{\rho(n),0}$ *and* $V(p_i) = b$ *if* $p_i b \in C_S \cup C_{\rho(n),S}$ *and (when applicable)* $V(p) = b$ *where* $p\,b$ *is the first element in lexicographic order of* $\Delta_1(A_n)$ *or* $\Delta_1(B_m)$ *is a countermodel for* A.

4.1 Examples

1. For the formula of example 1:
   (a) $C_0 = \{r\,0, s\,1\}$ (see Fig. 5),
   (b) $C_S = \Delta_+(\mathcal{F}(\widehat{T_{\neg A(+)}})) = \{q\,1\}$ (see example 3),
   Now, considering that TAS-D^{++} stops with the subtask (1), see Figs. 11 and 12,
   $$C_{\rho(1),0} = \{p\,\bar{0}\} = \{p\,1\}$$
   and since, $\Delta_1(\widehat{T}_{\neg t \vee \neg p}) = \{t\,0, p\,0\}$, TAS-D^{++} gives the following countermodel V for A:
   $$V(r) = V(q) = V(p) = 1 \qquad V(s) = V(t) = 0$$
2. For the formula in the example 2 of Sect. 3.5, since $T_{\neg A(+)}$ is finalizable with $\Delta_1(\neg A(+)) = \{s\,1\}$, every valuation V such that $V(s) = 1$ is a countermodel for A.

5 Extensions

The method TAS-D^{++} to verify the validity of a formula might be used in the usual manner to verify the validity of an inference:

Given an inference *Inf* with hypothesis $H_1, \ldots, H_n$ and conclusion C then,

Inf is valid if and only if TAS-D$^{++}(\Omega)$ = TAS-D$^{++}(T_{H_1 \wedge \cdots \wedge H_n \wedge \neg C})$ = "VALID"

the method of resolution and the method of semantic tableauxs use the equivalence shown above to study the validity of an inference. However TAS-D^{++}, to face efficiently the additional problem of the size (number of formulae of Ω), includes new strategies that allow to work with a "significative" subset, $\Omega_{T_F} \subset \Omega$ in order to either finish the algorithm or collect information enough to decrease the size of $\Omega - \Omega_{T_F}$. Therefore, not all the formulae in Ω are to be considered simultaneously, while we try to determine whether the structures of the hypothesis and the conclusion give either information about the unsatisfiability of the sequence Ω (and then TAS-D^{++} ends) or information which allow to transform the sequence Ω in a smaller-sized sequence Ω' simultaneously unsatisfiable with Ω.

The extension of TAS-D^{++} for first-order logic is given essentially as follows:

$$\text{TAS-D}^{++}_{FO} = \text{TAS-D}^{++}_{PROP} + \text{Unification in the definition of } \Delta_i$$

6 Conclusions

In this work a new ATP for classical logic named TAS-D^{++} has been presented; this is a refutation method that does not require the conversion to normal form. TAS-D^{++} has been implemented, see [1], and it has been tested and confronted with an implementation of the semantic tableauxs method using a random *wff* generator (the length of the formulae being a ramdom number between 1 and 100).

The implementation differs in some points with the method presented here:

- To make clearer the explanation of the method, the processes *sign, label* and *simplify* have been distinguished, but it is important to note that these three processes are executed in a single traverse of the syntactic tree.
- The sets Δ_0 and Δ_1 are defined as sets, but they have been implemented as lists.
- In the description of ($\wedge$-$\vee$) the distribution is made with respect to the leftmost branch, the implementation makes the distribution with respect to the smaller sized branch.

The following table shows the average ratio between the time consumed by the tableaux method and TAS-D^{++} depending the length of the formulae (the parenthesized number shows the number of formulae generated with that length):

Length (No. of formulae)	1–24 (43)	25–49 (38)	50–74 (52)	75–100 (47)
Ratio	1.01	6.61	26.02	71.81

Two important features of the method are the following: all of the processes involved in $\mathcal{F}$ and $\mathcal{S}$ are *linear* and, the use of the labels (Δ_0, Δ_1) in $\mathcal{E}$ allows one to improve outstanding algorithms for tree matching (see for example the introduced in [7]). The use of $\mathcal{F}$ and $\mathcal{S}$ inside ($\wedge$-$\vee$)-*par* allows one to improve the efficiency so we can affirm that TAS-D^{++} fulfills the goals sought after in this work.

In [2] can be found the extension of TAS-D^{++} to treat inferences and to first-order logic.

7 Acknowledgements

The authors would like to thank Jaume Agustí, John Darlington, Luis Fariñas del Cerro and Dov Gabbay for their invaluable comments in our discussions about this work.

References

1. G. Aguilera, I.P. de Guzmán and M. Ojeda. A graphical implementation of TAS-D^{++}. Submitted to GULP-PRODE'94, 1994.
2. G. Aguilera, I.P. de Guzmán and M. Ojeda. TAS-D^{++}: A parallel and computationally efficient ATP. Obtaining information from syntactic trees. Technical Report, Dept. Matemática Aplicada, 1993.
3. W. Bibel. Automated Theorem Proving. Vieweg & Sohn, 1987.
4. W. Bledsoe and D. Loveland, eds. Automated Theorem Proving after 25 years. Contemporary Mathematics, . **24** 1984.
5. S. A. Cook. The complexity of Theorem-Proving Procedures Proc. of 3rd annual ACM Symposium on the Theory of Computing. (1971) 151–158
6. R. Coferra and N. Zabel. Extending Resolution for Model Construction. Lect. Notes in Comp. Sci. **478** (1991)
7. Roberto Grossi. On finding common subtrees. Theor. Comp. Sci. **108** (1993)
8. Inma P. de Guzmán and M. Enciso. Topological semantics and ATP's for temporal logics. Technical Report, Dept. Matemática Aplicada, 1994.
9. F.R. Pelletier. Seventy-five problems for testing automatic theorem provers. Journal of Automated Reasoning **2**: (1986) 191–216
10. W.V. Quine. Methods of Logic. Henry Holt, New York, (1950)
11. R.M. Smullyan. First-Order Logic. Springer-Verlag, Berlin, (1968)
12. L.A. Wallen. Automated proof search in non-classical logics. The MIT Press, Cambridge MA, (1990)
13. S. Winker. Generation and verification of finite models and counter using an automated theorem prover answering two open questions. Journal of ACM, **29(2)** (1982) 273–284
14. L. Wos and S. Winker. Open questions solved with the assistance of AURA. Contemporary Mathematics **24** 1984

A Unification of Ordering Refinements of Resolution in Classical Logic

Hans de Nivelle
Department of Mathematics and Computer Science
Delft University of Technology
Julianalaan 132, 2628BL Delft, The Netherlands,
email: nivelle@cs.tudelft.nl

Abstract

We introduce a general notion of resolution calculus. We show that it is possible to obtain a resolution calculus from a semantic tableau calculus that can be described by rules of a certain type. It is also possible to obtain an ordered resolution calculus from a semantic tableau calculus. We apply this to classical logic to obtain simple proofs of the completeness of several refinements of resolution in classical logic.

1 Introduction

The purpose of this paper is 3-fold:

1. Introduce a general notion of resolution calculus, and show that it is possible to obtain a resolution calculus from any semantic tableau calculus. We give a general result about the existence of ordering refinements for resolution calculi.

2. Apply this general result to resolution in classical logic and show that it increases insight in ordering refinements for resolution in classical logic.

3. Introduce a new ordering refinement for resolution in classical logic, L-ordered resolution, and show that it subsumes A-ordered resolution, semantic resolution and A-ordered semantic resolution. The completeness of L-ordered resolution follows from the fact that ordering refinements exist for every resolution calculus.

(1) It has been stressed by various authors that there is a tight relation between Gentzen-sequent calculi, resolution calculi, and semantic tableaux calculi. ([Mints88]) We make this relation very precise, and show that it is possible to base a resolution calculus on the rules that specify a semantic tableau calculus. This method was implicitly used in ([Niv93b]). This method could also be used to obtain other resolution calculi, as defined in ([EnjFar89]).

It is possible to immediately obtain a resolution calculus from a semantic tableau calculus. We define a resolution calculus as a set of rules, which defines a semantic tableau calculus, and some structure that defines which literals and terms are allowed, so that it can handle predicate logic. We give the following 2 theorems:

1. For every resolution calculus, for every set of clauses: This set of clauses has a semantic tableau refutation iff it has a resolution refutation based on this set of rules.
2. For every resolution calculus, for any ordering on literals which satisfies some properties of well-behaviour, for any set of clauses: This set of calculi has a semantic tableau refutation iff it has an ordered resolution refutation based on this set of rules, and on this ordering.

Because in general the correctness (= soundness and completeness) of a semantic tableaux calculus, for a given logic is easier to prove than the correctness of a resolution calculus, we have simplified the task of proving the correctness of a resolution calculus.

(2) We will show that the second theorem mentioned above can be used to obtain the correctness of several known refinements for resolution in classical logic. These refinements are: **(a)** A-ordered resolution, **(b)** semantic resolution, **(c)** A-ordered semantic resolution, and **(d)** indexed (= lock)-resolution. We think that the proofs are simple and increase insight.

(3) The completeness of **(a)**, **(b)**, and **(c)** will be established by means of a new refinement: L-ordered resolution. It is known that A-ordered resolution is complete. In A-ordered resolution, the A-ordering is extended to an ordering on literals, by comparing the atoms of which the literals are constructed. However completeness is not lost, when an ordering on the literals that is not obtained from an A-ordering is used. Using an idea which is present in ([Lovlnd78], the example after 3.3.3), it is possible to order false literals before true literals with an L-ordering, to obtain semantic resolution.

2 Basic Definitions

Definition 1 A first order language is an ordered triple $\mathcal{L} = (F, P, V)$, in which:

- F is a finite set of function symbols and arities,
- P is a finite set of predicate symbols and arities,
- V is a countable set of variables.

The *terms* of $\mathcal{L}$ are defined as the set of objects that can be constructed from F and V. An *atom* of $\mathcal{L}$ is of the form $p(t_1, \ldots, t_n)$, where p has arity n, and $t_1, \ldots, t_n$ are terms of $\mathcal{L}$. A *literal* of $\mathcal{L}$ is either an atom (a *positive literal*), or the negation of an atom. (a *negative literal*). A literal is *ground* if it does not contain any variables. A *formula* is constructed in the usual way with the usual operators. ($\neg, \rightarrow, \wedge, \vee, \leftrightarrow, \forall$ and $\exists$). We define an *interpretation*, and a *model* of a first order language as usual. ([ChngLee73])

Definition 2 A *substitution* Θ (w.r.t. a first order language $\mathcal{L} = (F, P, V)$) is a finite set of the form $\{v_1 := t_1, \ldots, v_n := t_n\}$, where each $v_i \in V$, each t_i is a term of $\mathcal{L}$, no $v_i = t_i$, for no $i \neq j$, we have $v_i = v_j$. The *effect* of a substitution on a literal A is defined as usual: As the result of replacing simultaneously all v_i that occur in A by t_i. Notation: $A\Theta$. A literal A is an *instance* of B if there is a substitution Θ, such that $A = B\Theta$. A is a *ground instance* of B if A is an instance of B, and A is ground.

Let $\mathcal{E}$ be a set of equations of the form $A_1 = B_1, \ldots, A_m = B_m$. A *unifier* of $\mathcal{E}$ is a substitution Θ, such that $A_1\Theta = B_1\Theta, \ldots, A_m\Theta = B_m\Theta$. A *most general unifier* Θ is a unifier of $\mathcal{E}$, such that for any unifier Σ, of $\mathcal{E}$, there is a substitution Ξ, such that $\Sigma = \Theta \cdot \Xi$. Here $\cdot$ denotes the composition of Θ and Ξ.

If there exists a unifier, then there exists a most general unifier. ([Robins65])

Definition 3 A *clause* of a first order language $\mathcal{L}$ is a finite, possibly empty set of literals of $\mathcal{L}$. The *effect of a substitution* on a clause is obtained by applying the substitution memberwise. If C is a set of clauses, then ground(C) is the set of all ground instances of elements of C. An *interpretation* I of a set of clauses of a first order language is a set of ground literals of $\mathcal{L}$, not containing a complementary pair. I *satisfies*/is a *model* of a set of clauses, if for all $c \in \text{ground}(C)$, we have $c \cap I \neq \emptyset$.

There exist algorithms ([ChngLee73], [Lovlnd78]), that transform any formula F of any first order language into a set of clauses C, such that F is satisfiable iff C is satisfiable.

3 Resolution Systems

A semantic tableaux system is a system which tries to prove the unsatisfiability of a formula (or set of formulae) by adding formulae that necessarily hold in an interpretation where the original formula holds. These rules may be non-deterministic because it may be the case that the rule is of the form: It is a necessary condition for a certain formule F to hold in an interpretation that either F_1 or F_2 holds in that interpretation. When such a rule is applied the semantic tableau has to split. There are also rules that close a branch of the semantic tableau. So there are two types of rules:

1. Rules that can be used to add necessary conditions (i.e. new formulae) for a formula to be satisfied by an interpretation.

2. Rules that specify when a set of formulae can not be satisfied by an interpretation.

Both types of rules can be fitted into the following scheme:
$A_1, \ldots, A_p/B_1, \ldots, B_q$, to be read as: If all A_i hold, then one of the B_j holds. For $q > 0$, the rule has type 1, for $q = 0$, the rule has type 2. The rules of type 1 may take the form $A \wedge B/A$, or $A \wedge B/B$ in classical logic. A rule that forces a branch to split up is the rule $A \vee B/A, B$. The rules of type 2 may take the form $A, \neg A/$ in classical logic.

For a semantic tableaux system to work it must be sound and complete. The soundness is usually not problematic. It can be obtained by checking the rules. The completeness is also easily obtained usually. For the completeness the only necessary condition is that a branch that is closed under all rules of type 1, and from which no contradiction can be derived with a rule of type 2, has a model, i.e. there exists a (simultaneous) model of all formulae in the branch.

Example 4 For classical, propositional logic the following set of rules has this property:

$$A \rightarrow B/\neg A, B;\ A \leftrightarrow B/A \rightarrow B;\ A \leftrightarrow B/B \rightarrow A;\ A \wedge B/A;\ A \wedge B/B;$$

$$A \vee B/A, B;\ \neg(A \rightarrow B)/A \wedge \neg B;\ \neg(A \leftrightarrow B)/\neg(A \rightarrow B), \neg(B \rightarrow A);$$

$$\neg(A \wedge B)/\neg A \vee \neg B;\ \neg(A \vee B)/\neg A \wedge \neg B;\ \neg\neg A/A;$$

$$A, \neg A/;$$

Now it is the case that any set of rules of this type $A_1, \ldots, A_p/B_1, \ldots, B_q$ can be immediately used in a resulution calculus. It has been proven in ([Niv93a]) that from every set of rules that defines a sound and complete semantic tableau calculus, a sound and complete resolution calculus can be obtained as follows: For any rule $A_1, \ldots, A_p/B_1, \ldots, B_q$ let the resolution rule be $\{A_1\} \cup R_1, \ldots, \{A_p\} \cup R_p \vdash \{B_1, \ldots, B_q\} \cup R_1 \cup \cdots \cup R_p$. It is also proven that the resolution calculi that are thus obtained remain complete even when an ordering refinement is applied.

We will define a resolution calculus as a set of rules and sets of predicate names, function names and variables that specify which literals are allowed.

First we give an example that shows how the rules of Example 4 can be used as resolution calculus. This calculus is sound and complete for classical, propositional logic.

Example 5 We will prove that $\models (P \rightarrow Q) \rightarrow (\neg\, Q \rightarrow \neg\, P)$.

(1)	$\{\neg\,[(P \rightarrow Q) \rightarrow (\neg\, Q \rightarrow \neg\, P)]\}$	(negation of the conclusion)
(2)	$\{(P \rightarrow Q) \wedge \neg\,(\neg\, Q \rightarrow \neg\, P)\}$	(from 1, rule $\neg\,(A \rightarrow B)/A \wedge \neg\, B$)
(3)	$\{P \rightarrow Q\}$	(from 2, with rule $A \wedge B/A$)
(4)	$\{\neg(\neg\, Q \rightarrow \neg P)\}$	(from 2, with rule $A \wedge B/B$)
(5)	$\{\neg\, P, Q\}$	(from 3, with rule $A \rightarrow B/\neg\, A, B$)
(6)	$\{\neg\, Q \wedge \neg\,\neg\, P\}$	(from 4, rule $\neg\,(A \rightarrow B)/A \wedge \neg\, B$)
(7)	$\{\neg\, Q\}$	(from 6, with rule $A \wedge B/A$)
(8)	$\{\neg\,\neg\, P\}$	(from 6 with rule $A \wedge B/B$)
(9)	$\{P\}$	(from 8, with rule $\neg\,\neg\, A/A$)
(10)	$\{Q\}$	(from 5 and 9, with rule $A, \neg\, A/$)
(11)	$\{\}$	(from 7 and 10, with rule $A, \neg\, A/$)

Definition 6 A resolution system $S = (F, P, V, \mathcal{R})$ consists of the following parts:

- F is a finite set of function symbols and arities,
- P is a finite set of predicate symbols,
- V is a countable set of variables,
- $\mathcal{R}$ is a (not necessarily finite) set of rules of the form $A_1, \ldots, A_p/B_1, \ldots, B_q$, where the A_i and the B_j are atoms of S. (Atoms of S are defined below) It is possible that $q = 0$, but it must be the case that $p > 0$.

The *terms* of S, and the *atoms* of S are defined in the same manner as for first order languages, from F, P, and V. A *clause* of S is a finite set of atoms of S. A resolution system S, such that every member of P has arity 0, will be called *propositional*. In that case we omit F and V and write $S = (P, \mathcal{R})$. The *effect* of a substitution on a clause, or on a rule is defined memberwise:

$$\{A_1, \ldots, A_p\}\Theta = \{A_1\Theta, \ldots, A_p\Theta\},$$

$$(A_1, \ldots, A_p/B_1, \ldots, B_q)\Theta = A_1\Theta, \ldots, A_p\Theta/B_1\Theta, \ldots, B_q\Theta.$$

If C is a set of clauses, then ground(C) is the set of all ground instances of members of C. Similarly ground($\mathcal{R}$) is the set of all ground instances of members of $\mathcal{R}$. ground(S) can be seen as a propositional resolution system, and ground(C) is a clause of this system.

Intuitively a clause $\{A_1, \ldots, A_p\}$ should be read as: One of the A_i holds. A rule $A_1, \ldots, A_p/B_1, \ldots, B_q$ should be read as: If all A_i hold then a B_j holds. If a clause or a rule contains a variable then this variable should be read as universally quantified.

Definition 7 If $S = (P, \mathcal{R})$, is a propositional resolution system, then *an interpretation* I of S is a subset $I \subseteq P$, such that for every rule $A_1, \dots A_p / B_1, \dots, B_q$ for which $\{A_1, \dots, A_p\} \subseteq I$, we have $\{B_1, \dots, B_q\} \cap I \neq \emptyset$.

If c is a clause of S, then an *interpretation* of c is an interpretation of S. A model M of c is an interpretation of c, such that $I \cap c \neq \emptyset$. A model/interpretation of a set of clauses is a model/interpretation of every clause in the set.

A model/interpretation of a resolution system S that is not propositional is a model/interpretation of ground(S).

A model/interpretation of a set of non-propositional clauses C of a resolution system S is a model/interpretation of ground(C) of ground(S).

A set of clauses C of a resolution system S is *consistent* if it has a model. It is *inconsistent* if it has no model.

Models correspond to open branches in semantic tableaux.

Example 8 Take $S = (P, \mathcal{R})$, with $P = \{a, b, c, d\}$, $\mathcal{R} = \{(a, c/), (a/b, d)\}$. $\{a, b\}$ and $\{a, d\}$ are interpretations of S. $\emptyset$ is an interpretation of S. No interpretation of S contains both a and c. If $C = \{\ \{c\}, \{a, d\}\ \}$ is a set of clauses, then $\{c\}$ is not a model of C, but $\{c, d\}$ is a model. The set of clauses $\{\ \{a\}, \{c\}\ \}$ is inconsistent.

We now define the resolution operator:

Definition 9 Let $S = (F, P, V, \mathcal{R})$ be a resolution system. We define the resolution rule $\vdash$ and the factoring rule $\vdash_f$.

Resolution rule (In the following we assume that there are no overlapping variables in $c_1, \dots, c_p, A_1, \dots, A_p, B_1, \dots, B_q$. Variables in clauses and rules can be renamed freely) If there exist

1. a rule $A_1, \dots, A_p / B_1, \dots, B_q \in \mathcal{R}$,
2. clauses $c_1, \dots, c_p$ with atoms $a_1 \in c_1, \dots, a_p \in c_p$,
3. A most general unifier Θ of the set of equations

$$A_1 = a_1, \dots, A_p = a_p, \text{ then}$$

$$c_1, \dots, c_p \vdash (c_1\Theta)\backslash\{A_1\Theta\} \cup \cdots (c_p\Theta)\backslash\{A_p\Theta\} \cup \{B_1\Theta, \dots, B_q\Theta\}.$$

The atoms $a_1, \dots, a_p$ are called the *atoms resolved upon.*

Factoring rule If a clause c has two atoms $a_1 \in c$, and $a_2 \in c$, for which there exists a most general unifier Θ of $a_1 = a_2$, then

$$c \vdash_f c\Theta.$$

We define the generations that can be obtained from a set of clauses C as usual: The first generation of clauses equals C. Every next generation can be obtained from the previous generations, as the set of clauses that can be obtained in 1 step by $\vdash$, or $\vdash_f$.

Example 10 If $p(X), q(Y)/r(X,Y)$ is a rule, then

$$\{p(V), s(V)\},\ \{q(W), t(W)\} \vdash \{s(V), t(W), r(V,W)\},$$

$$\{p(0)\},\ \{q(1)\} \vdash \{r(0,1)\},$$

$$\{p(X,1), p(0,Y)\} \vdash_f \{p(0,1)\}.$$

Theorem 11 For any resolution system S and set C of clauses of this system, the following 3 statements are equivalent:

1. C is inconsistent with respect to S,
2. There is a resolution refutation of C, based on S,
3. There is a semantic tableaux refutation of C, based on S.

The correctness of this statement can be seen by noticing that resolution as defined in Definition 9, is in fact very similar to hyperresolution. The completeness proof can be obtained from the completeness of hyperresolution.

Now a resolution system can be applied to a certain logic when the following conditions are met:

1. There must be some algorithm that translates formulae F into sets of atoms, or sets of clauses.
   (a) If the algorithm translates into sets of clauses C, (as in resolution in classical logic) then it must be the case that F is satisfiable iff it is possible to choose a literal from every clause in ground(C), and find a simultaneous model (in the logic) for these literals.
   (b) If the algorithm translates into sets of literals L, (as in Example 5) then it must be the case that F is satisfiable iff ground(L) is satisfiable (in the logic).
2. A set of ground literals that satisfies all ground instances of rules (i.e. is a model of the resolution system) must have a model in the logic. (In most cases: Specifies a model)

It is also possible to obtain ordered resolution calculi.

Definition 12 Let $S = (F, P, V, \mathcal{R})$ be a resolution system. A *resolution ordering* of S is an ordering (i.e. an irreflexive, transitive relation), $\prec$ of the atoms of S, with the following additional property:

$$A \preceq B \Rightarrow A\Theta \preceq B\Theta, \text{ for any substitution } \Theta.$$

(Here $\preceq$ means: $\prec$ or $=$) This property is needed to ensure that lifting of ground deductions is possible.
We define the ordered factoring rule, and the ordered resolution rule:

- To the factoring rule we add the condition that one of the a_1, a_2 be $\prec$-maximal.(I.e., for no $\overline{a} \in c$, we have $a_1 \prec \overline{a}$)
- To the resolution rule we add in 2 the condition that a_1 is $\prec$-maximal in $c_1, \ldots, a_p$ is $\prec$-maximal in c_p.

Then Theorem 11 can be modified as follows:

Theorem 13 Let S be a resolution system, and let $\prec$ be a resolution ordering of S. Then, for any set of clauses C of S, the following 3 statements are equivalent:

1. C is inconsistent with respect to S,
2. there is an $\prec$-ordered resolution refutation of C, based on S.
3. there is a semantic tableaux refutation of C, based on S.

proof:
The remark under Theorem 11 also applies here. The theorem follows quickly from the completeness of ordered hyperresolution.

In the literature a distinction is made between applying orderings a priori and a posteriori. We apply the ordering a priori. When an ordering is applied a posteriori it is checked *after the resolvent is constructed* that the literal resolved upon is greater than every literal in the resulting clause.

This is essential for the application of A-orderings in decision procedures. ([Joy76], [FLTZ93]). It is possible to define a posteriori ordered resolution with resolution systems, but some care has to be taken.

4 Refinements of Classical Resolution

We will now apply the previous to classical logic to obtain the completenss of ordering refinements for resolution in classical logic. These refinements are:

1. A-ordered resolution,
2. semantic resolution,
3. A-ordered semantic resolution,
4. L-ordered resolution,
5. indexed resolution.

We will do this by proving the completeness of L-ordered resolution, and show that 1, 2 and 3 are a special form of L-ordered resolution. The completeness of L-ordered resolution follows immediately from the fact that resolution in classical logic is based on the rule $A, \neg A/$, and Theorem 13. The completeness of (5) can be obtained by making some small modifications in the proof of the completeness of (4).

For the propositional case the classical resolution rule is:

$$\{A\} \cup R_1, \{\neg A\} \cup R_2 \vdash R_1 \cup R_2.$$

It is easily verified that this is indeed equivalent to resolution using a rule $A, \neg A/$. In the non-propositional case, resolution in classical logic can be obtained with the rules $p(X_1, \ldots, X_n), \neg p(X_1, \ldots, X_n)/$ for every relevant predicate symbol. Indeed the conditions specified below Theorem 11 are satisfied:

Proposition 14 For every set of clauses C, every set of ground literals M that defines a model of C (in the sense of classical logic) is a model of C, in the sense of Definition 7. Also every set of ground literals M that defines a model of C, in the sense of Definition 7, can be extended to a set M' that defines a model of C, in the sense of classical logic.

We can now treat ordered resolution: An A-ordering is an irreflexive and transitive relation on the instances of the atoms that occur in a set of clauses, and that is preserved under substitutions: $A_1 \preceq A_2 \Rightarrow A_1\Theta \preceq A_2\Theta$. It is a well-known fact that resolution in classical logic remains complete when only the maximal elements under an A-ordering are used for resolution. The ordering on atoms is extended to an ordering on literals as follows: If $A \prec B$ then $A \prec \neg B, \neg A \prec B, \neg A \prec \neg B$. However it is also possible to define orderings on literals that are not the extension of an A-ordering. We will call such an ordering an L-ordering.

Definition 15 An L-ordering with respect to a set of clauses C is an ordering on the literals that can be constructed from the predicate symbols and function symbols in C, with the following additional property:

$$A \preceq B \Rightarrow A\Theta \preceq B\Theta, \text{ for any substitution } \Theta.$$

Theorem 16 *L*-ordered classical resolution is sound and complete.

proof:
It follows from Theorem 13

We give an example of an L-ordered refutation:

Example 17 Let $C = \{\ \{\neg\, a, \neg\, b\},\ \{a, \neg\, b\},\ \{b, c\},\ \{\neg\, c\}\ \}$. C is unsatisfiable. Let $\prec$ be the following L-ordering:
$b \prec \neg\, a \prec \neg\, b \prec c \prec \neg\, c \prec a$. An L-ordered refutation is:

(1)	$\{\neg\, a, \neg\, b\}$	(given)
(2)	$\{\neg\, b, a\}$	(given)
(3)	$\{b, c\}$	(given)
(4)	$\{\neg\, c\}$	(given)
(5)	$\{b\}$	(from 3 and 4)
(6)	$\{\neg\, a\}$	(from 1 and 5)
(7)	$\{\neg\, b\}$	(from 2 and 6)
(8)	$\{\}$	(from 5 and 7)

In ([Baum92]) it is shown that *L*-orderings can be combined with theory resolution. In ([FLTZ93]) it is shown that the completeness of *L*-ordered resolution can be obtained from the completeness of indexed resolution, but in neither of these places *L*-orderings are applied to other refinements.

Now that we have established the completeness of *L*-ordered resolution we can use it to prove the completeness of *A*-ordered resolution, semantic resolution, and *A*-ordered semantic resolution.

In ([Lovlnd78]) the concept of ordering O for a set of clauses S is used. He shows that semantic resolution can be described with an ordering by ordering all positive literals before all negative literals. Completeness proofs of resolution refinements consist of 2 parts in general: A proof of the propositional case and a proof of the lifting theorem. In ([Lovlnd78]) orderings are used to give a unified proof of the lifting theorem for all refinements that can be described with an ordering. His concept of ordering can not be used for the completeness proof of the propositional case, because it is too general: Ordered resolution is not complete. But *L*-orderings can still be used for describing semantic resolution, and *L*-ordered resolution is complete. So it is possible to give a unified proof. First we give some necessary definitions:

Definition 18 An interpretation I of a set of clauses C is a set of instances of literals of C, with the following properties:

1. For no instance of an atom A occurring in C, (either as A or as $\neg\, A$) both A and $\neg\, A$ are in I.

2. If both A and B are instances of literals in C, and B is an instance of A, then **(a)** $A \in I \Rightarrow B \in I$, and **(b)** $B \in I \Rightarrow A \in I$.

In ([Lovlnd78]) a set that satisfies (1) and (2a) is called a setting. A setting that also satisfies (2b) is called homogeneous there. In ([ChngLee73]) interpretations are defined as sets of (possibly negated) predicate symbols, that do not contain a complementary pair, so that the conditions 2 are automatically satisfied. In practice interpretations are mostly defined by generator sets, i.e. as $I = \{A\Theta \mid \Theta \text{ is a substitution, } A \in G\}$, for a finite G.

Definition 19 If I is an interpretation of C, and c is a clause, consisting of instances of literals occurring in C, then c is true in I iff $c \cap I \neq \emptyset$,
c is false in I iff $c \cap I = \emptyset$. A literal L is true in I iff $L \in I$.

Now *semantic resolution* based on I is defined by allowing resolution only between clauses c_1 and c_2, such that c_1 is true in I, and c_2 is false in I.

It is possible to combine semantic resolution with an A-ordering $\prec$ to obtain A-ordered semantic resolution: $c_1, c_2 \vdash c$ A-ordered semantically, iff c_1 is true in I, c_2 is false in I, and the literal resolved upon in c_2 is $\prec$-maximal in c_2.

We will now show that semantic resolution is complete by embedding semantic resolution into L-ordered resolution: Let I be an interpretation of a set of clauses C. Define the following L-ordering of C:

$$L_1 \prec L_2 \text{ iff } L_1 \text{ is false in } I \text{ and } L_2 \text{ is true in } I.$$

It is easily verified that this is an L-ordering.

Lemma 20 For any clauses c_1, c_2, and c, containing only instances of C: $c_1, c_2 \vdash c$ semantically based on I iff $c_1, c_2 \vdash c$, L-ordered, based on $\prec$.

proof:
Assume that $c_1, c_2 \vdash c$, L-ordered, based on $\prec$. Then literals $\pm A_1 \in c_1$, and $\mp A_2 \in c_2$, are resolved upon. Either $\pm A_1$, or $\mp A_2$ is true in I. Without loss of generality we can assume that $\pm A_1$ is true in I. (otherwise c_1 and c_2 can be exchanged) Then c_1 is true in I, because it contains $\pm A_1$. c_2 must be false in I, because $\mp A_2$ is false in I, and for any true literal L we would have $\mp A_2 \prec L$. But then $c_1, c_2 \vdash c$ semantically.

On the other hand assume that $c_1, c_2 \vdash c$ semantically. All literals in c_2 are $\prec$-maximal, because c_2 is false in I. This implies that the literal resolved upon in c_1 is true in I. But then this literal is $\prec$-maximal in c_1 and $c_1, c_2 \vdash c$, L-ordered, based on $\prec$.
end of proof
In the propositional case the ordering $\prec$ is highly partial. This suggests that better orderings are possible. A better ordering can be obtained by combining

$\prec$ with an A-ordering. Let I be an interpretation, and let $\sqsubset$ be an A-ordering. We define the following L-ordering:

$$L_1 \prec L_2 \text{ iff either } L_1 \text{ and } L_2 \text{ are false in } I, \text{ and } L_1 \sqsubset L_2, \text{ or}$$

$$L_1 \text{ is false in } I, \text{ and } L_2 \text{ is true in } I.$$

Lemma 21 For any clauses c_1, c_2 and c containing only instances of literals occurring in a set of clauses C, we have $c_1, c_2 \vdash c$, A-ordered semantically, based on $\sqsubset$ and I iff $c_1, c_2 \vdash c$, L-ordered, based on $\prec$.

proof:
Assume that $c_1, c_2 \vdash c$ L-ordered, based on $\prec$. Then literals $\pm A_1 \in c_1$, and $\mp A_2 \in c_2$ are being resolved upon. Exactly one of them is true in I. Without loss of generality assume that $\pm A_1$ is true in I, and that $\mp A_2$ is false in I. Then c_1 is true in I, and c_2 is false in c_2 because any true literal would be $\prec$-greater then c_2. Then also $\mp A_2$ is $\sqsubset$-maximal in c_2.

On the other hand assume that $c_1, c_2 \vdash c$, A-ordered semantically, based on I and $\sqsubset$. Then a true literal $\pm A_1$ from c_1, and a false literal $\mp A_2$ from c_2 is being resolved upon. Now $\pm A_1$ is $\prec$-maximal in c_1 because every true literal is $\prec$-maximal. $\pm A_1$ is $\prec$-maximal in c_2 because there are only false literals in c_2 and $\pm A_1$ is $\sqsubset$-maximal in c_2.

end of proof

This explains why semantic resolution and A-ordered resolution are compatible: Because the L-orderings that characterize semantic resolution and A-ordered resolution are compatible.

The ordering $\prec$, defined above to characterize A-ordered semantic resolution is still not total in the propositional case, because it does not specify any order between true literals. We can apply any A-ordering to the true literals.

For any set of clauses C, for every interpretation I of C, and for every two A-orderings $\sqsubset_1$ and $\sqsubset_2$, let $\prec$ be defined from:

$L_1 \prec L_2$ iff L_1 is false in I and L_2 is true in I,
L_1 and L_2 are true in I, and $L_1 \sqsubset_1 L_2$,
L_1 and L_2 are false in I, and $L_1 \sqsubset_2 L_2$.

This ordering corresponds to the following variant of semantic resolution:

$$c_1, c_2 \vdash c, \text{ semantically and } A\text{-ordered iff}$$

- c_1 is true in I and the literal resolved upon in c_1 is $\sqsubset_1$-maximal among the true literals in c_1, and
- c_2 is false in I and the literal resolved upon in c_2 is $\sqsubset_2$-maximal in c_2.

Finally we will show how Theorem 13 can be used to obtain the completeness of indexed resolution ([Boyer71]).

In indexed resolution, integers are attached to the literals in the initial clauses. When a resolution step is made the literals in the result inherit the indexes from the parent clauses. Only the literals in a clause with minimal index can be used in resolution. We will show that the completeness of indexed resolution can be obtained from Theorem 13, with a small adaptation of the construction.

Definition 22 Let C be an indexed set of clauses. C^i is obtained from C by replacing every positive, indexed literal $p(t_1, \ldots, t_n) : i$ by $(p, i)(t_1, \ldots, t_n)$, and by replacing every negative, indexed literal $\neg\, p(t_1, \ldots, t_n) : i$ by $(\neg\, p, i)(t_1, \ldots, t_n)$.
We define the indexed resolution system $S^i = (F^i, P^i, V^i, \mathcal{R}^i)$, where

- F^i is the set of function symbols that occur in C
- $P^i =$

$$\{(p, i) \text{ with arity } n \mid p \text{ occurs in } C, \text{ with arity } n, \text{ and } i \in \mathcal{Z}\} \cup$$

$$(\neg\, p, i) \text{ with arity } n \mid \neg\, p \text{ occurs in } C, \text{ with arity } n, \text{ and } i \in \mathcal{Z}\}.$$

- V^i is a countable set of variables at least containing the variables that occur in C.
- $\mathcal{R}^i = \{(p, i_1)(v_1, \ldots, v_n), (\neg\, p, i_2)(v_1, \ldots, v_n)/ \mid p$ occurs in C with arity n, and $i_1, i_2 \in \mathcal{Z}$ $\}$

The following is easy to see:

Lemma 23 Let C be an indexed set of clauses, let C^i be obtained from C, in the way that is described in Definition 22, and let S^i be the resolution system that is obtained in Definition 22. Then

$$C \text{ is satisfiable } \Leftrightarrow C^i \text{ is consistent w.r.t. } S^i.$$

It follows immediately, making use of Theorem 13 that there exists an indexed resolution refutation of C iff C is non-satisfiable. This is because the following ordering is a resolution ordering on the atoms of S^i :

$$(p, i_1)(t_1, \ldots, t_m) < (q, i_2)(u_1, \ldots, u_n) \text{ iff } i_1 > i_2.$$

The completeness of indexed resolution follows immediately from Theorem 13, when this ordering is used.

5 Acknowledgements

I would like to thank Ruud Sommerhalder for reading the initial version of this paper and Trudie Stoute for advice on English. I also would like to thank the anonymous referees for making some valuable suggestions.

References

[Baum92] P. Baumgartner, An ordered theory resolution calculus, in LPAR92, Springer Verlag, Berlin, 1992.

[Boyer71] R. S. Boyer, Locking: A restriction of resolution, Ph. D. Thesis, University of Texas at Austin, Texas 1971.

[ChngLee73] C-L. Chang, R. C-T Lee, Symbolic logic and mechanical theorem proving, Academic Press, New York, 1973.

[EnjFar89] P. Enjalbert, L. Fariñas, Modal resolution in clausal form, theoretical computer science 65, 1989.

[FLTZ93] C. Fermüller, A. Leitsch, T. Tammet, N. Zamov, Resolution Methods for the decision problem, LNAI 679, Springer Verlag, 1993.

[Joy76] W. H. Joyner, Resolution Strategies as Decision Procedures, Journal of the association for computing machinery, vol. 23, 1976.

[Lovlnd78] D. W. Loveland, Automated theorem proving: A logical basis, North Holland Publishing Company, Amsterdam, New York, Oxford 1978.

[Mints88] G. Mints, Gentzen-type systems and resolution rules, Part 1, Propositional logic, in COLOG-88, pp 198-231, Springer Verlag, 1988.

[Niv93a] H. de Nivelle, A general resolution scheme, reports of the faculty of technical mathematics and informatics, no. 93-62, Delft, 1993.

[Niv93b] H. de Nivelle, Generic resolution in propositional modal systems, in LPAR93, Springer Verlag, Berlin, 1993.

[Robins65] J.A. Robinson, A machine-oriented logic based on the resolution principle, Journal of the ACM, Vol. 12, No. 1, pp 23-41.

Two Logical Dimensions

Ewa Orlowska

Polish Academy of Sciences
Warsaw, Poland

Abstract

Relational formalization of nonclassical logics is provided by means of a relational semantics for the languages of these logics and relational proof systems. Relational semantics is determined by a class of algebras of relations, possibly with some nonstandard operations and constants, and by a meaning function that assigns relations to formulas. A formula is true under relational interpretation whenever its meaning is the unit element of the underlying algebra. Relational proof systems consist of Rasiowa-Sikorski style decomposition rules for every relational operator admitted in the respective algebras, and with specific rules that reflect properties of relational constants. Nonclassical logics are two-dimensional in the following sense: they consist of an extensional part and an intensional part. The extensional part carries a declarative information about knowledge states. The intensional part includes a procedural information, it represents transitions between knowledge states. Relational formalization exhibits these two dimensions in several nonclassical logics, including various modal logics, intuitionistic logic, Post logics, relevant logics. In the relational semantical structures of these logics the Boolean reducts of the underlying algebras of relations provide a formal counterpart of the first dimension. The monoid reducts are the counterpart of the second dimension. The two dimensions are also manifested in the relational proof systems. The decomposition rules for Boolean operators refer to the first dimension, and the rules for monoid operators correspond to the second dimension. In the paper the paradigm of two logical dimensions and their manifestation in relational formalisms is illustrated with examples of many-valued logics.

Prioritized Autoepistemic Logic

Jussi Rintanen *

Helsinki University of Technology
Department of Computer Science
Otakaari 1, 02150 ESPOO, Finland

Abstract. An important problem in data and knowledge representation is the possibility of default rules that conflict. If the application of both of two default rules leads to a contradiction, they cannot both be applied. Systems that support the use of default rules may either remain indifferent or prioritize one rule over the other. In this paper a prioritized version of autoepistemic logic is presented. Priorities determine a subset of all stable expansions of a set, the preferred stable expansions. The priority notion is declarative, unlike e.g. some recent approaches to priorities in default logic that modify the semi-constructive definition of extensions of Reiter. Computationally the new priority notion can nevertheless be seen as a mechanism for pruning search trees in procedures for autoepistemic reasoning, as demonstrated by procedures given in the paper.

1 Introduction

Default rules are used in many areas of knowledge representation, e.g. links in inheritance hierarchies [Touretzky *et al.*, 1987] and correctness assumptions in diagnostic reasoning [Reiter, 1987] can be seen as default rules. A problem in these and other areas of knowledge representation is how to deal with conflicting defaults. A safe way is to resort to skepticism and consider all possible ways of resolving the conflicts between defaults. In diagnostic reasoning this corresponds to the computation of *all* diagnoses. Another way is to bring more knowledge – priority information – to the system so that conflicts can be solved in a principled way: e.g. in single-inheritance the class-inclusion relation directly acts as priority information. Explicitly represented priority information is necessary e.g. for solving conflicts in multiple-inheritance.

Conflicts between defaults, as described above, show up in nonmonotonic logics and other formalizations of defeasible reasoning. Default logic [Reiter, 1980] and autoepistemic logic [Moore, 1985] generate several extensions and stable expansions for many sets of rules and facts. Similarly circumscription [McCarthy, 1980] sanctions distinct classes of minimal models. If conflicts are left unresolved, the convention is to take the intersection of the extensions, the stable expansions,

* This work was funded by Suomen Akatemia. Additional support from the following foundations is gratefully acknowledged: Emil Aaltosen säätiö, Heikki ja Hilma Honkasen säätiö, Alfred Kordelinin yleinen edistys- ja sivistysrahasto.

or the sets of formulae true in minimal models. The priorities can be coded in default rules. In default logic, for example, the justifications in semi-normal defaults can be used for blocking the application of a low-priority rule in favour of a higher-priority rule [Reiter and Criscuolo, 1981]. However, representing such dependencies explicitly in rules is inconvenient, and complicates the maintenance of sets of rules.

Another line of research distinct from autoepistemic logic and other nonmonotonic logics based on consistency uses model preference [Shoham, 1987; Kraus *et al.*, 1990]. Shoham defines a general framework of preferential logics in which both nonmonotonicity and resolution of conflicts between defaults are achieved using model preference. Kraus et al. [1990] use a framework similar to Shoham's for defining logics in which conflicts are solved according to the specificity of the defaults. The problem in these logics is that the addition of irrelevant premises cancels useful conclusions. Geffner and Pearl [1992] avoid these irrelevance problems in their conditional entailment which is obtained by modifying the logics of [Kraus *et al.*, 1990]. Conditional entailment can be seen as a variant of preferred subtheories of Brewka [1989] together with a fixed mechanism for computing priorities. In this paper we concentrate on consistency-based nonmonotonic logics.

There have been several proposals to incorporate priorities in consistency-based nonmonotonic logics in an abstract form, or to introduce related systems that involve explicit priorities, e.g. [Lifschitz, 1985; Konolige, 1989; Brewka, 1989; MacNish, 1991; Ryan, 1992; Tan and Treur, 1992; Brewka, 1992; Baader and Hollunder, 1993]. Some of the earlier proposals concern only simple defaults that correspond to prerequisite-free normal default rules of default logic [Brewka, 1989; Ryan, 1992]. The expressivity of these systems is insufficient. In the propositional case, prioritized circumscription [Lifschitz, 1985] is closely related to preferred subtheories of Brewka [1989], and hence the defaults expressible are essentially prerequisite-free normal. Konolige [1989] and Toyama et al. [1991] give variants of autoepistemic logic in which the syntactic form of defaults is not restricted. MacNish [1991] introduces priorities in default logic, but he encounters serious problems. For example, the extensions his system produces are not default extensions as defined by Reiter [1980]. Tan and Treur [1992] present a very general scheme for bringing priorities in default logic.

Baader and Hollunder [1993] consider arbitrary partial orders of unrestricted default rules of default logic. They give a version of Reiter's semi-constructive definition of default extensions, in which priorities restrict the order of application of default rules. Brewka [1992] has a similar definition. The generality of these systems is comparable to our prioritized autoepistemic logic. However, it seems that these approaches, like those of MacNish, and Tan and Treur, are likely to run into difficulties: the conclusions that are obtained are sometimes unintuitive, or can be explained only on the basis of the mechanism using which they were produced. The semi-constructive definition of extensions is the basis of these proposals, and the principle used is roughly the following: always apply a highest priority rule among all applicable rules that have not already been applied. Consider a default theory consisting of the following formulae and rules (Brewka has given an example similar to this).

$$a \qquad \frac{b:c}{c} \qquad \frac{a:\neg c}{\neg c} \qquad \frac{a:b}{b}$$

Let the three rules be prioritized in the order indicated, i.e. $\frac{b:c}{c}$ has the highest priority. With the above default theory, the second and third default rules are both initially applicable. Hence apply the second one and obtain $\neg c$. Now only the third rule is applicable and b – the prerequisite of the highest priority rule – is obtained. The unintuitivity is that the lowest priority rule is applied in all extensions of the theory, and hence the prerequisite of the highest priority rule $\frac{b:c}{c}$ belongs to all extensions. Despite this fact the highest priority rule does not become applicable because it is blocked by the the rule $\frac{a:\neg c}{\neg c}$. The problem in this example that shows up both in Baader and Hollunder's and Brewka's work, is that the application of high-priority defaults may depend on the application of low-priority rules in unobvious ways. It may be difficult to devise versions of Reiter's semi-constructive definition of extensions that avoid this kind of problems. Furthermore, there does not seem to be alternative, declarative ways of characterizing the preferred extensions in these approaches.

In this paper, we give a definition of priorities in autoepistemic logic. Priorities declaratively select a subset of all stable expansions of a set, the *preferred* stable expansions. Priorities on default rules can be defined in a way that properly fulfills the principle "apply the highest priority default if possible." For example, the problem illustrated in the above example is avoided.

2 Prioritized autoepistemic logic

Prioritized autoepistemic logic is defined as an extension of autoepistemic logic of Moore [1985]. Different versions of autoepistemic logic are obtained by basing it on different monotonic logics, like classical propositional logic or predicate logic. The language of the classical logic on which an autoepistemic logic is based on is denoted by $\mathcal{L}$. To obtain the language $\mathcal{L}_{ae}$ of an autoepistemic logic, $\mathcal{L}$ is extended with the operator L which is read "is believed". The definition of autoepistemic logic extends the notion of a model and the relation $\models$ to cover also formulae beginning with the operator L. These formulae are treated as atomic formulae.

Autoepistemic logics describe the reasoning capabilities of ideally rational agents. An agent is logically omniscient, i.e. it believes all logical consequences of its own beliefs, and it is capable of introspection, i.e. for each proposition ϕ either $L\phi$ or $\neg L\phi$ is its belief depending on whether ϕ is its belief. The beliefs of an autoepistemic agent are based on a set of initial premises. Given a set of initial premises Σ, possible states of belief T of an autoepistemic agent are the solutions of the following equation [Moore, 1985].

$$T = \{\phi \in \mathcal{L}_{ae} | \Sigma \cup \{L\phi | \phi \in T\} \cup \{\neg L\phi | \phi \notin T\} \models \phi\}$$

The sets T are *stable expansions of Σ*. When Σ contains conflicting defaults, there may be several stable expansions. If at least some of the preferences of the autoepistemic agent are available, the number of stable expansions that correspond to possible states of belief can be reduced, and by cautious reasoning, i.e. taking the intersection of the stable expansions, more formulae can be inferred. Prioritized autoepistemic logic is an explication of the preference mechanism used by an autoepistemic agent.

Definition 1. Let $\Phi \subseteq \mathcal{L}_{ae}$ be a set of formulae, and let $\mathcal{P} \subseteq \Phi \times \Phi$ be a transitive and asymmetric relation such that all strict total orders $\mathcal{T} \supseteq \mathcal{P}$ are well-orderings[2]. Then $P = \langle \Phi, \mathcal{P} \rangle$ is a *prioritization*.

The formulae in Φ represent those beliefs that are relevant in determining which state of belief the agent is going to choose. The relation $\phi \mathcal{P} \psi$ expresses that the agent is more reluctant to accept the belief ϕ than the belief ψ, i.e. when having to believe one of them the agent chooses ψ. We have chosen to prefer the absence of formulae of Φ in stable expansions rather than their presence. The opposite, i.e. preferring the presence of formulae in stable expansions, can be easily achieved by allowing formulae $+\phi$ in prioritizations, where $+\phi$ means simply $\neg L\phi$. We call formulae $+\phi$ in prioritizations *positive* and other formulae *negative*.

The relation $\mathcal{P}$ is asymmetric and hence it describes strict preferences. The transitivity property makes the prioritization an ordering relation. The well-ordering property rules out the existence of infinite chains of less and less believable beliefs, and thereby guarantees that any two stable expansions can be ordered. The partial order $\mathcal{P}$ is used lexicographically. Consider two stable expansions E_1 and E_2 given a strict total order $\mathcal{P}$ on Φ. If the least element of Φ belongs to E_1 but not to E_2, then prefer E_2 to E_1, and if it belongs to E_2 but not to E_1, then prefer E_1 to E_2. If the formula belongs to both of them or to none of them, the next formula of Φ is considered and so on. Hence in case of strict total orders a set of most preferred stable expansions naturally arises.

Sometimes not all preferences of an autoepistemic agent are known: the ordering in the prioritization may be properly partial. It seems that the most natural meaning for the partiality is obtained by extending the partial order to a total order, and then performing ordinary lexicographic comparison using the total order. Different notions of most preferred stable expansions are obtained depending on whether only one total order is considered in comparisons or whether different total orders may be used for different stable expansions. The first alternative, a variant of which is used by Brewka [1989], is to take the most preferred stable expansions to be the ones that are according to a single strict total order lexicographically preferred (not necessarily properly) to all other stable expansions. The second alternative, a variant of which is used

[2] An ordering $\mathcal{P}$ on a set D is a well-ordering if every non-empty subset S of D has an element σ such that $\tau \mathcal{P} \sigma$ for no $\tau \in S$.

by Ryan [1992][3], is to take the most preferred stable expansions to be the ones that are lexicographically preferred (not necessarily properly) to all other stable expansions according to possibly different total orders. The second definition essentially defines an ordering on stable expansions, the maximal elements of which are the most preferred stable expansions. The first definition – which we adopt – in general does not correspond to any such ordering.

Definition 2 P-preferredness. Let $P = \langle \Phi, \mathcal{P} \rangle$ be a prioritization and Σ a set of formulae. Then E is a *P-preferred* stable expansion of Σ if and only if it is a stable expansion of Σ and there is a strict total order T of Φ such that $T \supseteq \mathcal{P}$ and for all stable expansions E' of Σ the following holds.

For all $\phi \in \Phi (\phi \in E \backslash E'$ implies there is $\psi, \psi T \phi$ and $\psi \in E' \backslash E)$.

Example 1. The diagram below on the left depicts the prioritization $P = \langle \Phi, \mathcal{P} \rangle = \langle \{\alpha, \beta, \gamma\}, \{(\alpha, \beta), (\gamma, \beta)\} \rangle$. Whenever $a \mathcal{P} b$, a is depicted below b. The other two diagrams represent respectively the stable expansions E_1 and E_2 such that $E_1 \cap \Phi = \{\beta, \gamma\}$ and $E_2 \cap \Phi = \{\alpha\}$.

There are two strict total orders that extend $\mathcal{P}$, call them T_1 and T_2, which are shown below.

	T_1	E_1	E_2		T_2	E_1	E_2
	β	●	○		β	●	○
	↑	↑	↑		↑	↑	↑
	γ	●	○		α	○	●
	↑	↑	↑		↑	↑	↑
	α	○	●		γ	●	○

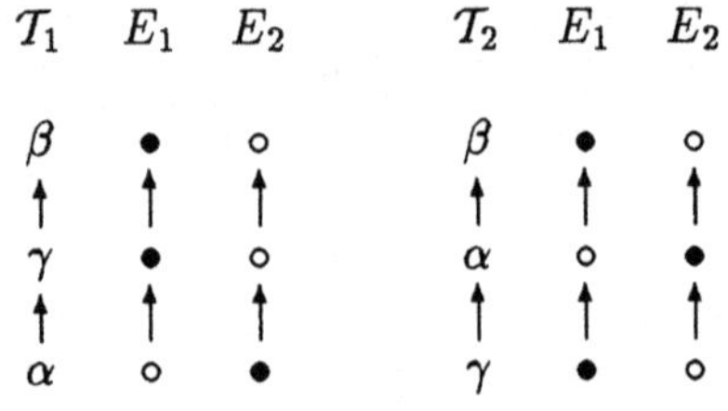

Because $\alpha \in E_2 \backslash E_1$ and there are no formulae $\psi such that \psi T_1 \alpha$, E_1 is a P-preferred stable expansion. Similarly, because $\gamma \in E_1 \backslash E_2$ and there are no formulae $\psi, \psi T_2 \gamma$, E_2 is a P-preferred stable expansion.

The decision problems of prioritized autoepistemic logic are *prioritized cautious reasoning* and *prioritized brave reasoning*, which correspond to the membership of formulae in all P-preferred stable expansions and the membership in some P-preferred stable expansions, respectively.

Theorem 3. When the underlying logic is the classical propositional logic, prioritized cautious and brave reasoning are respectively Π_2^p-hard and in Π_3^p, and Σ_2^p-hard and in Σ_3^p.

---

[3] Ryan does not mention the possibility of stating the meaning of the partiality in his priorities this way.

Proof. Outline: the hardness results of prioritized reasoning are immediate applications of the respective results of ordinary autoepistemic logic [Gottlob, 1992]. The membership in the third level of the polynomial hierarchy is because guessing a stable expansion (not) containing a formula takes nondeterministic polynomial time, and the test that the stable expansion is preferred (i.e. that there are no better stable expansions) uses a Π_2^p-oracle.

3 Relation to other formalisms

Brewka's [1989] preferred subtheories and Ryan's [1992] ordered theory presentations (OTPs) can be seen as prioritized versions of Reiter's default logic restricted to prerequisite-free normal defaults. OTPs are richer in using natural consequences of formulae. There is an exact translation of Brewka's preferred subtheories to prioritized autoepistemic logic. The formulae ϕ in a theory T translate simply into $\neg L \neg \phi \rightarrow \phi$, and the priorities into $\langle \{\neg \phi | \phi \in T\}, \{(\neg \phi, \neg \psi) | \phi < \psi\} \rangle$. The preference notion used by Ryan differs slightly from the one used by us and Brewka. Instead of requiring that a single total ordering of the prioritization partial order exists, it suffices that for every other stable expansion there is some total ordering of the prioritization using which preferredness is achieved. Hence the difference between these two preference notions is that of $\forall\exists$ and $\exists\forall$.

Definition 4. Let E be a stable expansion of Σ and $P = \langle \Phi, \mathcal{P} \rangle$ a prioritization. Then E is *P-maximal* if and only if for all stable expansions E' of Σ, there is a strict total order $\mathcal{T} \supseteq \mathcal{P}$ such that

$$\text{for all } \phi \in \Phi (\phi \in E \backslash E' \text{ implies there is } \psi \text{ such that } \psi \mathcal{T} \phi, \psi \in E' \backslash E).$$

With this preference notion the translation of OTPs to prioritized autoepistemic logic is almost like that of preferred subtheories. A formula ϕ of an OTP is translated to formulae in $\{\neg L \neg \nu \rightarrow \nu | \phi \breve{\models} \nu\}$ where each ν is a *natural consequence* of ϕ. Ryan's idea in defining natural consequences is the conjunctivity in formulae. Ryan can be seen – like Brewka – as constructing maximal consistent subsets of the formulae in an OTP except that if e.g. $p \wedge q$ cannot be consistently included in the set, some of its natural consequences – e.g. p or q – possibly can. Each formula has an infinite number of natural consequences, but Ryan [1992] conjectures that the conjuncts of a suitable conjunctive normal form of ϕ can be used instead of them. Complete proofs of these correspondences are given in [Rintanen, 1993].

Hierarchic autoepistemic logic [Konolige, 1989] and multi-agent autoepistemic logic of Toyama et al. [1991] support the expression of priorities, and do not make syntactic restrictions to the form of defaults. In hierarchic autoepistemic logic the introspection ability of an autoepistemic agent is restricted. Sets of formulae are divided into a number of layers. On layer n, the formula $L_m \phi$ ($m < n$) refers to believing ϕ on layer m. This mechanism effectively resolves conflicts between defaults: there is only one expansion for each layered set of formulae. A drawback of the logic is that priorities are obligatory: if conflicting defaults are on the same layer, an inconsistency results. This logic can be embedded in ordinary autoepistemic logic [Przymusinska, 1989]. Unlike Konolige's logic, our logic allows the expression of partial priorities. Partiality implies the possibility of several preferred stable expansions. Multi-agent autoepistemic logic generalizes autoepistemic logic to several agents. Each agent can access the beliefs of other agents. Toyama et al. demonstrate how specificity in inheritance reasoning can be conveniently represented in this logic. Priorities may be partial, in which case conflicts between defaults produce several expansions. The suitability of the multi-agent logic of Toyama et al. for other uses of priorities than expressing specificity is open. Multi-agent autoepistemic logic can be translated into Moore's autoepistemic logic [Rintanen, 1993].

The logics of Baader and Hollunder [1993] and Brewka [1992] bring priorities to default logic and allow prerequisites in default rules. These logics cannot be easily translated into prioritized autoepistemic logic. Our logic works properly for the example discussed in the introduction.

Example 2. The example given in the introduction can be expressed as formulae of autoepistemic logic as follows.

$$\Sigma = \{a, Lb \wedge \neg L\neg c \rightarrow c, La \wedge \neg Lc \rightarrow \neg c, La \wedge \neg L\neg b \rightarrow b\}$$

Three schemes of using priorities in our logic are obvious. Either give priorities on the justifications, on the conclusions, or on whole application conditions (prerequisites and justifications) of the defaults. That is, the above defaults can be prioritized in the order indicated using either $P = \langle \Phi, \mathcal{P} \rangle$ where $\Phi = \{\neg c, c, \neg b\}$ and $\mathcal{P}$ totally orders Φ in the order $\neg c, c, \neg b$, or P' that orders the formulae $+c, +\neg c, +b$ in this order, or P'' that orders the formulae $+(b \wedge \neg L\neg c), +(a \wedge \neg Lc), +(a \wedge \neg L\neg b)$ in this order. There are two stable expansions for the above set Σ.

$$E_1 = \{a, b, c, \neg L\neg c, \ldots\} \qquad E_2 = \{a, b, \neg c, \neg Lc, \ldots\}$$

According to our definition, E_1 is P-preferred and P'-preferred and P''-preferred, like expected, but E_2 is not. This is opposite to the result given by systems of Baader and Hollunder [1993] and Brewka [1992].

The reason for the unintuitive conclusions in logics of [Baader and Hollunder, 1993; Brewka, 1992] is that the applicability of high-priority rules may depend on the temporally prior application of lower-priority rules. In these logics, the order in which defaults become applicable is a significant factor in determining which extensions the priorities select. Our definition of prioritized autoepistemic logic does not depend on such procedural aspects of application of defaults.

4 Automated reasoning with priorities

In this section we give procedures for automated reasoning with priorities. Our procedures use *full sets* developed by Niemelä for handling stable expansions computationally. Alternatively we could use similar definitions by Shvarts [1990]. The relevant subset of a stable expansion of a set Σ is the set $Sf^L(\Sigma)$ of subformulae of Σ that begin with the L operator. The set $Sf^{qL}(\Sigma)$ is the subset of $Sf^L(\Sigma)$ consisting of those subformulae of Σ that begin with L and are not inside another L in Σ. The following definitions and theorems are from [Niemelä, 1990].

Definition 5. A set Λ is Σ-full if it satisfies the following conditions.

$$\Lambda \subseteq Sf^L(\Sigma) \cup \{\neg L\chi | L\chi \in Sf^L(\Sigma)\}$$
$$\text{for all } L\chi \in Sf^L(\Sigma),\ L\chi \in \Lambda \text{ iff } \Sigma \cup \Lambda \models \chi$$
$$\text{for all } L\chi \in Sf^L(\Sigma),\ \neg L\chi \in \Lambda \text{ iff } \Sigma \cup \Lambda \not\models \chi$$

Theorem 6. *For a set of sentences Σ there is a bijective mapping from the Σ-full sets to the stable expansions of Σ.*

Definition 7. Given a set of sentences Σ and a sentence ϕ

$$\Sigma \models_L \phi \text{ iff } \Sigma \cup SB_\Sigma(\phi) \models \phi$$

where $SB_\Sigma(\phi) = \{L\chi \in Sf^{qL}(\phi) | \Sigma \models_L \chi\} \cup \{\neg L\chi | L\chi \in Sf^{qL}(\phi), \Sigma \not\models_L \chi\}$.

Theorem 8. *Let Λ be a Σ-full set. Then $\Delta = \{\phi | \Sigma \cup \Lambda \models_L \phi\}$ is the unique stable expansion of Σ for which $\Lambda \subseteq \{L\phi | \phi \in \Delta\} \cup \{\neg L\phi | \phi \notin \Delta\}$.*

Definition 9. Let Σ be a set of formulae and Λ a Σ-full set. Define $SE_\Sigma(\Lambda) = \{\phi | \Sigma \cup \Lambda \models_L \phi\}$.

This way of representing stable expansions immediately suggests a decision procedure: generate all candidate full sets, test whether they are indeed full, and if they are, test the membership of a formula in the respective stable expansion.

The number of candidate full sets for a set of formulae Σ is exponential on the size of Σ: there are $2^{|Sf^L(\Sigma)|}$ such sets. The problem that arises in these decision procedures is how to effectively reduce the size of the search space formed by the candidate full sets. In the stratified case [Gelfond, 1987; Marek and Truszczyński, 1991] there is an efficient way of reducing this search space as shown in [Niemelä and Rintanen, 1992]. In fact, search is completely avoided.

With prioritized autoepistemic reasoning, an obvious algorithm for automated reasoning is to construct all stable expansions (their full sets), test the preferredness condition for each of them, and then test the membership of a formula in the preferred ones. This however is inefficient, and can be improved at least in cases where the prioritizations are on formulae ϕ for which $L\phi \in Sf^L(\Sigma)$. The procedures that use full sets (or something similar) produce search trees where the edges of the tree correspond to the inclusion of a formula $L\phi$ or $\neg L\phi$ in the full set, and the leafs of the tree correspond to candidate full sets, i.e. the set of formulae on the path from the root of the tree to a leaf. The idea is to use priorities for pruning the search tree. Roughly, the basic reduction step is to ignore a subtree T of the search tree that corresponds to full sets containing $L\phi$ (respectively $\neg L\phi$ for a *positive* ϕ), whenever there was a full set containing $\neg L\phi$ (respectively $L\phi$) and all full sets corresponding to T contain exactly the same formulae $L\psi$ or $\neg L\psi$ for formulae ψ that are more important than ϕ. In this case, the subtree T is guaranteed *not to* contain full sets of preferred stable expansions.

It is essential for our procedures that the order in which full sets are generated in our algorithms fulfills the following: the case $L\phi \in \Lambda$ is considered before the case $\neg L\phi \in \Lambda$ for positive formulae $+\phi$, and the case $\neg L\phi \in \Lambda$ before the case $L\phi \in \Lambda$ for negative formulae ϕ. Two subprocedures and the main procedure of our algorithms are shown in Figure 1. For the clarity of presentation, the handling of positive formulae $+\phi$ in prioritizations is not discussed. The modifications required for them are straightforward. The reason for giving them a separate treatment in the procedures instead of using them as a shorthand for $\neg L\phi$, is that this way the constraint $L\phi \in Sf^L(\Sigma)$ works symmetrically for positive $+\phi$.

The main procedure traverses the search space of $2^{|Sf^L(\Sigma)|}$ full sets of a set Σ. The pruning of the search tree is performed by the procedure *extendible*, and the order in which the formulae are chosen in full sets is determined by the procedure *next*. In the main procedure, the variable Q contains the full sets of all preferred stable expansions found so far. The procedure *next* selects elements ϕ of $\Phi \subseteq \{\phi | L\phi \in Sf^L(\Sigma)\}$ in some total order that extends the relation $\mathcal{P}$ in the prioritization P. This total order fulfills a further condition that is taken advantage of in the procedure *extendible*. At each node of the search tree the procedure *extendible* is invoked to detect whether the full sets to be found in the respective subtree can correspond to preferred stable expansions. The procedure attempts to construct a strict total order $\mathcal{T} \supseteq \mathcal{P}$ so that the preferredness condition of Definition 2 would be fulfilled for the stable expansions of the current subtree. It turns out, that for the fulfillment of the condition it suffices to look at the preferred stable expansions found so far (the full sets in Q). The possibility that a preferred stable expansion found later would be "better" than the stable expansions corresponding to the current subtree is ruled out by the way the procedure *next* selects elements: the order in which elements $\phi, L\phi \in Sf^L(\Sigma)$ are chosen is acceptable as the strict total order $\mathcal{T}$. Hence, all full sets Λ found later contain $L\phi$ for the most important formula ϕ (according to $\mathcal{T}$) in which Λ differs from the full sets of the current subtree, and all full sets of the current subtree contain $\neg L\phi$. Therefore the stable expansions of the current subtree are no "worse" than those found later.

We briefly discuss properties of our algorithm. First, priorities are usefully taken advantage of when they are on formulae ϕ such that $L\phi \in Sf^L(\Sigma)\}$ [4]: the number of candidate subsets considered is smaller than in decision procedures of Moore's autoepistemic logic [5]. In cases where prioritizations are strict total

[4] Extending Σ with tautologies $L\phi \rightarrow L\phi$ for formulae ϕ such that $L\phi \notin Sf^L(\Sigma)$ makes our algorithm applicable for all prioritizations.

[5] Improvements to the trivial algorithms that use full sets, as proposed by Niemelä [1994], can be easily incorporated in our algorithm.

PROCEDURE extendible(P, Λ, Q)
BEGIN
  $\langle \Phi, \mathcal{P} \rangle := P$;
  A $:= \{\phi \in \Phi | \phi$ is $\mathcal{P}$-minimal in Φ, $L\phi \in \Lambda$ implies $L\phi \in \bigcap Q\}$;
  IF $A = \emptyset$ *THEN RETURN* false;
  $\phi :=$ any member of A;
  $P' := \langle \Phi \backslash \{\phi\}, \mathcal{P} \cap (\Phi \backslash \{\phi\} \times \Phi \backslash \{\phi\}) \rangle$;
  IF $L\phi \in \Lambda$ *THEN RETURN* extendible(P', Λ, Q)
  ELSE RETURN extendible($P', \Lambda, \{\Lambda' \in Q | \neg L\phi \in \Lambda'\}$)
END

PROCEDURE next(P, Σ, Λ, Q)
BEGIN
  $\langle \Phi, \mathcal{P} \rangle := P$;
  A $:= \{\phi \in \Phi | \phi$ is $\mathcal{P}$-minimal in Φ, $L\phi \in \Lambda$ implies $L\phi \in \bigcap Q\}$;
  IF $A = \emptyset$ *THEN RETURN* any ϕ such that $L\phi \in Sf^L(\Sigma) \backslash Sf^{qL}(\Lambda)$;
  B $:= \{\phi \in A | L\phi \notin \Lambda, \neg L\phi \notin \Lambda\}$;
  IF $B \neq \emptyset$ *THEN RETURN* any member of B;
  $\phi :=$ any member of A;
  $P' := \langle \Phi \backslash \{\phi\}, \mathcal{P} \cap (\Phi \backslash \{\phi\} \times \Phi \backslash \{\phi\}) \rangle$;
  IF $L\phi \in \Lambda$ *THEN RETURN* next(P', Σ, Λ, Q)
  ELSE RETURN next($P', \Sigma, \Lambda, \{\Lambda' \in Q | \neg L\phi \in \Lambda'\}$)
END

PROCEDURE decide($P, \Sigma, \Lambda, \phi, Q$)
BEGIN
  IF extendible(P, Λ, Q) = false *THEN RETURN* ($\emptyset$,false);
  IF for some $\neg L\chi \in \Lambda, \Sigma \cup \Lambda \models \chi$ *THEN RETURN* ($\emptyset$,false);
  IF $Sf^L(\Sigma) \subseteq Sf^{qL}(\Lambda)$ *THEN*
    IF for all $L\chi \in \Lambda$, $\Sigma \cup \Lambda \models \chi$ *THEN RETURN* ($\{\Lambda\}$,test(Σ, Λ, ϕ))
    ELSE RETURN ($\emptyset$,false);
  $\chi :=$ next(P, Σ, Λ, Q);
  $(S, b) :=$ decide($P, \Sigma, \Lambda \cup \{\neg L\chi\}, \phi, Q$);
  $(S', b') :=$ decide($P, \Sigma, \Lambda \cup \{L\chi\}, \phi, S \cup Q$);
  RETURN ($S \cup S', b$ or b')
END

Fig. 1. The procedure for prioritized autoepistemic logic

orders on $\{\phi | L\phi \in Sf^L(\Sigma)\}$, our algorithm finds the unique stable expansion and all subsequent computation is avoided. Second, more trivial ways of using priorities in the computation may require considering $n!$ strict total orders for a prioritization with n elements. The definition of Brewka's preferred subtheories [1989] and priorities in default logic [1992] suggest such algorithms. In our algorithms the consideration of $\mathcal{O}(2^n)$ different cases suffices. This seems to be the case also with the algorithm for default logic by Baader and Hollunder [1993].

The running time of our algorithm is exponential on the size of Σ for two reasons: the classical reasoning component (e.g. for propositional logic) takes exponential time, and there may be an exponential number of candidate full sets despite the reduction due to priorities. In the special case of a tractable subset of a classical logic (e.g. propositional Horn clauses), autoepistemic formulae of the form $\neg L\neg\phi \rightarrow \phi$ (prerequisite-free normal defaults), and strict total prioritizations on $\{\phi|L\phi \in Sf^L(\Sigma)\}$, both these causes for exponentiality disappear. The unique preferred stable expansion can be immediately found without search, and the classical theorem-prover – that runs in polynomial time – is called only a polynomial number of times, hence resulting in a tractable decision procedure.

The correctness proof of the algorithm uses the following definitions and lemmata that are proved in [Rintanen, 1993] (In the definitions and lemmata Σ is a finite set of formulae, and P is a finite prioritization).

Definition 10 Preferred in. A stable expansion E of Σ is *P-preferred in* a set of stable expansions $X = \{E_1, \ldots, E_n\}$ of Σ if and only if there is a strict total order $\mathcal{T} \supseteq \mathcal{P}$ on Φ such that for all $E' \in X$,

$$\text{for all } \phi \in \Phi(\phi \in E \backslash E' \text{ implies there is } \psi \in E' \backslash E \text{ such that } \psi \mathcal{T} \phi).$$

Lemma 11. *Let X be a set of stable expansions of Σ such that each $E \in X$ is P-preferred in X, and all P-preferred stable expansions of Σ are in X. Then X is exactly the set of P-preferred stable expansions of Σ.*

***Lemma 12 Procedure* extendible.** *Let $P = \langle \Phi, \mathcal{P} \rangle$ be a prioritization such that $\Phi \subseteq \{\phi|L\phi \in Sf^L(\Sigma)\}$, Λ a Σ-full set, and Q a finite set of Σ-full sets. Let b be the value returned by the procedure call extendible(P, Λ, Q). Then b is* true *if and only if $SE_\Sigma(\Lambda)$ is P-preferred in $SE_\Sigma(Q)$.*

Let Λ' be a subset of a Σ-full set Λ and let the procedure call extendible(P, Λ', Q) return false. *Then $SE_\Sigma(\Lambda)$ is not P-preferred in $SE_\Sigma(Q)$.*

***Lemma 13 Procedure* next.** *Let $P = \langle \Phi, \mathcal{P} \rangle$ be a prioritization such that $\Phi \subseteq \{\phi|L\phi \in Sf^L(\Sigma)\}$, Λ a consistent subset of $Sf^L(\Sigma) \cup \{\neg L\phi|L\phi \in Sf^L(\Sigma)\}$ such that $Sf^L(\Sigma)\backslash Sf^{qL}(\Lambda) \neq \emptyset$, and Q a finite set of Σ-full sets.*

Then next(P, Σ, Λ, Q) returns a formula χ such that $L\chi \in Sf^L(\Sigma)\backslash Sf^{qL}(\Lambda)$ and the following holds. If S is a set of Σ-full sets $\Lambda' \supseteq \Lambda \cup \{\neg L\chi\}$ such that each $SE_\Sigma(\Lambda')$ is P-preferred in $SE_\Sigma(S \cup Q)$, and S' is a set of Σ-full sets $\Lambda' \supseteq \Lambda \cup \{L\chi\}$ such that each $SE_\Sigma(\Lambda')$ is P-preferred in $SE_\Sigma(S \cup S' \cup Q)$, then each $SE_\Sigma(\Lambda'), \Lambda' \in S$ is P-preferred in $SE_\Sigma(S \cup S' \cup Q)$.

***Lemma 14 Procedure* decide.** *Let $P = \langle \Phi, \mathcal{P} \rangle$ be a prioritization such that $\Phi \subseteq \{\phi|L\phi \in Sf^L(\Sigma)\}$, Λ a consistent set such that $\Lambda \subseteq (Sf^L(\Sigma) \cup \{\neg L\phi|L\phi \in Sf^L(\Sigma)\})$, $\phi \in \mathcal{L}_{ae}$ a formula, and Q a finite set of Σ-full sets.*

The procedure call decide($P, \Sigma, \Lambda, \phi, Q$) returns (S,b). The set S consists of Σ-full sets $\Lambda' \supseteq \Lambda$ such that each $SE_\Sigma(\Lambda')$ is P-preferred in $SE_\Sigma(S \cup Q)$, and each P-preferred stable expansion E of Σ such that $E = SE_\Sigma(\Lambda')$ for some $\Lambda' \supseteq \Lambda$, is contained in S. The value b is true *if and only if for some $\Lambda' \in S$, test(Σ, Λ', ϕ) returns* true.

Theorem 15 Correctness. *Let $\phi \in \mathcal{L}_{ae}$ be a formula. The procedure call decide($P, \Sigma \cup \{L\phi \rightarrow L\phi | \phi \in \Phi, L\phi \notin Sf^L(\Sigma)\}, \emptyset, \phi, \emptyset$) returns (S, b), where b is* true *if and only if there is a P-preferred stable expansion E of Σ such that $E = SE_\Sigma(\Lambda)$ and the procedure call test(Σ, Λ, ϕ) returns* true.

Proof. The idea in extending Σ to $\Sigma' = \Sigma \cup \{L\phi \rightarrow L\phi | \phi \in \Phi, L\phi \notin Sf^L(\Sigma)\}$ is to fulfill the requirement that for each formula $\phi \in \Phi$, $L\phi \in Sf^L(\Sigma)$. The procedures *next*, *extendible* and *decide* depend on this requirement. By Lemma 14 $SE_{\Sigma'}(S)$ contains all P-preferred stable expansions. Because every member of $SE_{\Sigma'}(S)$ is P-preferred in $SE_{\Sigma'}(S)$, every member of $SE_{\Sigma'}(S)$ is P-preferred by Lemma 11. The value of b is *true* if and only if for some stable expansion $SE_{\Sigma'}(\Lambda), \Lambda \in S$ test(Σ', Λ, ϕ) returns *true*.

Procedures for different decision problems are obtained by supplying different procedures *test*. For prioritized brave reasoning the procedure *test* simply tests $\Sigma \cup \Lambda \models_L \phi$, and for prioritized cautious reasoning it tests $\Sigma \cup \Lambda \not\models_L \phi$ and the value returned by *decide* is negated. We have implemented all these procedures in an automatic theorem-proving system for autoepistemic and default logics.

Decision procedures for several nonmonotonic modal logics ("nonmonotonic versions" of N, K, T, S4, S4F, KD45, SW5, W5) are given in [Marek *et al.*, 1993]. These procedures use the finite characterization of expansions developed by [Shvarts, 1990]. Similar procedures based on full sets – with some improvements – for Moore's autoepistemic logic and enumeration-based autoepistemic logics are given in [Niemelä, 1994]. Our techniques can be applied in computing expansions for the prioritized versions of these logics. Using prioritized versions of any of the nonmonotonic logics N, K, T, S4, S4F or the L-hierarchic autoepistemic logic of Niemelä [1994] we can avoid the groundedness problems that distinguish default logic from Moore's autoepistemic logic. Using the translation $L\alpha \wedge L\neg L\neg\beta \rightarrow \gamma$ for defaults $\frac{\alpha:\beta}{\gamma}$ [Truszczyński, 1991], a version of prioritized default logic can be embedded in any of these logics.

5 Conclusions

In this work we have presented a formalization of priorities within autoepistemic logic. The work generalizes earlier work on priorities and nonmonotonicity, e.g. [Brewka, 1989; Ryan, 1992], by allowing unrestricted defaults. Comparable generalizations have been proposed by Brewka [1992] and Baader and Hollunder [1993]. We believe that because of the declarativity of our priority notion unintuitive conclusions that are likely to arise in more procedural approaches to default priorities [Baader and Hollunder, 1993; Brewka, 1992; MacNish, 1991; Tan and Treur, 1992] are avoided in our system. For example, logics of Baader and Hollunder as well as Brewka prefer some defaults on the basis that their prerequisites are more directly derivable. In some cases, this preference takes precedence over preference indicated by priorities, as demonstrated by the example in the introduction.

Automated nonmonotonic reasoning with priorities has been investigated earlier in [Baker and Ginsberg, 1989; Junker and Brewka, 1991]. Both approaches essentially restrict to prerequisite-free normal defaults. Baker and Ginsberg work only with layered partial orders. Our work does not make these restrictions. Reasoning with priorities can be more efficient than without. We have demonstrated this for an important class of priorities that are on formulae that occur inside L in the premises: priorities justify a principle for pruning search trees in decision procedures.

Through translations of autoepistemic logic to other formalisms, our priorities and the associated reasoning procedures can be brought to e.g. default logic, justification-based TMSs, and the theory of diagnosis of Reiter [1987].

Future investigations on priorities concern the computation of priorities automatically. An important source of priorities is rule specificity. A method for computing priorities for conditional entailment was presented by Geffner and Pearl [1992]. However, their method involves several prioritizations instead of only one, and hence these priorities are not convenient for automated reasoning. Default rules expressible in autoepistemic logic are more general than in conditional entailment, and Geffner and Pearl's method does not immediately generalize. The problem of rule specificity in the general context of default rules is a challenging subject for further research.

Acknowledgements

I would like to thank Dr I. Niemelä for advice, discussions, and for proposing this research subject.

References

Baader, F. and B. Hollunder: How to prefer more specific defaults in terminological default logic. In Ruzena Bajcsy, editor, *Proceedings of the 13th International Joint Conference on Artificial Intelligence*, volume 1, pages 669–674, Chambery, France, 1993.

Baker, A. B. and M. L. Ginsberg: A theorem prover for prioritized circumscription. In *Proceedings of the 11th International Joint Conference on Artificial Intelligence*, pages 463–467, Detroit, 1989. Morgan Kaufmann Publishers.

Brewka, G.: Preferred subtheories: an extended logical framework for default reasoning. In *Proceedings of the 11th International Joint Conference on Artificial Intelligence*, pages 1043–1048, Detroit, 1989. Morgan Kaufmann Publishers.

Brewka, G.: Specificity in default logic and its application to formalizing obligation. Unpublished, 1992.

Geffner, H. and Judea Pearl: Conditional entailment: bridging two approaches to default reasoning. *Artificial Intelligence*, 53:209–244, 1992.

Gelfond, M.: On stratified autoepistemic theories. In *Proceedings of the 6th National Conference on Artificial Intelligence*, pages 207–211, Seattle, July 1987. American Association for Artificial Intelligence.

Gottlob, G.: Complexity results for nonmonotonic logics. *Journal of Logic and Computation*, 2(3):397–425, June 1992.

Junker, U. and G. Brewka: Handling partially ordered defaults in TMS. In *Symbolic and quantitative approaches to uncertainty. European Conference*, number 548 in Lecture Notes in Computer Science, pages 211–218, Marseille, 1991. Springer-Verlag.

Konolige, K.: Hierarchic autoepistemic theories for nonmonotonic reasoning: Preliminary report. In *Proceedings of the 2nd Workshop on Non-Monotonic Reasoning*, number 346 in Lecture Notes in Artificial Intelligence, pages 42–59, Grassau, Germany, June 1989. Springer-Verlag.

Kraus, S., D. Lehmann, and M. Magidor: Nonmonotonic reasoning, preferential models and cumulative logics. *Artificial Intelligence*, 44:167–207, 1990.

Lifschitz, V.: Computing circumscription. In A. Joshi, editor, *Proceedings of the 9th International Joint Conference on Artificial Intelligence*, pages 121–127, Los Angeles, August 1985. Morgan Kaufmann Publishers.

MacNish, C.: Hierarchical default logic. In *Symbolic and quantitative approaches to uncertainty. European Conference*, number 548 in Lecture Notes in Computer Science, pages 246–253, Marseille, 1991. Springer-Verlag.

Marek, W. and M. Truszczyński: Autoepistemic logic. *Journal of the ACM*, 38:588–619, 1991.

Marek, V. W., G. F. Schwarz, and M. Truszczyński: Modal nonmonotonic logics: ranges, characterization, computation. *Journal of the ACM*, 40(4):963–990, September 1993.

McCarthy, J.: Circumscription – a form of non-monotonic reasoning. *Artificial Intelligence*, 13:27–39, 1980.

Moore, R. C.: Semantical considerations on nonmonotonic logic. *Artificial Intelligence*, 25:75–94, 1985.

Niemelä, I. and J. Rintanen: On the impact of stratification on the complexity of nonmonotonic reasoning. In B. Nebel, C. Rich, and W. Swartout, editors, *Proceedings of the 3rd International Conference on Principles of Knowledge Representation and Reasoning*, pages 627–638, Cambridge, MA, October 1992. Morgan Kaufmann Publishers.

Niemelä, I.: A decision method for nonmomotonic reasoning based on autoepistemic reasoning. In *Proceedings of the 4th International Conference on Principles of Knowledge Representation and Reasoning*, Bonn, May 1994. To appear.

Niemelä, I.: Towards automatic autoepistemic reasoning. In J. Van Eijck, editor, *Proceedings of the European Workshop on Logics in Artificial Intelligence—*

JELIA'90, number 478 in Lecture Notes in Artificial Intelligence, pages 428–443, Amsterdam, September 1990. Springer-Verlag.

Przymusinska, H.: The embeddability of hierarchic autoepistemic logic in autoepistemic logic. In Z. W. Ras, editor, *Methodologies for Intelligent Systems, 4: Proceedings of the Fourth International Symposium on Methodologies for Intelligent Systems*, pages 485–493, New York, 1989. North-Holland.

Reiter, R. and G. Criscuolo: On interacting defaults. In *Proceedings of the 7th International Joint Conference on Artificial Intelligence*, pages 270–276, 1981.

Reiter, R.: A logic for default reasoning. *Artificial Intelligence*, 13:81–132, 1980.

Reiter, R.: A theory of diagnosis from first principles. *Artificial Intelligence*, 32:57–95, 1987.

Rintanen, J.: Priorities and nonmonotonic reasoning. Research report A 28, Helsinki University of Technology, Digital Systems Laboratory, December 1993.

Ryan, M.: Representing defaults as sentences with reduced priority. In B. Nebel, C. Rich, and W. Swartout, editors, *Proceedings of the 3rd International Conference on Principles of Knowledge Representation and Reasoning*, pages 649–660, Cambridge, MA, October 1992. Morgan Kaufmann Publishers.

Shoham, Y.: Nonmonotonic logics: meaning and utility. In D. McDermott, editor, *Proceedings of the 10th International Joint Conference on Artificial Intelligence*, pages 388–393, Milano, 1987.

Shvarts, G.: Autoepistemic modal logics. In *Proceedings of the 3rd Conference on Theoretical Aspects of Reasoning about Knowledge*, pages 97–109, Pacific Grove, CA, March 1990. Morgan Kaufmann Publishers.

Tan, Y. H. and J. Treur: Constructive default logic and the control of defeasible reasoning. In B. Neumann, editor, *Proceedings of the 10th European Conference on Artificial Intelligence*, pages 299–303, Vienna, 1992. John Wiley & Sons.

Touretzky, D. S., J. F. Horty, and R. H. Thomason: A clash of intuitions: the current state of nonmonotonic multiple inheritance systems. In D. McDermott, editor, *Proceedings of the 10th International Joint Conference on Artificial Intelligence*, pages 476–482, Milano, August 1987. Morgan Kaufmann Publishers.

Toyama, K., Y. Inagaki, and T. Fukumura: Knowledge representation based on autoepistemic logic for multiple agents. Technical report 90-1-06, Chukyo University, School of Computer and Cognitive Sciences, March 1991.

Truszczyński, M.: Embedding default logic into modal nonmonotonic logics. In *Proceedings of the 1st International Workshop on Logic Programming and Non-monotonic Reasoning*, pages 151–165, Washington, DC, July 1991. The MIT Press.

Adding Priorities and Specificity to Default Logic

Gerhard Brewka

GMD, Schloss Birlinghoven
53757 Sankt Augustin, Germany

Abstract. Reiter's Default Logic (DL) is one of the most popular formalizations of default reasoning. Nevertheless, the logic has a serious deficiency: the specificity principle, i.e. the commonly accepted idea that in case of a conflict more specific defaults should be preferred over more general ones, is lacking. In this paper we show how this principle can be added to Default Logic. We first present a prioritized version of DL for normal defaults, called PDL. Adapting ideas underlying Geffner and Pearl's conditional entailment we then show how the priorities needed to handle specificity can be defined. This treatment of specificity avoids two serious problems in Pearl's system Z and leads to stronger conclusions than conditional entailment in many cases.

1 Introduction

Reiter's Default Logic [17], DL for short, is one of the most popular nonmonotonic formalisms. In DL default theories consist of a set of facts W and a set of defaults D. Each default is of the form $A : B_1, \ldots, B_n/C$. A default theory generates extensions which are defined as fixed points of an operator Γ. Γ maps an arbitrary set of formulas S to the smallest deductively closed set S' that contains W and satisfies the condition: if $A : B_1, \ldots, B_n/C \in D$, $A \in S'$ and for all i $(1 \leq i \leq n)$ $B_i \notin S$ then $C \in S'$.

The popularity of DL is basically due to two reasons. Firstly, although the technical definition of DL, in particular that of an extension, is rather tricky, the way defaults are represented is natural and intuitive: defaults are inference rules with an additional consistency check. Secondly, DL is of greater expressiveness than many of its competitors. This expressiveness makes it, for instance, possible to use DL for the specification of the semantics of logic programs with negation as failure [2].

Nevertheless, DL suffers from a serious deficiency: it does not prefer more specific defaults over more general ones. [1] Specificity is a fundamental principle of commonsense reasoning, and it is commonly agreed that conclusions based

[1] We consider this deficiency as much more serious than other problems of DL discussed in the literature. In [5] we showed how *cumulativity* and *joint consistency of justifications* can be achieved in DL. A way of guaranteeing *existence of extensions* for finite DL theories is discussed in [6]. *Reasoning by cases* is possible in DL if the defaults are represented without prerequisites, as discussed in [8].

on more general default rules should be given up when more specific conflicting rules are available. DL does not obey to this principle, that is, if two applicable conflicting defaults, say about birds and penguins, are given, then DL will generate two extensions instead of preferring the more specific one. The goal of this paper is to provide a solution to this problem, i.e. to add the specificity principle to DL.

There is a tremendous amount of literature on the notion of specificity in AI. Based on his notion of inferential distance, Touretzky has developed a mathematical theory of inheritance [21]. This work has initiated a whole subarea of research in AI dealing with inheritance networks. Different intuitions and options modelled in various approaches are described in [22] and [23]. All these approaches impose rather severe restrictions on the syntax of the represented theories and are for this reason of little help for our problem.

Rule based systems that prefer more specific inference rules have, for instance, been developed by Nute [12] and Simari and Loui [19]. The latter system is based on the notion of specificity developed by Poole [14]. Poole has later refined his notion of specificity [15]. For a critical analysis of this treatment of specificity and of the results provided by Nute's approach see [10].

In recent years a number of approaches to default reasoning based on conditionals have been developed [7, 11, 3, 13]. These approaches handle specificity particularly well. On the other hand, they have immense difficulties to deal with irrelevant information. For instance, if the default "birds fly" is given, this does not sanction the conclusion that a particular green bird flies since the property of being green might be relevant to flying. The conditional approaches therefore have to make additional, often rather tedious, meta-theoretic assumptions that make it possible not only to reason about, but also with defaults in a satisfactory way. DL is one of the logics that handle irrelevant information appropriately. This is the reason why it seems worthwhile to combine this logic with a treatment of specificity.

Two approaches are of particular interest here: Pearl's system Z [13] and Geffner and Pearl's conditional entailment [10]. Both extract priorities from implicit specificity information, the former ranks defaults according to a specific criterion, the latter uses partial orderings to enforce that a default "If a then normally b" leads to the conclusion b given the evidence at hand is a. System Z has two well-known problems: property inheritance is not handled adequately, and sometimes unwanted priorities are introduced. These problems do not arise in conditional entailment. Nevertheless, as noted by Geffner and Pearl, "the conclusions sanctioned by conditional entailment remain too weak". We will discuss these problems in more detail in Section 2.

Unfortunately, we cannot adopt the techniques developed in these approaches for our purposes in a straighforward way. The main reason is that they are based on a fundamentally different conception of a default. In default logic a default is viewed as a directed inference rule that does not allow for contraposition. If default contraposition is wanted one has to explicitly add the contraposed default, or use a representation without prerequisite. This inference rule conception of defaults also underlies logic programming and the standard treatment of inheritance networks.

In conditional entailment and in system Z defaults are viewed differently, namely as particular formulas. This does not mean that using a default is equivalent to using its contraposed form. In both systems the syntactic form of a default has an influence on the priority ordering that is generated. Nevertheless, unless there is information of higher priority to the contrary a default $a \rightarrow b$ will sanction the conclusion $\neg a$ given $\neg b$ as a premise. In Geffner and Pearl's words [10]:

> In conditional entailment the only difference between a default $p \rightarrow q$ and a default $true \rightarrow (p \Rightarrow q)$, which clearly allows contraposition, is that the former provides a more specific context in which the material $p \Rightarrow q$ can be asserted, and *thus* is stronger than the latter.

There does not seem to be an a priori reason for preferring one view over the other, but choosing one for determining priorities and the other for default inferencing seems inappropriate.[2] We will therefore define our own method for generating priority information. This method shares intuitions with the one underlying conditional entailment but differs in two respects:

- it treats defaults as inference rules, not as formulas
- it produces a single partial priority ordering, not possibly several ones as in conditional entailment.

Our treatment of specificity proceeds in two steps: in a first step (Sect. 3) we define a prioritized, constructive version of DL where, in addition to the set of defaults D and facts W a strict partial order on D can be specified. A similar, but not equivalent approach has independently been developed in [1].[3]

In a second step (Sect. 4) we show how the priorities needed to model specificity can be generated automatically. For that purpose we have to split the facts W of a default theory into a set of formulas T representing the background theory, and a set of formulas C representing the particular case at hand. Only the background theory will be used to determine specificity. The priorities will be obtained from minimal conflicting sets of defaults as in Delgrande and Schaub's approach [9] yet we use a somewhat different notion of conflict. We demonstrate that the problems of system Z and the weakness of conditional entailment do not arise in our approach.

An alternative way of handling priorities is considered in Section 5. It models somewhat different intuitions about the right behaviour of prioritized defaults

[2] Such a "mixing" of views underlies an approach recently proposed by Delgrande and Schaub [9] where the priorities obtained from system Z are coded into non-normal defaults. This approach thus differs from ours in two respects: we treat defaults as rules also when determining specificity and we use a modified version of DL with explicit priorities.

[3] Baader and Hollunder are dealing with terminological systems and assume to get the specificity information for their prioritized default logic from the terminological reasoner.

which can only be captured using a non-constructive definition, even for normal defaults.

Throughout the paper we will restrict ourselves to normal default theories, i.e. theories where all defaults are of the form $a{:}b/b$. We will use a somewhat simplified notation and write this default in the form a/b. As discussed in [18, 5] one of the main reasons for using non-normal defaults in the first place is their ability to encode priorities between defaults. Since we will introduce other explicit means of representing such priorities non-normal defaults won't be relevant for our purposes.

For sake of simplicity we will also assume that the set of defaults is finite. This second restriction merely simplifies the presentation in this paper.

2 System Z and conditional entailment

Pearl's system Z [13] partitions a set of default rules R into sets $R_0, R_1, \ldots$. A rule $a \rightarrow b \in R$ is in R_0 if adding its prerequisite a to R does not lead to inconsistency. For the consistency check the rules are treated like ordinary implications. Similarly, a rule is in R_i if it is not in any of the sets R_{i-k}, and adding its prerequisite does not render $R \setminus \{R_0, \ldots, R_{i-1}\}$ inconsistent. Here is a simple example. Let R consist of:

1) $bird \rightarrow flies$	3) $bird \rightarrow wings$
2) $penguin \rightarrow \neg flies$	4) $penguin \rightarrow bird$

The partition contains two sets: $R_0 = \{1, 3\}$, and $R_1 = \{2, 4\}$, those in the latter are considered of higher priority. Note that the more specific defaults get a higher rank.

The ranking of rules obtained this way is used to rank models according to the highest rank of a rule that they violate. In our example, the model

$$\{penguin, bird, wings, \neg flies\}$$

is considered better than the model

$$\{penguin, bird, wings, flies\}$$

since the former only violates a rule of rank 0, whereas the latter violates a rule of rank 1. A formula p is considered a consequence of a formula q if the rank of the best $p \wedge q$ model is lower than the rank of the best $p \wedge \neg q$ model.

There are two well-known problems with this notion of consequence, however: 1) property inheritance does not work adequately, and 2) rankings are too inflexible and introduce unwanted priorities. For the first problem consider our example again. System Z does not infer $wings$ from $penguin$, contrary to intuition. The reason is that the best model of $penguin \wedge wings$ is not considered better than the best model of $penguin \wedge \neg wings$ since both violate a rule of rank 0, namely rule 1.

For the second problem consider the following example:

1) $a \rightarrow s_1$
2) $b \rightarrow s_2$
3) $c \rightarrow s_3$
4) $a \rightarrow b$

Assume the s_i are mutually inconsistent. Intuitively, rule 4 tells us that a is a more specific class than b, therefore 1 should get priority over 2. But since there is no information about the relative specificity of a and c there should be no priority relationship among 1 and 3. Z, however, ranks 1 higher than 3, i.e. one would derive s_3 from $a \wedge c$ instead of remaining agnostic.

Conditional entailment as defined by Geffner and Pearl in [10] does not suffer from these two problems of system Z. The exact definition of conditional entailment is too complex to be reproduced here, but we want to emphasize that, as our approach, it is based on arbitrary partial orderings used to define a notion of preferred models. The orderings are defined in such a way that they enforce the conclusion b if the default $\delta = a \rightarrow b$ is given and the evidence is a. To achieve this the sets of defaults conflicting with δ for evidence a are identified, and at least one default in each such set is given lower priority than δ. In general, multiple *admissible* orderings are obtained this way.

As noted by Geffner and Pearl conditional entailment sometimes is too weak. As an example they discuss an inheritance hierarchy consisting of the following defaults[4]:

1) $a \rightarrow b$
2) $b \rightarrow c$
3) $c \rightarrow \neg d$
4) $a \rightarrow d$

Conditional entailment sanctions the conclusion d, but not c from premise a. The reason, intuitively, is that the contraposition of 3 can be used as an argument for $\neg c$. Also, if the evidence is $\{a, c\}$ then d is no longer conditionally entailed. This is at odds with the standard treatment of inheritance in inheritance systems where one would expect c to be provable, and d to remain provable if c is added as additional evidence.

Since our default logic approach views defaults as inference rules it will be much more in accordance with the treatment of defaults in inheritance systems.

3 Prioritized Default Logic

In this section we introduce PDL (Prioritized Default Logic), a version of DL where in addition to facts and defaults the user can specify an explicit priority ordering on the defaults.[5]

Our definition of prioritized extensions is closely related, but not equivalent to an independently developed definition in [1]. Baader and Hollunder show that the two approaches are in fact orthogonal, although based on similar ideas.

[4] The actual representation of defaults in conditional entailment is more complex but irrelevant for the discussion here.

[5] An "ancestor" of this definition appeared in [5]. This version, however, did not handle arbitrary partial orderings and lead to reasonable results for prerequisite-free normal defaults only.

The presentation of our definition in this paper is strongly influenced by their terminology.

Our definition of extensions for such theories is based on Reiter's quasi-inductive characterization of extensions [17]. The basic idea is to use the priority ordering as follows: during the generation of an extension defaults with higher priorities whose prerequisite has been derived are always considered before defaults of lower priority. This guarantees that defaults of lower priority are applied only if this does not lead to a conflict with a 'better' default. Since the partial ordering usually leaves some priorities undecided we have to take every total ordering of the defaults into account that respects the original partial order. Every such total order will give rise to one extension. Different orderings may generate the same extension.

A (prioritized) default theory is a triple $(D, W, <)$. D is a finite set of normal defaults of the form a/b, W a set of formulas, and $<$ a strict partial order on D. Note that we follow a tradition in nonmonotonic reasoning and consider the smaller defaults as the better ones, i.e., δ_1 has higher priority than δ_2 if $\delta_1 < \delta_2$.

Definition 1. Let E be a set of formulas, $\delta = a/c$ a default. We say δ is active in E iff

1. $a \in E$,
2. $c \notin E$,
3. $\neg c \notin E$.

Definition 2. Let $\Delta = (D, W, <)$ be a (prioritized) default theory, $\ll$ a strict total order containing $<$. We say E is the (prioritized) extension of Δ generated by $\ll$ iff $E = \bigcup E_i$, where $E_0 := Th(W)$, and

$$E_{i+1} = \begin{cases} E_i & \text{if no default is active in } E_i \\ Th(E_i \cup \{c\}) & \text{otherwise, where } c \text{ is the consequent of the} \\ & \ll\text{-minimal default that is active in } E_i. \end{cases}$$

Definition 3. Let $\Delta = (D, W, <)$ be a (prioritized) default theory. E is a prioritized extension of Δ iff there is a strict total order containing $<$ that generates E.

As an example consider the following default theory

1) b/f	3) $p/\neg f$
2) p/b	4) p

In DL we obtain two extensions, namely

$$E = Th(\{p, b, f\})$$

and

$$E' = Th(\{p, b, \neg f\}).$$

Now assume we define $3 < 1$. There are exactly three total orderings of the defaults respecting $<$, namely

$$2 < 3 < 1$$
$$3 < 2 < 1$$
$$3 < 1 < 2$$

It is easy to verify that in each case we obtain E' as the generated extension. Consider as an example the first of the three total orderings. We obtain the following sequence of sets

$$E_0 = Th(\{p\}), E_1 = Th(\{p, b\}), E_2 = Th(\{p, b, \neg f\}), E_3 = E_2, \ldots$$

The two other orderings lead to the same extension. The single generated extension thus is the one where the preferred default 3 is applied.

Our definition of prioritized extensions is fully constructive: during the inductive step no reference to the final result of the induction is made. That this can be done is not too surprising due to our restriction to normal defaults.[6] The following Lemma therefore follows immediately:[7]

Lemma 4. *The existence of extensions for prioritized default theories is guaranteed.*

Our definition generalizes the notion of preferred subtheories as defined in [4] for prerequisite-free normal defaults in a natural way. The difference between our definition and the one independently proposed by Baader and Hollunder is that we apply a single default in each inductive step whereas they apply all $<$-minimal active defaults at the same time. The different behaviour of the two approaches is discussed in [1].

There is also a close relationship between our definition and recent work by Tan and Treur [20]. These authors use an arbitrary selection function to pick out the defaults applied in each step of the induction and are thus more general than we are. On the other hand, there will be many selection functions that lead to counterintuitive results, and it is difficult to characterize those functions where this does not happen in a constructive way. For our purposes the generality of selection functions is not necessary.

In the remainder of this section we discuss several properties of our approach. The following lemma follows immediately from our definition:

Lemma 5. *Let $<_1$ and $<_2$ be strict partial orderings on a set of defaults D such that $<_1 \subseteq <_2$. The set of extensions of $(D, W, <_2)$ is a subset of the set of extensions of $(D, W, <_1)$.*

We next demonstrate that the introduction of priorities is an extension, not a modification of DL. We show that in the limiting case where the set of priorities is empty, PDL extensions and DL extensions coincide.

[6] A constructive characterization of extensions for normal default theories - without priorities - is already contained in Reiter's original paper.

[7] To generalize this result to theories with infinitely many defaults one would have to make appropriate well-foundedness assumptions for the involved priority orderings.

Proposition 6. *Let $T = (D, W, <)$ be a prioritized default theory such that $<$ is empty. The prioritized extensions of T are exactly the Reiter extensions of (D, W).*

Proof. Sketch:

"$\leftarrow$". Let E be a Reiter extension of (D, W), $GD(E)$ the set of generating defaults of E, that is the defaults whose prerequisite and consequent both are in E. Every total ordering satisfying the condition

$$d_i < d_j \text{ whenever } d_i \in GD(E) \text{ and } d_j \notin GD(E)$$

generates E.

"$\rightarrow$". Let $E = \bigcup Th(E_i)$ where the E_i are defined as in Def. 2 be a prioritized extension of T. E is an extension of (D, W) iff it is a fixed point of Reiter's Γ-operator, i.e. if it is the smallest set which 1) contains W, 2) is deductively closed, and 3) satisfies the condition: if $a/b \in D$, $a \in \Gamma(E)$, and $\neg b \notin E$, then $b \in \Gamma(E)$. We show by induction that, for all i, $Th(E_i) \subseteq \Gamma(E)$ and thus $E \subseteq \Gamma(E)$. Moreover, since E satisfies the 3 conditions for $\Gamma(E)$ we have that E is the minimal such set and thus an extension of (D, W).

From this proposition and Lemma 5 it follows immediately that every prioritized extension of a default theory $(D, W, <)$ must be a Reiter extension of (D, W).

4 Specificity in DL

In this section we will use the ability to explicitly represent priorities in the extended default logic PDL in order to handle specificity. We will show how the priorities needed to add this principle can be defined. For this purpose we first split the certain knowledge W in our default theories into two separate parts, as is common in conditional approaches (see e.g. [7, 3, 10]):

- a set T representing background knowledge, and
- a set C representing the contingent facts.

The background knowledge contains general rules, definitions, and taxonomic relationships, like 'penguins are birds', whereas C represents what is known about the current case or situation. Only the background knowledge and defaults will be used to establish the specificity of defaults.

We say that E is a (prioritized) extension of $(D, T, C, <)$ iff E is a (prioritized) extension of $(D, T \cup C, <)$. Similarly, we say E is a Reiter extension of (D, T, C) iff E is a Reiter extension of $(D, T \cup C)$.

Now what are the right priorities for modeling specificity? As mentioned in the introduction we cannot simply apply the techniques used in system Z or conditional entailment for this purpose because of their different underlying conception of a default. Nevertheless, our definition is influenced by Geffner/Pearl's definition of admissible orderings for conditional entailment.

To define the needed priorities we will identify minimal conflicting sets of defaults. This notion has first been used in [9] where a combination of system Z and semi-normal defaults is proposed which, however, mixes the inference rule and the formula view of defaults. Our notion of conflicting defaults views defaults as inference rules. Basically, a set D' will be called p-conflicting if D' generates a conflict when one of the prerequisites in D' is given. Consider the inheritance example discussed in Section 2:

1) a/b	3) $c/\neg d$
2) b/c	4) a/d

Let us assume T is empty. Now the four defaults are minimally p-conflicting since a) we obtain an inconsistent conclusion from a (viewing the defaults as inference rules), and b) no subset of the defaults is p-conflicting.

In this case we want to give the defaults with more specific prerequisite, here 1 and 4, priority over the less specific defaults 2 and 3. Generally, a default a_1/c_1 in a minimal p-conflicting set D' is considered more specific than a_2/c_2 in D' if a_1 is inconsistent under D' wheras a_2 is not.

Some useful notation: for a given set of defaults D we use $P \vdash_D q$ to express that q is contained in the smallest deductively closed set of formulas that contains P and is closed under D, where D is interpreted as a set of inference rules.

Definition 7. Let $\Delta = (D, T, C)$ be a default theory, $D' \subseteq D$ is p-conflicting (in Δ) iff for some $\delta = pre/cons \in D'$ we have

$$T \cup \{pre\} \vdash_{D'} false.$$

A p-conflicting set D' is resolvable if it contains at least one default $\delta = pre/cons$ such that

$$T \cup \{pre\} \not\vdash_{D'} false.$$

Definition 8. Let $\Delta = (D, T, C)$ be a default theory. $<_\Delta$ is the specificity ordering associated with Δ iff $<_\Delta$ is the smallest strict partial ordering on D satisfying the condition that $\delta_1 = pre_1/cons_1 <_\Delta \delta_2 = pre_2/cons_2$ whenever

1. δ_1, δ_2 are contained in a minimal set D' of defaults p-conflicting in Δ, and
2. $T \cup \{pre_1\} \vdash_{D'} false$, yet $T \cup \{pre_2\} \not\vdash_{D'} false$, or
   D' is not resolvable.

If Δ is clear from context we will often omit the suffix and simply write $<$ to denote the specificity ordering. Note that the purpose of the second alternative in clause 2 of this definition is to render the specificity ordering non-existent whenever D contains non-resolvable subsets, e.g., the two defaults a/b and $a/\neg b$. This is in full accordance with Geffner and Pearl: no admissible ordering (and no Z-ranking) exists in that case in their approach either.

There are further similarities between our definition and admissible orderings. In particular, both derive preferences from conflicting default sets and default prerequisites. Nevertheless, there are important differences, and in fact the two approaches are orthogonal.

The most obvious difference is that the specificity ordering is unique if it exists whereas in conditional entailment there may be multiple admissible orderings. In many cases the single specificity ordering is stronger than any of the admissible orderings. For instance, in the inheritance example we have that 1 and 4 both have priority over 2 *and* 3. For the corresponding admissible orderings it is only required that 1 has priority over 2 *or* 3, and 4 priority over 2 *or* 3. Note that the stronger ordering is needed in order to prove d from evidence a as well as from evidence a, c in this example.

There are also cases where the specificity ordering is weaker. This is due to the fact that defaults are considered as inference rules. For instance, the set of defaults

1) a/c
2) $b/\neg c$
3) $\neg b/\neg a$

is not p-conflicting and does not lead to any preferences at all. The corresponding set of defaults in conditional entailment would lead to a preference of 1 over 2. The intuitive reason for this different behaviour is that conditional entailment uses the contraposition of 3 to prove b from a which is unintended in the view of defaults as directed inference rules underlying our approach.

As we have seen the existence of the specificity ordering is not guaranteed. Like Geffner and Pearl we take this as a hint that there are problems in the represented knowledge and leave the notion of specificity-respecting extension, s-extension for short, undefined if no specificity ordering exists:

Definition 9. E is an s-extension of $\Delta = (D, T, C)$ iff the specificity ordering $<_\Delta$ of Δ exists and E is a prioritized extension of $(D, T, C, <_\Delta)$.

In the rest of this section we will show that the problems of system Z and the weakness of conditional entailment as discussed in Section 2 do not arise in our approach. Let us first continue the inheritance example. We have seen that the ordering prefers the 2 defaults with prerequisite a over the two others. From this it is obvious that the single s-extension for $C = \{a\}$ is the set $Th(\{a, b, c, d\})$. Adding c to C does not change this result.

Now consider the two examples illustrating the difficulties of system Z:

1) $bird/flies$
2) $penguin/\neg flies$
3) $bird/wings$
4) $penguin/bird$

Assume T is empty. The single minimal p-conflicting set of defaults is $\{1, 2, 4\}$ and leads to the priorities $2 < 1$, $4 < 1$. For $C = \{peng\}$ the s-extension thus contains $peng, \neg flies$ and $wings$, Z's property inheritance problem is thus avoided.

Here is the second example:

1) a/s_1
2) b/s_2
3) c/s_3
4) a/b

Again the different s_i are mutually conflicting and T is empty. The single minimal p-conflicting set is $\{1, 2, 4\}$. We thus obtain $1 < 2$ and $4 < 2$. Given $C = \{a, c\}$ we get 2 s-extensions, one containing s_1, the other s_3. No unintended priority is introduced.

5 An alternative treatment of priorities

In our treatment of priorities it may be the case that a default of low priority leads to the derivation of the prerequisite of a default with high priority which nevertheless is not applied because of a conflicting default of medium priority. Consider the following example:

1) a/c 3) $true/a$
2) $true/\neg c$

Assume $W = \emptyset$ and the priority ordering is $1 < 2 < 3$. Since this ordering is total there is only one extension generated, namely $Th(\{a, \neg c\})$. This is also the extension obtained by Baader and Hollunder in their approach [1].

In his dissertation Prakken [16] criticizes this treatment of priorities. He argues that in such an example default 1 should have been applied instead of default 2, since its prerequisite a is contained in the final extension.

Intuitions seem to differ here. On one hand we can justify the behaviour of our original formalization as follows: since the priority of 3 is low it can be considered questionable whether 1 really should have been applied since there is only weak evidence that a is true. On the other hand one might argue that if the extension contains a anyway there is no reason not to apply the default with higher priority.

The Prakken intuition can be modeled using a slight modification of our definition of prioritized extensions (we call the new extensions PDL2 extensions). However, the price to pay is rather high: constructiveness and existence of extensions have to be given up. The basic idea is to introduce a form of lookahead into the construction of extensions. The role of this lookahead is to guess whether a default prerequisite will be derivable in a later step in the inductive construction of the extension.

To achieve this we have to change the notion of active default used in Def. 2. A default $\delta = a/c$ must be considered active if

1. $a \in E$ (the result of the induction!),
2. $c \notin E_i$,
3. $\neg c \notin E_i$.

Now the most preferred default whose prerequisite is proven or becomes provable in a later inductive step will be applied. This modification as it stands, however, leads to ungrounded extensions. For instance, a default a/a can be used to derive a. This shows that an additional groundedness check has to be added.

The simplest way to achieve this is to eliminate by definition all extensions that are not Reiter extensions.[8]

With these modifications the extension obtained for our example is $Th(\{a, c\})$ which is the result desired by Prakken. Unfortunately, this alternative notion of extensions for prioritized default logic looses some of the nice features of the original one. In particular, existence of extensions is not guaranteed. This can be illustrated using the following slight modification of our example:

1) a/c 3) $\neg c/a$
2) $true/\neg c$

Again assume the priority ordering $1 < 2 < 3$. Now $Th(\{a, \neg c\})$ is not a PDL2 extension since 1 has not been applied. Similarly $Th(\{a, c\})$ is not an extension since a cannot be reproduced during the construction of the E_i's.

This lack of constructiveness is the reason why we favour our original approach. Nevertheless, our method of automatically generating the specificity ordering does not depend on the form of the underlying prioritized default logic and can be combined with the second, modified version as well.

6 Conclusions

In this paper we have defined two versions of prioritized normal default logic, a constructive, more well-behaved one, and a non-constructive one. These logics formalize somewhat different ways of handling prioritized defaults. We further developed a method of automatically generating priorities that model the specificity principle. Our method is related to the definition of admissible orderings in conditional entailment, but it is based on a different view of defaults where, as in default logic, defaults are conceived as directed inference rules. We have thus shown that techniques for handling specificity can also be defined for rule-based nonmonotonic logics. Additionally, we have demonstrated that our approach does not suffer from some weaknesses of Pearl's system Z and conditional entailment.

The notion of specificity is an important preference criterion for defaults. Nevertheless, it is not the only such criterion. For instance, in legal reasoning a more general rule may override a more specific rule if the general rule was created more recently. This shows that it must be possible to combine specificity with other forms of preferences. Since in our approach the generation of the specificity ordering and the definition of prioritized default logic are independent from each other we can easily add methods of combining specificity with, say, temporal preference criteria. A simple way of doing this would be to use a lexicographical ordering based on the different available criteria and to use this combined ordering as input for PDL. It is much more difficult to do this in approaches where specificity is deeply built into the logical machinery, e.g., in conditional nonmonotonic logics.

[8] Note also that even the grounded extensions generated by a single total ordering may be non-unique.

The computation of our specificity ordering is obviously extremely expensive as it involves the generation of all minimal p-inconsistent subsets. From the practical point of view there is some hope, however. It seems reasonable to assume that the background knowledge changes much less frequently than the case knowledge. Since the specificity ordering only depends on the background knowledge there is a good chance that we can compute it once and then, hopefully, apply it to many different cases.

Acknowledgements

I would like to thank Jim Delgrande and Torsten Schaub for interesting discussions on the topic of this paper.

References

1. Baader, Franz, Hollunder, Bernhard, How to Prefer More Specific Defaults in Terminological Default Logic, Proc. IJCAI-93, Chambery, France, 1993
2. Bidoit, Nicole, Froidevaux, Christine, Negation by Default and Non-Stratifiable Logic Programs, Research Report 437, Universite Paris Sud, Centre d'Orsay, LRI, 1988
3. Boutilier, Craig, Conditional Logics for Default Reasoning and Belief Revision, PhD thesis, Dep. of Computer Science, Univ. of Toronto, 1992
4. Brewka, Gerhard, Preferred Subtheories - An Extended Logical Framework for Default Reasoning, Proc. IJCAI-89, Detroit, 1989
5. Brewka, Gerhard: Cumulative Default Logic - In Defense of Nonmonotonic Inference Rules, Artificial Intelligence 50, 1991
6. Brewka, Gerhard, Konolige, Kurt, An Abductive Framework for General Logic Programs and Other Nonmonotonic Systems, Proc. IJCAI-93, Chambery, France, 1993
7. Delgrande, J.P., A First-Order Conditional Logic for Prototypical Properties, Artificial Intelligence, 33(1), 1987
8. Delgrande, J.P., Jackson, W.K., Default Logic Revisited, Proc. Second International Conference on Principles of Knowledge Representation and Reasoning, 1991
9. Delgrande, J.P., Schaub, T., On Using System Z to Generate Prioritised Default Theories: Extended Abstract, unpublished manuscript
10. Geffner, Hector, Pearl, Judea, Conditional Entailment: Bridging Two Approaches to Default Reasoning, Artificial Intelligence 53, 1992
11. Lehmann, D., What Does a Conditional Knowledge Base Entail?, Proc. First International Conference on Principles of Knowledge Representation and Reasoning, 1990
12. Nute, D., Defeasible Reasoning: A Philosophical Analysis in Prolog, in: J. Fetzer (ed), Aspects of Artificial Intelligence, Kluwer, Boston, 1988
13. Pearl, Judea, System Z: A Natural Ordering of Defaults with Tractable Applications to Nonmonotonic Reasoning, Proc. Third Conference on Theoretical Aspects of Reasoning About Knowledge, 1990
14. Poole, David, On the Comparison of Theories: Prefering the Most Specific Explanation, Proc. IJCAI-85, Los Angeles, 1985

15. Poole, David, Dialectics and Specificity: Conditioning in Logic-Based Hypothetical Reasoning (Preliminary Report), Proc. 3rd Intl. Workshop on Nonmonotonic Reasoning, South Lake Tahoe, 1990
16. Prakken, Henry, Logical Tools for Modelling Legal Argument, dissertation, VU Amsterdam, 1993
17. Reiter, Raymond, A Logic for Default Reasoning, Artificial Intelligence 13 (1980) 81-132.
18. Reiter, Raymond, Criscuolo, G., On Interacting Defaults, Proc. IJCAI 81, 1981
19. Simari, G., Loui, R., A Mathematical Treatment of Defeasible Reasoning and its Implementation, Artificial Intelligence 53 (1992)
20. Tan, Yao Hua, Treur, Jan, Constructive Default Logic and the Control of Defeasible Reasoning, Proc. ECAI 92, Vienna, 1992
21. Touretzky, David S., The Mathematics of Inheritance, Pitman Research Notes in Artificial Intelligence, London, 1986
22. Touretzky, David S., Horty, John F., Thomason, Richmond H., A Clash of Intuitions: The Current State of Nonmonotonic Multiple Inheritance Systems, Proc. IJCAI-87, Milan, 1987
23. Touretzky, David S., Thomason, Richmond H., Horty, John F., A Skeptic's Menagerie: Conflictors, Preemptors, Reinstaters, and Zombies in Nonmonotonic Inheritance, Proc. 12th IJCAI, Sydney, 1991

Viewing Hypothesis Theories as Constrained Graded Theories

Philippe CHATALIC

Laboratoire de Recherche en Informatique - UA CNRS 410
Bat 490 - Université Paris-Sud - 91405 ORSAY Cedex - FRANCE
email: chatalic@lri.lri.fr

Abstract

Modeling expert tasks often leads to consider uncertain and/or incomplete knowledge. This generally requires reasoning about uncertain beliefs and sometimes making additional hypotheses. While numerical models are often used to model uncertainty, the estimation of precise and meaningful values for certainty degrees is sometimes problematic. Moreover, the use of a numerical scale implies that any two certainty degrees are comparable. This paper presents a qualitative approach, where uncertainty is represented by means of partially ordered symbolic grades. The framework is a multimodal logic in which each grade is expressed as a modal operator. An extension of this framework is proposed which makes it possible to state additional hypotheses in the spirit of Siegel and Schwind's hypothesis theory. We show that such hypotheses may be interpreted as constraints on the set of possible beliefs. We thus obtain a very natural integration of multimodal graded logic and hypothesis theory. The resulting framework allows for the simultaneous representation of uncertain and/or incomplete information. Some correspondence results between extensions of graded default logic and those of such new *graded hypothesis theories* are established.

KEYWORDS: UNCERTAIN AND INCOMPLETE KNOWLEDGE REPRESENTATION AND SEMANTICS, PARTIALLY ORDERED GRADES, MODAL LOGIC, HYPOTHESIS THEORIES.

1 Introduction

To perform successful reasoning under incomplete information it is generally necessary to take advantage of any kind of available knowledge. This includes true facts as well as less certain pieces of information, which may be considered as uncertain beliefs. Generally such beliefs do not all have the same strength and when using such beliefs in deductions, the characterization of the strength of the derived conclusions is an important issue. In the case of missing knowledge, it also happens that some facts are deduced under some hypotheses. It is then generally assumed that such conclusions only hold as far as they do not contradict the premises from which they are drawn.

Many approaches for handling uncertain beliefs are based on numerical settings , e.g., Probability theory [Ni 86, Pe 88], Dempster-Shafer theory [Sh 76] and Possibility theory [Za 78, DP 88]. However, experts do not always feel very comfortable when they have to give precise estimations of certainty degrees. In such cases, instead of giving precise values, they often prefer using qualitative estimates to express that a statement is considered as more certain than another. Several attempts have proposed qualitative views of numerical settings (e.g., [Fi 73], [Gä 75], [HR 87], [WL 91]). Still these proposals can be found to be too constraining, because of the (often implicit) assumption that the certainty degrees of any two pieces of knowledge are comparable. But there might be circumstances under which experts do not want to compare uncertain beliefs, either because it does not make sense, or because not enough information is available. A possible way to overcome such a limitation is to express uncertain beliefs by means of a partially ordered set of qualitative grades. This idea has been already explored in [CF 91, 92]. The approach followed in [CF 92] is a multimodal framework in which grades are represented by modal operators. It is however purely monotonic and gives no possibility of reasoning in presence of incomplete information.

In this paper we show how this framework may be augmented in order to allow for the representation of incomplete knowledge. The proposed extension gives the possibility to deduce some facts under some additional assumptions. The approach we propose here is very close to Siegel & Schwind's hypothesis theory [SS 91,93]. We show that by considering hypotheses as constraints on the set of possible beliefs we obtain a very natural integration of multimodal graded logic and hypothesis theory. In this new framework, called *graded hypothesis theory* it is possible to handle simultaneously uncertain and/or incomplete information. Imperfect knowledge may thus be represented and dealt with in different ways to perform either monotonic or non-monotonic graded deductions.

The remainder of the paper is organized as follows. Section 2 presents the framework of multimodal graded logic, its syntax, semantics and some useful properties. Section 3 extends the previous framework to introduce hypotheses, as constraints on a set of graded beliefs, and formalizes the notion of graded hypothesis theory. In Section 4 we show how it is possible to map a graded default theory [CF 91] into a corresponding graded hypothesis theory and establish correspondence results between extensions in both formalisms. Section 5 presents a comparison of this work with other related approaches. Eventually, we conclude with some directions for further research.

2 Multimodal graded logic

2.1 Expressing uncertain beliefs

Representing uncertain beliefs requires putting together two components, one for characterizing what is believed and another for characterizing how much it is believed. In multimodal graded logic [CF 92, 93] uncertain beliefs are expressed by means of modal formulas of the form $[\alpha]$f. The formula f denotes what is believed and may be any formula of the language. The modal operator $[\alpha]$ is used to express that the formula f is supported with some grade α. Note that such a grade α is not meant to represent the very certainty degree of f. When reasoning under uncertainty any piece of information may prove to be useful and it often happens that a given formula is supported in different ways. For instance, it might be supported by different sources of evidence or it might be obtained as the conclusion of different deductions. In a graded theory, a formula may thus be supported with different grades. We argue that in such a case, its certainty degree should be greater than or equal to each of these grades. As a consequence, grades are to be considered as lower bounds of certainty degrees of beliefs. From this it results that if a formula is believed with different grades, it is also believed with the least upper bound of all these grades (principle 1).

Another important concern, when considering uncertain beliefs, is the way levels of support are combined during deductions. It is generally considered as a good knowledge representation principle that a belief obtained as the conclusion of some deduction should not be more supported that each of the beliefs used in the deduction (principle 2). When considering qualitative values for grades, this leads to consider the greatest lower bound of the grades used in the deduction.

Since, according to these two basic principles, we have to deal with least upper bounds and greatest lower bounds of sets of grades, we use for representing grades a finite distributive lattice structure $(\Gamma, \wedge, \vee, \preccurlyeq)$, where $\preccurlyeq$ denotes the partial order relation on the set of grades Γ. For any grades $\alpha, \beta \in \Gamma$, the expressions $\alpha \vee \beta$ and $\alpha \wedge \beta$ denote respectively the least upper bound (*lub*) and the greatest lower bound (*glb*) of α and β. Note that all the grades of Γ are supposed to denote a strictly positive level of support. The greatest element, denoted by $\top$, corresponds to the highest possible level of support i.e. total certainty.

In the following we also denote by $(\Gamma_\wedge, \wedge, \preccurlyeq)$ the lower semi-lattice of Γ. This corresponds to elements of Γ that can be described by expressions which do not use the operator $\vee$ [Bi 73]. An important property of distributive lattices is that any grade may be described by an expression under disjunctive normal form. Formally this means that $\forall \alpha \in \Gamma, \exists\, \alpha_1, \ldots, \alpha_p \in \Gamma_\wedge$ such that $\alpha = \alpha_1 \vee \ldots \vee \alpha_p$ and $\forall i, j$ α_i and α_j are not comparable.

2.2 Syntax of multimodal graded logic

Let $P = \{p_0, p_1, \ldots, p_n\}$ be a finite set of atomic propositions. Let $(\Gamma, \wedge, \vee, \preccurlyeq)$ be a distributive lattice. As usual $\neg, \wedge, \vee, \rightarrow, \leftrightarrow$ will denote the boolean connectives[1]. We also introduce true and false as constants. For each grade α of Γ, we define a corresponding parameterized modal operator $[\alpha]$. The propositional multimodal graded language based on P and Γ is defined as follows :

Definition 2.1 : *The multimodal graded language ML_Γ induced by P and Γ is the least set satisfying the following conditions:*

- $P \subset ML_\Gamma$, *true* $\in ML_\Gamma$, *false* $\in ML_\Gamma$,
- •• $\forall p, q \in ML_\Gamma$, $\neg p, p \vee q, p \wedge q, p \rightarrow q, p \leftrightarrow q$ *are formulas of* ML_Γ
- ••• $\forall p \in ML_\Gamma$, $\forall \alpha \in \Gamma, [\alpha]p \in ML_\Gamma$

Axiom System

We now introduce the system Σ_Γ, which is based on the modal system K, as a syntactical characterization of our basic principles.

Axiom schemes:

(C) Classical Axioms

(K) $[\alpha](A \rightarrow B) \rightarrow ([\alpha]A \rightarrow [\alpha]B)$

---

[1] To preserve usual notations, we use the same symbols $\wedge$ and $\vee$ to denote logical connectors as well as meet and join operations on Γ. The interpretation of these symbols thus depends on the context in which they are used.

(D_T) $\neg[T]false$

(A_1) $([\alpha]A \wedge [\beta]A) \rightarrow [\alpha \vee \beta]A$

(A_2) $[\alpha]A \rightarrow [\beta]A$ $\quad \forall \alpha, \beta \in \Gamma$ such that $\beta < \alpha$.

Inference rules:

(MP) $\dfrac{\vdash_\Sigma A \quad \vdash_\Sigma A \rightarrow B}{\vdash_\Sigma B}$ (modus ponens rule)

(NR) $\dfrac{\vdash_\Sigma A}{\vdash [T]A}$ (necessitation rule)

In this system, the symbols α and β refer to any element of Γ, while symbols A and B refer to any formulas. Theorems p of Σ_Γ are denoted by $\vdash_\Sigma$ p. As in [Ch 80] we say that a formula p is *deducible* or *derivable* from a set S of formulas (written S $\vdash_\Sigma$ p) in Σ_Γ if and only if we can find a finite subset $\{q_1, ..., q_n\} \subseteq S$ such that $(q_1 \wedge ... \wedge q_n) \rightarrow p$ is a theorem of Σ_Γ. We denote by $Th_\Sigma(S)$ the set of formulas that may be derived from S. The parameterized modal operators $[\alpha]$ correspond to necessity operators in the K system [Ch 80], except $[T]$ which corresponds to the necessity operator of the D system . This results from the axiom D_T, which expresses that any theory from which it is possible to derive the contradiction with total certainty is inconsistent. Note that the notion of consistency in Σ_Γ corresponds to the usual notion of consistency in modal logic. A set of S of formulas is *inconsistent* if and only if the false formula is deducible from S. However the consistency of a set S of formulas does not imply the consistency of the set of beliefs expressed by S. Particularly, S may be consistent while allowing the formula $[\alpha]$false (with $\alpha \neq T$) to be derivable from S. This is rather satisfactory from the knowledge representation point of view, since with uncertain knowledge we often have to deal with conflicting pieces of information [DL 91]. We may call such theories as *α-inconsistent* theories.
In Σ_Γ, the partial order on Γ is expressed by means of the weakening axiom schemes A_2 (strict ordering is sufficient since $p \rightarrow p$ is a theorem of classical logic). Using this axiom scheme, K and modus ponens we may obtain as a derived axiom scheme :

$$\vdash ([\alpha]A \wedge [\beta](A \rightarrow B)) \rightarrow [\alpha \wedge \beta]B \quad (A_{gmp})$$

This axiom scheme clearly expresses our second basic principle and is related to the *graded modus ponens* rule introduced in [CF 91]. It allows for an easy formalization of deductions in theories with uncertain knowledge.

Axiom scheme A_1 expresses our first basic principle which states that, if there are two ways to deduce a given proposition with different grades, the level of support for this proposition should be at least as high as any of those grades. More generally, in the case where there are several derivations of a same formula with different grades, A_1 will be used to obtain the *best* possible grade, i.e. the *greatest lower bound* of the certainty degree of this formula. The following property may be established :

Property 2.1: *Let S be a set of graded formulas over Γ, let $\alpha \in \Gamma$ and $\alpha_1 \vee ... \vee \alpha_p$ be its disjunctive normal form, where $\forall i=1..p$, $\alpha_i \in \Gamma_\wedge$.*

$S \vdash_\Sigma [\alpha]p$ iff $\forall i=1..p$, $S \vdash_\Sigma [\alpha_i]p$

2.3 Semantics

In this section we recall the semantics introduced in [CF 93] for multimodal graded logic and recall useful properties. We define the meaning of formulas in terms possible worlds semantics [Ch 80] involving families of accessibility relations. With each grade $\alpha \in \Gamma$, we associate an accessibility relation R_α. Our intuition is that the higher the grade associated with a formula, the more constraining the corresponding accessibility relation should be. We express this idea by the fact that if $\alpha \preccurlyeq \beta$ then $R_\alpha \subseteq R_\beta$. In such a case, if, from a given possible world w, there are more possible worlds accessible thru R_β than thru R_α, it will be more difficult to satisfy the formula $[\beta]p$ than the formula $[\alpha]p$ in w. Additional constraints insure a minimal correspondence between the structure of the family of accessibility relations and the structure of Γ.

Definition 2.2: *A Γ-**interpretation** is defined as a triple $I = \langle W, (R_\alpha)_{\alpha\in\Gamma}, s \rangle$ where:*

- *W is a set of worlds*
- *$(R_\alpha)_{\alpha\in\Gamma}$ is a family of accessibility relations verifying:*
  - *i) $\forall \alpha,\beta \in \Gamma$, if $\alpha \preccurlyeq \beta$ then $R_\alpha \subseteq R_\beta$*
  - *ii) $\forall \alpha,\beta \in \Gamma$, $R_{\alpha\vee\beta} \subseteq R_\alpha \cup R_\beta$*
  - *iii) R_T is serial (i.e., $\forall w \in W$, $\exists w' \in W$ such that $R_T(w,w')$)*
- *s is a classical truth value assignment $s: P \times W \rightarrow \{true, false\}$*

The definition of a formula being true at a world is the standard one used in modal logic.

Definition 2.3: *Let $I = \langle W, (R_\alpha)_{\alpha\in\Gamma}, s \rangle$ be a Γ-interpretation. A formula f of ML_Γ is said to be **true at a world** w of I (written $I,w \models_m f$) iff:*

- *$I,w \models_m true$*
- *$I,w \models_m f$ iff $s(f,w) = true$, for $f \in P$*
- *$I,w \models_m f \vee g$ iff $I,w \models_m f$ or $I,w \models_m g$, for $f, g \in ML_\Gamma$*
- *$I,w \models_m f \wedge g$ iff $I,w \models_m f$ and $I,w \models_m g$, for $f, g \in ML_\Gamma$*
- *$I,w \models_m \neg f$ iff not $I,w \models_m f$ (that will be denoted by $I,w \not\models_m f$), for $f \in ML_\Gamma$*
- *$I,w \models_m [\alpha]f$ iff $\forall w' \in W$, $R_\alpha(w,w')$ implies $I,w' \models_m f$, for $f \in ML_\Gamma$*

In the following, when there is no ambiguity, the interpretation I will not be mentioned any more and we shall merely write $w \models_m f$ instead of $I,w \models_m f$.

Definition 2.4:

- *A formula f is **satisfiable** iff there exists a Γ-interpretation $I = \langle W, (R_\alpha)_{\alpha\in\Gamma}, s \rangle$ and a world $w \in W$, such that $I,w \models_m f$.*
- *A formula f is **valid in an interpretation I** $= \langle W, (R_\alpha)_{\alpha\in\Gamma}, s \rangle$ (written $I \models f$) iff $\forall w \in W$, $I, w \models_m f$. We also say that I is a Γ-**model** of the formula f.*
- *A formula f is **valid** (written $\models_m f$) iff every Γ-interpretation is a Γ-model of f.*

The previous definitions may be extended to sets of formulas.

Property 2.2 : *Let $I = \langle W, (R_\alpha)_{\alpha\in\Gamma}, s\rangle$ be a graded interpretation, and let $\alpha, \beta \in \Gamma$*

a) $\models_m [\beta]f \rightarrow [\alpha]f$ *if* $\alpha \preccurlyeq \beta$

b) $\models_m [\alpha]f \wedge [\beta]f \rightarrow [\alpha\vee\beta]f$

c) If $\alpha_1 \vee ... \vee \alpha_p$ *is the disjunctive normal form of* α*, with* $\forall i=1..p,\ \alpha_i \in \Gamma_\wedge$*, then* $\models_m [\alpha]p$ *iff* $\forall i=1..p,\ \models_m [\alpha_i]p$

The system Σ_Γ has been proved to be sound and complete with respect to this semantics [CF 93].

Theorem 2.1 : $\vdash_\Sigma f$ *iff* $\models_m f$.

3. Hypotheses in graded multimodal logic

The multimodal logic presented so far is monotonic. The purpose of this section is to extend the current formalism in order to allow for defeasible reasoning. Our aim is to integrate similar ideas to those developed in Siegel and Schwind's Hypothesis Theory [SS 91,93] to allow for making additional hypotheses in the case of incomplete information. Our approach essentially differs from the one of [SS 93] on two points. While Siegel and Schwind use a single modal operator L to characterize what is known, we use our family of modal operators $[\alpha]$ to denote what is believed. The second difference is that since Siegel and Schwind are concerned with what is known, their necessity operator L follows the rule of the T system [Ch 80]. This is justified by the fact that what is known should be consistent with the true facts. Because in our approach we agree to have partially inconsistent beliefs, our parameterized modal operators $[\alpha]$ (resp. $[\top]$) follow the rules of the modal system K (resp. D).

3.1 Viewing hypotheses as constraints on beliefs

In this section, we introduce a new grade $\perp$. In contrast with the grades of Γ, which are supposed to characterize strictly positive amount of support, $\perp$ will be used to characterize the lowest possible level of support, i.e. empty support. The new grade $\perp$ is thus a universal lower bound for the elements of Γ and we have $\forall\alpha\in\Gamma,\ \perp < \alpha$, $\alpha\wedge\perp = \perp$ and $\alpha\vee\perp = \alpha$. The set $\Gamma\cup\{\perp\}$ is also a distributive lattice and in the following of the paper, we consider that this new grade is now an element of Γ and denote respectively by $\Gamma^+_\wedge$ and Γ^+ the sets $\Gamma_\wedge \setminus \{\perp\}$ and $\Gamma \setminus \{\perp\}$.

Since the grade $\perp$ corresponds to empty support, a formula like $[\perp]$p may be considered as a vacuous piece of information. At a first glance, such a formula does not look very interesting. However, $[\perp]$p may be considered as a way to express that, although we do not have any particular evidence in favor of p or ¬p, we still have some *preference* for p rather than for ¬p.

A formulas like $\neg[\perp]$p is also interesting. In a consistent theory, such a formula expresses that the formula p is not supported by any belief. Otherwise from $[\alpha]$p it would be possible to derive $[\perp]$p by weakening, since $\forall\alpha\in\Gamma\ \perp \preccurlyeq \alpha$, and this would lead to an inconsistency. Turning to a negative formula, $\neg[\perp]\neg$p says that the formula ¬p is not supported by any belief... which is to say that having such beliefs would lead to an inconsistency. Thus, although it does not imply any positive belief in favor of the formula p, it rejects any possible belief in favor of ¬p. This can be considered as a way of *assuming* p although not giving any support to p.

Similar behavior may be encountered in Siegel and Schwind's hypothesis theory [SS 91, 93]. The main idea of hypothesis theory is to try to augment a set of non logical axioms by adding hypotheses while preserving consistency. Similarly, hypotheses are considered as propositions that are assumed though not known. The consistency of added assumptions with what is known is insured by means of a set of additional constraints of the form $Hp \rightarrow \neg L \neg p$. This states that it is only possible to assume a formula p (i.e. to make the hypothesis Hp) if the formula $\neg p$ is not known.

In our multimodal approach such constraints naturally result from the axiom (A_2) and the fact that $\perp$ is a universal lower bound of Γ. For any formula p and any grade $\alpha \in \Gamma^+$ we have by (A_2) $[\alpha]\neg p \rightarrow [\perp]\neg p$ and thus by contraposition $\neg[\perp]\neg p \rightarrow \neg[\alpha]\neg p$. As a consequence, adding a formula $\neg[\perp]\neg p$ to a multimodal graded theory is clearly similar to adding a hypothesis in hypothesis theory. It can be considered as a *constraint* that is added to the initial theory and that rejects any belief supporting the proposition $\neg p$.

In modal logic, it is quite common to associate to a modal operator of necessity $\Box$, a dual modal operator of possibility , denoted by $\Diamond$ and defined by $\Diamond = \neg\Box\neg$. Similarly, we may define in our framework a possibility operator $\langle\perp\rangle$ as $\neg[\perp]\neg$. A *hypothesis* then corresponds to a formula of the form $\langle\perp\rangle p$. From the semantical point of view, a Γ-interpretation is thus a model of an hypothesis if from any possible world, there exists at least one possible world, accessible by $R_\perp$, in which p is true. This seems intuitively satisfactory. In the following, for convenience, we shall also use the notation Hp to represent the formula $\langle\perp\rangle p$, as originally introduced in Siegel and Schwind's work.

3.2 Formalizing graded hypothesis theories

In the rest of this section we show how multimodal graded logic may be extended in order to incorporate this notion of hypotheses and show that most results of hypothesis theory are preserved in their graded version.

Definition 3.1: *A **graded hypothesis theory** is a pair HT = (S, Hyp) where :*

- *S is a set of formulas of ML_Γ*
- *Hyp is a set of hypotheses $Hp = \neg[\perp]\neg p$, $p \in ML_\Gamma$*

Given a graded hypothesis theory (S, Hyp), we are interested in the characterization of sets of hypotheses from Hyp that are consistent with S. An extension is thus obtained by successive additions to S of new hypotheses from Hyp, until no more hypothesis may be added while preserving consistency.

Definition 3.2: *An **extension** of a graded hypothesis theory HT = (S, Hyp) is a set $E = Th_\Sigma(S \cup H)$ where $H \subseteq Hyp$ is a maximal subset of Hyp consistent with S.*

Example :

Let us consider a given student, supposed to be a good student. We also think that, if this student is good and if it is possible to make the hypothesis that he has worked hard to prepare his exams, then he is likely to pass his exams. This can be expressed by the hypothesis theory HT = (S, Hyp) defined as follows:

S = { ([T]good_student∧Hworked_hard) → [α]pass_exams, [T]good_student }
Hyp = {Hworked_hard}

Since nothing contradicts the hypothesis Hworked_hard, the theory HT has exactly one extension E which contains the formula [α]pass_exams.
Now, let us suppose we have also heard from another person that this student spent all the days before the exam at the swimming pool. We might then reasonably believe (but we cannot be certain of that) that he did not worked hard. This can be expressed by adding to S the formula [β]¬worked_hard. But from [β]¬worked_hard, it is possible to derive [⊥]¬worked_hard (since $\perp < \beta$) and recall that Hworked_hard ≡ ¬[⊥]¬worked_hard. Therefore it is no more possible to make the hypothesis Hworked_hard, since this would lead to an inconsistency. Thus the only possible extension of HT is reduced to the set of theorems of S.

As it can be seen in this example, adding a new formula f to the set S may have an incidence on the set of possible extensions. It may cause some hypothesis, that was previously in some extension of S, to be no more consistent with the new set of beliefs $S \cup \{f\}$. This exhibits the nonmonotonic behavior of this approach.

As for hypothesis theories, we may state a sufficient condition for the existence of extensions.

Theorem 3.1: *Let HT = (S, Hyp) be an hypothesis theory.*
If S is consistent then HT has an extension.

The proof is analogous to the one of [SS 91]. As pointed in [SS 93] this implies that extending the set of Hyp of hypotheses of a graded hypothesis theory can augment the content of extensions, yield more extensions but will never eliminate any of the previous extensions.

Among other results of [SS 93] which still hold in the graded version are the compactness property and the fixpoint characterization of extensions that follow from the definition.

Theorem 3.2: *Let HT = (S, Hyp) be a graded hypothesis theory and let f be a formula of* ML_Γ.

(1) let $E = Th_\Sigma(S \cup H)$ *be an extension of HT.*
Then $f \in E$ *iff* $\exists\{h_1, \ldots, h_n\} \subseteq H$ *such that* $f \in Th_\Sigma(S \cup \{h_1, \ldots, h_n\})$.

(2) f is a theorem of some extension E of HT iff $\exists\{h_1, \ldots, h_n\} \subseteq Hyp$ *such that* $f \in Th_\Sigma(S \cup \{h_1, \ldots, h_n\})$ *and* $S \cup \{h_1, \ldots, h_n\}$ *is consistent.*

Theorem 3.3: *Let HT=(S, Hyp) be a graded hypothesis theory and E an extension of HT. Then E is a solution of the recursive equation :*

$$E = Th_\Sigma(S \cup \{h \in Hyp / \neg h \notin E\})$$

Corollary 3.1: *If E is an extension of S in (S,Hyp), then for all* $h \in Hyp$, *either* $h \in E$ *or* $\neg h \in E$

Again proofs are analogous to those of [SS 93]. The only difference is that the proof of the compactness property relies on the compactness of K instead of the compactness of T.

4. Relation with graded default logic

In [CF 91, CF 93] another formalism has been introduced that makes it possible to handle simultaneously uncertain and incomplete knowledge. A graded formula is then represented by a pair (p α), where p is a classical propositional formula and α is a grade on the lattice Γ. In the following, to avoid ambiguities we shall refer this former work as the *classical graded logic* approach, as opposed to the *multimodal graded logic* approach. In [CF 91], nonmonotonic behavior is achieved by associating grades with classical defaults [Re 80, Be 89] in a similar way and applying the principle of graded inference to default inference. The purpose of this section is to establish some correspondence results between the notions of extension, for such graded defaults theories and for graded hypothesis theories.

4.1 Classical graded logic

We first compare the monotonic parts of each formalism, namely classical graded logic (without graded defaults) and multimodal graded logic (without hypotheses). The axiom system of classical graded logic (that will be denoted here by Ξ_Γ) is composed of classical axioms schemes of propositional logic (graded by ⊤) and three inference rules denoted respectively by:

$$\frac{(p\ \ \alpha)\quad (p \rightarrow q\ \ \beta)}{(q\ \ \alpha\wedge\beta)}\,(\text{MPG}) \qquad \frac{(p\ \ \alpha)}{(p\ \ \beta)\ \forall\,\beta < \alpha}\,(\text{WR}) \qquad \frac{(p\,\alpha)\quad (p\ \ \beta)}{(p\ \ \alpha\vee\beta)}\,(\text{SR})$$

graded modus ponens rule — weakening rule — strengthening rule

Multimodal graded logic clearly encompasses classical graded logic. The multimodal language is more expressive. For instance, it makes it possible to have complex graded formulas of the form [α]f where f is itself made of graded formulas. Such nested use of grades is not allowed in the classical approach. It is also possible to use negation (as any other logical connector) with graded formulas, which is impossible in the classical approach. In fact classical graded logic can be seen to some extent as a fraction of multimodal graded logic, where only modal formulas of the form [α]p are considered, such that p is a classical propositional formula. From the semantical point of view, although classical graded logic is not a multimodal language its semantics is also based on Kripke like interpretations with a family of accessibility relations. Actually the only difference between the two semantics is that instead of being serial, as in the multimodal framework, the relation $R_\top$ is equal to W×W. Classical graded logic has been proved to be correct and complete with respect to this semantics. Let us notice that the class of Γ-interpretations for the classical graded logic approach is strictly included in the class of Γ-interpretations for the multimodal graded logic approach.

To establish a formal correspondence between both formalisms, we may define a mapping Θ, which transforms a classical graded theory S into a corresponding multimodal graded theory MS, by translating every graded proposition (p α) of S into the corresponding modal formula [α]p. In this section, to avoid possible confusions, we denote respectively by $\vdash_\Xi$ and $\vdash_\Sigma$ the derivability in classical graded logic and in multimodal graded logic. Similarly we use respectively $\models_\Xi$ and $\models_\Sigma$ to denote validity.

Theorem 4.1 : *Let S be a set of classical graded formulas and MS = $\Theta(S)$ its translation into multimodal graded logic:*

$$\text{if } S \vdash_{\Xi} (p \quad \alpha) \text{ then } MS \vdash_{\Sigma} [\alpha]p.$$

PROOF:

We show that all the components of Ξ_Γ may be recovered in Σ_Γ. Axiom schemes of Ξ_Γ correspond to axiom schemes of classical propositional logic graded by $\top$. The corresponding axiom schemes may be derived in Σ_Γ using the necessitation rule. The three inference rules of Ξ_Γ have corresponding axioms schemes in Σ_Γ and thus may be obtained as derived inference rule in Σ_Γ. Indeed, strengthening and weakening rules correspond exactly to axiom schemes (A_1) and (A_2), while graded modus ponens corresponds to the derived axiom scheme (A_{gmp}). As a consequence the multimodal system Σ_Γ subsumes Ξ_Γ. From this it follows it follows that if S $\vdash_{\Xi}$ (p α) then MS $\vdash_{\Sigma}$ [α]p.

It is also worth considering the reverse mapping Θ^{-1}. But in this case, it can only be defined for theories containing formulas of the form [α]p, where p does not contain itself any modal operator.

Theorem 4.2 : *Let MS be a theory containing only formulas of the form [α]p, such that p does not contain itself any modal operator, and let S = Θ^{-1}(MS) be its reverse translation:*

$$\text{if } MS \vdash_{\Sigma} [\alpha]p \quad \text{then} \quad S \vdash_{\Xi} (p \quad \alpha)$$

PROOF:

The proof may be done in a semantic way. We know that MS $\vdash_{\Sigma}$ [α]p iff $\exists f_1, \ldots, f_n \in$ MS such that $(f_1 \wedge \ldots \wedge f_n) \rightarrow [\alpha]p$ is a theorem. Since MS contains only formulas of the form [α]p, we may write each f_i as $[\alpha_i]p_i$. By correction if $([\alpha_1]p_1 \wedge \ldots \wedge [\alpha_n]p_n) \rightarrow [\alpha]p$ is a theorem, it is a tautology. Thus any Γ-interpretation is a model of this formula. This means that for any interpretation $I = \langle W, (R_\alpha)_{\alpha \in \Gamma}, s \rangle$, and any $w \in W$ we have $w \models ([\alpha_1]p_1 \wedge \ldots \wedge [\alpha_n]p_n) \rightarrow [\alpha]p$. This means that $\forall w \in W$, if $w \models [\alpha_1]p_1$ and ... and $w \models [\alpha_n]p_n$ then $w \models [\alpha]p$. In both approaches, the semantics of the formula [α]p and (p α) (where p is a classical propositional formula) is defined exactly in the same way. Moreover, the class of interpretations considered in classical graded logic is strictly included in the class of interpretation considered in multimodal graded logic. Thus, this implies that from the classical graded logic point of view, any Γ-model of$\{(p_1 \ \alpha_1) \ldots (p_n \ \alpha_n)\}$ is also a Γ-model of (p α). Then, by completeness of classical graded logic this implies that S $\vdash_{\Xi}$ (p α).

4.2 Correspondence between graded default theories and hypothesis theories

In [SS 93] a correspondence between Reiter's default logic and hypothesis theories is established. Given the links that we have established between classical graded logic and multimodal graded logic, this suggests a similar correspondence between graded default theories [FG 90, CF 91] and graded hypothesis theories. In the following we show that Siegel and Schwind's translation of classical default theories into corresponding hypothesis theories may be adapted to the graded case and point out the links between the notion of extension of a graded default theory and of its translation in graded hypothesis theory.

Let us recall that in Reiter's approach [Re 80] a default d is a specific nonmonotonic inference rule of the form (p : q / r), where p, q and r are elements of P ; p is called the *prerequisite* of d, q its *justification* and r its *consequent.* We use Pre(D), Just(D) and Cons(D) to denote respectively the set of prerequisites, justifications and consequents of a set of defaults D. Just as a graded formula, *a graded default* is defined as a pair (d α) where d is a classical default and α a grade. As for a classical default, p and q are called respectively the prerequisite and the justification of d. A *graded default theory* Δ is then defined as a pair (W, D) where W is a set of graded formulas and D is a set of graded defaults.

The approach followed by [CF 91] consists in generalizing the principle of graded inference (principle 1) to the case of default inference. A default (p : q / r β) may be triggered if its prerequisite p is believed with some grade α and if nothing contradicts q (i.e. if $\neg$q is not believed at all). Then r is inferred with the grade $\alpha \wedge \beta$. Note that in contrast with classical defaults, if the prerequisite of a graded default is believed with different grades, this default may produce different consequents according to the grades of its prerequisite. Extensions of a graded default theory are then characterized by the following fixpoint definition :

Definition 4.1: *Let $\Delta = (W, D)$ be a graded default theory. Let E be a set of graded formulas. The sequence $(E_i)_{i\geq 0}$* is defined as follows :

$E_0 = W$ *and for* $i\geq 0$,

$E_{i+1} = Th_{\Sigma}(E_i) \cup \{ (r \quad \alpha \wedge \beta) \ / \ (p : q / r \quad \beta) \in D, (p \quad \alpha) \in E_i \text{ and } \neg q \notin \overline{E} \}$.

E is a graded extension for Δ iff $E = \bigcup_{i\geq 0} E_i$.

In this definition $\overline{S}$ denotes the *support* of a set of graded formulas S, i.e. the set of formulas of S without their grades. It has been shown in [CF 91] that graded extensions are closely related to classical extensions of a non-graded default theory. In particular, we have the following result :

Theorem 4.3: *Let $\Delta = (W, D)$ be a graded default theory, F be a set of formulas and E be a set of graded formulas. If E is a graded extension for Δ then $\overline{E}$ is an extension for ($\overline{W}$, $\overline{D}$) and conversely, if F is an extension for $\overline{\Delta}$ then there exists some graded extension E for Δ such that $F = \overline{E}$.*

We now extend the mapping Θ in order to translate any graded default theory into a corresponding multimodal graded theory. Following [SS 93], we suggest the following transformation :

Let Δ = (W, D) be a graded default theory, the translation $\Theta(\Delta)$ of Δ into hypothesis logic is defined by $\Theta(\Delta)$ = (S,Hyp) where :

$$\bullet\ S = \Theta(W) \cup \Theta(D) \text{ where } \begin{cases} \Theta(W) = \{[\alpha]p \ / \ (p \quad \alpha) \in W\} \\ \Theta(D) = \{[\gamma]p \wedge Hq \rightarrow [\gamma]r \ / \ (\frac{p:q}{r} \quad \alpha) \in D \text{ and } \bot \preccurlyeq \gamma \preccurlyeq \alpha\} \end{cases}$$

• Hyp = {Hp / p ∈ just(D)}

In this mapping, to preserve the principle of graded inference, a graded default $(p : q / r \quad \alpha)$ is translated into a set of formulas of the form $[\gamma]p \wedge Hq \rightarrow [\gamma]r$ with $\gamma \prec \alpha$. Note that the formula $[\alpha]p \wedge Hq \rightarrow [\alpha]r$ alone would not be sufficient since in such a case it would be impossible to infer anything from $[\beta]p$ if $\alpha \nprec \beta$. The set of hypotheses of the translation corresponds to the set of justifications of the defaults of D.

Let us note that there is a little difference between the sets of grades used in both approaches. While Δ is characterized with only grades of Γ^+ (i.e. strictly positive), its translation is expressed using the grades of Γ, including $\perp$. However it is possible to overcome this little problem by considering that, although the set of grades used in Δ does not contain $\perp$, we still express Δ as a graded default theory based on a set of grades Γ.

By introducing such a new universal lower bound in the set of grades causes new theorems of the form $(p \quad \perp)$ to be derivable. But it may be shown that as far as $\perp$ is not used in the characterization of a set of formulas S, a formula p is derivable from S with the grade $\perp$ if and only if it may be derived with some grade $\alpha \in \Gamma^+$. More generally it may be shown that :

Property 4.1: Let S be a set of graded formulas such that $\forall (p \quad \alpha) \in S,\ \alpha \in \Gamma^+$,
$Th_{\equiv\Gamma^+}(S) = Th_{\equiv\Gamma}(S)^+$ where $Th_{\equiv\Gamma}(S)^+ = \{(p \quad \alpha) \in Th_{\equiv\Gamma}(S) / \alpha \in \Gamma^+\}$

In the following, to avoid ambiguities, we denote by Δ_+ the initial graded default theory based on the set of grades Γ^+ and by Δ the same graded default theory described on the extended set of grades Γ. A consequence of property 4.1 is that a set E_+ of Γ^+-formulas is a Γ^+-extension of Δ_+ iff there exists a Γ-extension E of Δ such that $E_+ = (E)^+$.

Now we are able to state the correspondence theorems. Roughly the idea is that to any consistent extension E of a graded default theory Δ_+, corresponds some extensions E' of $\Theta(\Delta_+)$ such that the set of hypotheses characterizing the extension E' corresponds precisely to the set of justifications of the generating defaults of E.

Theorem 4.4 : *Let $\Delta_+ = (W,D)$ be a graded default theory and HT = (S,Hyp) be its translation into hypothesis theory. Let E_+ be a consistent extension of Δ_+, and let $H(E_+) = \{Hp / p \in Just(D)$ and $\forall \gamma \in \Gamma^+, (\neg p \quad \gamma) \notin E_+\}$*

1) $E' = Th_\Sigma(S \cup H(E_+))$ is an extension of HT

2) $E_+ = \{(p \quad \alpha) / \alpha \in \Gamma^+,\ [\alpha]p \in E'$ and p is a non-modal formula$\}$

3) If $\neg Hp \in E'$ then $\exists \alpha \in \Gamma^+$ such that $[\alpha]\neg p \in E'$

Theorem 4.5: *Let $\Delta_+ = (W,D)$ be a default theory and HT = (S,Hyp) be its translation.*

Let E' be an extension of HT such that if $\neg Hq \in E'$ then $\exists \gamma \in \Gamma^+, [\gamma]\neg q \in E'$.

Then there exists an extension E_+ of Δ_+ such that :

$E_+ = \{(p \quad \alpha) / \alpha \in \Gamma^+, [\alpha]p \in E'$ and p is a non-modal formula$\}$.

The reader is referred to the appendix for the proofs of these theorems.

5 Related work

The expression *graded modal logic* has also been used by VanDerHoek in [VH 92]. Although this work is also motivated by the representation of uncertain knowledge by means of modal operators, it differs from ours in a fundamental way, since it aims at counting the number of exceptional situations where some proposition p does not hold. For this, an infinity of necessity operators [n] ($n \in \mathbb{N}$) is introduced, such that [n]p is satisfied by a (possible) world w if and only if there are at most n worlds w' that are reachable from w and that satisfy formula $\neg p$.

There has been another attempt at translating graded default theories into graded hypothesis theories [Pg 92]. However it is not based on a multimodal approach. The basic idea of this work is to encode any grade α of Γ by a chain of modalities of the form $S_\alpha = \Box S \Box$, where S is a sequence of modal operators $\Box$ and $\Diamond$, containing a fixed number of $\Box$ and possibly some $\Diamond$ provided they are not placed at consecutive places in the chain. Such an encoding is made possible by the fact that the logic T contains an infinity of distinct modalities represented by such chains. The author shows that for a given set of grades Γ, it is possible to define a mapping $\phi: \alpha \mapsto S_\alpha$ such that all chains S_α contain the same number n of $\Box$, and such that $\forall \alpha, \beta \in \Gamma$, $\alpha \lessdot \beta$ iff $S_\beta p \rightarrow S_\alpha p$ is an axiom of the system T. Uncertain beliefs are then represented by formulas of the form $S_\beta p$. A noticeable drawback of this approach is that it is not incremental. If new grades are introduced, it is necessary to reconstruct the encoding. To perform graded inference in the same way as graded modus ponens does, the author proposes to use the inference rule : $\dfrac{S_\alpha \Box p \quad S_\alpha \Box (p \rightarrow q)}{S_\alpha \Box q}$. A graded default theory $\Delta = (W, D)$, is then translated into the hypothesis theory HT = (S, Hyp) where :

$$S = \{ S_\alpha \Box p \,/\, (p \;\; \alpha) \in W \} \cup \{ S_\alpha (\Box p \wedge Hq \rightarrow \Box r \;\; / \, (p : q / r \;\; \alpha) \in D \}$$
$$Hyp = \{ \Box^n Hq \,/\, q \in Just(D) \}$$

The author justifies the introduction of an extra necessity operator $\Box$ after any chain S_α by technical reasons. In our formalism (apart from the fact that we use the system K), the corresponding translation would require us to replace any chain S_α by a modal operator $[\alpha]$, and to use an extra modal operator of necessity $\Box$. This would give :

$$S = \{ [\alpha] \Box p \,/\, (p \;\; \alpha) \in W \} \cup \{ [\alpha] (\Box p \wedge Hq \rightarrow \Box r \;\; / \, (p : q / r \;\; \alpha) \in D \}$$
$$Hyp = \{ [\top] Hq \,/\, q \in Just(D) \}$$

6. Conclusion and perspectives

In this paper we have presented a formalism which allows for the simultaneous handling of uncertain and incomplete information. It is based on a qualitative approach, where the uncertainty is represented by means of partially ordered symbolic grades attached to logical formulas and expressed as modal operators. The use of a partially set of grades makes it possible to state that some beliefs are more supported than others and are to be interpreted in term of level of certainty. However it is not intended to be used for expressing preferences or priorities among the set of beliefs.

We have shown that Siegel and Schwind's notion of hypothesis may naturally be integrated into the framework of multimodal graded logic. We have thus extended this framework and shown that most properties of hypothesis theory still hold in graded hypothesis theory. The semantics of modal formulas this approach differs from the original one of Siegel and Schwind's approach. They are interpreted as graded beliefs instead of known formulas. As a consequence, the modal operator used to state hypotheses is similar to the other and may be given a clear semantics. Making an hypothesis may be considered as setting a constraint on the set of possible belief. More precisely, the addition of the hypothesis Hq excludes the possibility of having any support in q (i.e. to have $[\alpha]q$, for any $\alpha \in \Gamma$). The extended framework offers two non-exclusive possibilities to express imperfect knowledge. By using graded implications, it is possible to perform graded deductions, combining the various levels of support involved in the reasoning processes. The conclusion of such deductions may be uncertain beliefs but they are derived in a monotonic way. Defeasible (and possibly uncertain) conclusions may be obtained by introducing hypothesis into the involved formulas.

Another idea to be explored could be to consider less restrictive constraints. Let us recall that Hq actually corresponds to the modal operator $\langle \bot \rangle$ defined as $\neg[\bot]\neg$. One possible way to set weaker constraints may be to consider modal operators of the form $\langle \alpha \rangle$ corresponding to $\neg[\alpha]\neg$. Then, by making an hypothesis of the form $\langle \alpha \rangle q$, we merely state that we do not have $[\alpha]\neg q$, which implies that we do not have either $[\beta]\neg q$, for any $\alpha \preccurlyeq \beta$. But this leaves some possibility of having $[\beta]\neg q$ if $\alpha \not\preccurlyeq \beta$. Such a point of view is very close to the notion of relaxed condition of applicability for graded defaults as introduced in [Ng 92, CF 93b]. In this variant of graded default logic, a graded default may still be triggered if the negation of its justification is derivable, but only with a grade that falls under some *ε-bar* $B_\varepsilon = \{\varepsilon_1, \ldots, \varepsilon_n\}$. The ε-bar is used to partition the whole set of grades Γ in two parts, one of which is considered as a set of *negligible grades*, with respect to the rest of Γ. The translation of such a graded default rule $(p : q / r \quad \alpha)$ could then be expressed in hypothesis theory by a set of formulas of the form $[\gamma]p \wedge \langle \varepsilon_1 \rangle q \wedge \ldots \wedge \langle \varepsilon_n \rangle q \rightarrow [\gamma]r \ \forall \gamma \preccurlyeq \alpha$. The hypotheses to be considered would be all the formulas of the form $\langle \varepsilon_j \rangle q_i$ such that q_i is the justification of some graded default. Such a translation would amount to setting a uniform condition of applicability for each default. Of course graded hypothesis theory allows much more precision and it would be possible to make the criteria of applicability to vary according to the default considered. This topic is currently under investigation, as well as the extension of the formalism to the first order case [CSF 94].

Another direction worth to be considered is to look for alternative translations of graded defaults into hypothesis theory. For instance we might consider the translation of a graded default $(p : q / r \quad \alpha)$ into a formula of the form $Hq \rightarrow [\alpha](p \rightarrow r)$ or $[\alpha](p \wedge Hq \rightarrow r)$. Such translations would lead to different notions of extensions and might prove to have other interesting properties.

Acknowledgments

I would like to thank Christine Froidevaux for many fruitful discussions and valuable comments on this paper. This work has been supported by DRUMS II Esprit Basic Research Action project n°6156.

References

[Be 89] Besnard P. (1989), *An introduction to default logic*, Springer Verlag, Heidelberg.

[Bi 73] Birkhoff G. (1973), *Lattice Theory*, American Mathematical Society Colloquium Publications, vol. XXV.

[CF 91] Chatalic P. and Froidevaux C. (1991), *Graded logics: A framework for uncertain and defeasible knowledge*, *in* Methodologies for Intelligent Systems, (Ras Z.W. & Zemankova M. eds.), Proc. of ISMIS-91, Lecture Notes in Artificial Intelligence, 542, 479-489.

[CF 92] Chatalic P. and Froidevaux C. (1992), *Lattice based Graded logics: A multimodal Approach*, Proc. Uncertainty in AI, Stanford, CA, USA, 33-40.

[CF 93] Chatalic P. and Froidevaux C. (1993), *A multimodal Approach to Graded Logic*, L.R.I. Tech Report n° 808, L.R.I. Tech. Report, Université Paris-Sud, Orsay, France.

[CF 93b] Chatalic P. and Froidevaux C. (1993), *Weak inconsistency in graded default logic and hypothesis theory*, DRUMS II Tech. Report 3.1.1. BRA n°6156

[CFS 94] Chatalic P., Froidevaux C. and Schwind C. (1994), *A logic with graded hypotheses* (forthcoming)

[Ch 80] Chellas B. (1980), *Modal logic - an introduction*, Cambridge University Press, New York.

[DL 91] Dubois D., Lang J. and Prade H., (1991), *Inconsistency in knowledge bases - to live or not live with it -*, Fuzzy Logic for the management of Uncertainty (Zadeh L.A., Kacprzyk J. eds) J. Wiley.

[DP 88] Dubois D. and Prade H. (with the collaboration of Farreny H., Martin-Clouaire R., Testemale C.) (1988), Possibility Theory: An approach to computerized processing of uncertainty. Plenum Press, New-York.

[Fi 73] Fine T.L. (1973) *Theories of Probability: An Examination of Foundations.* Academic Press, New York.

[FG 90] Froidevaux C. and Grossetête C. (1990), Graded default theories for uncertainty, Proc. of the 9th European Conference on Artificial Intelligence, Stockholm, 283-288.

[Gä 75] Gärdenfors P. (1975) *Qualitative probability as an intensional logic*, J. Phil. Logic 4, 171-185.

[HR 87] Halpern J. and Rabin M. (1987) A logic to reason about likelihood, Artificial Intelligence 32, 379-405.

[Ng 92] Nguyen F. (1992) *Towards the introduction of inconsistency in the extensions of graded default theory.* Research Note, LRI, Orsay. France.

[Ni 86] Nilsson N.J. (1986) *Probabilistic logic*, Artificial Intelligence 28, 71-87.

[Pe 88] Pearl J. (1988) *Probabilistic Reasoning in Intelligent Systems - Networks of plausible Inference.* Morgan Kaufmann Pub., San Matheo, Cal, USA.

[Pg 92] Pereira Gonzalez W. (1992) *Une logique modale pour le raisonnement dans l'incertain.* PhD thesis. University of Rennes I, France.

[Re 80] Reiter R. (1980) A logic for default reasoning, Artificial Intelligence 13, 81-132.

[Re 87] Reiter R. (1987), Nonmonotonic reasoning, Annual Reviews Computer Science 2, 147-186.

[SS 91] Siegel P. and Schwind C. (1991) *Hypothesis Theory for Nonmonotonic Reasoning*, 2nd Int. Workshop on Non-monotonic and Inductive Logic, Reinhardsbrunn Castle.

[SS 93] Siegel P. and Schwind C. (1993) *Modal logic based theory for non-monotonic reasoning*, Journal of Applied Non-Classical Logics, Vol 3-1, pp 73-92

[Sh 76] Shafer G. (1976) *A mathematical theory of evidence*, Princeton University Press, NJ, USA.

[Vh 92] Van der Hoek W. (1992) *On the Semantics of Graded Modalities,* Journal of Applied Non Classical Logics, Vol. II n°1, pp. 81-123.

[WL 91] Wong S.K., Lingras P. and Yao Y. (1991) *Propagation of Preference relations in qualitative inference networks,* Proc. IJCAI 91, Sydney, Australia, 1204-1209.

[Za 78] Zadeh L.A. (1978) *Fuzzy Sets as a basis for a theory of possibility.* Fuzzy Sets and Systems 1, 3-28.

Appendix

We give here the main lines of the proofs of theorems 4.4 and 4.5.

SKETCH OF PROOF FOR THEOREM 4.4 :

Let $\Delta_+ = (W, D)$ be a graded default theory based on the set of grades Γ^+ and let HT = (S,Hyp) be its translation into hypothesis theory. Let E_+ be a consistent extension of Δ_+ and let $H(E_+) = \{Hp \,/\, p \in Just(D)$ and $\forall\gamma\in\Gamma^+, (\neg q \quad \gamma) \notin E_+\}$. We first consider the graded default theory Δ similar to Δ_+ but based on the extended set of grades Γ. We know that Δ has a consistent extension E such that $E_+ = (E)^+$. Let $H(E) = \{Hp \,/\, p\in Just(D)$ and $\forall\gamma\in\Gamma^+, (\neg q \quad \gamma)\notin E+\}$. By construction we have $H(E^+) = H(E)$. Let $E' = Th_\Sigma(S \cup H(E))$ and let $E'' = \{(p \quad \alpha) \,/\, [\alpha]p \in E'$ and p is a non-modal formula$\}$. The proof proceeds in four steps :

a) We first show that $E \subseteq E''$: If A is a set of classical graded formulas, $A \subseteq E''$ implies $\Theta(A) \subseteq E'$ and $Th_\Xi(A) \subseteq E''$. If $A \subseteq E''$, A contains only formulas of the form $(p \quad \alpha)$, such that $[\alpha]p\in E'$ and p is a non-modal formula. Hence $\Theta(A) \subseteq \{[\alpha]p \in E' \,/\, p$ is a non-modal formula$\}$ and thus $\Theta(A) \subseteq E'$. By theorem 4.1 this implies $\Theta(Th_\Xi(A)) \subseteq Th_\Sigma(\Theta(A)) \subseteq E'$. Thus $Th_\Xi(A) \subseteq E''$. Since E is an extension of Δ, it may be expressed as $E = \bigcup_{i\geq 0} E_i$, where $E_0 = W$ and for $i\geq 0$ and $E_{i+1} = Th_\Xi(E_i) \cup \{ (r \quad \alpha\wedge\beta) \,/\, (p{:}q/r \quad \beta)\in D, (p \quad \alpha) \in E_i$ and $\forall\gamma\in\Gamma^+, (\neg q \quad \gamma) \notin E\}$. Then the proof proceeds by induction to prove that $\forall i$ we have $E_i \subseteq E''$, which entails $E = \bigcup_{i\geq 0} E_i \subseteq E''$.

b) The next step is to show that E' is an extension of HT, which means that E' is consistent and that the set of hypotheses H(E) is maximal.

b.1) We prove the consistency of E' by exhibiting a Γ-model of $S\cup H(E)$. If E is a consistent extension of Δ, it has a Γ^+-model. Particularly it has a canonical model $I_E = \langle W, (R_\alpha)_{\alpha\in\Gamma}, s\rangle$ defined by :

- W is the set of all (classical) models of $\overline{E_T}$,
- $(R_\alpha)_{\alpha\in\Gamma}$ is such that $\begin{cases} \text{(i) } R_T = W\times W \\ \text{(ii) } \forall\beta\in\Gamma_\wedge \setminus\{T\},\ R_\beta = \{(w,w')\in W\times W \,/\, w' \models \overline{E_\beta}\ \} \\ \text{(iii) } \forall\beta,\ \delta\in\Gamma\quad R_{\beta\vee\delta} = R_\beta \cup R_\delta \end{cases}$

where $\forall\alpha\in\Gamma, E_\alpha = \{(p \quad \beta)\in E \,/\, \alpha \lessdot \beta\}$ and $\overline{E} = \{p \,/\, \exists\alpha\in\Gamma\ (p \quad \alpha)\in E\}$

The proof first establishes that I_E is a Γ-interpretation and that I_E is such that $\forall\alpha\in\Gamma, \forall w, w'\in W, (p \;\; \alpha)\in E$ iff $I_E \models (p \;\; \alpha)$ iff $\forall w\in W\ w \models (p \;\; \alpha)$ iff $\exists w\in W\ w \models (p \;\; \alpha)$. The end of the proof for this point consists in showing that this interpretation is also a Γ-model of E'.

b.2) To prove the maximality of H(E), let us consider $Hq\in Hyp \setminus H(E)$. Then $\exists\gamma\in\Gamma^+, (\neg q \;\; \gamma)\in E$, which implies $[\gamma]\neg q \in E'$ and thus $[\bot]\neg q\in E'$. Adding Hq to $S\cup H(E)$ would mean adding $\neg[\bot]\neg q$, which would not be consistent with $[\bot]\neg q$. Thus H(E) is maximal.

As a consequence the set E' is an extension of the graded hypothesis theory HT.

c) Now we prove that if $\neg Hq\in E'$ then $\exists\gamma\in\Gamma^+, [\gamma]\neg q\in E'$. Since E' is an extension of HT, we know that for any Hq of Hyp we have either $Hq\in E'$ or $\neg Hq\in E'$. If $\neg Hq\in E'$ then $Hq\notin H(E)$, i.e. $\exists\gamma\in\Gamma^+, (\neg q \;\; \gamma)\in E$. But this implies that $[\gamma]\neg q\in E'$. Thus if $\neg Hq\in E'$ then $\exists\gamma\in\Gamma^+, [\gamma]\neg q\in E'$.

d) Eventually, we prove that $E'' \subseteq E$. Let $(p \;\; \alpha)\in E''$, i.e. such that $\alpha\in\Gamma^+, [\alpha]p\in E'$ and p is a non-modal formula. By property 2.1 it is sufficient to restrict to the case where $\alpha\in\Gamma_\wedge$. Since $[\alpha]p\in E'$ it is valid in any Γ-model of E'. In particular, it is valid in the Γ-model I_E that we have seen earlier in this proof. In I_E we have, $\forall w\in W, w \models [\alpha]p$. But since $\alpha\in\Gamma_\wedge$, this implies that $\forall w\in W, w \models (p \;\; \alpha)$ and thus $(p \;\; \alpha)\in E$. Hence $E'' \subseteq E$.

SKETCH OF PROOF FOR THEOREM 4.5 :

Let $\Delta_+ = (W, D)$ be a graded default theory based on the set of grades Γ^+ and let HT = (S,Hyp) be its translation $\Theta(\Delta_+)$ into hypothesis theory. Let E' be an extension of HT such that if $\neg Hq\in E'$ then $\exists\gamma\in\Gamma^+, [\gamma]\neg q\in E'$. Let $E'' = \{[\alpha]p\in E' \,/\, p$ is a non-modal formula$\}$.

a) We first prove that the set E'' may be characterized as the union $E = \bigcup_{i\geq 0} E_i$ where the sets E_i are defined by : $E_0 = \Theta(W)$ and for $i\geq 0$,

$$E_{i+1} = \{[\alpha]p\in Th_\Sigma(E_i)\}\cup\{[\alpha]r \,/\, [\alpha]p\wedge Hq \rightarrow [\alpha]r \in\Theta(D),$$
$$[\alpha]p\in E_i \text{ and } Hq\in E'\}.$$

a.1) To show that $E \subseteq E''$, we prove by induction that $\forall i\geq 0$ we have $E_i \subseteq E''$. For i=0, $E_0 = \Theta(W)$, and $\Theta(W) \subseteq S \subseteq E'$. Since $\Theta(W)$ contains only formulas of the form $[\alpha]p$ we have $E_0 \subseteq E''$. Let us suppose that $\forall i\leq n, E_i \subseteq E''$. Let $[\alpha]p\in E_{n+1}$, by construction either $[\alpha]p\in Th_\Sigma(E_n)$ or $[\alpha]p\in \{[\alpha]r \,/\, [\alpha]p\wedge Hq \rightarrow [\alpha]r\in\Theta(D), [\alpha]p\in E_n$ and $Hq\in E'\}$. In the first case, we know by hypothesis that $E_n \subseteq E' \subseteq E''$, thus $Th_\Sigma(E_n) \subseteq Th_\Sigma(E') = E'$. This implies $\{[\alpha]p\in Th_\Sigma(E_n)\} \subseteq E''$. In the second case, $[\alpha]p$ is of the form $[\alpha_i]r_i$ where $[\alpha_i]p_i\wedge Hq_i \rightarrow [\alpha_i]r_i\in\Theta(D)$, $[\alpha_i]p_i\in E_n$ and $Hq_i\in E'$. But if $[\alpha_i]p_i\in E_n$ and $Hq_i\in E'$, we have $[\alpha_i]p_i\in E'$ and thus $[\alpha_i]r_i\in E'$. Again this implies that $[\alpha_i]r_i\in E''$. As a consequence we have $E_{n+1} \subseteq E''$. By induction we obtain that $\forall i\ E_i \subseteq E''$, i.e. $E \subseteq E''$.

a.2) Now we show that $E'' \subseteq E$. Since E' is an extension, it is consistent as well as E'' and E. It is then possible to define a canonical model in a similar way as for classical graded theories. Since E contains only formulas of the form $[\alpha]p$

where p is a non-modal formula, for any grade α of Γ, we may consider the set $\overline{E_\alpha}$ defined as $\{p / \exists\gamma\in\Gamma, [\gamma]p\in E\ \alpha \preccurlyeq \gamma\}$.

Let us consider the interpretation $I_C = \langle W, \underline{(R_\alpha)_{\alpha\in\Gamma}}, s\rangle$ defined by :

- W is the set of all (classical) models of $\overline{E_T}$,
- $(R_\alpha)_{\alpha\in\Gamma}$ is such that $\begin{cases} \text{(i) } R_T = W\times W \\ \text{(ii) } \forall\beta\in\Gamma_\wedge\setminus\{T\},\ R_\beta = \{(w,w')\in W\times W / w'\models \overline{E_\beta}\ \} \\ \text{(iii) } \forall\beta,\delta\in\Gamma\ \ R_{\beta\vee\delta}=R_\beta\cup R_\delta \end{cases}$

The proof first establishes that I_C is a Γ-model of E such that for any non-modal formula p and any grade α of Γ, we have the following property :

$[\alpha]p\in E$ iff $I_C \models [\alpha]p$ iff $\forall w\in W\ w \models [\alpha]p$ iff $\exists w\in W\ w \models [\alpha]p$

Then we prove that I_C is a also Γ-model of E". Thus if $[\alpha]p\in$ E" then $I_C \models [\alpha]p$, which is equivalent to say that $[\alpha]p\in E$. Hence E"$\subseteq$E.

As a consequence we have E" = E.

b) In the second part of the proof, we consider the graded default theory Δ similar to Δ_+ but based on the extended set of grades Γ, and show that Δ has an extension corresponding to $\Theta^{-1}(E)$. We consider the sequence defined by E^Δ_0 = W and for i≥0, $E^\Delta_{i+1} = Th_{\equiv}(E^\Delta_i) \cup \{(r\ \ \alpha\wedge\beta) / (p{:}q/r\ \ \beta)\in D, (p\ \alpha)\in E^\Delta_i$ and $\forall\gamma\in\Gamma^+, (\neg q\ \ \gamma)\notin\Theta^{-1}(E)\}$. Let E^Δ be the set $\bigcup_{i\geq0}E^\Delta_i$.

It may be proved by induction that $\forall i$, $\Theta(E^\Delta_i) = E_i$. Hence we have $\Theta(E^\Delta) = E$ which also means $E^\Delta = \Theta^{-1}(E)$. Thus E^Δ is a Γ-extension of the graded default theory Δ defined on the set of grades Γ. This implies that Δ_+ has an extension E_+ such that $E_+ = (E)^+$. Hence the graded default theory Δ_+ has an extension $E_+ = (E)^+ = \{(p\ \ \alpha) / \alpha\in\Gamma^+, [\alpha]p\in E'$ and p is a non-modal formula $\}$.

Temporal Theories of Reasoning*

Joeri Engelfriet and Jan Treur

Free University Amsterdam
Department of Mathematics and Computer Science
De Boelelaan 1081a, 1081 HV Amsterdam
The Netherlands
email: {joeri, treur}@cs.vu.nl

1 Introduction

In practical reasoning usually there are different patterns of reasoning behaviour possible, each leading to a distinct set of conclusions. In logic one is used to express semantics in terms of models that represent descriptions of (conclusions about) the world and in terms of semantic entailment relations based on a specific class of this type of models. In the (sound) classical case each reasoning pattern leads to conclusions that are true in all of these models: each line of reasoning fits to each model. However, for the non-classical case the picture is quite different. For example, in default reasoning conclusion sets can be described by (Reiter) extensions. In common examples this leads to a variety of mutually contradictory extensions. It depends on the chosen line of reasoning which one of these extensions fits the pattern of reasoning.

The general idea underlying our approach is that a particular reasoning pattern can be formalized by a sequence of *information states* $\mathbf{M}_0$, $\mathbf{M}_1$, ... Here any $\mathbf{M}_t$ is a description of the (partial) knowledge that has been deduced up to the moment in time $\mathbf{t}$. An inference step is viewed as a transition $\mathbf{M}_t \text{ -> } \mathbf{M}_{t+1}$ of the current information state $\mathbf{M}_t$ to the next information state $\mathbf{M}_{t+1}$. In the current paper we formalize the information states $\mathbf{M}_t$ by partial models (although other choices are possible). A particular reasoning pattern is formalized by a sequence $(\mathbf{M}_t)_{t \in T}$ of subsequent partial models labelled by elements of a flow of time $\mathbf{T}$; such a sequence is interpreted as a partial temporal model. A transition relating a next information state to the current one can be formalized by temporal formulae the partial temporal model has to satisfy. So, inference rules will be translated into temporal rules to obtain a temporal theory describing the reasoning behaviour. Each possible pattern of the reasoning process can be described by a model of this theory (in temporal partial logic).

Using our techniques the semantics of reasoning can be viewed as a set of (intended) partial temporal models. The branching character of these reasoning

* This work has been carried out in the context of SKBS and the ESPRIT III Basic Research project 6156 DRUMS II.

processes can be described by branching time partial temporal models. One (strict) line of reasoning corresponds to a linear time model: a branch in the tree of all possibilities. In one branching time model more than one line of reasoning (and the resulting conclusion sets) can be represented (even when they are mutually contradictory).

In this paper we will present temporal axiomatizations of three types of reasoning: reasoning based on a classical proof system, default reasoning and meta-level reasoning. We show that it is possible to define formal semantics where (temporal) aspects of the process of reasoning and the resulting conclusions are both integrated in an explicit manner.

In Section 2 we introduce temporal partial logic. In Section 3 it is pointed out that under some conditions a temporal theory has a unique (up to isomorphism) final branching time model that covers all possible lines of reasoning. The semantics of such a temporal theory can be defined on the basis of this unique final model: a form of final model semantics.

In Section 4 we show how proof rules in a classical proof system can be represented by temporal rules. Thus a temporal theory is provided that has a final model where all possible classical proofs are represented as branches. In Section 5 we show how default reasoning can be formalized in temporal partial logic. In Section 6 we treat the reasoning of a meta-level architecture. We show how also in this case a temporal axiomatization of the reasoning can be obtained.

The approach as worked out here can be viewed as a generalization of the manner in which modal and temporal semantics can be given to intuitionistic logic (see [13], [17]). One of the differences is that we use partial (information) states; another one is that we apply the approach to a wider class of reasoning systems.

In practical reasoning systems for complex tasks of the type analysed in [27] often the dynamics of the reasoning is subject of reasoning itself (using strategic knowledge to control the reasoning). Therefore for the overall semantics of this type of reasoning system it is hard to make a distinction between static aspects and dynamic aspects. In particular, it is impossible to provide independent declarative semantics for such systems without taking into account the dynamics of the reasoning. For the overall reasoning system a formal semantical description is needed that systematically integrates both views. The lack of such (overall) semantics for complex reasoning systems (with meta-level reasoning capabilities) was one of the major open problems that were identified during the ECAI-92 workshop on Formal Specification Methods for Complex Reasoning Systems where 8 formal specification languages for reasoning systems for complex tasks where analysed and compared (see [16], p. 280). The approach introduced in the current paper can be considered as a first step to provide semantics of formal specification languages of this type.

2 Temporal Partial Logic

In this section we introduce our temporal partial logic, based on branching time structures. Our approach is in line with what in [12] is called temporalizing a given logic; in our case the given logic is partial logic with the Strong Kleene semantics. We shall start by defining the time structures, then we will define the models based on them and finally we will define temporal formulae and their interpretation.

Definition 2.1 (Flow of Time)
A *flow of time* is a pair $(T, <)$ where T is a set of time points and $<$ is a binary relation over T, called the *immediate successor relation*. We want to consider forward branching structures: $(T, <)$ viewed as a graph has to be a *forest*, that is a disjoint union of trees. Furthermore the transitive (but not reflexive) closure $\ll$ of $<$ is introduced.

Definition 2.2 (Partial Temporal Model)
a) A (propositional) *partial temporal model* $\mathbb{M}$ of signature Σ is a triple $(M, T, <)$, where $(T, <)$ is a flow of time and M is a mapping

$$M : T \times At(\Sigma) \rightarrow \{0, 1, u\}$$

If a is an atom, and t is a time point in T, and $M(t, a) = 1$, then we say that in this model M *at time point* t *the atom* a *is true*. Similarly we say that *at time point* t *the atom* a is *false*, respectively *undefined*, if $M(t, a) = 0$, respectively $M(t, a) = u$. We will sometimes leave out the flow of time and denote a partial temporal model by M only.
b) If M is a partial temporal model, then for any fixed time point t the partial model $M_t : At(\Sigma) \rightarrow \{0, 1, u\}$ (the *snapshot at time point* t) is defined by

$$M_t\!: \; a \; \mapsto \; M(t, a)$$

We will sometimes use the notation $(M_t)_{t \in T}$ where each M_t is a partial model as an equivalent description of a partial temporal model M.
The *ordering of truth values* is defined by $u \leq 0$, $u \leq 1$, $u \leq u$, $0 \leq 0$, $1 \leq 1$. We call the partial model N a *refinement* of the model M, denoted by $M \leq N$, if for all atoms a it holds: $M(a) \leq N(a)$. A partial temporal model M is called *conservative* if for all time points s and t with $s \ll t$ it holds $M_s \leq M_t$.
c) The refinement relation $\leq$ between partial temporal models based on the same flow of time is defined by: $M \leq N$ if for all time points t and atoms a it holds $M(t, a) \leq N(t, a)$.

Because our partial temporal models based on forests have a more differentiated structure towards the future than towards the past, we assume the following temporal operators. In the standard manner we build temporal formulae. In these definitions, $(M, t) \models^{+} \alpha$ means that in the model M at time point t the formula α is true, $(M, t) \models^{-} \alpha$ that it is false and $(M, t) \models^{u} \alpha$ that it is undefined. Furthermore $(M, t) \not\models^{+} \alpha$ denotes that $(M, t) \models^{+} \alpha$ is not the case.

Definition 2.3 (Interpretation of Temporal Formulae)
Let a (temporal) formula α, a partial temporal model M, and a time point $t \in T$ be given, then:

a)

$$(M, t) \models^{+} P\alpha \quad \Leftrightarrow \quad \exists s \in T \; [\, s \ll t \;\&\; (M, s) \models^{+} \alpha \,]$$

$$(M, t) \models^{+} C\alpha \quad \Leftrightarrow \quad (M, t) \models^{+} \alpha$$

$$(M, t) \models^{+} \exists X\alpha \quad \Leftrightarrow \quad \exists s \in T \; [\, t < s \;\&\; (M, s) \models^{+} \alpha \,]$$

$$(M, t) \models^{+} \exists F\alpha \quad \Leftrightarrow \quad \exists s \in T \; [\, t \ll s \;\&\; (M, s) \models^{+} \alpha \,]$$

$(M, t) \models^{+} \forall F\alpha \quad \Leftrightarrow \quad$ **for all branches including t there exists an s in that branch such that** $[\, t \ll s \;\&\; (M, s) \models^{+} \alpha \,]$

b) The temporal operators are defined in a two-valued manner, so for every $\mathbf{O} \in \{\mathbf{P}, \mathbf{C}, \exists\mathbf{X}, \exists\mathbf{F}, \forall\mathbf{F}\}$:

$$(\mathbf{M}, \mathbf{t}) \models^{-} \mathbf{O}\alpha \quad \Leftrightarrow \quad (\mathbf{M}, \mathbf{t}) \not\models^{+} \mathbf{O}\alpha$$

c) The connectives are evaluated according to the strong Kleene semantics and the atoms according to Definition 2.2.

d) For a partial temporal model $\mathbf{M}$, by $\mathbf{M} \models^{+} \alpha$ we mean $(\mathbf{M}, \mathbf{t}) \models^{+} \alpha$ for all $\mathbf{t} \in \mathbf{T}$ and by $\mathbf{M} \models^{+} \mathbf{K}$ we mean $\mathbf{M} \models^{+} \varphi$ for all $\varphi \in \mathbf{K}$, where $\mathbf{K}$ is a set of formulae possibly containing any of the defined operators. We will say that $\mathbf{M}$ is a *model of* the theory $\mathbf{K}$.

e) A partial temporal model $\mathbf{M}$ of a theory $\mathbf{K}$ is called a *minimal model* of $\mathbf{K}$ if for every model $\mathbf{C}$ of $\mathbf{K}$ with $\mathbf{C} \leq \mathbf{M}$ it holds $\mathbf{C} = \mathbf{M}$.

If in a model $\mathbf{M}$ the formula $\mathbf{P}(\mathrm{T})$ is true at time point $\mathbf{t}$ then $\mathbf{t}$ must have a predecessor, and therefore $\neg\mathbf{P}(\mathrm{T})$ will be true exactly in the time points which are minimal with respect to $<$.

From now on the word (temporal) formula will be used to denote a formula possibly containing any of the new operators, unless stated otherwise. If a formula contains no operators it is called *objective*. We call a subformula *guarded* if it is in the scope of an operator. A *purely temporal* formula is one in which all objective subformulae are guarded.

What is most interesting about a reasoning process, is of course its set of final conclusions. Talking about *final* conclusions, we will assume that the reasoning is *conservative*, which means that once a fact is established, it will remain true in the future of the reasoning process. In that case a fact is a final conclusion of a process if it is established at some point in time in the branch representing the process. So besides reasoning paths also the conclusions they result in are defined in a branching time model in the following manner:

Definition 2.4 (Limit Models of a Conservative Model)
Let $\mathbf{M}$ be a conservative partial temporal model. The *limit model of a branch* $\mathbf{B}$ of $\mathbf{M}$ based on flow of time $(\mathrm{T}', <')$, denoted by $\mathbf{lim_B\,M}$, is the partial model with for all atoms $\mathbf{p}$:

(i) $\mathbf{lim_B\,M} \models^{+} \mathbf{p} \quad \Leftrightarrow \quad \exists\, \mathbf{t} \in \mathbf{T}'\colon (\mathbf{M}, \mathbf{t}) \models^{+} \mathbf{p}$

(ii) $\mathbf{lim_B\,M} \models^{-} \mathbf{p} \quad \Leftrightarrow \quad \exists\, \mathbf{t} \in \mathbf{T}'\colon (\mathbf{M}, \mathbf{t}) \models^{-} \mathbf{p}$

Notice that $\mathbf{p}$ is undefined in $\mathbf{lim_B\,M}$ if and only if $\mathbf{p}$ is undefined in $\mathbf{B_t}$ for all $\mathbf{t} \in \mathbf{T}'$.

3 Final Models

In this section $\mathbf{M}$ and $\mathbf{M}'$ denote partial temporal models based on the flows of time $(\mathrm{T}, <)$ and $(\mathrm{T}', <')$ respectively.

Definition 3.1 (Homomorphism)
A mapping $\mathrm{f} : \mathrm{T} \to \mathrm{T}'$ is called a *homomorphism* of $\mathbf{M}$ to $\mathbf{M}'$ if

(i) $\mathrm{s} < \mathrm{t} \Rightarrow \mathrm{f(s)} <' \mathrm{f(t)}$

(ii) $\mathbf{M(s)} = \mathbf{M'(f(s))}$
(iii) If $\mathbf{s}$ is a minimal element of $\mathbf{T}$ then $\mathbf{f(s)}$ is minimal element of $\mathbf{T'}$

Definition 3.2 (Persistency under Homomorphisms)
Let $\mathbf{f : M \to M'}$ be a homomorphism. The formula α is called *forward persistent* (under $\mathbf{f}$) if for all time points $\mathbf{t}$ in $\mathbf{T}$:

$$(\mathbf{M}, \mathbf{t}) \models^{+} \alpha \;\Rightarrow\; (\mathbf{M'}, \mathbf{f(t)}) \models^{+} \alpha$$

The following proposition is an immediate consequence of [10].

Proposition 3.3
Let α be any formula containing at most the operators $\mathbf{C}$ and $\mathbf{P}$ and β any objective formula. Then the formulae α and $\alpha \to \exists\mathbf{X}(\beta)$ are forward persistent under any homomorphism.

Definition 3.4 (Final Model)
The model $\mathbf{F}$ of a temporal theory $\mathbf{Th}$ is called a *final model* of $\mathbf{Th}$ if for each model $\mathbf{M}$ of $\mathbf{Th}$ there is a unique homomorphism $\mathbf{f : M \to F}$. The model $\mathbf{F}$ is called a *final minimal model* of $\mathbf{Th}$ if $\mathbf{F}$ is a minimal model of $\mathbf{Th}$ and for each minimal model $\mathbf{M}$ of $\mathbf{Th}$ there is a unique homomorphism $\mathbf{f : M \to F}$.

The following result shows the existence of final models for a certain class of theories (see [10]).

Theorem 3.5
If a final (minimal) model of a temporal theory $\mathbf{Th}$ exists, then it is unique (up to isomorphism). If all formulae in $\mathbf{Th}$ are forward persistent under surjective homomorphisms then there exists a (unique) final model $\mathbf{F_{Th}}$ of $\mathbf{Th}$.

4 Temporal Axiomatization of a Classical Proof System

In this section we will apply our approach to a relatively simple type of reasoning: based on a classical proof system. We will show how proof rules can be represented by temporal formulae. As an example, consider modus ponens:

$$\frac{\mathbf{A} \quad \mathbf{A} \to \mathbf{B}}{\mathbf{B}}$$

Here $\mathbf{A}$ and $\mathbf{B}$ are meta-variables ranging over the set of formulae, and $\mathbf{A} \to \mathbf{B}$ is a term structure built from them using the logical connective $\to$. We want the partial temporal models to reflect the proof process, such that a partial model at a certain point in time reflects what has been derived up to that moment. The temporal interpretation of such a proof rule we have in mind is then the following.

if for any formulae A *and* B
in the current information state both A *and* A→B *have been derived*
then in a next information state B *has been derived*

This interpretation of modus ponens is formalized by the following temporal axiom scheme (for all formulae **A** and **B**):

$$C(A) \wedge C(A \rightarrow B) \rightarrow \exists X(B)$$

However, the truth of the formulae **A** and **A→B** in a current state already implies the truth of **B** in the same state, due to the compositional truth definition in the Strong Kleene semantics. As we want to describe the steps of reasoning by time steps this is undesirable. A solution for this is to extend the notion of partial model to the notion of valuation of all formulae, in a manner similar to [3], also see [21]. For each formula φ of the original language we define a new atom at_φ, and then we take the propositional language induced by these new atoms as our new language. So if $FORM(\Sigma)$ denotes the set of formulae based on the signature Σ, then we define a new signature Σ' based on the set of atoms $At(\Sigma') = \{ at_\varphi \mid \varphi \in FORM(\Sigma) \}$. So we have a natural bijection $\varphi \text{ -> } at_\varphi$ between $FORM(\Sigma)$ and $At(\Sigma')$. Notice that $At(\Sigma)$ is embedded in $At(\Sigma')$ by $At(\Sigma) \ni p \text{ -> } at_p \in At(\Sigma')$.

After this change of language has been accomplished, we can describe any instance of the proof rule modus ponens by a temporal formula as follows:

$$C(at_\varphi) \wedge C(at_{\varphi \rightarrow \psi}) \rightarrow \exists X(at_\psi)$$

This allows us to give a temporal axiomatization of a proof system. In addition we need a temporal translation of the initial axioms: the theory from which conclusions are to be drawn. Suppose **K** is any set of formulae of signature Σ. Let $at(K)$ be the set of atoms corresponding to the formulae in **K**. We require that these atoms are true at each moment of time. Therefore for any such formula φ we can simply add the formulae $C(at_\varphi)$ to our temporal theory.

After these preparations we are ready to formalize the translation of the proof rules into temporal formulae:

Definition 4.1
a) By **Forterm** we denote the set of term structures built up from (meta-) variables, ranging over $FORM(\Sigma)$, by use of the logical connectives. A *proof system* **PS** is a set of *proof rules* of type $(A_1, ..., A_k) / B$ where the $A_i, B \in$ **Forterm**. Let a proof rule **PR**: $(A_1, ..., A_k) / B$ be given en let MV_{PR} be the set of *meta-variables* occurring in $A_1, ..., A_k$ and **B**. A mapping $\sigma: MV_{PR} \text{ -> } FORM(\Sigma)$ is called a *meta-variable assignment*. Any meta-variable assignment σ can be extended in a canonical manner to a substitution mapping

$$\sigma^*: \mathbf{Forterm} \text{ -> } FORM(\Sigma)$$

such that σ^* substitutes formulae for the meta-variables of MV_{PR} in any term structure of **Forterm**.
The *temporal translation* of a proof rule **PR** of the form $(A_1, ..., A_k) / B$ is the set T_{PR} of instances of temporal formulae defined by:

$$\{ C(at_{\sigma^* A_1}) \wedge .... \wedge C(at_{\sigma^* A_k}) \rightarrow \exists X(at_{\sigma^* B}) \mid \sigma \text{ meta-variable assignment for PR} \}$$

The *temporal translation* T_{PS} of **PS** is defined by: $T_{PS} = \bigcup_{PR \in PS} T_{PR}$.

b) Let **K** be any set of objective formulae of signature **Σ**. The *temporal translation* $\mathbf{T_K}$ of **K** is defined by: $\mathbf{T_K} = \{ \mathbf{C(at_\varphi)} \mid \varphi \in \mathbf{K} \}$.
c) We have to make sure that once a fact has been established, it remains known at all later points (conservativity); this can be axiomatized by the temporal theory

$$\mathbf{C'} = \{ \mathbf{P(a)} \rightarrow \mathbf{C(a)} \mid \mathbf{a} \in \mathbf{At(\Sigma')} \}$$

The overall translation of proof rules and theory is defined by:

$$\mathbf{Th_{PS,K}} = \mathbf{T_{PS}} \cup \mathbf{T_K} \cup \mathbf{C'}$$

Some proof systems may consist of both proof rules and axioms; these may be incorporated by adding them to the theory **K**.

The first observation about this temporal theory $\mathbf{Th_{PS,K}}$ is that there exist partial temporal models of it. Such a model could be constructed incrementally, starting with a root, adding its successor partial models in the next step, and any time the model has been constructed up until a certain level, one can construct the next level by adding successor partial models to those at the current level. This is possible since the formulae of $\mathbf{Th_{PS,K}}$ prescribe existence of successors, obeying certain properties. It is easy to see that these properties are never contradictory since only truth of certain atoms is prescribed. Taking such a model and changing the truth value of atoms which are not prescribed to be true by $\mathbf{Th_{PS,K}}$ to undefined, points out a manner to establish the existence of minimal models of $\mathbf{Th_{PS,K}}$. We have the following theorem.

Theorem 4.2
Let **PS** be any proof system and **K** any set of objective formulae of signature **Σ** and let $\mathbf{Th_{PS,K}}$ be the temporal theory $\mathbf{T_{PS}} \cup \mathbf{T_K} \cup \mathbf{C'}$. Let **M** be a minimal partial temporal model of $\mathbf{Th_{PS,K}}$. For any formula φ of signature **Σ** it holds

$$\mathbf{K} \vdash_{\mathbf{PS}} \varphi \Leftrightarrow \mathbf{M} \models^{+} \neg\mathbf{P(T)} \rightarrow \exists\mathbf{F(at_\varphi)}$$

Note that for a minimal partial temporal model of $\mathbf{Th_{PS,K}}$ the partial models of time points which are minimal, are the same (atoms corresponding to formulae of **K** are true, other atoms are undefined). In this way a semantics is defined which can be seen as a generalization of the manner in which modal and temporal semantics can be given to intuitionistic logic (see [13], [17]). Apart from the use of partial models, our approach can be used for any proof system.

Proposition 4.3
The temporal theory $\mathbf{Th_{PS,K}}$ has a final model $\mathbf{F_{PS,K}}$.

If we have a proof $\varphi_1, \ldots, \varphi_n$ of which (only) the first **k** formulae are axioms from **K**, then a proof trace is a sequence $(\mathbf{M_i})_{\mathbf{i=0..n-k}}$ of partial models such that $\mathbf{Lit(M_i)} = \{\mathbf{at_{\varphi_j}} \mid \mathbf{j} = 1, .., \mathbf{k+i}\}$. In such a trace the partial model $\mathbf{M_i}$ reflects exactly the formulae which have been derived up until the $\mathbf{i^{th}}$ step of the proof. It is easy to see that, although such a proof trace itself is in general not a model of $\mathbf{Th_{PS,K}}$, it can always be embedded in the final model $\mathbf{F_{PS,K}}$. Note that for a branch **B** of the final model the limit model $\mathbf{lim_B\,F_{PS,K}}$ corresponds to the set of all conclusions drawn in that reasoning pattern; this is a subset of the deductive closure of **K** under **PS** (since we allow non-exhaustive reasoning patterns).

5 Temporal Theories of Default Reasoning

In this section we will show that default reasoning patterns based on normal defaults can be captured by temporal theories. The main point is how to interpret a default reasoning step

if α *and it is consistent to assume* β
then β *can be assumed*

in a temporal manner. We will view the underlying default $(\alpha : \beta)/\beta$ as a (meta-level) proof rule stating that if the formula α has already been established in the past, and there is a possible future reasoning path where the formula β remains consistent, then β can be assumed to hold in the current time point. As partial models can only be used to describe literals, instead of arbitrary formulae, we restrict our default rules to ones which are *based on literals*, which means that α and β have to be literals. As in the previous section, this is not an important hindrance, since for an arbitrary formula α we can add a new atom $\mathbf{at}_\alpha$ to our signature, adding the formula $\alpha \leftrightarrow \mathbf{at}_\alpha$ to our (non-default) knowledge in $\mathbf{W}$. The translation of the rule $(\alpha : \beta)/\beta$ in temporal partial logic will be:

$$\mathbf{P}\alpha \wedge \neg \forall\mathbf{F}\neg \beta \rightarrow \mathbf{C}\beta$$

Here $\neg \forall\mathbf{F}\neg \beta$ is true at a point in time if not for all future paths, the negation of β becomes true at some point in that branch, which is equivalent to: there is a branch, starting at the present time point, on which β is always either true or undefined. To ensure that formulae which are true at a certain point in time, will remain true in all of its future points (once a fact has been established, it remains established), we will add for each literal $\mathbf{L}$ a rule $\mathbf{P(L)} \rightarrow \mathbf{C(L)}$. Furthermore, if we have an additional (non-default) theory $\mathbf{W}$ it can be shown (see [9], [11]) that there exists a temporal theory which ensures that all conclusions which can be drawn using the theory $\mathbf{W}$ and the default conclusions at a certain time point, are true in the partial model at that time point. The complete translation of a normal default theory is as follows:

Definition 5.1 (Temporal Interpretation of a Normal Default Theory)

Let $\Delta = \langle \mathbf{W}, \mathbf{D} \rangle$ be a normal default theory of signature Σ. Define

$$\begin{aligned}
\mathbf{C}' &= \{ \mathbf{P(L)} \rightarrow \mathbf{C(L)} \mid \mathbf{L} \in \mathbf{Lit}(\Sigma) \} \cup \{ \exists \mathbf{F}(\top) \} \\
\mathbf{D}' &= \{ \mathbf{P}\alpha \wedge \neg \forall \mathbf{F} \neg \beta \rightarrow \mathbf{C}\beta \mid (\alpha : \beta) / \beta \in \mathbf{D} \} \\
\mathbf{W}' &= \{ \mathbf{C(L)} \mid \mathbf{L} \textbf{ literal}, \ \mathbf{W} \models \mathbf{L} \} \cup \\
&\quad\ \{ \mathbf{C(con(F))} \rightarrow \mathbf{C(L)} \mid \mathbf{L} \textbf{ literal}, \ \mathbf{F} \neq \varnothing \textbf{ a finite set of literals with } \mathbf{F} \cup \mathbf{W} \models \mathbf{L} \}
\end{aligned}$$

The *temporal interpretation* of Δ is the temporal theory

$$\mathbf{Th}_\Delta = \mathbf{C}' \cup \mathbf{D}' \cup \mathbf{W}'$$

Minimal temporal models of $\mathbf{Th}_\Delta$ describe the possible reasoning paths when reasoning with defaults from Δ. It turns out that there is a nice connection between the branches of such a minimal temporal model and Reiter extensions of the default theory. We will first give a definition of a Reiter extension of a default theory, equivalent to Reiter's original definition (in [20]):

Definition 5.2 (Reiter Extension)
Let $\Delta = \langle W, D \rangle$ be a default theory of signature Σ, and let E be a consistent set of sentences for Σ. Then E is a Reiter extension of Δ if $E = \bigcup_{i=0}^{\infty} E_i$ where $E_0 = Th(W)$, and for all $i \geq 0$

$$E_{i+1} = Th(E_i \cup \{ \beta \mid (\alpha : \beta) / \beta \in D, \alpha \in E_i \text{ and } \neg\beta \notin E \})$$

If E is a Reiter extension, then by E_i we will denote the subsets of E as defined in this definition. The following theorem shows that in the case of linear models there is a clear correspondence between extensions of a default theory Δ and the minimal linear time models of the temporal theory Th_Δ (also see [11]). For a linear model we can always assume that it is based on the flow of time $(\mathbb{N}, <')$ with $s <' t$ iff $t = s + 1$. For a consistent set of literals S by $< S >$ we denote the unique partial model M with $Lit(M) = S$.

Theorem 5.3
Let $\Delta = \langle W, D \rangle$ be a normal default theory.
a) If M is a minimal linear time temporal model of Th_Δ, then $Th(Lit(lim_M\, M) \cup W)$ is a Reiter extension E of Δ.
Moreover, $E_t = Th(Lit(M_t) \cup W)$ for all $t \in \mathbb{N}$.
b) If W is consistent and E a Reiter extension of Δ, then the partial temporal model M defined by $M = (< Lit(E_t) >)_{t \in \mathbb{N}}$ is a minimal linear time temporal model of Th_Δ with $Lit(lim_M\, M) = Lit(E)$.

There is the following connection between the branches of (minimal) temporal models of Th_Δ and the linear (minimal) temporal models of Th_Δ:

Proposition 5.4
Let Δ be a normal default theory and M a temporal model of Th_Δ.
a) Every maximal branch of M is a linear time model of Th_Δ.
b) M is a minimal temporal model of T_Δ if and only if every maximal branch of M is a (linear) minimal temporal model of Th_Δ.

What we would now hope is that for a default theory a final model would exist which captures all of the linear partial temporal models, and thus captures all of the extensions in one model. This is however not in general the case, but there is a category of theories (which includes all the theories with a finite number of default rules) for which such a final model always exists. To identify this category we need the following notion:

Definition 5.5 (Extension Complete)
Let Δ be a default theory.
a) We call a chain of sets of formulae

$$S_0 \subseteq S_1 \subseteq S_2 \subseteq \ldots$$

approximated by a (Reiter) extension E of Δ *up to depth* n if for all $i \leq n$ it holds $S_i = E_i$ (where the E_i are as in Definition 5.2). The chain $(S_k)_{k \in \mathbb{N}}$ is called *approximated by a set* of Reiter extensions R of Δ if for every $n \in \mathbb{N}$ there is an extension $E \in R$ such that $(S_k)_{k \in \mathbb{N}}$ is approximated by E up to depth n.

b) We call Δ (Reiter) *extension complete* if for any chain of sets of formulae $(S_k)_{k \in \mathbb{N}}$ that is approximated by a set of (Reiter) extensions **R** of Δ, its union $\bigcup_{k \in \mathbb{N}} S_k$ is a Reiter extension **E** with (where the E_i are as in Definition 5.2) $E_i = S_i$ for all i.

In [11] an example is given of a default theory which is not extension complete. For the default theories which are extension complete, we have the following (see [11]):

Theorem 5.6
Let Δ be a normal default theory.
a) If Δ is extension complete, then there exists a (unique) final minimal temporal model $\mathbf{FM}_\Delta$ of $\mathbf{Th}_\Delta$.
b) Suppose a final minimal temporal model $\mathbf{FM}_\Delta$ of $\mathbf{Th}_\Delta$ exists.
Then there is a one to one correspondence between the set $\mathbb{LT}(\mathbf{FM}_\Delta)$ of maximal branches **B** of $\mathbf{FM}_\Delta$ and the set $\mathbb{E}(\Delta)$ of all Reiter extensions **E** of Δ.
Here **B** and **E** correspond to each other if and only if $\mathbf{B} = (< \mathrm{Lit}(E_t) >)_{t \in \mathbb{N}}$ and $\mathbf{E} = \mathbf{Th}(\mathrm{Lit}(\lim_B \mathbf{M}) \cup \mathbf{W})$.

6 Temporal Axiomatization of a Meta-level Architecture

In this section we apply our approach to a third kind of reasoning patterns: generated by a meta-level architecture reasoning system. Meta-level architectures form the basis of quite powerful reasoning systems: they have been applied for example to non-monotonic reasoning and reasoning about control (e.g., [6], [7], [8], [14], [19], [22], [23], [28]). A meta-level architecture consists of two separate reasoning levels or components: the object level component and the meta-level component. The connections between the components are defined by so called upward and downward reflections.

As an example, suppose the meta-level reasoning component has (meta-) knowledge by which it can be deduced in which state what goal is adequate for the reasoning of the object level component:

if the atom a *is unknown, then the atom* b *is proposed as a goal*

If we assume that after downward reflection indeed the proposed goal has been chosen (in the literature this is called the causal connection assumption; [19]), this meta-knowledge can be interpreted in a temporal manner:

if in the current state the atom a *is unknown*	*(in the object level reasoning component)*
then in a next state the atom b *is a goal*	*(for the object level reasoning process)*

Thus meta-level reasoning implies a shift in time, replacing the goals at the object-level by new goals (the ones proposed by the meta-level). We will formalize these notions in subsequent (sub)sections.

6.1 Formalizing the Object Level Component

In the sequel by $\vdash$ we will denote any sound inference relation that is not necessarily complete (e.g., one of: natural deduction, chaining, full resolution, SLD resolution, unit resolution, etc.). A partial model is complete if it does not assign the truth value undefined to any atom. For a consistent set of formulae $\mathbf{K}$, of signature Σ, by $\mathbf{IS_K(\Sigma)}$ we denote the set of partial models which have a complete refinement (with respect to $\leq$) which is a model of $\mathbf{K}$. This set can be seen as the set of partial models which are "consistent" with $\mathbf{K}$.

Definition 6.1 (Deductive Closure)
Let $\mathbf{K}$ be a consistent set of objective formulae of signature Σ.
For $\mathbf{M} \in \mathbf{IS_K(\Sigma)}$ we define the partial model $\mathbf{dc_K}^{\vdash}\mathbf{(M)}$ by

$$\mathbf{dc_K}^{\vdash}\mathbf{(M)} \models^{+} \mathbf{L} \iff \mathbf{K} \cup \mathbf{Lit(M)} \vdash \mathbf{L}$$

for any literal $\mathbf{L}$. This model is called the *deductive closure* of $\mathbf{M}$ under $\mathbf{K}$.
We call $\mathbf{M}$ *deductively closed* under $\mathbf{K}$ if $\mathbf{M} = \mathbf{dc_K}^{\vdash}\mathbf{(M)}$.

Definition 6.2 (Conservation, Monotonicity, Idempotency)
Let $\mathbf{K}$ be a consistent set of objective formulae of signature Σ.
The mapping $\alpha : \mathbf{IS_K(\Sigma)} \rightarrow \mathbf{IS_K(\Sigma)}$ is called:

(i) *conservative*	if	$\mathbf{M} \leq \alpha(\mathbf{M})$	for all $\mathbf{M} \in \mathbf{IS_K(\Sigma)}$
(ii) *monotonic*	if	$\alpha(\mathbf{M}) \leq \alpha(\mathbf{N})$	for all $\mathbf{M}, \mathbf{N} \in \mathbf{IS_K(\Sigma)}$ with $\mathbf{M} \leq \mathbf{N}$
(iii) *idempotent*	if	$\alpha(\alpha(\mathbf{M})) = \alpha(\mathbf{M})$	for all $\mathbf{M} \in \mathbf{IS_K(\Sigma)}$

Proposition 6.3
Let $\mathbf{K}$ be a consistent set of objective formulae of signature Σ. Then the mapping $\mathbf{dc_K}^{\vdash}: \mathbf{IS_K(\Sigma)} \rightarrow \mathbf{IS_K(\Sigma)}$ is conservative, monotonic and idempotent.
Moreover, for any $\mathbf{M} \in \mathbf{IS_K(\Sigma)}$ and any complete model $\mathbf{N}$ of $\mathbf{K}$ with $\mathbf{M} \leq \mathbf{N}$ it holds $\mathbf{dc_K}^{\vdash}\mathbf{(M)} \leq \mathbf{N}$.

In case of controlled non-exhaustive reasoning, mappings are involved (depending on certain control settings) that in principle are not idempotent. However, we still require these mappings (called controlled inference functions) to be conservative and monotonic. Now we can formalize the object level component as follows.

Definition 6.4 (Object Level Component)
The *object-level reasoning component* $\mathbf{OC}$ is defined by a tuple

$$\mathbf{OC} = \langle \langle \Sigma_o, \mathbf{OT}, \vdash \rangle, \langle \Sigma_c, \mu_{\mathbf{OT}}{}^{\vdash}, \upsilon_{\mathbf{OT}}{}^{\vdash} \rangle \rangle$$

with

Σ_o a signature, called the *object-signature*
$\mathbf{OT}$ a set of ground formulae expressed in terms of Σ_o : the *object theory*
$\vdash$ a classical inference relation (assumed sound but not necessarily complete)
Σ_c a signature, called the *control signature* and
$\mu_{\mathbf{OT}}{}^{\vdash}$ $: \mathbf{IS_{OT}(\Sigma_o)} \times \mathbf{IS(\Sigma_c)} \rightarrow \mathbf{IS_{OT}(\Sigma_o)}$
$\upsilon_{\mathbf{OT}}{}^{\vdash}$ $: \mathbf{IS_{OT}(\Sigma_o)} \times \mathbf{IS(\Sigma_c)} \rightarrow \mathbf{IS(\Sigma_c)}$

We call $\mu_{\mathbf{OT}}{}^{\vdash}$ the *(controlled) inference function* for the object-level, and $\upsilon_{\mathbf{OT}}{}^{\vdash}$ the *process state update function*. For any $\mathbf{N} \in \mathbf{IS(\Sigma_c)}$ the mapping

$$\mu_{\mathbf{OT}}{}^{\mathbf{N}} : \mathbf{IS_{OT}(\Sigma_o)} \rightarrow \mathbf{IS_{OT}(\Sigma_o)}$$

is defined by $\mu_{OT}^{N}(M) = \mu_{OT}^{\vdash}(M, N)$. We assume that for any $N \in IS(\Sigma_c)$ this μ_{OT}^{N} is conservative and monotonic and satisfies $\mu_{OT}^{N}(M) \leq dc_{OT}^{\vdash}(M)$ for all $M \in IS_{OT}(\Sigma_o)$. When no confusion is expected, we will leave out the subscript and superscript of $\mu_{OT}^{\vdash}$ and $\upsilon_{OT}^{\vdash}$ and write shortly μ and υ.

The control-information state **N** specifies at a high level of abstraction all information relevant to the control of the (future) reasoning behaviour; i.e., the object-information state **M** and the control-information state **N** together determine in a deterministic manner the behaviour of the object-level reasoning component during its next activation. The process state update function expresses what the process brings about with respect to the descriptors in the control-information state. Examples: an object-atom was unknown, but becomes known during the reasoning; an object atom that was a goal has failed to be found.

6.2 Formalizing the Meta-level Component

In the meta-reasoning we distinguish two special types of (meta-)information: a) information on relevant aspects of the *current* (control-)state of the object-level reasoning process (possibly also including facts inherited from the past), and b) information on proposals for control parameters that are meant to guide the object-level reasoning process in the next activation. Therefore we assume that in the meta-signature two copies of the control signature of the object-level component are included as a subsignature: one that refers to the current state and another one referring to the proposed truth values for the next state of the object-level reasoning process. For example, if **goal(h)** is an atom of the control signature, then there are copies **current_goal(h)** and **proposed_goal(h)** in the set of atoms for the meta-signature. Two syntactic functions **c** and **p** transforming a meta-atom into a current variant and a proposed variant of it can simply be defined; e.g.,

c(goal(h)) = current_goal(h), p(goal(h)) = proposed_goal(h)

We assume that the reasoning of the meta-level itself has no sophisticated control: for simplicity we assume that it concerns taking deductive closures with respect to the inference relation used at the meta-level.

Definition 6.5 (Meta-level Architecture)

a) The *meta-level component* **MC** related to Σ_c is defined as a tuple

$$MC = \langle\langle \Sigma_m, MT, \vdash_m \rangle, \langle c, p \rangle\rangle$$

with

Σ_m a signature, called the *meta-signature* related to Σ_c

MT a set of objective ground formulae of signature Σ_m: the *meta-theory*

$\vdash_m$ a classical inference relation (assumed sound but not necessarily complete)

c, p $: At(\Sigma_c) \rightarrow At(\Sigma_m)$ injective mappings

It is assumed that every information state in $M \in IS(\Sigma_m)$ with $M(a) = u$ for all $a \in At(\Sigma_m)\backslash c(At(\Sigma_c))$ is consistent with **MT**, i.e. $M \in IS_{MT}(\Sigma_m)$.

The *inference function of the meta-level* $\mu_{MT}^{\vdash_m}$ (or shortly μ^*)

$$\mu_{MT}^{\vdash_m}: IS_{MT}(\Sigma_m) \rightarrow IS_{MT}(\Sigma_m)$$

is defined by $\mu_{MT}^{\vdash_m}(N) = dc_{MT}^{\vdash_m}(N)$.

b) A *meta-level architecture* is defined as a pair

$$MA = \langle OC; MC \rangle$$

with **OC** an object level component and **MC** a related meta-level component.

c) For a meta-level architecture **MA** the *upward reflection function* α_u: $IS(\Sigma_c) \rightarrow IS(\Sigma_m)$ is defined for $N \in IS(\Sigma_c)$ and $b \in At(\Sigma_m)$ by

$$\alpha_u(N)(b) = \begin{cases} N(a) & \text{if } b = c(a) \text{ for some } a \in At(\Sigma_c) \\ u & \text{otherwise} \end{cases}$$

The *downward reflection function* α_d: $IS(\Sigma_m) \rightarrow IS(\Sigma_c)$ is defined for $N \in IS(\Sigma_m)$ and $a \in At(\Sigma_c)$ by

$$\alpha_d(N)(a) = \begin{cases} 1 & \text{if } N(p(a)) = 1 \\ 0 & \text{otherwise} \end{cases}$$

6.3 Formalizing the Overall Reasoning Behaviour

Four types of actions take place as depicted in Fig. 1. For a formal description, see the following definition.

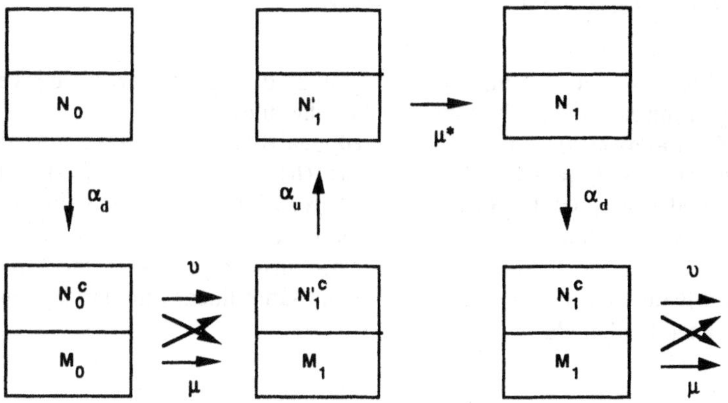

Fig 1 Reasoning pattern in a meta-level architecture

For a meta-level architecture with propositional signatures Σ_0 and Σ_m we denote the combined signature (based on the disjoint union of their sets of atoms) by $\Sigma_0 \oplus \Sigma_m$. We denote partial models for this combined signature as a pair of partial models $M \oplus N$ for Σ_0 and Σ_m.

Definition 6.6 (Semantics Based on Overall Traces)
a) An *overall trace* for the meta-level architecture **MA** is a linear partial temporal model $(M_t \oplus N_t)_{t \in \mathbb{N}}$ of signature $\Sigma_0 \oplus \Sigma_m$ satisfying for each $t \in \mathbb{N}$:

$$M_{t+1} = \mu(M_t, \alpha_d(N_t))$$
$$N_{t+1} = \mu^*(\alpha_u(\upsilon(M_t, \alpha_d(N_t))))$$

b) The *(intended) trace semantics* of **MA** is the set of overall traces.

The following theorem (also see [26]) shows that for any meta-level architecture **MA** with finite sets of atoms there exists a temporal theory such that a specific class of its models are precisely the overall traces of **MA**. For any $\langle M', N' \rangle \in IS(\Sigma_0) \times IS(\Sigma_m)$ let the temporal theory $T_{\langle M', N' \rangle}$ be given that defines all partial temporal models with initial state $\langle M', N' \rangle$, i.e. with $(M_t \oplus N_t)_{t \in \mathbb{N}}$ is a minimal model of $T_{\langle M', N' \rangle}$ iff $M_0 = M'$ and $N_0 = N'$.

Theorem 6.7 (Temporal Theory of a Meta-level Architecture)
Let **MA** be a meta-level architecture with finite sets of atoms.
There exists a temporal theory Th_{MA} (consisting of formulae of the form **A** or $A \rightarrow \exists X(L)$ where **L** is some objective literal and **A** a formula only referring to the past and current state) such that:
a) For any linear partial temporal model $(M_t \oplus N_t)_{t \in \mathbb{N}}$ of signature $\Sigma_o \oplus \Sigma_m$ the following are equivalent:
(i) $(M_t \oplus N_t)_{t \in \mathbb{N}}$ is a minimal model of $Th_{MA} \cup T_{\langle M', N' \rangle}$.
(ii) $(M_t \oplus N_t)_{t \in \mathbb{N}}$ is an overall trace for **MA** with initial state $\langle M', N' \rangle$.
In other words: the intended semantics of **MA** is described by the linear models of Th_{MA} that are minimal in the set of models with fixed initial state.
b) The theory Th_{MA} has a final model F_{MA}.

To get an idea how such a temporal theory can be defined, the example given in the beginning of this section can be formalized by the rules

$C(\neg\, known(a)) \rightarrow C(proposed\_goal(b))$ (meta-knowledge)

$C(proposed\_goal(b)) \rightarrow \exists X(goal(b))$ (downward reflection)

The final model F_{MA} contains all overall traces as linear submodels (branches), but in general it will contain other, less useful branches as well. Another variant can be obtained by using the temporal operator $\forall X$ instead of $\exists X$ in the formulae of Th_{MA} ; this leads to a temporal theory with the same linear time models but with a more restricted final model.

7 Conclusions

Partial temporal models can be used to describe the behaviour of dynamic reasoning processes, such as those performed by reasoning agents. The linear models usually describe a particular reasoning pattern, and a set of such models can be used to describe all possible patterns. These models may be described by a temporal theory. Another way of describing possible behaviour is by a branching time model which is branching at any time a pattern can continue in more than one way. These models can also often be axiomatized by a temporal theory. In this fashion we can use branching time temporal partial logic to obtain semantics for a variety of reasoning patterns including regular monotonic logics, default logic and for all patterns a meta-level architecture can perform. In these patterns one can often identify object level reasoning (by means of classical logic) and meta-level reasoning which complements it. The inferences on the object level can be axiomatized by a theory which restricts the (current) partial models at each time point, whereas the meta-level inferences (potentially introducing non-monotonicity) can be axiomatized by a theory which restricts the successor partial models at any point in time. The theory axiomatizing object level inference consists of formulae containing no operators but the **C** operator, whereas the theory for meta-level inference will consist of formulae containing at least one of the other operators.

In default logic the classical propositional logic is used for the object level inferences (axiomatized by the theory **W'**), whereas the meta-level inferences are performed by the default rules (axiomatized by the theory **D'**). The truth of formulae in the theory **W'** at a certain point depends only on the partial model at that point, whereas the theory **D'** restricts successor partial models.

In the example of the classical proof system, the proof rules have been lifted to the meta-level so that "proving" a formula is an explicit temporal process. Note that the partiality in this case is not explicitly needed: in a minimal model the truth value false will never occur. In the meta-level architecture some aspects of the object inferences have been lifted to the meta-level to allow for explicit control of the reasoning process.

In a number of cases it is possible to identify a final model, that is a "biggest" branching time model in which all possible patterns are incorporated in the most compact manner. Not only do we then have one structure which holds all information about a reasoning process, also the points in a process where a choice has to be made are explicitly identified. Based on this final model one can define a number of entailment relations, depending for instance on whether conclusions have to be established in all possible patterns, or if it is enough if there is at least one possibility to establish the conclusion.

We feel that temporal partial logic and other temporalized logics are a powerful way of describing complex reasoning patterns as they can be used to model a variety of reasoning patterns in a clear fashion. This approach contributes to a better integration of dynamic aspects in logical systems, as, for instance, advocated in [2]. It can be used as a basis for providing formal semantics of (specification languages for) complex reasoning systems, where often control of reasoning itself is a subject of reasoning: one of the open problems formulated in [16].

Acknowledgements

This work has been carried out in the context of SKBS and the ESPRIT III Basic Research project 6156 DRUMS II.

References

1. J.F.A.K. van Benthem, The logic of time: a model-theoretic investigation into the varieties of temporal ontology and temporal discourse, Reidel, Dordrecht, 1983.

2. J.F.A.K. van Benthem, Logic and the flow of information, in: D. Prawitz, B. Skyrms, D. Westerstahl (eds.), Proc. 9th Int. Congress of Logic, Methodology and Philosophy of Science, North Holland, 1991

3. P. Besnard, R.E. Mercer, Non-monotonic logics: a valuations-based approach, In: B. du Boulay, V. Sgurev (eds.), Artificial Intelligence V: Methodology, Systems, Applications, Elsevier Science Publishers, 1992, pp. 77-84

4. H. Bestougeff, G. Ligozat, Logical tools for temporal knowledge representation, Ellis Horwood, 1992

5. S. Blamey, Partial logic, in: D. Gabbay and F. Guenthner (eds.), Handbook of philosophical logic, Vol. III, 1-70, Reidel, Dordrecht, 1986.

6. K.A. Bowen, R. Kowalski, Amalgamating language and meta-language in logic programming. In: K. Clark, S. Tarnlund (eds.), Logic programming. Academic Press, 1982.

7. W.J. Clancey, C. Bock, Representing control knowledge as abstract tasks and metarules, in: Bolc, Coombs (Eds.), Expert system applications, 1988.

8. R. Davis, Metarules: reasoning about control, Artificial Intelligence 15 (1980), pp. 179-222.

9. J. Engelfriet, J. Treur, A temporal model theory for default logic, in: M. Clarke, R. Kruse, S. Moral (eds.), Proc. 2nd European Conference on Symbolic and Quantitative Approaches to Reasoning and Uncertainty, ECSQARU '93, Springer Verlag, 1993, pp. 91-96. Extended version: Report IR-334, Vrije Universiteit Amsterdam, Department of Mathematics and Computer Science, 1993, pp. 38

10. J. Engelfriet, J. Treur, Relating linear and branching time temporal models, Report, Free University Amsterdam, Department of Mathematics and Computer Science, 1994

11. J. Engelfriet, J. Treur, Final model semantics for normal default theories, Report, Free University Amsterdam, Department of Mathematics and Computer Science, 1994

12. M. Finger, D.M. Gabbay, Adding a temporal dimension to a logic system, Journal of Logic, Language and Information 1 (1992), pp. 203-233

13. D.M. Gabbay, Intuitionistic basis for non-monotonic logic, In: G. Goos, J. Hartmanis (eds.), 6th Conference on Automated Deduction, Lecture Notes in Computer Science, vol. 138, Springer Verlag, 1982, pp. 260-273

14. E. Giunchiglia, P. Traverso, F. Giunchiglia, Multi-context systems as a specification framework for complex reasoning systems, In: J. Treur, Th. Wetter (eds.), Formal specification of complex reasoning systems, Ellis Horwood, 1993.

15. R. Goldblatt, Logics of time and computation. CSLI Lecture Notes, vol. 7. 1987, Center for the Study of Language and Information.

16. F. van Harmelen, R. Lopez de Mantaras, J. Malec, J. Treur, Comparing formal specification languages for complex reasoning systems. In: J. Treur, Th. Wetter (eds.), Formal specification of complex reasoning systems, Ellis Horwood, 1993, pp. 257-282

17. S. Kripke, Semantical analysis of intuitionistic logic, In: J.N. Crossley, M. Dummett (eds.), Formal systems and recursive function theory, North Holland, 1965, pp. 92-129

18. T. Langholm, Partiality, truth and persistance, CSLI Lecture Notes No. 15, Stanford University, Stanford, 1988.

19. P. Maes, D. Nardi (eds.), Meta-level architectures and reflection, Elsevier Science Publishers, 1988.

20. R. Reiter, A logic for default reasoning, Artificial Intelligence 13, 1980, pp. 81-132

21. E. Sandewall, A functional approach to non-monotonic logics, Computational Intelligence 1 (1985), pp. 80-87

22. Y.H. Tan, J. Treur, A bi-modular approach to nonmonotonic reasoning, In: De Glas, M., Gabbay, D. (eds.), Proc. World Congress on Fundamentals of Artificial Intelligence, WOCFAI-91, 1991, pp. 461-476.

23. Y.H. Tan, J. Treur, Constructive default logic and the control of defeasible reasoning, B. Neumann (ed.), Proc. 10th European Conference on Artificial Intelligence, ECAI-92, Wiley and Sons, 1992, pp. 299-303.

24. E. Thijsse, Partial logic and knowledge representation, Ph.D. Thesis, Tilburg University, 1992

25. J. Treur, Declarative functionality descriptions of interactive reasoning modules, In: H. Boley, M.M. Richter (eds.), Processing Declarative Knowledge, Proc. of the International Workshop PDK-91, Lecture Notes in Artificial Intelligence, vol. 567, Springer Verlag, 1991, pp. 221-236.

26. J. Treur, Temporal semantics of meta-level architectures for dynamic control, Report, Free University Amsterdam, Department of Mathematics and Computer Science, 1994. Shorter version in: Proc. 4th International Workshop on Meta-programming in Logic, META '94, 1994.

27. J. Treur, Th Wetter (eds.), Formal specification of complex reasoning systems , Ellis Horwood, 1993, p. 282

28. R.W. Weyhrauch, Prolegomena to a theory of mechanized formal reasoning, Artificial Intelligence 13 (1980), pp. 133-170.

Appendix A to Section 2

Flows of Time

Definition 2.1' (Flow of Time)
A *flow of time* (T, <) is a pair consisting of a non-empty set T of time points, and a binary relation < on T x T, called the *immediate successor relation* which has to be:
(i) *Antitransitive*: for no a, b and c in T it holds a < b, b < c and a < c.
(ii) *A-cyclic*: there is no $n \geq 0$ and $a_0, ..., a_n$ in T with for $0 \leq i \leq n-1$: $a_i < a_{i+1}$ and $a_n < a_0$. (This implies < is irreflexive and antisymmetrical).
Here for s, t in T the expression s < t denotes that t is an *immediate successor* of s, and that s is an *immediate predecessor* of t.
We also introduce the transitive (but not reflexive) closure « of this binary relation: $« = <^+$. A flow of time is called *linear* if « is a total ordering.

Note that with this definition T together with « is a discrete time structure.
We will further limit our flows of time to be forests or trees:

Definition 2.2' (Tree and Forest)
a) The following properties are defined:
 (i) *Successor existence*
  Every time point has at least one successor:
  for all $s \in T$ there exists a $t \in T$ such that $s < t$.
 (ii) *Rooted*
  A flow of time is rooted with root r if r is a (unique) smallest element:
  for all t it holds r = t or r « t.
 (iii) *Left linear*
  For all t the set of s with s « t is totally ordered by «.
 (iv) *Well-founded*
  There are no infinite descending chains of elements $s_i < s_{i-1}$
b) A flow of time is called a *tree* if it is rooted and left linear.
c) A flow of time is called a *forest* if it is well-founded, left linear.

Note that a forest is just a disjoint union of trees. We will assume all flows of time to be forests.

Definition 2.3' (Sub-ft and Branch)
a) A flow of time (T', <') is called a *sub-ft (sub-flow of time)* of a flow of time (T, <) if $T' \subseteq T$ and $<' = < \cap T' \times T'$. It is also called the sub-ft of (T, <) *defined by* T', or the *restriction of* (T, <) to T'.
b) A *branch* in a flow of time T is a sub-ft B = (T', <') of T such that:
(i) «' (the transitive closure of <') is a total ordering on $T' \times T'$
(ii) Every $t \in T'$ with a successor in T also has a successor in T':
for all $s \in T', t \in T : s < t \Rightarrow$ there is a $t' \in T' : s < t'$
(iii) Every element of T that is in between elements of T' is itself in T':
for all $s \in T', t \in T, u \in T' : s « t « u \Rightarrow t \in T'$
A branch is called *maximal* if every t in T' with a predecessor in T also has a predecessor in T': for all $s \in T, t \in T' : s < t \Rightarrow$ there is an $s' \in T' : s' < t$.

Partial Temporal Models

By a signature Σ for convenience we mean a sequence of proposition symbols (propositional atom names). What counts is the set of atoms $At(\Sigma)$ and the set of literals $Lit(\Sigma)$ based on this signature.

Definition 2.4' (Partial Model)
Let Σ be a signature.
a) A *partial model* **M** for the signature Σ is an assignment of a truth value from $\{0, 1, u\}$ to each of the atoms of Σ, i.e. $M: At(\Sigma) \rightarrow \{0, 1, u\}$.
We say an atom **a** is *true* in **M** if **1** is assigned to it, and *false* if **0** is assigned; else it is called *undefined* (or *unknown*). In the same way we say a literal $\neg p$ is *true* in **M** if **M** assigns **0** to **p** and it is *false* if **M** assigns **1** to **p**. Otherwise it is *undefined.*
By $Lit(M)$ we denote the set of literals (atoms or negations of atoms) with truth value *true* in **M**.
b) The truth, falsity or undefinedness of any formulae in a partial model is evaluated according to the Strong Kleene semantics (e.g., [5], [18]).
c) The *ordering of truth values* is defined by $u \leq 0, u \leq 1, u \leq u, 0 \leq 0, 1 \leq 1$.
We call the model **N** a *refinement* of the model **M**, denoted by $M \leq N$, if for all atoms **a** it holds: $M(a) \leq N(a)$.
d) For a consistent set of literals **S** the unique partial model **M** with $Lit(M) = S$ is denoted by $< S >$.

Definition 2.5'
The partial temporal model **M'** is *sub-model* of **M** if $(T', <')$ is a sub-flow of time of $(T, <)$ with $M(t) = M'(t)$ for all **t** in **T'**. We also call **M'** the *restriction* of **M** to **T'**, denoted by $M|T'$.
Also the other notions defined in the above subsection for flows of time inherit to models.

Definition 2.6' (Temporal Operators and Their Semantics)
Let a formula α, a partial temporal model **M**, and a time point $t \in T$ be given, then:

a) $(M, t) \models^+ \exists F\alpha \quad \Leftrightarrow \quad \exists\, s \in T\ [\, t \ll s\ \&\ (M, s) \models^+ \alpha\,]$
$(M, t) \models^- \exists F\alpha \quad \Leftrightarrow \quad (M, t) \not\models^+ \exists F\alpha$

b) $(M, t) \models^+ \forall F\alpha \quad \Leftrightarrow$ **for all branches including t there exists an s in that branch such that** $[\, t \ll s\ \&\ (M, s) \models^+ \alpha\,]$
$(M, t) \models^- \forall F\alpha \quad \Leftrightarrow \quad (M, t) \not\models^+ \forall F\alpha$

c) $(M, t) \models^+ \forall G\alpha \quad \Leftrightarrow \quad \forall\, s \in T\ [\, t \ll s \Rightarrow (M, s) \models^+ \alpha\,]$
$(M, t) \models^- \forall G\alpha \quad \Leftrightarrow \quad (M, t) \not\models^+ \forall G\alpha$

d) $(M, t) \models^+ \exists G\alpha \quad \Leftrightarrow$ **there exists a branch including t such that for all s in that branch** $[\, t \ll s \Rightarrow (M, s) \models^+ \alpha\,]$
$(M, t) \models^- \exists G\alpha \quad \Leftrightarrow \quad (M, t) \not\models^+ \exists G\alpha$

e) $(M, t) \models^+ P\alpha \quad \Leftrightarrow \quad \exists\, s \in T\ [\, s \ll t\ \&\ (M, s) \models^+ \alpha\,]$
$(M, t) \models^- P\alpha \quad \Leftrightarrow \quad (M, t) \not\models^+ P\alpha$

f)	$(M, t) \models^+ H\alpha$	$\Leftrightarrow$	$\forall s \in T\ [s \ll t \Rightarrow (M, s) \models^+ \alpha]$
	$(M, t) \models^- H\alpha$	$\Leftrightarrow$	$(M, t) \not\models^+ H\alpha$
g)	$(M, t) \models^+ C\alpha$	$\Leftrightarrow$	$(M, t) \models^+ \alpha$
	$(M, t) \models^- C\alpha$	$\Leftrightarrow$	$(M, t) \not\models^+ C\alpha$
h)	$(M, t) \models^+ \exists X\alpha$	$\Leftrightarrow$	$\exists s \in T\ [t < s\ \&\ (M, s) \models^+ \alpha]$
	$(M, t) \models^- \exists X\alpha$	$\Leftrightarrow$	$(M, t) \not\models^+ \exists X\alpha$
i)	$(M, t) \models^+ \forall X\alpha$	$\Leftrightarrow$	$\forall s \in T\ [t < s \Rightarrow (M, s) \models^+ \alpha]$
	$(M, t) \models^- \forall X\alpha$	$\Leftrightarrow$	$(M, t) \not\models^+ \forall X\alpha$

Definition 2.7' (Interpretation of Temporal Formulae)
Let Σ be a signature, let **M** be a partial temporal model for Σ, and $t \in T$ a time point.

a) For any propositional atom $p \in At(\Sigma)$:

$$(M, t) \models^+ p \quad \Leftrightarrow \quad M(t, p) = 1$$
$$(M, t) \models^- p \quad \Leftrightarrow \quad M(t, p) = 0$$

b) For a formula of the form $\exists F\alpha$, $\forall F\alpha$, etcetera, see Definition 2.6'

c) For any two temporal formulae φ and ψ:

(i)	$(M, t) \models^+ \varphi \wedge \psi$	$\Leftrightarrow$	$(M, t) \models^+ \varphi$ and $(M, t) \models^+ \psi$
	$(M, t) \models^- \varphi \wedge \psi$	$\Leftrightarrow$	$(M, t) \models^- \varphi$ or $(M, t) \models^- \psi$
(ii)	$(M, t) \models^+ \varphi \rightarrow \psi$	$\Leftrightarrow$	$(M, t) \models^- \varphi$ or $(M, t) \models^+ \psi$
	$(M, t) \models^- \varphi \rightarrow \psi$	$\Leftrightarrow$	$(M, t) \models^+ \varphi$ and $(M, t) \models^- \psi$
(iii)	$(M, t) \models^+ \neg\varphi$	$\Leftrightarrow$	$(M, t) \models^- \varphi$
	$(M, t) \models^- \neg\varphi$	$\Leftrightarrow$	$(M, t) \models^+ \varphi$

d) For any temporal formula φ:

$(M, t) \not\models^+ \varphi$	$\Leftrightarrow$	$(M, t) \models^+ \varphi$ does not hold
$(M, t) \not\models^- \varphi$	$\Leftrightarrow$	$(M, t) \models^- \varphi$ does not hold
$(M, t) \models^u \varphi$	$\Leftrightarrow$	$(M, t) \not\models^+ \varphi$ and $(M, t) \not\models^- \varphi$

e) For a partial temporal model **M**, by $M \models^+ \varphi$ we mean $(M, t) \models^+ \varphi$ for all $t \in T$ and by $M \models^+ K$ we mean $M \models^+ \varphi$ for all $\varphi \in K$, where **K** is a set of formulae possibly containing any of the defined operators. We will say that **M** is a model of the theory **K**.

f) A partial temporal model **M** of a theory **K** is called a *minimal* model of **K** if for every model **C** of **K** with $C \leq M$ it holds $C = M$.

Appendix B Proofs for Section 4

In this Appendix we give proofs of results in Section 4. Proofs of results in the other sections can be found in [9], [10], [11], [26].

Theorem 4.2
Let **PS** be any proof system and **K** any set of formulae of signature Σ and let $Th_{PS,K}$ be the temporal theory $T_{PS} \cup T_K \cup C'$. Let **M** be a minimal partial temporal model of $Th_{PS,K}$. For any formula φ of signature Σ it holds

$$K \vdash_{PS} \varphi \Leftrightarrow M \models^+ \neg P(T) \rightarrow \exists F(at_\varphi)$$

Proof

"$\Rightarrow$" Suppose $K \vdash_{PS} \varphi$ and suppose that $\psi_1, \ldots, \psi_{n-1}, \psi_n$, with $\psi_n = \varphi$, is a proof for φ. For a non-minimal element t in T it holds trivially that $(M, t) \models^+ \neg P(T) \rightarrow \exists F(at_\varphi)$, so let r be a minimal element in T. We shall prove the following by induction:
For every $1 \leq i \leq n$ there is a time point s reachable from r such that $at_{\psi 1}, .., at_{\psi i}$ are true in M at time point s.
$i = 1$: ψ_1 has to be an element of K and as M is a model of T_K, at_φ has to be true in M at time point r.
$i \rightarrow i + 1$: suppose that s is a time point reachable from r and that $at_{\psi 1}, .., at_{\psi i}$ are true in M at time point s. If ψ_{i+1} is an element of K then the same argument as above yields that $at_{\psi i+1}$ must be true in M at point s, so assume that ψ_{i+1} is the result of applying a proof rule PR to a subset of the formulae $\psi_1, .., \psi_i$ (say $\alpha_1, .., \alpha_k$). Then there is a rule $C(at_{\alpha 1}) \wedge \ldots \wedge C(at_{\alpha k}) \rightarrow \exists X(at_{\psi i+1})$ in T_{PS} which has to be true in M at point s. As $at_{\alpha 1}, .., at_{\alpha k}$ are true in M at point s, there has to be a successor t to s in which $at_{\psi i+1}$ is true. The rules in C' ensure that $at_{\psi 1}, .., at_{\psi i}$ have to be true in M at point t too.
Taking n for i we have that there must be a point s reachable from r such that $at_{\psi n}$ is true in M at point s. It follows that $(M, s) \models^+ \neg P(T) \rightarrow \exists F(at_\varphi)$.
"$\Leftarrow$" Suppose there is a formula φ and a minimal element r such that $(M, r) \models^+ \exists F(at_\varphi)$ although $K \not\vdash_{PS} \varphi$. Take the formulae φ at minimal depth, i.e. if s is a point at minimal depth for which $(M, s) \models^+ at_\varphi$, then there is no formula α such that there is a point t at smaller depth than s with $(M, t) \models^+ at_\alpha$ but $K \not\vdash_{PS} \alpha$. As M is a minimal model of Th, if at_φ were undefined in M at point s, a formula in Th would become false. If this is a formula from T_K then it has to be the formula $C(at_\varphi)$, but then φ is in K and therefore $K \vdash_{PS} \varphi$. If it is a formula in C' then it must be the rule $P(at_\varphi) \rightarrow C(at_\varphi)$ at time point s. This means that at_φ is true in a point at smaller depth, which was not the case. Therefore it must be a rule of T_{PS}, say $C(at_{\alpha 1}) \wedge \ldots \wedge C(at_{\alpha k}) \rightarrow \exists X(at_\varphi)$ which will become false in a point t with $t < s$. But as $at_{\alpha 1}, .., at_{\alpha k}$ have to be true in M at point t and t is at smaller depth than s, we must have that $K \vdash_{PS} \alpha_1, .., K \vdash_{PS} \alpha_k$. But there is a proof rule in PS which can be applied to $\alpha_1, .., \alpha_k$ yielding φ, and therefore $K \vdash_{PS} \varphi$. This shows that such a formula can not exist.

Proposition 4.3
The temporal theory $Th_{PS,K}$ has a final model $F_{PS,K}$.

Proof
The theory $Th_{PS,K}$ consists of formulae which are forward persistent under any homomorphism (Proposition 3.3) and therefore by Theorem 3.5 a final model exists.

Reasoning about Knowledge on Computation Trees

Konstantinos Georgatos*

Department of Mathematics
Hunter College
City University of New York
695 Park Avenue
New York, NY 10021
U.S.A.

Abstract. This paper presents a bimodal logic for reasoning about knowledge during knowledge acquisition. One of the modalities represents (effort during) non-deterministic time and the other represents knowledge. The semantics of this logic are tree-like spaces which are a generalization of semantics used for modelling branching time and historical necessity. A finite system is shown to be canonically complete for the formentioned spaces. A characterization of the satisfaction relation implies the small model property and decidability for this system.

1 Introduction

The classical models of reasoning about knowledge are *possible worlds*. Restricting ourselves to the single knower case, the actual world or state of the knower is related to a set of alternative states. This set represents his or her view and the alternativeness relation reflects the basic properties of knowledge. Therefore, in most cases, it is taken to be reflexive, symmetric and transitive, i.e. an equivalence relation. This treatment of knowledge agrees with the traditional one ([Hin62], [HM84], [PR85], [CM86]) expressed in a variety of contexts (artificial intelligence, distributed processes, economics, etc).

We are interested in formulating a basic logical framework for reasoning about a resource-conscious acquiring of knowledge. Such a framework can be applied to many settings such as the ones involving time, computation, physical experiments or observations. In these settings an (discrete or continuous) increase of information available to us takes place and results in an increase of our knowledge. How could this simple idea be embodied in the formentioned semantical framework? An increase of knowledge can be represented with a restriction of the knower's view, i.e. of the equivalence class of the alternative worlds. This restriction is nondeterministic (we do not know what kind of additional information will be available to us if at all) but not arbitrary: it will always contain

* Author's current address: The Institute of Mathematical Sciences, C.I.T. Campus, Madras 600 113, India; email: geo@imsc.ernet.in.

the actual state of the knower, i.e. it is a *neighborhood restriction* of the actual state. In this way, set-theoretic considerations come in.

A discrete version of our epistemic framework can arise in scientific experiments or tests. We acquire knowledge by "a step-by-step" process, each step being an experiment or test. The outcome of such an experiment or test is unknown to us beforehand, but after being known it restricts our attention to a smaller set of possibilities. A sequence of experiments, test or actions comprises a *strategy of knowledge acquisition*. This model is in many respects similar to Hintikka's "oracle" (see [Hin86]). In Hintikka's model the "inquirer" asks a series of questions $Q_1, Q_2, \ldots, Q_n, \ldots$ to an external information source, called "oracle" (can be thought as a knowledge base). The oracle answers yes or no and the inquirer increases his or her knowledge by this piece of additional evidence. At any point of this process the inquirer follows a branch of a tree determined by the possible answers to his or her series of questions. Such an interrogative model is recognized by Gadamer ([Gad75]) as an important part of the epistemic process. To illustrate better these simple but fundamental ideas we present the following examples:

Example 1. Suppose that our view, the set of possible worlds, is $\{q_1, q_2, q_3, q_4\}$ and our query consists of two questions Q_1, Q_2, in that order. The answer to Q_1 is yes in q_1, q_2 and no in q_3, q_4. The answer to Q_2 is yes in q_1, q_2, q_3 and no in q_4. Then the possible sequences of knowledge states comprise a tree of subsets as shown in Fig. 1. The space of subsets labelling the nodes of the *computation tree* will be called a *tree-like space*.

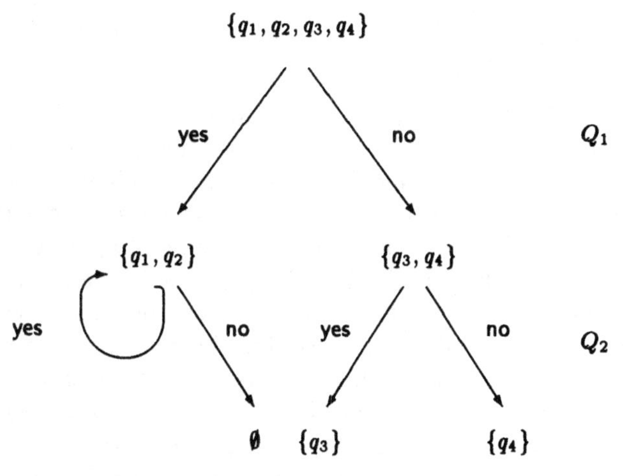

Fig. 1.

Example 2. Suppose that a machine emits a stream of binary digits representing the output of a recursive function f. After time t_1 the machine emitted the stream 111. The only information we have about the function being computed at this time on the basis of this (finite) observation is that

$$f(1) = f(2) = f(3) = 1.$$

As far as our knowledge concerns, f is indistinguishable from the constant function $\mathbf{1}$, where $\mathbf{1}(n) = 1$ for all n. After some additional time t_2, i.e. spending more time and resources, 0 might appear and thus we could be able to distinguish f from $\mathbf{1}$. In any case, each binary stream will be an initial segment of f and this initial segment is a neighborhood of f. In this way, we can acquire better knowledge of the function the machine computes. The space of finite binary streams is a structure which models computation. The sets of binary streams under the initial segment ordering is an example of a computation tree.

To express this framework we use two modalities K for knowledge and $\Box$ for effort, i.e. spending of resources. Moss and Parikh observed in [MP92] that if the formula

$$A \to \Diamond \mathsf{K} A$$

is valid, where A is an atomic predicate and $\Diamond$ is the dual of the $\Box$, i.e. $\Diamond \equiv \neg\Box\neg$, then the set which A represents is an open set of the topology where we interpret our systems. Under the reading of $\Diamond$ as "possible" and K as "is known", the above formula says that

> "if A is true then it is possible for A to be known",

i.e. A is *affirmative*. Vickers defines similarly an affirmative assertion in [Vic89]

> "an assertion is affirmative iff it is true precisely in the circumstances when it can be affirmed."

Affirmative assertions are closed under infinite disjunctions and refutative assertions are closed under infinite conjunctions. Smyth in [Smy83] observed first these properties in semi-decidable properties. Semi-decidable properties are those properties whose truth set is r.e. and are a particular kind of affirmative assertions. In fact, changing our power of affirming or computing we get another class of properties with a similar knowledge-theoretic character. For example, using polynomial algorithms affirmative assertions become polynomially semi-decidable, i.e. NP properties. If an object has this property then it is possible to know it with a polynomial algorithm even though it is not true we know it now.

Our approach has an independent theoretical interest. A new family of Kripke frames, called *subset frames*, arises. These are Kripke frames which are equivalent to sets of subsets. In particular, we have identified those which are equivalent to (complete) lattices of subsets and topologies (see [Geo93]). In this paper, we shall identify those which correspond to the above interrogative model, called *tree-like spaces*. Tree-like spaces have a particular interest; they correspond to an indeterminist's theory of time called *Ockhamism* (see [Pri67]), which gives rise to branching time. The models used for treating historical necessity according to an Ockhamist's notion of time are examples of tree-like spaces (see [Bur79] or [Tho84]).

A family of logics for knowledge and time is studied in [HV89] and various complexity results are established. However the framework of the above logics is restricted to distributed systems and their interpretation differs significantly from ours.

Interpreting the knowledge modal operator as a universal quantifier we present a novel way of understanding the meaning of quantifiers in varying domains (which can be ordered). This is one of the main difficulties in formulating a meaningful first-order system for modal logic (see [Fit93] for a discussion).

The language and semantics of our logical framework is presented in Sect. 2. In the same section we present two systems which belong to the same family of logics, studied in [MP92], [Geo93] and [Geo94]. In Sect. 3 we present an axiomatization, called **MPT** for our semantics and we prove completeness, small model property and decidability.

Due to space limitations we shall omit most of the lemmas which lead to the main results of the present paper. However, the overall structure of the proofs is described in sufficient detail.

2 Two Systems: MP and MP*

2.1 Language and Semantics

We follow the notation of [MP92].

We construct a bimodal propositional modal logic. Formally, we start with a countable set A of *atomic formulae*, then the *language* $\mathcal{L}$ is the least set such that $\mathsf{A} \subseteq \mathcal{L}$ and closed under the following rules:

$$\frac{\phi, \psi \in \mathcal{L}}{\phi \wedge \psi \in \mathcal{L}} \qquad \frac{\phi \in \mathcal{L}}{\neg\phi, \Box\phi, \mathsf{K}\phi \in \mathcal{L}}$$

We abreviate, as usual, $\phi \wedge \neg\phi$ with $\bot$ and $\neg\bot$ with $\top$. The language $\mathcal{L}$ can be interpreted inside any spatial context as follows.

Definition 1. Let X be a set and $\mathcal{O}$ a subset of the powerset of X, i.e. $\mathcal{O} \subseteq \mathcal{P}(X)$ such that $X \in \mathcal{O}$. We call the pair $\langle X, \mathcal{O}\rangle$ a *subset space*. A *model* is a triple $\langle X, \mathcal{O}, i\rangle$, where $\langle X, \mathcal{O}\rangle$ is a subset space and i a map from A to $\mathcal{P}(X)$ with $i(\top) = X$ and $i(\bot) = \emptyset$ called *initial interpretation*.

We denote the set $\{(x, U) : x \in X, U \in \mathcal{O}, \text{ and } x \in U\} \subseteq X \times \mathcal{O}$ with $X \dot{\times} \mathcal{O}$. For each $U \in \mathcal{O}$ let $\downarrow U$ be the set $\{V : V \in \mathcal{O} \text{ and } V \subseteq U\}$, i.e. the lower closed set generated by U in the partial order $(\mathcal{O}, \subseteq)$.

Definition 2. The *satisfaction relation* $\models_{\mathcal{M}}$, where $\mathcal{M}$ is the model $\langle X, \mathcal{O}, i\rangle$, is a subset of $(X \dot{\times} \mathcal{O}) \times \mathcal{L}$ defined recursively by (we write $x, U \models_{\mathcal{M}} \phi$ instead of

$((x, U), \phi) \in \models_{\mathcal{M}}$):

$$\begin{array}{lll} x, U \models_{\mathcal{M}} A & \text{iff} & x \in i(A), \text{ where } A \in \mathsf{A} \\ x, U \models_{\mathcal{M}} \phi \wedge \psi & \text{if} & x, U \models_{\mathcal{M}} \phi \text{ and } x, U \models_{\mathcal{M}} \psi \\ x, U \models_{\mathcal{M}} \neg\phi & \text{if} & x, U \not\models_{\mathcal{M}} \phi \\ x, U \models_{\mathcal{M}} \mathsf{K}\phi & \text{if} & \text{for all } y \in U, \quad y, U \models_{\mathcal{M}} \phi \\ x, U \models_{\mathcal{M}} \Box\phi & \text{if} & \text{for all } V \in \downarrow U \text{ such that } x \in V, \quad x, V \models_{\mathcal{M}} \phi. \end{array}$$

If $x, U \models_{\mathcal{M}} \phi$ for all (x, U) belonging to $X \dot{\times} \mathcal{O}$ then ϕ is *valid* in $\mathcal{M}$, denoted by $\mathcal{M} \models \phi$.

We abbreviate $\neg\Box\neg\phi$ and $\neg\mathsf{K}\neg\phi$ with $\Diamond\phi$ and $\mathsf{L}\phi$ respectively. We have that

$x, U \models_{\mathcal{M}} \mathsf{L}\phi$ if there exists $y \in U$ such that $y, U \models_{\mathcal{M}} \phi$

$x, U \models_{\mathcal{M}} \Diamond\phi$ if there exists $V \in \mathcal{O}$ such that $V \subseteq U$, $x \in V$, and $x, V \models_{\mathcal{M}} \phi$.

Definition 3. A *tree-like space* is a subset space $\langle X, \mathcal{O} \rangle$ where for all $U, V \in \mathcal{O}$, either $U \subseteq V$, or $V \subseteq U$, or $U \cap V = \emptyset$. A model induced by a tree space will be called a *tree-like model.*

It is clear that in the countable case the set of subsets of a tree-like space forms a tree under the subset ordering.

Example 3. Let

$$X = \{f \mid f \text{ recursive}\}.$$

Now, let

$$[a_1 a_2 \ldots a_n] = \{f \mid f(k) = a_k, \text{ for } k = 1, 2, \ldots, n\} \subseteq X,$$

where $a_1, a_2, \ldots, a_n$ are natural numbers, and

$$\mathcal{O} = \{[a_1 a_2 \ldots a_n] \mid n = 1, 2, \ldots\} \cup X.$$

Then it is easily verified, using Definition 3, that $\langle X, \mathcal{O} \rangle$ is a tree-like space.

Now let $\mathbf{1}$ be a predicate with

$$i(\mathbf{1}) = \{f \mid \text{ there exists } n \text{ such that } f(n) = 1\}.$$

Then the formula

$$\Box \mathsf{L} \mathbf{1}$$

which translates to "it will never be known that 1 appears infinitely often", is valid in the tree-like model $\langle X, \mathcal{O}, i \rangle$. This comes with no surprise since the knowledge of "infinitely often" requires an infinite amount of resources. This formula is an example of a refutative assertion (see Sect. 1).

Tree-like spaces are $T \times W$ frames in disguise. $T \times W$ frames have appeared as semantics for the Ockhamist's concept of non-deterministic time and been used for treating necessity and conditionals (see [Tho84] and [VF81]). The validity on these frames, called $T \times W$ *validity* differs from ours in that satisfaction of atomic formulae is time dependent, while in our case it depends only in the choice of the world. So the assumption we make is that basic properties are time invariant and such an assumption is crucial during a knowledge query. As far as we know $T \times W$ validity has not been given a reasonable axiomatization.

2.2 MP and MP*

We saw that the semantics of the bimodal language is interpreted in any pair $\langle X, \mathcal{O} \rangle$. What happens when we allow $\mathcal{O}$ to be any class of sets of subsets? If $\mathcal{O}$ is an arbitrary set of subsets then the system **MP** is complete for such subset spaces. The axiom system **MP** consists of Axiom schemes 1 through 10 and rules of Table 1 (see page 7) and appeared first in [MP92].

The following was proved in [MP92].

Theorem 4. *The axioms and rules of* **MP** *are sound and complete with respect to subset spaces.*

If $\mathcal{O}$ is a complete lattice under set-theoretic union and intersection then the system **MP*** is canonically complete for this class of subset spaces. The axiom system **MP*** consists of the axiom schemes and rules of **MP** plus the following two additional axiom schemes:

$$\Diamond\Box\phi \to \Box\Diamond\phi$$

and

$$\Diamond(\mathsf{K}\phi \wedge \psi) \wedge \mathsf{L}\Diamond(\mathsf{K}\phi \wedge \chi) \to \Diamond(\mathsf{K}\Diamond\phi \wedge \Diamond\psi \wedge \mathsf{L}\Diamond\chi).$$

The first axiom is a well-known formula which characterizes *incestual* frames, i.e. if two points β and γ in a frame can be accessed by a common point α then there is a point δ which can be accessed by both β and γ. The second characterizes union.

The following was proved in [Geo93].

Theorem 5. *The axioms and rules of* **MP*** *are sound and canonically complete with respect to subset spaces, which are complete lattices.*

3 The System MPT

We add the Axioms 11 and 12 to form the system **MPT** for the purpose of axiomatizing tree-like spaces.

A word about the axioms (most of the following facts can be found in any introductory book about modal logic, e.g. [Che80] or [Gol87]). Axiom 2 expresses the fact that the truth of atomic formulae is independent of the choice of subset

Table 1. Axioms and Rules of **MPT**

Axioms

1. All propositional tautologies
2. $(A \rightarrow \Box A) \wedge (\neg A \rightarrow \Box \neg A)$, for $A \in \mathsf{A}$
3. $\Box(\phi \rightarrow \psi) \rightarrow (\Box\phi \rightarrow \Box\psi)$
4. $\Box\phi \rightarrow \phi$
5. $\Box\phi \rightarrow \Box\Box\phi$
6. $\mathsf{K}(\phi \rightarrow \psi) \rightarrow (\mathsf{K}\phi \rightarrow \mathsf{K}\psi)$
7. $\mathsf{K}\phi \rightarrow \phi$
8. $\mathsf{K}\phi \rightarrow \mathsf{KK}\phi$
9. $\phi \rightarrow \mathsf{KL}\phi$
10. $\mathsf{K}\Box\phi \rightarrow \Box\mathsf{K}\phi$
11. $\Box(\Box\phi \rightarrow \psi) \vee \Box(\Box\psi \rightarrow \phi)$
12. $\Box\mathsf{K}\phi \wedge \mathsf{K}(\Box\phi \rightarrow \Box\psi) \rightarrow \Box\mathsf{K}(\Box\phi \rightarrow \Box\psi)$

Rules

$$\frac{\phi \rightarrow \psi, \phi}{\psi}\ \mathrm{MP}$$

$$\frac{\phi}{\mathsf{K}\phi}\ \text{K-Necessitation} \qquad \frac{\phi}{\Box\phi}\ \Box\text{-Necessitation}$$

and depends solely on the choice of point. Axioms 3 through 5 and Axioms 6 through 9 are used to axiomatize the normal modal logics **S4** and **S5** respectively. The former group of axioms expresses the fact that the passage from one subset to a restriction of it is done in a constructive way as actually happens to an increase of information or resource spending (the classical interpretation of necessity in intuitionistic logic is axiomatized in the same way). The latter group is generally used for axiomatizing logics of knowledge.

Axiom 10 expresses the fact that if a formula holds in arbitrary subsets will hold as well in the ones which are neighborhoods of a point. The converse of this axiom is not sound.

Axiom 11 is a well-known axiom which characterizes reflexive, transitive and *connected* frames, i.e. if two points β and γ in a frame can be accessed by a common point α then either β accesses γ or γ accesses β (or both).

Soundness of Axioms 1 through 10 has already been established for arbitrary subset spaces (see [MP92]). The soundness of Axiom 11 is easy to see since the *subset frame* (see [Geo93]) of a tree-like model is connected.

Proposition 6. *Axiom 12 is sound.*

Proof. We shall show soundness for the following formula

$$\Box \mathsf{K}\phi \wedge \Diamond \mathsf{L}(\psi \wedge \Box\phi) \rightarrow \mathsf{L}(\Diamond\psi \wedge \Box\phi)$$

which can be easily shown to be equivalent to Axiom 12.

Let $x, U \models \Box\mathsf{K}\phi \wedge \Diamond\mathsf{L}(\psi \wedge \Box\phi)$. Then there exists $V \subseteq U$ such that $x, V \models \mathsf{L}(\psi \wedge \Box\phi)$. This implies that there exists $y \in V$ such that $y, V \models \psi \wedge \Box\phi$. We must have that $y \in U$ and $y, U \models \Box\phi$ since all subsets between U and V are linearly ordered and x belongs to all of them. Therefore $y, U \models \Diamond\psi \wedge \Box\phi$. □

3.1 Completeness

Our proof of completeness is based on a construction of a tree-like model which is (strongly) equivalent to each generated submodel of the canonical model of **MPT**.

The *canonical model* of **MPT** is the structure

$$\mathcal{C} = \left(S, \{\stackrel{\Diamond}{\rightarrow}, \stackrel{\mathsf{L}}{\rightarrow}\}, v\right),$$

where

$$S = \{s \subseteq \mathcal{L} | s \text{ is } \mathbf{MPT}\text{-maximal consistent}\},$$
$$s \stackrel{\Diamond}{\rightarrow} t \text{ iff } \{\phi \in \mathcal{L} | \Box\phi \in s\} \subseteq t,$$
$$s \stackrel{\mathsf{L}}{\rightarrow} t \text{ iff } \{\phi \in \mathcal{L} | \mathsf{K}\phi \in s\} \subseteq t,$$
$$v(A) = \{s \in S | A \in S\},$$

along with the usual satisfaction relation (defined inductively):

$$\begin{array}{ll} s \models_{\mathcal{C}} A & \text{iff } s \in v(A) \\ s \not\models_{\mathcal{C}} \bot & \\ s \models_{\mathcal{C}} \neg\phi & \text{iff } s \not\models_{\mathcal{C}} \phi \\ s \models_{\mathcal{C}} \phi \wedge \psi & \text{iff } s \models_{\mathcal{C}} \phi \text{ and } s \models_{\mathcal{C}} \psi \\ s \models_{\mathcal{C}} \Box\phi & \text{iff for all } t \in S,\ s \stackrel{\Diamond}{\rightarrow} t \text{ implies } t \models_{\mathcal{C}} \phi \\ s \models_{\mathcal{C}} \mathsf{K}\phi & \text{iff for all } t \in S,\ s \stackrel{\mathsf{L}}{\rightarrow} t \text{ implies } t \models_{\mathcal{C}} \phi. \end{array}$$

We write $\mathcal{C} \models \phi$, if $s \models_{\mathcal{C}} \phi$ for all $s \in S$.

A canonical model exists for all consistent bimodal systems with the normal axiom scheme for each modality (as **MPT**). We have the following well known theorems (see [Che80], or [Gol87]).

Theorem 7 Truth Theorem. *For all $s \in S$ and $\phi \in \mathcal{L}$,*

$$s \models_{\mathcal{C}} \phi \quad \textit{iff} \quad \phi \in s.$$

Theorem 8 Completeness Theorem. *For all $\phi \in \mathcal{L}$,*

$$\mathcal{C} \models \phi \quad \textit{iff} \quad \vdash_{\mathbf{MPT}} \phi.$$

We shall now state some properties of $\mathcal{C}$.

Proposition 9. *1. The relation $\xrightarrow{\mathsf{L}}$ is an equivalence relation.*

2. For all $s, s', t \in S$, if $s \xrightarrow{\diamond} s' \xrightarrow{\mathsf{L}} t$ then there exists $t' \in S$ such that $s \xrightarrow{\mathsf{L}} t' \xrightarrow{\diamond} t$.

3. For all $s, s' \in S$, if $s \xrightarrow{\mathsf{L}} s'$ and $s \xrightarrow{\diamond} s'$ then $s = s'$.

4. The canonical frame is reflexive, antisymmetric, transitive (i.e. partially ordered) and connected with respect to the relation $\xrightarrow{\diamond}$.

Proof. For (1), K is axiomatized with the **S5** axioms. (2) is an immediate consequence of Axiom 10. (3) is shown by induction on the complexity of ϕ. For (4), we use (3) for antisymmetry, and the fact that Axioms 3 through 5 and Axiom 11 characterize reflexive, transitive and connected frames (these axioms comprise the system **S4.3**). □

For all $t \in S$, let $[t] = \{s \in S \mid s \xrightarrow{\mathsf{L}} t\}$, i.e. the equivalence class under $\xrightarrow{\mathsf{L}}$ where t belongs. Let $\mathcal{C}_{\mathsf{K}} = \{[t] \mid t \in S\}$. We define the following relation on $\mathcal{C}_{\mathsf{K}}$

$$[t_1] \leq [t_2] \quad \text{iff} \quad \text{there exist } s_1, s_2 \in S \text{ such that } s_1 \in [t_1], s_2 \in [t_2] \text{ and } s_2 \xrightarrow{\diamond} s_1.$$

We have the following proposition

Proposition 10. *The relation $\leq$ is a partial order.*

A subset X of S, the domain of the canonical model $\mathcal{C}$, is called K□*-closed* whenever

$$\text{if } s \in X, \text{ and } s \xrightarrow{\diamond} t \text{ or } s \xrightarrow{\mathsf{L}} t, \quad \text{then} \quad t \in X.$$

The intersection of K□-closed sets is still closed, therefore we can define the smallest K□-closed containing t, for all $t \in S$. We shall denote this set by S^t. Fix $t_0 \in S$. We define the model

$$\mathcal{C}^{t_0} = \left(S^{t_0}, \xrightarrow{\diamond}|_{S^{t_0}}, \xrightarrow{\mathsf{L}}|_{S^{t_0}}, v^{t_0}\right),$$

where $\xrightarrow{\diamond}|_{S^{t_0}}$, $\xrightarrow{\mathsf{L}}|_{S^{t_0}}$ and v^{t_0} are the restrictions of $\xrightarrow{\diamond}$, $\xrightarrow{\mathsf{L}}$ and v to S^{t_0}. We shall call this model *the submodel of $\mathcal{C}$ generated by t_0*.

Observe that if we restrict the partial order $\leq$ to $\mathcal{C}^{t_0}$ then $[t_0]$ is the greatest element under $\leq$.

We are now ready to construct the tree-like model.

Let

$$[\![s]\!] \quad = \quad \{t \in [t_0] \mid \text{there exists } t' \in [s] \text{ such that } t \xrightarrow{\diamond} t'\}.$$

Observe that $[\![s]\!] \subseteq [t_0]$.

For each $s \in S_{t_0}$ we define the following relation $\sim_s$ on $[\![s]\!]$

$$t_1 \sim_s t_2 \quad \text{iff} \quad \text{for all } [s] \leq [s'],\ t_1 \in [\![s']\!] \text{ iff } t_2 \in [\![s']\!].$$

Proposition 11. *For all $s \in \mathcal{C}^{t_0}$, the relation $\sim_s$ is an equivalence relation.*

We denote the equivalence class of t under $\sim_s$ with $[t]_s$. We have $[t]_s \subseteq \mathbf{[}s\mathbf{]} \subseteq [t_0]$.

Let $\langle X, \mathcal{O}^{t_0}\rangle$ be the subset space where

$$X = \{t \mid t \in [t_0]\}$$

and

$$\mathcal{O}^{t_0} = \{[t]_s \mid t \in \mathbf{[}s\mathbf{]} \text{ and } s \in \mathcal{C}^{t_0}\}.$$

Note here, of course, that $\mathcal{O}^{t_0} \subseteq \mathcal{P}(X)$.

Lemma 12. *If* $[s_1] \leq [s_2]$ *and* $t \in \mathbf{[}s_1\mathbf{]} \cap \mathbf{[}s_2\mathbf{]}$ *then* $[t]_{s_1} \subseteq [t]_{s_2}$.

Proposition 13. *The subset space* $\langle X, \mathcal{O}^{t_0}\rangle$ *is a tree-like space.*

Proof. Suppose $[t_1]_{s_1} \cap [t_2]_{s_2} \neq \emptyset$. Let $t \in [t_1]_{s_1} \cap [t_2]_{s_2}$. Since $t \overset{\diamond}{\rightarrow} s_1$, $t \overset{\diamond}{\rightarrow} s_2$ and the canonical frame is connected we have either $s_1 \overset{\diamond}{\rightarrow} s_2$ or $s_2 \overset{\diamond}{\rightarrow} s_1$ which implies that, either $[s_1] \leq [s_2]$ and thus by Lemma 12 $[t_1]_{s_1} = [t]_{s_1} \subseteq [t]_{s_2} = [t_2]_{s_2}$, or $[s_2] \leq [s_1]$ and thus by Lemma 12 $[t]_{s_2} \subseteq [t]_{s_1}$. □

Let $\langle X, \mathcal{O}^{t_0}, i\rangle$ be the tree-like model where X and $\mathcal{O}^{t_0}$ are as above and $i(A) = v^{t_0}(A)$ where v^{t_0} the initial interpretation $\mathcal{C}^{t_0}$. The proof of the main theorem (Theorem 15) of this section is based on the bollowing property of the canonical model.

Lemma 14. *Let* $t \in [t_0]$ *and* $s \in \mathcal{C}^{t_0}$ *such that* $t \overset{\diamond}{\rightarrow} s$*. Then for all* $s' \in [s]$ *there exists* $t' \in [t_0]$ *such that* $t' \overset{\diamond}{\rightarrow} s'$ *and* $t \sim_s t'$*, i.e.* $t' \in [t]_s$.

We now have the following theorem

Theorem 15. *For all* $s \in \mathcal{C}^{t_0}$ *and* $t \in X$ *such that* $t \overset{\diamond}{\rightarrow} s$,

$$\phi \in s \quad \textit{iff} \quad t, [t]_s \models \phi.$$

Combining now Proposition 13 and Theorem 15 we have the following

Corollary 16. *The system* **MPT** *is complete with respect tree-like spaces.*

3.2 Decidability

For each tree-like model and formula ϕ, we shall construct an equivalent finite subset space of bounded size with respect to the complexity of ϕ. To this end, we shall prove a more general property of tree-like spaces along the lines of the Partition theorem in [Geo94].

First we need some definitions. In the following we assume that $\langle X, \mathcal{O}\rangle$ is a tree-like space. Our aim is to find a partition of $\mathcal{O}$, where a given formula ϕ "retains its truth value" for each point throughout a member of this partition. It turns out that there exists a finite partition of this kind.

Note that the following hold, although we refer to a tree-like space $\mathcal{O}$, for an arbitrary family of subsets of X and their proofs can be found in [Geo94].

Definition 17. Given a finite family $\mathcal{F} = \{U_1, \ldots, U_n\} \subseteq \mathcal{P}(X)$, i.e. of subsets of X, we define the *remainder* of (the principal ideal in $(\mathcal{O}, \subseteq)$ generated by) U_k by

$$\mathsf{Rem}^{\mathcal{F}} U_k \quad = \quad \downarrow U_k - \bigcup_{U_k \not\subseteq U_i} \downarrow U_i,$$

where $\downarrow U_k = \{V \in \mathcal{O} \mid V \subseteq U_k\}$. Note that $\mathsf{Rem}^{\mathcal{F}} U_k \subseteq \mathcal{O}$.

Proposition 18. *If $\mathcal{F} = \{U_1, \ldots, U_n\}$ is a finite family of subsets of X, closed under intersection, then*

1. $\mathsf{Rem}^{\mathcal{F}} U_i = \downarrow U_i - \bigcup_{U_j \subset U_i} \downarrow U_j$, *for $i = 1, \ldots, n$. We shall denote $\bigcup_{U_i \in \mathcal{F}} \downarrow U_i$ with $\downarrow \mathcal{F}$.*
2. $\mathsf{Rem}^{\mathcal{F}} U_i \cap \mathsf{Rem}^{\mathcal{F}} U_j = \emptyset$, *for $i \neq j$,*
3. $\bigcup_{i=1}^{n} \mathsf{Rem}^{\mathcal{F}} U_i = \downarrow \mathcal{F}$, *i.e. $\{\mathsf{Rem}^{\mathcal{F}} U_i\}_{i=1}^{n}$ is a partition of $\downarrow \mathcal{F}$. We call such a* $\mathcal{F}$ finite splitting (of $\downarrow \mathcal{F}$),
4. *if $V_1, V_2 \in \mathcal{O}$, $V_1 \in \mathsf{Rem}^{\mathcal{F}} U_i$ and $V_1 \subseteq V_2 \subseteq U_i$ then $V_2 \in \mathsf{Rem}^{\mathcal{F}} U_i$, i.e. $\mathsf{Rem}^{\mathcal{F}} U_i$ is convex,*
5. *if $\{V_j\}_{j \in J} \subseteq \mathsf{Rem}^{\mathcal{F}} U_i$ then $\bigcup_{j \in J} U_j \subseteq U_i$.*

Every partition of a set induces an equivalence relation on this set. The members of the partition comprise the equivalence classes. We denote the equivalence relation induced by $\mathcal{F}$ by $\sim_{\mathcal{F}}$.

Definition 19. Given a set of subsets $\mathcal{G}$, we define the relation $\sim'_{\mathcal{G}}$ on $\mathcal{O}$ with $V_1 \sim'_{\mathcal{G}} V_2$ if and only if $V_1 \subseteq U \Leftrightarrow V_2 \subseteq U$ for all $U \in \mathcal{G}$.

We have the following

Proposition 20. *1. The relation $\sim'_{\mathcal{G}}$ is an equivalence.*

2. *Given a finite splitting $\mathcal{F}$, $\sim'_{\mathcal{F}} = \sim_{\mathcal{F}}$ i.e. the remainders of $\mathcal{F}$ are the equivalence classes of $\sim'_{\mathcal{F}}$.*
3. *If $\mathcal{G}$ is a finite set of subsets of X then $\mathsf{Cl}(\mathcal{G})$, its closure under intersection, is a finite splitting for $\downarrow \mathcal{G}$.*

The last proposition enables us to give yet another characterization of remainders: every family of points in a complete lattice closed under arbitrary joins comprises a *closure system*, i.e. a set of fixed points of a closure operator of the lattice (cf. [GHK+80]). Here, the lattice is the powerset of X. If we restrict ourselves to a finite number of fixed points then we just ask for a finite set of subsets closed under intersection i.e. Proposition 20(3). Thus a closure operator in the lattice of the powerset of X induces an equivalence relation to any family of subsets of X – two subsets being equivalent if they have the same closure – and the equivalence classes of this relation are just the remainders of the subsets which are fixed points of the closure operator.

We now introduce the notion of stability corresponding to what we mean by "a formula retains its truth value on a set of subsets".

Definition 21. Let $\mathcal{G} \subseteq \mathcal{O}$ then $\mathcal{G}$ is *stable for* ϕ, if for all x, either $x, V \models \phi$ for all $V \in \mathcal{G}$, or $x, V \models \neg\phi$ for all $V \in \mathcal{G}$.

Proposition 22. *Let* $\mathcal{G}_1, \mathcal{G}_2 \subseteq \mathcal{O}$ *then*

1. *if* $\mathcal{G}_1 \subseteq \mathcal{G}_2$ *and* $\mathcal{G}_2$ *is stable for* ϕ *then* $\mathcal{G}_1$ *is stable for* ϕ ,
2. *if* $\mathcal{G}_1$ *is stable for* ϕ *and* $\mathcal{G}$ *is stable for* χ *then* $\mathcal{G}_1 \cap \mathcal{G}_2$ *is stable for* $\phi \wedge \chi$.

Definition 23. A finite partition $\mathcal{F} = \{U_1, \ldots, U_n\}$ is called a *stable partition for* ϕ, if $\mathsf{Rem}^{\mathcal{F}} U_i$ is stable for ϕ for all $U_i \in \mathcal{F}$.

Proposition 24. *If* $\mathcal{F} = \{U_1, \ldots, U_n\}$ *is a stable partition for* ϕ, *so is*

$$\mathcal{F}' = \mathsf{Cl}(\{U_0, U_1, \ldots, U_n\}),$$

where $U_0 \in \downarrow\mathcal{F}$.

The above proposition tells us that if there is a finite stable splitting for a tree-like space $\mathcal{O}$ then there is a closure operator with finite fixed points whose associated equivalence classes are stable subsets of $\mathcal{O}$.

The following is the main theorem of this section.

Theorem 25 Partition Theorem. *Let* $\mathcal{M} = \langle X, \mathcal{O}, i \rangle$ *be a e tree-like model. Then there exists a a set* $\{\mathcal{F}^{\psi}\}_{\psi \in \mathcal{L}}$ *of finite stable splittings such that if* ϕ *is a subformula of* ψ *then* $\mathcal{F}^{\phi} \subseteq \mathcal{F}^{\psi}$ *and* $\mathcal{F}^{\psi}$ *is a finite stable splitting for* ϕ.

Proof. By induction on the structure of the formula ψ. In each step we take care to refine the partition of the induction hypothesis. For each $U \in \mathcal{F}^{\psi}$ we let $U^{\psi} = \{x \in U | x, U \models \psi\}$ which determines completely the satisfaction of ψ on $\mathsf{Rem}^{\mathcal{F}^{\psi}} U$ whenever $\mathcal{F}^{\psi}$ is stable.

If $\psi = A$ is an atomic formula, then $\mathcal{F}^A = \{X\} = \{i(\top)\}$, since $\mathcal{O}$ is stable for all atomic formulae. We have $X^A = i(A)$.

If $\psi = \neg\phi$ then let $\mathcal{F}^{\psi} = \mathcal{F}^{\phi}$, since the statement of the theorem is symmetric with respect to negation. We also have that for an arbitrary $U \in \mathcal{F}^{\psi}$, $U^{\psi} = (X - U^{\phi}) \cap U$.

If $\psi = \chi \wedge \phi$, let

$$\mathcal{F}^{\psi} = \mathsf{Cl}(\mathcal{F}^{\chi} \cup \mathcal{F}^{\phi}).$$

Observe that $\mathcal{F}^{\chi} \cup \mathcal{F}^{\phi} \subseteq \mathcal{F}^{\chi \wedge \phi}$. Now, $\mathcal{F}^{\psi}$ is a stable splitting for $\chi \wedge \phi$ containing X, since it is a refinement of both $\mathcal{F}^{\chi}$ and $\mathcal{F}^{\phi}$. Thus $\mathcal{F}^{\psi}$ is a finite stable splitting for ψ containing X.

Suppose $\psi = \mathsf{K}\phi$. Then, by induction hypothesis, there exists a finite stable splitting $\mathcal{F}^{\phi} = \{U_1, \ldots, U_n\}$ for ϕ containing X.

Now, if $V \in \mathsf{Rem}^{\mathcal{F}^{\phi}} U_i \cap \downarrow U_i^{\phi}$, for some $i \in \{1 \ldots, n\}$, then $x, V \models \phi$ for all $x \in V$, by definition of U_i^{ϕ}, hence $x, V \models \mathsf{K}\phi$ for all $x \in V$.

On the other hand, if $V \in \mathsf{Rem}^{\mathcal{F}^\phi} U_i - {\downarrow}U_i^\phi$ then there exists $x \in V$ such that $x, V \models \neg\phi$ (otherwise $V \subseteq U_i^\phi$). Thus we have $x, V \models \neg \mathsf{K}\phi$ for all $x \in V$. Hence $\mathsf{Rem}^{\mathcal{F}^\phi} U_i \cap {\downarrow}U_i^\phi$ and $\mathsf{Rem}^{\mathcal{F}^\phi} U_i - {\downarrow}U_i^\phi$ are stable for $\mathsf{K}\phi$. Thus, the set

$$F = \{\mathsf{Rem}^{\mathcal{F}} U_i \mid U_i^\phi \notin \mathsf{Rem}^{\mathcal{F}} U_i\} \cup \\ \{\mathsf{Rem}^{\mathcal{F}} U_j - {\downarrow}U_j^\phi, \mathsf{Rem}^{\mathcal{F}} U_j \cap {\downarrow}U_j^\phi \mid U_j^\phi \in U_j\}$$

is a partition of $\mathcal{O}$ and its members are stable for $\mathsf{K}\phi$. Let

$$\mathcal{F}^{\mathsf{K}\phi} = \mathsf{Cl}(\mathcal{F}^\phi \cup U_i^\phi).$$

We have that $\mathcal{F}^{\mathsf{K}\phi}$ is a finite set of opens and $\mathcal{F}^\phi \subseteq \mathcal{F}^{\mathsf{K}\phi}$. Thus, $\mathcal{F}^{\mathsf{K}\phi}$ is finite and contains X. We have only to prove that $\mathcal{F}^{\mathsf{K}\phi}$ is a stable splitting for $\mathsf{K}\phi$, i.e. every remainder of an open in $\mathcal{F}^{\mathsf{K}\phi}$ is stable for $\mathsf{K}\phi$. But for that observe that $\mathcal{F}^{\mathsf{K}\phi}$ is a refinement of F. Therefore $\mathcal{F}^{\mathsf{K}\phi}$ is a finite stable splitting for $\mathsf{K}\phi$ using Proposition 22(1).

Now if $U \in \mathcal{F}^\psi$ then either $U^{\mathsf{K}\phi} = U$ or $U^{\mathsf{K}\phi} = \emptyset$.

Suppose $\psi = \Diamond\phi$. Then, let

$$\mathcal{F}^{\Diamond\phi} = \mathcal{F}^\phi,$$

where $\mathcal{F}^\phi$ is a finite stable splitting for ϕ by induction hypothesis.

We shall show that $\mathcal{F}^\phi$ is also a finite stable spitting for $\Diamond\phi$. Pick $U \in \mathcal{F}^\phi$ and $x \in U$. If $x, V \models \neg\phi$ for all $V \subseteq U$ such that $x \in V$ we are done because in this case $x, V \models \neg\Diamond\phi$. If $x, V \models \phi$ for some $V \in \mathsf{Rem}^{\mathcal{F}^\phi} U$ we have $x, W \models \phi$ for all $W \in \mathsf{Rem}^{\mathcal{F}^\phi} U$ because $\mathcal{F}^\phi$ is stable for ϕ. Therefore $x, W \models \Diamond\phi$ for all $W \in \mathsf{Rem}^{\mathcal{F}^\phi} U$. If $x, V \models \phi$ for some $V \subseteq U$ but $V \notin \mathsf{Rem}^{\mathcal{F}^\phi} U$ then since the set of subsets containing x is linearly ordered and $\mathsf{Rem}^{\mathcal{F}^\phi} U$ is stable and convex we have $V \subseteq W$ for all $W \in \mathsf{Rem}^{\mathcal{F}^\phi} U$, therefore $x, W \models \Diamond\phi$ for all $W \in \mathsf{Rem}^{\mathcal{F}^\phi} U$. □

Given a tree-like model $\langle X, \mathcal{O}, i\rangle$ and a formula ϕ there exists a finite splitting $\mathcal{F}^\phi$ on $\mathcal{O}$, stable for ϕ, by the Partition Theorem. For each $U \in \mathcal{F}^\phi$, let

$$\overline{U} \quad = \quad \left\{\bigcup \mathsf{Rem}^{\mathcal{F}^\phi} U \mid U \in \mathcal{F}^\phi\right\}$$

and now let

$$\overline{\mathcal{F}^\phi} \quad = \quad \{U \mid U \in \mathcal{F}^\phi \text{and } \overline{U} \neq \emptyset\}.$$

Let $<$ be the following relation on $\overline{\mathcal{F}^\phi}$

$U_1 < U_2$ iff $\overline{U_1} \cap \overline{U_2} \neq \emptyset$ and
for all $x \in \overline{U_1} \cap \overline{U_2}$ $V_1 \in \mathsf{Rem}^{\mathcal{F}^\phi} U_1$ such that $x \in V_1$
and $V_2 \in \mathsf{Rem}^{\mathcal{F}^\phi} U_2$ such that $x \in V_2$, $V_1 \subset V_2$.

Clearly we cannot have $U_1 < U_2$ and $U_2 < U_1$. Let $U_1 \leq U_2$ if $U_1 = U_2$ or $U_1 < U_2$.

Lemma 26. *$\leq$ is reflexive and antisymmetric.*

Instead of transitivity we have the following property of $\leq$

Lemma 27. *Let $U_1, U_2, U_3 \in \overline{\mathcal{F}^\phi}$. If $U_1 \leq U_2$, $U_2 \leq U_3$ and $\overline{U_1} \cap \overline{U_2} \cap \overline{U_3} \neq \emptyset$ then $U_1 \leq U_3$.*

Since $\overline{\mathcal{F}^\phi} \subseteq \mathcal{F}^\phi$ then $\overline{\mathcal{F}^\phi}$ is finite. Let $\overline{\mathcal{F}^\phi} = \{U_1, U_2, \ldots, U_n\}$ for some finite number n. Now let $\sim_i$ be the following equivalence relation on $\overline{U_i}$

$$x \sim_i y \text{ iff for all } \overline{U_j}, j \in \{1, 2, \ldots, n\} \text{ such that } U_i \leq U_j, \\ x \in \overline{U_j} \text{ iff } y \in \overline{U_j}.$$

We denote the equivalence of x under $\sim_i$ with $[x]_i$.
Now let

$$[\overline{\mathcal{F}^\phi}] = \{[x]_i \mid x \in \overline{U_i},\ i \in \{1, 2, \ldots, n\}\}.$$

Proposition 28. *The subset space $\langle X, [\overline{\mathcal{F}^\phi}]\rangle$ is a tree-like space.*

Let $\overline{\mathcal{M}} = \langle X, [\overline{\mathcal{F}^\phi}], \bar{i}\rangle$ be the tree-like model where $\bar{i}(A) = \{[x]_i \mid x \in i(A)\}$.
The following proposition shows that the model we constructed is equivalent to the initial one with respect to the satisfaction of ϕ.

Proposition 29. *For all $x \in X$, $V \in \mathcal{O}$ and $\psi \in \mathcal{L}$ such that ψ is a subformula of ϕ, if $V \in \mathsf{Rem}^{\mathcal{F}^\phi} U_i$ for some $i \in \{1, 2, \ldots, n\}$ then*

$$x, V \models_{\mathcal{M}} \psi \qquad \textit{iff} \qquad x, [x]_i \models_{\overline{\mathcal{M}}} \psi.$$

Constructing the above model is not adequate for generating a finite model; there may still be an infinite number of points. It turns out that we only need a finite number of them.

Let $\mathcal{M} = \langle X, \mathcal{O}, i\rangle$ be a tree-like model, and define an equivalence relation $\sim$ on X by $x \sim y$ iff

1. for all $U \in \mathcal{O}$, $x \in U$ iff $y \in U$, and
2. for all atomic A, $x \in i(A)$ iff $y \in i(A)$.

Further, denote by x^* the equivalence class of x, and let $X^* = \{x^* : x \in X\}$. For every $U \in \mathcal{O}$ let $U^* = \{x^* : x \in X\}$, then $\mathcal{O}^* = \{U^* : U \in \mathcal{O}\}$ is a tree-like space on X^*. Define a map i^* from the atomic formulae to the powerset of X^* by $i^*(A) = \{x^* : x \in i(A)\}$. The entire model $\mathcal{M}$ lifts to the model $\mathcal{M}^* = \langle X^*, \mathcal{O}^*, i^*\rangle$ in a well-defined way.

Lemma 30. *For all x, U, and ϕ,*

$$x, U \models_{\mathcal{M}} \phi \qquad \textit{iff} \qquad x^*, U^* \models_{\mathcal{M}^*} \phi\,.$$

Theorem 31. *If ϕ is satisfied in any tree-like space then ϕ is satisfied in a finite tree-like space.*

Proof. Let $\mathcal{M} = \langle X, \mathcal{O}, i \rangle$ be such that for some $x \in U \in \mathcal{O}$, $x, U \models_{\mathcal{M}} \phi$. Let $\mathcal{F}^{\phi}$ be a finite stable splitting (by Theorem 25) for ϕ and its subformulae with respect to $\mathcal{M}$. By Proposition 29, $x, U \models_{\mathcal{N}} \phi$, where $\mathcal{N} = \langle X, \mathcal{F}, i \rangle$. We may assume that $\mathcal{F}$ is a tree-like space, and we may also assume that the overall language has only the (finitely many) atomic symbols which occur in ϕ. Then the relation $\sim$ has only finitely many classes. So the model $\mathcal{N}^*$ is finite. Finally, by Lemma 30, $x^*, U^* \models_{\mathcal{N}^*} \phi$. □

Observe that the finite tree-like space is a quotient of the initial one under two equivalences. The one equivalence is on the elements of the tree-like space and its number of equivalence classes is a function of the complexity of ϕ. The other equivalence is on the points of the tree-like space and its number of equivalence classes is a function of the atomic formulae appearing in ϕ. So the overall size of the (finite) tree-like space is bounded by a function of the complexity of ϕ. Thus if we want to test if a given formula is invalid we have a finite number of finite tree-like spaces where we have to test its validity. Thus we have the following

Theorem 32. *The theory of tree-like spaces is decidable.*

Aknowledgements: The author wishes to thank Larry Moss and Rohit Parikh for helpful comments and suggestions.

References

[Bur79] John P. Burgess. Logic and time. *The Journal of Symbolic Logic*, 44:566–582, 1979.

[Che80] Brian F. Chellas. *Modal Logic: An Introduction*. Cambridge University Press, Cambridge, 1980.

[CM86] M. Chandy and J. Misra. How processes learn. *Distributed Computing*, 1(1):40–52, 1986.

[Fit93] Melvin C. Fitting. Basic modal logic. In D. M. Gabbay, C. J. Hogger, and J. A. Robinson, editors, *Handbook of Logic in Artificial Intellingence and Logic Programming*, volume 1. Oxford University Press, 1993.

[Gad75] Hans Georg Gadamer. *Truth and Method*. Continuum, New York, 1975.

[Geo93] Konstantinos Georgatos. Modal logics for topological spaces. Ph.D. Dissertation. City University of New York, 1993.

[Geo94] Konstantinos Georgatos. Knowledge theoretic properties of topological spaces. In Michael Masuch and Polos Laszlo, editors, *Knowledge Representation and Uncertainty*, number 808 in Lecture Notes in Computer Science, pages 147–159, Springer-Verlag, Berlin, 1994.

[GHK+80] Gerhard Gierz, Karl Heinrich Hoffman, Klaus Keimel, James D. Lawson, Michael W. Mislove, and Dana S. Scott. *A Compendium of Continuous Lattices*. Springer-Verlag, Berlin, 1980.

[Gol87] Robert Goldblatt. *Logics of Time and Computation*. Number 7 in CSLI Lecture Notes. CSLI, Stanford, 1987.

[Hin62] Jaakko Hintikka. *Knowledge and Belief*. Cornell University Press, Ithaca, New York, 1962.

[Hin86] Jaakko Hintikka. Reasoning about knowledge in philosophy. The paradigm of epistemic logic. In J. Y. Halpern, editor, *Theoretical Aspects of Reasoning about Knowledge: Proceedings of the First Conference (TARK 1986)*, pages 63–80, 1986. Morgan Kaufmann.

[HM84] Joseph Y. Halpern and Yoram Moses. Knowledge and common knowledge in a distributed environment. In *Proceedings of the Third ACM Symposium on Principles of Distributed Computing*, pages 50–61, 1984.

[HV89] Joseph Y. Halpern and Moshe Y. Vardi. The complexity of reasoning about knowledge and time. I. Lower bounds. *Journal of Computer and System Sciences*, 38:195–237, 1989.

[MP92] Lawrence S. Moss and Rohit Parikh. Topological reasoning and the logic of knowledge. In Yoram Moses, editor, *Theoretical Aspects of Reasoning about Knowledge: Proceedings of the Fourth Conference (TARK 1992)*, pages 95–105, 1992. Morgan Kaufmann.

[PR85] Rohit Parikh and R. Ramanujam. Distributed computing and the logic of knowledge. In Rohit Parikh, editor, *Logics of Programs*, number 193 in Lecture Notes in Computer Science, pages 256–268, Springer-Verlag, Berlin, 1985.

[Pri67] Arthur Prior. *Past, Present and Future*. Oxford University Press, London, 1967.

[Smy83] M. B. Smyth. Powerdomains and predicate transformers: a topological view. In J. Diaz, editor, *Automata, Languages and Programming*, number 154 in Lecture Notes in Computer Science, pages 662–675, Springer-Verlag, Berlin, 1983.

[Tho84] Richmond H. Thomason. Combinations of tense and modality. In D. Gabbay and F. Guenthner, editors, *Handbook of Philosophical Logic*, volume II, pages 135–165. D. Reidel Publishing Company, 1984.

[VF81] B. Van Fraassen. A temporal framework for conditionals and chance. In W. Harper, R. Stalnaker, and G. Pearce, editors, *Ifs: Conditionals, Belief, Decision, Chance, and Time*, pages 323–340. D. Reidel, Dordrecht, 1981.

[Vic89] Steven Vickers. *Topology via Logic*. Cambridge Studies in Advanced Computer Science. Cambridge University Press, Cambridge, 1989.

Propositional State Event Logic

Gerd Große*

Intellektik,TH Darmstadt, Alexanderstr. 10, 64289 Darmstadt, Germany

Abstract. In this paper we propose the representation of concurrent events and causality between events in modal logic. This approach differs from previous approaches in the following directions: first, events enjoy the same attention as states. In the same way as states can be viewed as models of the formulae describing the facts that hold in them we think of events as models of the formulae describing the subevents. Second, instead of postulating just one set of states as primitive objects we use two sets, a set of states and a set of events. In terms of modal logic, the universe then becomes a set of pairs in which one component is a state and the other is one of the events following the state. The connection between two subsequent pairs is expressed by an accessibility relation.

1 Introduction

It is quite common to represent the behavior of dynamic systems by a sequence of states. These states are considered to be models of some set of logical formulae. Two subsequent states are connected by the occurrence of some event[2]. Syntactically each event is represented by a constant on the term level. Semantically, they are often identified by the pairs of states between which the event might happen (most obvious in dynamic logic [11], [7]). Thus there is a strong asymmetry in handling states and events.

Recently, there has been a growing interest in forming a theory that allows for the concurrent execution of actions [3]. The handling of concurrent events is important since in any bigger domain a number of events happen at the same time. However, a proper treatment of concurrency has to solve the following difficulties: Starting from the axioms describing the effects of the atomic events

1. we want to derive the effects of a set of atomic events occurring in parallel. This point can be divided into the questions of what has changed after the simultaneous occurrence of a set of events and what remains unchanged.
2. We do not want to be able to derive plans which are not executable. This might happen if our formalism cannot handle conflicts between actions.

* This research was supported by the German Research Council under grant no. HE 1170/5 - 1.

[2] We will use the term event as a generic term for actions and other changes

[3] see Hendrix [8], Georgeff [4], Pednault [10], Sandewall and Rönnquist [12], Wilkins [14], Belgrinos and Georgeff [2], Gelfond *etal.* [3], Große and Waldinger [5], Lin and Shoham [9] or Baral and Gelfond [1]

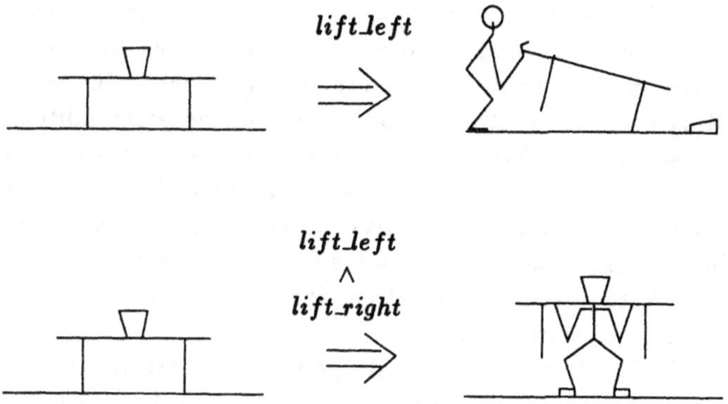

Fig. 1. Lifting a table

As a typical example see Figure 1. The first part of Figure 1 describes the single action of lifting the table on the left side. As an effect we would expect that the table is lifted on the left side and that the vase dropped down. The effect of lifting on the right side would be similar. However, lifting both sides of the table concurrently will lead to a state in which the vase has not dropped down (depicted in the second part of Figure 1). Thus the position of the vase is not affected and belongs therefore to the frame of the concurrent action. As we showed in Figure 1 the agent is able to lift the table simultaneously on both sides. Furthermore, suppose the agent were not as strong then our theory should somehow guarantee that we cannot form a plan in which the agent lifts the table on both sides simultaneously.

In this paper we propose a method for parallelizing events. The four main ideas are as follows: Firstly, much in the line of Lin and Shoham [9] we establish a symmetry of states and events. Analogously to the description of states by formulas over the set of primitive state symbols we describe events by formulas over a set of primitive event symbols. A possible description of the event happening in the first part of Figure 1 would be the formula $lift\_left \wedge \neg lift\_right$. Secondly, the effects of an event are divided into direct and indirect effects. Direct effects are effects which necessarily hold after the event (described by some event formula) has occurred. For instance, if we want to describe the effect of lifting the left side of the table we would state

> *if* down_left_side *holds in a state and if* lift_left *holds in the event occurring in this state then* ¬down_left_side *holds in the next state*

The position of the vase or of the right-hand side of the table is not included in the postcondition, since they might depend on other properties of the event. In Figure 1 both events make *lift_left* true. However, the position of the vase is differently affected. This representation of events is in contrast to the common description of events. There we have effect axioms in which we describe the effect of an event if it occurs in isolation. Here *lift_left* means all events for which the formula *lift_left* holds. Thus a conjunction like *lift_left* ∧ *lift_right* reduces the number of possible events to the ones which satisfy both conjuncts. The distinction between direct and indirect effects enables us to combine two (or more) effect axioms and to describe more specific events. The direct effect of such a resulting event is exactly the conjunction of the postconditions.

Secondly, we provide frame axioms which look similar to Haas' domain-specific frame axioms [6] and Schubert's explanation closure [13], i.e., " if P is true in a state and the occurring event does not satisfy E, F or G, then P is true in the successor state". The combination of direct effects and this frame handling allows us to infer the propositions that change and the ones that persist as a result of some concurrent event.

Thirdly, indirect effects such as the dropping of the vase are treated as events. The movement of the vase cannot be seen as a direct effect of *lift_left*$\wedge\neg$*lift_right*. The vase falls if and only if the table is sloped, i.e., which is here a consequence of the direct effect of *lift_left*. Moreover, indirect effects such as a movement take time like any other event. This view results in the sequence of events as depicted in Figure 2. The fact that such an event automatically occurs can be

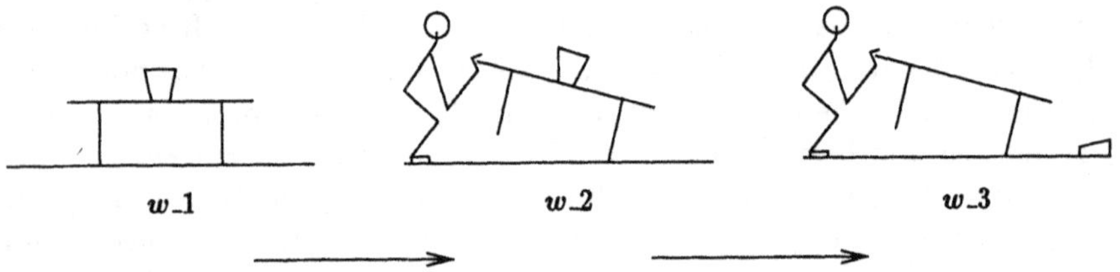

Fig. 2. Lifting a table with a vase on top of it described by a sequence of lifting and falling

represented by sentences of the form "if P holds in a state, then event E will necessarily occur". The dropping of the vase in Figure 2 could thus be formalized by

if $\neg$down_left_side $\wedge$ down_right_side $\wedge$ vase_on_table *holds in a state*
then vase_falls *holds in the event occurring in this state*

Thus when we state that the vase will drop down we restrict the possible events happening in the given state to the models of *vase_falls*. It is clear that treating indirect effects as events implies that we have to provide effect axioms for them similar to the ones which we introduced for *lift_left*.

Finally, we want our formalism to handle the conflict problem properly. By regarding events as models of a set of subevents we can achieve that. If two subevents, say E and F, cannot occur in parallel then there exists no event which is a model of both. Thus, although we can form a statement "if event E and event F happen, then ...", no trouble arises. On the one hand, the statement is still true, because the precondition is false. On the other hand, for proving that the goal condition holds in some future state we have to prove that the applied actions can be executed simultaneously.

These four points enable us to reason about the effects of the parallel occurrence of events. We realized our theory in modal logic K. However, since we are exclusively dealing with states and events we developed a more intuitive semantics. In this semantics we have a complete symmetry between states and events. For this purpose the semantics provides two sets of primitive objects, a set of states and a set of events. Analogously, the syntactical expressions are built up from two sorts of atomic formulae, a set of state formulae and a set of event formulae. From the set of states and the set of events we form pairs which relate a state to one of the events applicable in it. The pairs are in turn related to each other by an accessibility relation. Thus, instead of reasoning about states which are connected by events we reason about pairs which are related by an accessibility relation. We shall refer to this approach as "state event logic". The intention of the present article is to introduce state event logic and to demonstrate its expressiveness by examples from the planning area. The question how to derive plans within this framework is not in the scope of this paper.

The next section contains the formal description of propositional state event logic. In Section 3 we illustrate the theory by three examples. The last section will relate the work to other approaches.

2 Propositional State Event Logic

In the following part we embed our formalism in modal logic K. The central difference is that we provide two sets for building the universe, a set of states W and a set of events E. The universe is a set $T \subseteq W \times E$. Each element of T is a pair which has a state and an event component as depicted in Figure 3. We

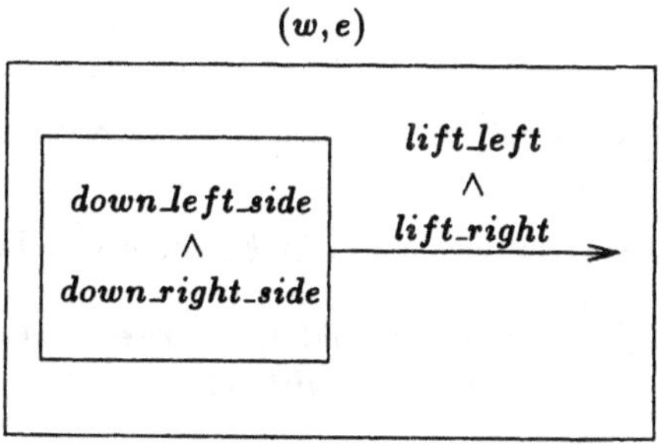

Fig. 3. A pair consisting of the state w and the event e.

provide two sorts of atomic formulae, i.e., state and event formulae, to describe the respective components of pairs. Thus for the pair depicted in Figure 3 the formula

$$down\_left\_side \land down\_right\_side \land \; lift\_left \land lift\_right$$

holds, because *down_left_side*∧*down_right_side* holds for the state component and *lift_left*∧*lift_right* holds for the event component. The expressions of our language are built up from the two sorts of formulae by using the standard connectives. This is sufficient to describe pairs as depicted in Figure 3. The conjunction of two subevents means that the entire event is a model of both subevents, which we want to interpret as parallel occurrence. The disjunction of two subevents means that the entire event is a model of at least one of the two events. Finally the negation of a subevent means that the entire event is not a model of the respective event formula.

We are also free to build expressions combining both sorts. For instance, the formula

$$(\textit{down\_left\_side} \wedge \neg \textit{down\_right\_side} \wedge \textit{vase\_on\_table}) \rightarrow \textit{vase\_falls}$$

describes the union of the set of pairs for which antecedent and conclusion hold and the set of pairs which are not models of the antecedent. If we take this formula as an axiom we express that in every pair for which the antecedent holds the event formula *vase_falls* holds for the subsequent event.

In order to describe the course of the world we have to connect the pairs by an accessibility relation $R \subseteq T \times T$. Statements about subsequent pairs will be expressed by using a modal operator $\Box$. The formula $\Box\ p$ can be read as 'formula p holds in the next pair'. $\Box\ p$ is true in a pair (w, e), if p is true in all pairs $(w', e') \in T$ with $[(w, e), (w', e')] \in R$. In Figure 4 we show a relation between two pairs.

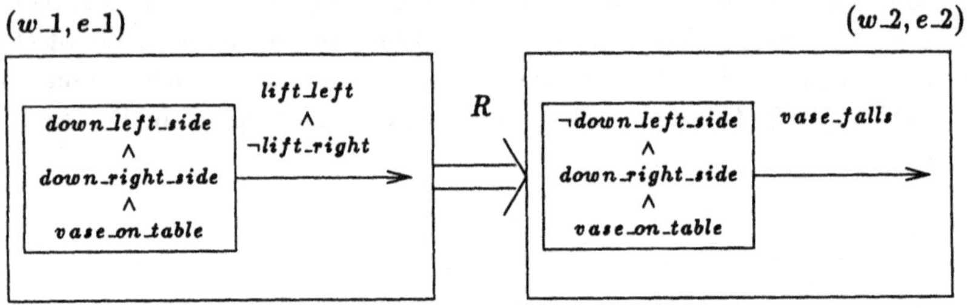

Fig. 4. The accessibility relation R between two pairs

The transition between the two pairs in Figure 4 can be expressed by the formula

$$\begin{gathered} \textit{down\_left\_side} \wedge \textit{down\_right\_side} \wedge \textit{vase\_on\_table} \\ \wedge\ \textit{lift\_left} \wedge \neg \textit{lift\_right} \\ \rightarrow \\ \Box\ (\neg \textit{down\_left\_side} \wedge \textit{down\_right\_side} \\ \wedge\ \textit{vase\_on\_table} \wedge\ \textit{vase\_falls}) \end{gathered}$$

Observe that we use the events in the precondition and include them into the effects. Our transition cannot be seen as an event connecting two states, the connection is between the pairs.

Observe also that the use of the $\Box$ operator represents the direct effects of an event. The statement

$$(\textit{down\_left\_side} \wedge \textit{lift\_left}) \rightarrow \Box \neg \textit{down\_left\_side}$$

expresses that if *lift_left* is part of the event component and if the state component satisfies *down_left_side*, then the formula $\neg$*down_left_side* holds in the state component of each possible next pair.

This view on states and events is expressive enough to handle simultaneously occurring events, causality between events, and conflicting events. This we illustrate by the examples in section 3. But before that we set state event logic on a formal ground.

Syntax: We have two sets of symbols, Φ_0 and Π_0, standing for atomic state formulae and atomic event formulae. We use $s, t, \ldots$ to denote the elements of Φ_0, and $a, b, \ldots$ to denote the elements of Π_0.

The set of well-formed formulae Σ is defined as follows:

$$\begin{array}{l} \Phi_0, \Pi_0 \subseteq \Sigma \\ \text{if } p, q \in \Sigma \text{ then } \neg p, p \vee q, \Box p \in \Sigma \end{array}$$

We use $\wedge$, $\rightarrow$, $\leftrightarrow$ as abbreviations in the standard way. In addition, we abbreviate $p \vee \neg p$ to *True*, $p \wedge \neg p$ to *False*, and as in modal logic $\neg\Box\neg\, p$ to $\Diamond\, p$.

Semantics: The structure underlying the state event logic is basically adopted from modal logic K. The difference is that we included events in the universe by forming pairs of states and events. For this purpose we add a second sort of primitives, a set of events E. As the universe serves a set $T \subseteq W \times E$. Extending the universe to state-event pairs implies that we also require two assignment functions, M_1 and M_2. M_1 assigns to each atomic state formula some subset of the set of states W. In addition, the mapping M_2 assigns to each atomic event formulae some subset of the set of events E. Based on these two mappings we will form a mapping M for assigning well-formed formulae to state-event pairs. Finally we take the relation R to be a subset of $T \times T$. Thus we use the modal logic system to reason about the relation of state/event pairs.

The semantics of state event logic logic is defined relative to a given structure $\mathcal{S} = \langle W, E, T, R, M_1, M_2 \rangle$. We postulate

1. a set of states W.
2. a set of events E.
3. a set of pairs $T \subseteq W \times E$,
4. a relation $R \subseteq T \times T$, and
5. two mappings M_1 and M_2:

$$\begin{array}{l} M_1 : \Phi_0 \rightarrow 2^W \\ M_2 : \Pi_0 \rightarrow 2^E \end{array}$$

Finally we need a mapping from the set of expressions Σ to the universe T. It can be inductively created with the mappings M_1 and M_2:

$$M : \Sigma \rightarrow 2^T$$

as follows

$$M(p) = \{(w,e) \in T \mid \quad [p \in \Phi_0 \wedge w \in M_1(p)] \quad \vee \quad [p \in \Pi_0 \wedge e \in M_2(p)] \quad \}$$
$$M(\neg p) = T - M(p)$$
$$M(p \vee q) = M(p) \cup M(q)$$
$$M(\Box\, p) = \{(w,e) \in T \mid \quad \forall (w',e') \in T: \quad (w,e)R(w',e') \rightarrow (w',e') \in M(p) \quad \}$$

The last equation asserts that $\Box\, p$ is true in those pairs (w,e) from which only pairs (w',e') are accessible such that p holds in (w',e').

We want to denote $(w,e) \in M(p)$ by $\models_{(w,e)} p$. Given a structure $\mathcal{S} = < W, E, T, R, M_1, M_2 >$. A formula p is *S-valid* (written $\models_{\mathcal{S}} p$) if $\models_{(w,e)} p$ is true for every $(w,e) \in T$. The formula p is *valid* (written $\models p$) if it is *S-valid* for every structure S.

The axiom system is the one of modal logic K:

Axiom schemata:

PL. All instances of tautologies of the propositional calculus

K. $\qquad \Box(p \rightarrow q) \rightarrow (\Box p \rightarrow \Box q)$

Inference rules:

$$\textbf{RN} \quad \frac{p}{\Box p} \qquad\qquad \frac{p \qquad p \rightarrow q}{q} \quad \textbf{MP}$$

Observe that a structure $\mathcal{S} = < W, E, T, R, M_1, M_2 >$ of state-event logic is logically identical to a structure $\mathcal{S}' = < T', R, M' >$ in normal modal logic K. The system K can be obtained by taking $\Phi_0 \cup \Pi_0$ as the set of atomic expressions. On the semantical side we postulate the universe T' in which each pair $(w,e) \in T$ of state-event logic is represented by a corresponding $t \in T'$. In addition we merge M_1 and M_2 to a mapping M'. We can state that a formula is a theorem in state event logic if and only if it is a theorem in its corresponding modal logic system (see long version for the proof). Consequently, propositional state event logic inherits the properties of modal logic.

The advantage of state event logic is that we can access the components of the state/event pairs. Semantically, we can express that different events can follow the same state, i.e., $(w,e_1) \in T$ and $(w,e_2) \in T$. In the corresponding system K these two pairs would be represented by two different elements $t_1, t_2 \in T$ which had no relationship with each other. Moreover, the effort of separating the two different concepts of states and events from each other is comparatively small to the gain in clarification.

In the sequel we will make extensive use of the following theorem

R. $\qquad \Box(p \wedge q) \leftrightarrow (\Box p \wedge \Box q)$

Let us now illustrate our logic with a number of examples.

3 Examples

In the first example we explain how we can reason about the effects of simultaneously occurring events. This includes the changes caused by the events, the question of what remains unchanged, and our way of handling conflicting events. The second example is an extension of the first one. There we demonstrate how indirect effects are treated. We give a number of axioms describing this scenario, in order to show how our approach is used. Finally, we prove one sentence by using the axiom system given in the previous section. In third example we discuss a tricky example in which an agent wants to interchange the position of two blocks by moving them both in parallel.

3.1 Lifting a Table

The scenario in Figure 5 consists of a table which can be lifted on two sides by an agent. The goal of the agent is to lift the table on both sides. There are three ways for it to achieve its goal: first, it can lift on the left side and then on the right side expressed by the event sequence e_1 and e_2. Second, the agent might lift the right side first and then the left side, i.e., first e_2 and then e_1. And third, the agent is strong enough to lift on both sides simultaneously, which is represented by event e_3. Our agent cannot wait or lower the table.

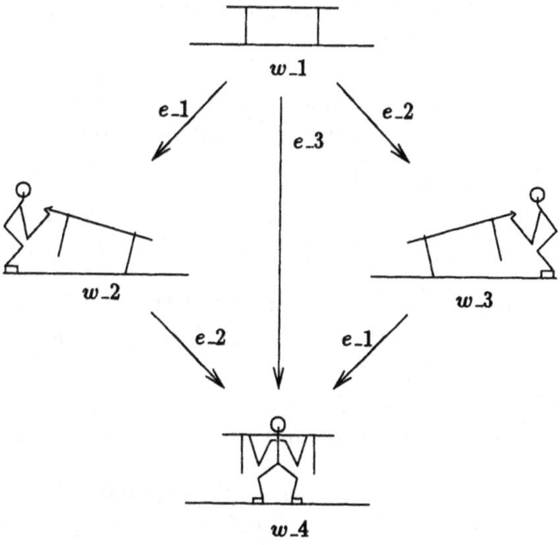

Fig. 5. Simultaneous lifting of the table

The atomic states and events will be described by the sets:

$\Phi_0 = \{$*down_left_side*, *down_right_side*$\}$ $\quad\quad$ $\Pi_0 = \{$*lift_left*, *lift_right*$\}$

In addition we provide a structure $S = (W, E, T, R, M_1, M_2)$ with

$$W = \{w_1, w_2, w_3, w_4\}$$
$$E = \{e_1, e_2, e_3, e_4\}$$
$$T = \{(w_1, e_1), (w_1, e_2), (w_1, e_3), (w_2, e_2), (w_3, e_1), (w_4, e_4)\}$$
$$R = \left\{ \begin{array}{l} [(w_1, e_1), (w_2, e_2)], [(w_1, e_2), (w_3, e_1)], [(w_1, e_3), (w_4, e_4)], \\ [(w_2, e_2), (w_4, e_4)], \\ [(w_3, e_1), (w_4, e_4)] \end{array} \right\}$$

$$M_1(\mathit{down\_left\_side}) = \{w_1, w_3\}$$
$$M_1(\mathit{down\_right\_side}) = \{w_1, w_2\}$$
$$M_2(\mathit{lift\_left}) = \{e_1, e_3\}$$
$$M_2(\mathit{lift\_right}) = \{e_2, e_3\}$$

From the mappings M_1 and M_2 we can form the mapping M for the atomic formulae as follows:

$$M(\mathit{down\_left\_side}) = \{(w_1, e_1), (w_1, e_2), (w_1, e_3), (w_3, e_1)\}$$
$$M(\mathit{down\_right\_side}) = \{(w_1, e_1), (w_1, e_2), (w_1, e_3), (w_2, e_2)\}$$
$$M(\mathit{lift\_left}) = \{(w_1, e_1), (w_1, e_3), (w_3, e_1)\}$$
$$M(\mathit{lift\_right}) = \{(w_1, e_2), (w_1, e_3), (w_2, e_2)\}$$

Let us examine how we reason about the simultaneous occurrence of events and the conflict problem: We provide descriptions of the events in Figure 5 in the common fashion, a precondition implies a postcondition:

lift_left: $(\mathit{down\_left\_side} \wedge \mathit{lift\_left}) \rightarrow \Box(\neg \mathit{down\_left\_side})$
lift_right: $(\mathit{down\_right\_side} \wedge \mathit{lift\_right}) \rightarrow \Box(\neg \mathit{down\_right\_side})$

It is easy to see that our example is a model of these two sentences. An implication is true in each pair of the universe, if the conclusion is true in each pair which satisfies the antecedent.

By applying the tautologies of propositional calculus and the rule R of the given axiom system we can derive from these two sentences the formula:

$$(\mathit{down\_left\_side} \wedge \mathit{down\_right\_side} \wedge \mathit{lift\_left} \wedge \mathit{lift\_right}) \longrightarrow \Box\ (\neg \mathit{down\_left\_side} \wedge \neg \mathit{down\_right\_side})$$

It states that, if there exists a pair which satisfies the antecedent

$$\mathit{down\_left\_side} \wedge \mathit{down\_right\_side} \wedge \mathit{lift\_left} \wedge \mathit{lift\_right}$$

then the next pair will satisfy the conjunction of both postconditions

$$\Box\ (\neg \mathit{down\_left\_side} \wedge \neg \mathit{down\_right\_side})$$

The conclusion that the next pair is a model of both postconditions is correct, because of the meaning of the □ operator. The single event descriptions express what necessarily happens after the event has occurred independent of other simultaneously occurring events.

The proper handling of conflicts is also a consequence of treating events on the predicate level. Consider the structure depicted in Figure 5 but without event e_3. Obviously this structure is also a model of both event descriptions. The derived sentence about the simultaneous occurrence is also correct. There is no pair in the structure which can satisfy the antecedent of the theorem. Gelfond *etal.* [3] present another example of conflicting actions in which they want to open and close the same door simultaneously. They propose that two actions are in conflict if the postconditions are inconsistent (see also Belgrinos and Georgeff [2]). This, however, is not a sufficient condition for conflicts. In our case the axiom for the parallel occurrence of the two actions would not be false, because the conjunction of the two actions makes the antecedent false. For the derivation of conflicts we propose to express that the conjunction of the two actions never holds.

In addition to describing the necessary effects caused by an event, we also want to express what remains unchanged. For instance, in Figure 5 we would like to express that the position of the right side of the table is not affected by lifting the left side. In situation calculus we would add an axiom

$$\mathit{down\_right\_side}(w) \rightarrow \mathit{down\_right\_side}(\mathit{result}(\mathit{lift\_left}, w))$$

to the system. However, this cannot be translated to

$$(\mathit{down\_right\_side} \wedge \mathit{lift\_left}) \rightarrow \Box \ \mathit{down\_right\_side}$$

The latter statement would be false for the pair (w_1, e_3). It is not necessarily true that the formula *down_right_side* holds in the next pair, because *lift_left* might be executed in parallel to *lift_right*. The reason is that in the traditional approach the formula *lift_left* denotes the entire event, thus a "closed event assumption" is applied. In our approach we just state that *lift_left* is satisfied in the upcoming event. The formula $\mathit{down\_right\_side} \wedge \mathit{lift\_left}$ is mapped by M to all pairs (w, e) where *down_right_side* is true in the state component and *lift_left* is a subevent of the next event.

However, viewing events like states and applying the standard connectives has very nice advantages. For instance, we can use the axiom

$$(\mathit{down\_right\_side} \wedge \neg \mathit{lift\_right}) \rightarrow \Box \ \mathit{down\_right\_side}$$

instead. This sentence makes a statement for the case that some event will not happen, i.e., if the table is on the ground on the right side and nobody lifts it there, then it will keep its position. This is similar to Haas' domain-specific frame axioms [6] or Schubert's explanation closure [13].

A further advantage is that we can combine frame axioms arbitrarily. For instance from the two sentences

$$(p \wedge a) \rightarrow \Box \ p$$
$$(r \wedge \neg c) \rightarrow \Box \ r$$

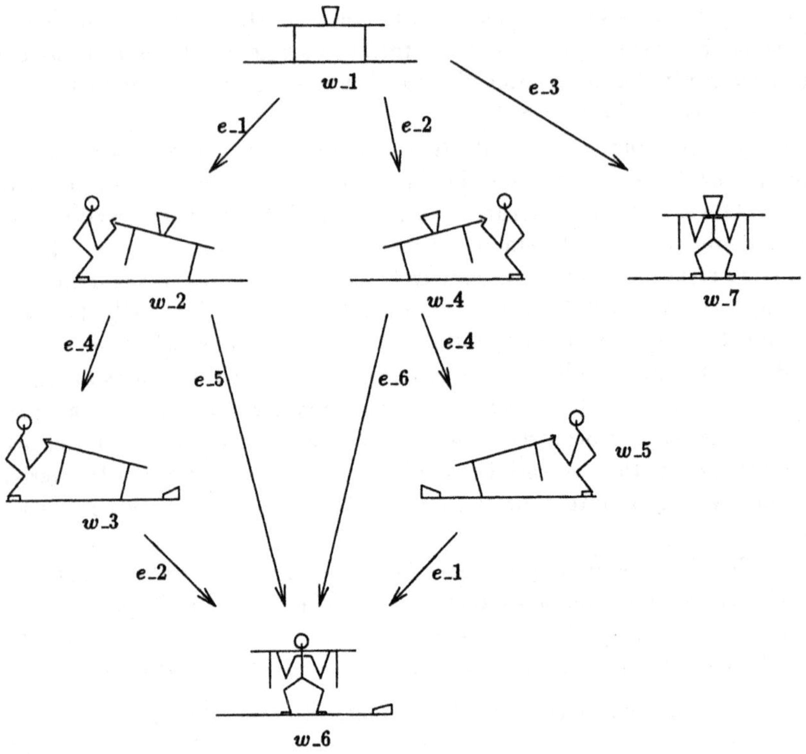

Fig. 6. Caused Events in the lifting scenario

we can conclude by using rule R and the laws of propositional logic

$$(p \wedge a \wedge r \wedge \neg c) \rightarrow \Box\ (p \wedge r)$$

Let us continue with a larger example in which the simultaneous occurrence of events has a side effect which is represented by a new event.

3.2 Lifting a Table with a Vase on Top of it

We turn to an example in which the parallel occurrence of events has an indirect effect. As we have mentioned before we can achieve such a case by putting a vase on the table. Lifting the table now on either side without lifting on the opposite side would cause the vase to fall as depicted in Figure 2. We extended the small example to a structure as depicted in Figure 6. The sets Φ_0 and Π_0 are as follows:

$$\Phi_0 = \{down\_left\_side,\ down\_right\_side,\ vase\_on\_table\}$$
$$\Pi_0 = \{lift\_left,\ lift\_right,\ vase\_falls\}$$

We take a structure $S = < W, E, T, R, M_1, M_2 >$ with

$$W = \{w_1, w_2, w_3, w_4, w_5, w_6, w_7\}$$
$$E = \{e_1, e_2, e_3, e_4, e_5, e_6, e_7\}$$
$$T = \left\{ \begin{array}{l} (w_1, e_1), (w_1, e_2), (w_1, e_3), \\ (w_2, e_4), (w_2, e_5), \\ (w_3, e_2), \\ (w_4, e_4), (w_2, e_6), \\ (w_5, e_1), \\ (w_6, e_7) \\ (w_7, e_7) \end{array} \right\}$$
$$R = \left\{ \begin{array}{l} [(w_1, e_1), (w_2, e_4)],\ [(w_1, e_1), (w_2, e_5)],\ [(w_1, e_2), (w_4, e_6)], \\ [(w_1, e_2), (w_4, e_4)],\ [(w_1, e_3), (w_7, e_7)], \\ [(w_2, e_4), (w_3, e_2)],\ [(w_2, e_5), (w_6, e_7)], \\ [(w_3, e_2), (w_6, e_7)], \\ [(w_4, e_6), (w_6, e_7)],\ [(w_4, e_4), (w_5, e_1)], \\ [(w_5, e_1), (w_6, e_7)] \end{array} \right\}$$

$$\begin{aligned} M_1(\mathit{down\_left\_side}) &= \{w_1, w_4, w_5\} \\ M_1(\mathit{down\_right\_side}) &= \{w_1, w_2, w_3\} \\ M_1(\mathit{vase\_on\_table}) &= \{w_1, w_2, w_4, w_7\} \\ M_2(\mathit{lift\_left}) &= \{e_1, e_3, e_6\} \\ M_2(\mathit{lift\_right}) &= \{e_2, e_3, e_5\} \\ M_2(\mathit{vase\_falls}) &= \{e_4, e_5, e_6\} \end{aligned}$$

From the mappings M_1 and M_2 we can form the mapping M for the atomic formulae as follows:

$$\begin{aligned} M(\mathit{lift\_left}) &= \{(w_1, e_1), (w_1, e_3), (w_4, e_6), (w_5, e_1)\} \\ M(\mathit{lift\_right}) &= \{(w_1, e_2), (w_1, e_3), (w_2, e_5), (w_3, e_2)\} \\ M(\mathit{vase\_falls}) &= \{(w_2, e_4), (w_2, e_5), (w_4, e_6), (w_4, e_4)\} \end{aligned}$$

$$\begin{aligned} M(\mathit{down\_left\_side}) &= \{(w_1, e_1), (w_1, e_2), (w_1, e_3), (w_4, e_4), (w_4, e_6), (w_5, e_1)\} \\ M(\mathit{down\_right\_side}) &= \{(w_1, e_1), (w_1, e_2), (w_1, e_3), (w_2, e_4), (w_2, e_5), (w_3, e_2)\} \\ M(\mathit{vase\_on\_table}) &= \left\{ \begin{array}{l} (w_1, e_1),\ (w_1, e_2),\ (w_1, e_3),\ (w_2, e_4),\ (w_2, e_5), \\ (w_4, e_4),\ (w_4, e_6),\ (w_7, e_7) \end{array} \right\} \end{aligned}$$

We represent the events shown in Figure 6 in a form slightly different than before. We break up the implication describing an event as a precondition implying a postcondition. Instead we provide two sorts of axioms: one sort for describing when events can or must occur and one sort for describing the effects of an event.

The following sentences express some conditions which hold for the occurrence of events. They can be divided into three groups. Firstly, we can restrict the events following a state by axioms such as

$$(down\_left\_side \wedge down\_right\_side) \rightarrow (lift\_left \vee lift\_right) \quad (I)$$

Secondly, we can specify the precondition in the traditional way, e.g.,

$$\begin{array}{ll} lift\_left \rightarrow down\_left\_side & (II) \\ lift\_right \rightarrow down\_right\_side & (III) \end{array}$$

And finally we can give sufficient preconditions as in the following example

$$\begin{array}{c} [\; vase\_on\_table \;\wedge \\ (\; (down\_left\_side \wedge \neg down\_right\_side) \\ \vee\; (\neg down\_left\_side \wedge down\_right\_side)\;)\;]\; (IV) \\ \leftrightarrow \\ vase\_falls \end{array}$$

The first three sentences contain conditions which have to hold for the occurrence of the agent's actions. Sentence *IV* determines exactly the states in which the vase will fall down. There is no $\Box$ operator in the formulae, because we formulate conditions on pairs.

The necessary effects of the events are described by the following sentences:

$$\begin{array}{ll} lift\_left \rightarrow \Box\; \neg down\_left\_side & (V) \\ lift\_right \rightarrow \Box\; \neg down\_right\_side & (VI) \\ vase\_falls \rightarrow \Box\; \neg vase\_on\_table & (VII) \end{array}$$

The falling of the vase is described as any other event. The $\Box$ operator is used, because we make a statement about the next pair.

The last category of sentences is concerned with describing the frame.

$$\begin{array}{lr} (down\_left\_side \wedge \neg lift\_left) \rightarrow \Box\; down\_left\_side & (VIII) \\ (down\_right\_side \wedge \neg lift\_right) \rightarrow \Box\; down\_right\_side & (IX) \\ (vase\_on\_table \wedge \neg vase\_falls) \rightarrow \Box\; vase\_on\_table & (X) \\ \neg down\_left\_side \rightarrow \Box\; \neg down\_left\_side & (XI) \\ \neg down\_right\_side \rightarrow \Box\; \neg down\_right\_side & (XII) \\ \neg vase\_on\_table \rightarrow \Box\; \neg vase\_on\_table & (XIII) \end{array}$$

The first three sentences describe which state formulae remain true unless the specified event happens. The last three implications express that the named atomic formulae do not change anymore.

Let us illustrate the theory by proving the following theorem:

$$\begin{array}{c} [lift\_left \wedge lift\_right \\ \wedge \;\; down\_left\_side \wedge down\_right\_side \wedge vase\_on\_table] \\ \rightarrow \\ \Box\; (\neg down\_left\_side \wedge \neg down\_right\_side \wedge vase\_on\_table) \end{array}$$

With respect to Figure 6 this sentence expresses that the occurrence of event e_3 leads from state w_1 to w_7.
Proof:

$$(1) \quad (down\_left\_side \wedge down\_right\_side \wedge vase\_on\_table) \rightarrow \neg vase\_falls \qquad (PL; IV)$$

$$(2) \quad (down\_left\_side \wedge down\_right\_side \wedge vase\_on\_table) \rightarrow (\neg vase\_falls \wedge vase\_on\_table) \qquad (PL; 1)$$

$$(3) \quad (down\_left\_side \wedge down\_right\_side \wedge vase\_on\_table) \rightarrow \Box\ vase\_on\_table \qquad (PL; 2, X)$$

$$(4) \quad (lift\_left \wedge lift\_right) \rightarrow \Box\ (\neg down\_left\_side \wedge \neg down\_right\_side) \qquad (PL, R; V, VI)$$

$$(5) \quad [lift\_left \wedge lift\_right \wedge down\_left\_side \wedge down\_right\_side \wedge vase\_on\_table] \rightarrow \Box\ (\neg down\_left\_side \wedge \neg down\_right\_side \wedge vase\_on\_table) \qquad (PL, R; 3, 4)$$

This proof shall give us a flavor of how proofs look like in our logic. Because of space limitations we have refer the reader to the long version of the paper. However, we want to mention another theorem which can be proved:

$$(lift\_left \wedge \neg lift\_right \wedge down\_left\_side \wedge down\_right\_side \wedge vase\_on\_table) \rightarrow \Box\ [lift\_right \rightarrow \Box\ (\neg down\_left\_side \wedge \neg down\_right\_side \wedge \neg vase\_on\_table)]$$

In Figure 6 this sentence expresses that the subsequent occurrence of the events e_1 and e_5 leads from state w_1 to w_6.

3.3 Interchanging Blocks

Suppose, we had two blocks A and B being at the positions f and g, respectively. The problem is now to interchange the positions of the two blocks. There are (at least) two ways to accomplish this: In the first case we move one of the blocks (say A) to some free intermediate position h and afterwards we move block B to position f. Finally, we move A to g. In the second case we know how to do it without using an intermediate position. For instance, we could take one block in each hand and change their positions simultaneously.

The first solution can be represented in any previous formalism. We provide axioms for movement of blocks in which we require that the location to which we want to move the block is free. The second method is more difficult. We cannot require anymore that the location to which we want to move is free beforehand. Because of the parallel movement the destinations become free. Thus for solving this problem we have to refer to the exact composition of subevents. Formally,

$$(move\_A\_from\_f\_to\_g \wedge move\_B\_from\_g\_to\_f) \rightarrow (A\_at\_f \wedge B\_at\_g)$$

If the agent is able to execute more actions than the two described, then we have to extend the description appropriately. For illustrating our solution the present description is sufficient.

The effects of each action can be separately described in two general axioms.

$$move\_A\_from\_f\_to\_g \rightarrow \Box \; (A\_at\_g \wedge \neg A\_at\_f)$$
$$move\_B\_from\_g\_to\_f \rightarrow \Box \; (B\_at\_f \wedge \neg B\_at\_g)$$

4 Related Work

The work mostly related to ours seems to be the one of [9]. They established the symmetry between states and events in the situation calculus (which is first-order predicate logic). It seems to us that most of our sentences can be expressed in their approach. However, it is difficult to keep the simplicity and elegance of our language. For instance, the expression "event E or F" (formally $E \vee F$), will turn into "primitive action E is in global action G or primitive action F is in global action G" (formally $In(E, G) \vee In(F, G)$). One real difference, however, is our handling of indirect effects. Lin and Shoham need to adapt their theorem prover for including the effects of caused events. We can use a common modal logic theorem prover.

From the logical point of view, dynamic logic is closely related. Whereas all statements which are interesting with respect to dynamic worlds can be transformed from dynamic logic to state event logic, the other direction is not possible. For instance, consider the expression "Event E" (formally E). It means in our logic that E is satisfied in all events of the structure. This statement is reasonable when we want to express the progress of time. At least in propositional dynamic logic this is not expressible, maybe not even in first-order dynamic logic.

5 Acknowledgments

I would like to thank Johanna Wiesmet, Ute Sigmund and Hesham Khalil for many enlightening comments and improvements on previous versions of this paper.

References

1. C. Baral and M. Gelfond. Representing Concurrent Actions in Extended Logic Programming. International Joint Conference on Artificial Intelligence, Morgan Kaufmann Publishers, Inc., 1993.

2. P. Belegrinos and M. Georgeff. A Model of Events and Processes. International Joint Conference on Artificial Intelligence, p. 506–511. Morgan Kaufmann Publishers, Inc., August 1991.
3. M. Gelfond, V. Lifschitz, and A. Rabinov. What are the Limitations of the Situation Calculus. Automated Reasoning — Essays in Honor of Woody Bledsoe, p. 167–179, Kluwer Academic Publishers, 1991.
4. M. P. Georgeff. Actions, Processes, and Causality. Proceedings of the Workshop on Reasoning about Actions and Plans, eds. M. P. Georgeff and A. L. Lansky, p. 99–122, 1986.
5. G. Große and R. Waldinger. Towards a Theory of Simultaneous Actions. Proceedings of the European Workshop on Planning, ed. J. Hertzberg, p. 78–87, Springer LNAI 522, 1991.
6. A. Haas. The Case for Domain-Specific Frame Axioms. Proceedings of the Workshop on The Frame Problem in Artificial Intelligence, ed. F. Brown, 1987.
7. D. Harel. Dynamic Logic. Handbook of Philosophical Logic, eds. F. Günthner and D. Gabbay, D. Reidel, 1984.
8. G. Hendrix. Modeling Simultaneous Actions and Continuous Processes. Artificial Intelligence 4, p. 145–180, 1973.
9. F. Lin and Y. Shoham. Concurrent actions in the situation calculus. Proceedings of the AAAI National Conference on Artificial Intelligence, 1992.
10. E. P. D. Pednault. Solving Multi-Agent Dynamic World Problems in the Classical Planning Framework. Proceedings of the Workshop on Reasoning about Actions and Plans, eds. M. P. Georgeff and A. L. Lansky, 1986.
11. V. R. Pratt. Semantical Considerations on Floyd-Hoare Logic. Proceedings of the 17th IEEE Symposium on Foundations of Computer Science, 1976.
12. E. Sandewall and R. Rönnquist. A Representation of Action Structures. Proceedings of the AAAI National Conference on Artificial Intelligence, p. 89–97, 1986.
13. L. Schubert. Monotonic Solution of The Frame Problem in The Situation Calculus. Knowledge Representation and Defeasible Reasoning, eds. Jr. etal H.E. Kyberg, pages 23–67, Kluwer Academic Publishers, 1990.
14. D. E. Wilkins. Practical Planning: Extending the Classical AI Planning Paradigm. Morgan Kaufmann Publishers, Inc., San Mateo, CA, 1988.

Description Logics with Inverse Roles, Functional Restrictions, and N-ary Relations

Giuseppe De Giacomo and Maurizio Lenzerini

Dipartimento di Informatica e Sistemistica, Università di Roma "La Sapienza"
Via Salaria 113, 00198 Roma, Italia
e-mail: [degiacom,lenzerini]@assi.dis.uniroma1.it

Abstract. Description Logics (DLs) are used in Artificial Intelligence to represent knowledge in terms of objects grouped into classes, and offer structuring mechanisms for both characterizing the relevant properties of classes in terms of binary relations, and establishing several interdependencies among classes. One of the main themes in the area of DLs has been to identify DLs that are both very expressive and decidable. This issue can be profitably addressed by relying on a correspondence between DLs and propositional dynamic logics (PDLs). In this paper, we exploit the correspondence as a framework to investigate the decidability and the complexity of a powerful DL, in which functional restrictions on both atomic roles and their inverse are expressible. We then show that such DL is suitable to represent n-ary relations, as needed in the applications of class-based formalisms to databases. The PDL that we use in this work is a proper extension of Converse Deterministic PDL, and its decidability and complexity is established contextually.

1 Introduction

The research in Artificial Intelligence and Computer Science has always paid special attention to formalisms for the structured representation of classes and relations. In Artificial Intelligence, the investigation on such formalisms began with semantic networks and frames, which have been influential for many formalisms proposed in the areas of knowledge representation, data bases, and programming languages, and developed towards formal logic-based languages, that will be called here *description logics*[1] (DLs). Generally speaking, DLs are decidable logics specifically designed for allowing the representation of knowledge in terms of objects (individuals) grouped into classes (concepts), and offer structuring mechanisms for characterizing the relevant properties of classes in terms of relations (roles).

Description logics have been the subject of many investigations in the last decade. It is our opinion that the main reason for investigating such logics is that they offer a clean, formal and effective framework for analyzing several important issues related to class-based representation formalisms, such as expressive power, deduction algorithms, and computational complexity of reasoning. This

[1] Terminological logics, and concept languages are other possible names.

is confirmed by the fact that the research in DLs both produced several theoretical results (see [22] for an overview), and has been influential for the design of knowledge representation systems, like LOOM [19], CLASSIC [6], and KRIS [2].

In order to use DLs as abstract formalisms for addressing diverse issues related to class-based representation schemes, they should be sufficiently general, and, at the same time, sufficiently simple so as to not fall into undecidability of reasoning. Currently, however, those DLs that have been studied from a formal point of view suffer from several limitations:

1. Relationships between classes are modeled by binary relations (roles), while n-ary relations are not supported.
2. They do not allow the modeler to refer to the inverse of a binary relation; or, if they do, they impose several restrictions in the usage of inverse relations (for example, although in general one can state that a relation is actually a function, there is no possibility to state that the inverse of a relation is a function).
3. While they offer a rich variety of constructs for building class descriptions (i.e. expressions denoting classes), they do not generally allow one to represent universal properties of classes (such as: all instances of class A must be related to at least another instance of A by means of the relation R).

All the above limitations prevent one to consider DLs general enough to capture a sufficiently broad family of class-based representation formalisms. Indeed:

1. Nonbinary relations are important in general and in particular for capturing conceptual and semantic database models (see [16]).
2. Inverse relations are essential in domain modeling (see [29]), for example, without the possibility of referring to the inverse of a relation, we are forced to use two unrelated relations *child* and *parent*, with no chance of stating their mutual dependency; also, in case inverse relations can be used in the DL, they should be used as any other relation (for example, it should be possible to state that the inverse of a relation is actually a function, analogously to the case of direct relations).
3. The possibility of expressing universal properties of classes is a basic feature for capturing both conceptual and object-oriented data models (see [9]).

Our goal in this paper is to propose a very expressive DL that both supports all the above features, and such that reasoning in the logic is decidable. To this end, we resort to the work by Schild [26], which singled out an interesting correspondence between DLs and several propositional dynamic logics (PDL), which are modal logics specifically designed for reasoning about program schemes. The correspondence is based on the similarity between the interpretation structures of the two logics: at the extensional level, objects in DLs correspond to states in PDLs, whereas connections between two objects correspond to state transitions. At the intensional level, classes correspond to propositions, and roles corresponds to programs. The correspondence is extremely useful mainly for two reasons. On one hand, it makes clear that reasoning about assertions on classes is equivalent to reasoning about dynamic logic formulae. On the other hand, the large body of research on decision procedures in PDL (see, for example, [17]) can be exploited in the setting of DLs, and, on the converse, the various works on tractability/intractability of DLs (see for example [14]) can be used in the setting of PDL.

We argue that the work on PDLs is a good starting point for our investigation, because it provides us with:

- a general method for reasoning with universal properties of classes;
- a general method for reasoning with inverses of relations (indeed, several PDLs proposed in the literature, include a construct that exactly correspond to the inverse of relations).

However, looking carefully at the expressive power of PDLs, it turns out that the following problems need to be addressed:

1. No existing PDL allows one to impose that the inverse of relation is functional.
2. No existing PDL provides a construct that can be directly used to model nonbinary relations.

In this paper we present a solution to such problems. The solution is based on a particular methodology, which we believe has its own value: the inference in DLs is formulated in the setting of PDL, and in order to represent functional restrictions on relations (both direct and inverse), special "constraints" are added to the PDL formulae. The solution to the problem of expressing functional restrictions on both direct and inverse roles directly leads to a method for incorporating n-ary relation in DLs.

The results have a twofold significance. From the standpoint of DLs, we derive decidability and complexity results for one of the most expressive DLs appeared in the literature, and from the standpoint of PDLs, we define a very powerful PDL (it subsumes Converse Deterministic PDL), and establish its decidability and complexity by a methodology that can be exploited to derive reasoning procedures for many extensions of known PDLs - e.g. PDLs including several forms of program determinism.

The paper is organized as follows. In Section 2, we recall the basic notions of both DLs and PDLs. In Section 3, we present the result on functional restrictions, showing that Converse PDL is powerful enough to allow the representation of functional restrictions on both direct and inverse roles. In Section 4, we deal with the problem of representing n-ary relations in DLs. Finally, Section 5 ends the paper with some conclusions. For the sake of brevity, all proves are omitted.

2 Preliminaries

We base our work on two logics, namely the DL $\mathcal{CI}$, and the PDL $\mathcal{DI}$ (traditionally called Converse PDL), whose basic characteristics are recalled in this section.

The formation rules of $\mathcal{CI}$ are specified by the following abstract syntax

$$C \longrightarrow \top \mid \bot \mid A \mid C_1 \sqcap C_2 \mid C_1 \sqcup C_2 \mid C_1 \Rightarrow C_2 \mid \neg C \mid \exists R.C \mid \forall R.C$$
$$R \longrightarrow P \mid R_1 \sqcup R_2 \mid R_1 \circ R_2 \mid R^* \mid R^- \mid id(C)$$

where A denotes an atomic concept, C (possibly with subscript) denotes a concept, P denotes an atomic role, and R (possibly with subscript) denotes a role. The semantics of concepts is the usual one: an interpretation $\mathcal{I} = (\Delta^{\mathcal{I}}, \cdot^{\mathcal{I}})$ consists of a domain $\Delta^{\mathcal{I}}$, and an interpretation function $\cdot^{\mathcal{I}}$ that assigns subsets of $\Delta^{\mathcal{I}}$ to concepts and binary relations over $\Delta^{\mathcal{I}}$ to roles as follows:

$$\begin{array}{l}
A^{\mathcal{I}} \subseteq \Delta^{\mathcal{I}}, \\
\top^{\mathcal{I}} = \Delta^{\mathcal{I}}, \\
\bot^{\mathcal{I}} = \emptyset, \\
(\neg C)^{\mathcal{I}} = \Delta^{\mathcal{I}} - C^{\mathcal{I}}, \\
(C_1 \sqcap C_2)^{\mathcal{I}} = C_1^{\mathcal{I}} \cap C_2^{\mathcal{I}}, \\
(C_1 \sqcup C_2)^{\mathcal{I}} = C_1^{\mathcal{I}} \cup C_2^{\mathcal{I}}, \\
(C_1 \Rightarrow C_2)^{\mathcal{I}} = (\neg C_1)^{\mathcal{I}} \cup C_2^{\mathcal{I}}, \\
(\exists R.C)^{\mathcal{I}} = \{d \in \Delta^{\mathcal{I}} \mid \exists d'.(d,d') \in R^{\mathcal{I}} \text{ and } d' \in C^{\mathcal{I}}\}, \\
(\forall R.C)^{\mathcal{I}} = \{d \in \Delta^{\mathcal{I}} \mid \forall d'.(d,d') \in R^{\mathcal{I}} \text{ implies } d' \in C^{\mathcal{I}}\}, \\
P^{\mathcal{I}} \subseteq \Delta^{\mathcal{I}} \times \Delta^{\mathcal{I}}, \\
(R_1 \sqcup R_2)^{\mathcal{I}} = R_1^{\mathcal{I}} \cup R_2^{\mathcal{I}}, \\
(R_1 \circ R_2)^{\mathcal{I}} = R_1^{\mathcal{I}} \circ R_2^{\mathcal{I}}, \\
(R^*)^{\mathcal{I}} = (R^{\mathcal{I}})^*, \\
(R^-)^{\mathcal{I}} = \{(d_1,d_2) \in \Delta^{\mathcal{I}} \times \Delta^{\mathcal{I}} \mid (d_2,d_1) \in R^{\mathcal{I}}\}, \\
id(C)^{\mathcal{I}} = \{(d,d) \in \Delta^{\mathcal{I}} \times \Delta^{\mathcal{I}} \mid d \in C^{\mathcal{I}}\}.
\end{array}$$

Note that $\mathcal{CI}$ is a very expressive language, comprising all usual concept constructs, and a rich set of role constructs, namely: union of roles $R_1 \sqcup R_2$, chaining of roles $R_1 \circ R_2$, transitive closure of roles R^*, inverse roles R^-, and the identity role $id(C)$ projected on C.

A ($\mathcal{CI}$) TBox (i.e., intensional knowledge base in $\mathcal{CI}$) is defined as a finite set $\mathcal{K}$ of inclusion assertions of the form $C_1 \sqsubseteq C_2$, where C_1, C_2 are $\mathcal{CI}$-concepts. An assertion $C_1 \sqsubseteq C_2$ is satisfied by an interpretation $\mathcal{I}$ if $C_1^{\mathcal{I}} \subseteq C_2^{\mathcal{I}}$, and $\mathcal{I}$ is a model of $\mathcal{K}$ if every assertion of $\mathcal{K}$ is satisfied by $\mathcal{I}$. A TBox $\mathcal{K}$ logically implies an assertion $C_1 \sqsubseteq C_2$, written $\mathcal{K} \models C_1 \sqsubseteq C_2$, if $C_1 \sqsubseteq C_2$ is satisfied by every model of $\mathcal{K}$.

There is a direct correspondence between $\mathcal{CI}$ and the PDL $\mathcal{DI}$, whose syntax is as follows:

$$\phi \longrightarrow \mathit{true} \mid \mathit{false} \mid A \mid \phi_1 \wedge \phi_2 \mid \phi_1 \vee \phi_2 \mid \phi_1 \Rightarrow \phi_2 \mid \neg\phi_1 \mid < r > \phi_1 \mid [r]\phi_1$$
$$r \longrightarrow P \mid r_1 \cup r_2 \mid r_1 ; r_2 \mid r^* \mid r^- \mid \phi?$$

where A denotes a propositional letter, ϕ (possibly with subscript) denotes a formula, P denotes an atomic program, and r (possibly with subscript) denotes a program. The semantics of $\mathcal{DI}$ is based on the notion of (Kripke) structure, which is defined as a triple $M = (\mathcal{S}, \{\mathcal{R}_P\}, \Pi)$, where $\mathcal{S}$ denotes a set of states, $\{\mathcal{R}_P\}$ is a family of binary relations over $\mathcal{S}$ such that each atomic program P is given a meaning through R_P, and Π is a mapping from $\mathcal{S}$ to propositional letters such that $\Pi(s)$ determines the letters that are true in the state s. Given M, the family $\{\mathcal{R}_P\}$ can be extended in the obvious way so as to include, for every program r, the corresponding relation $\mathcal{R}_r$ (for example, $\mathcal{R}_{r_1;r_2}$ is the composition of $\mathcal{R}_{r_1}$ and $\mathcal{R}_{r_2}$). For this reason, we often denote a structure by $(\mathcal{S}, \{\mathcal{R}_r\}, \Pi)$, where $\{\mathcal{R}_r\}$ includes a binary relations for every program (atomic or non-atomic). A structure M is called a model of a formula ϕ if there exists a state s in M such that $M, s \models \phi$. A formula ϕ is satisfiable if there exists a model of ϕ, unsatisfiable otherwise.

The correspondence between $\mathcal{CI}$ and $\mathcal{DI}$, first pointed out by Schild [26], is based on the similarity between the interpretation structures of the two logics: at the extensional level, individuals (members of $\Delta^{\mathcal{I}}$) in DLs correspond to states in PDLs, whereas connections between two individuals correspond to state transitions. At the intensional level, classes correspond to propositions, and roles corresponds to programs. The correspondence is realized through a (one-to-one and onto) mapping δ from $\mathcal{CI}$-concepts to $\mathcal{DI}$-formulae, and from $\mathcal{CI}$-roles to $\mathcal{DI}$-programs. The mapping δ is defined inductively as follows (we assume $\sqcup, \Rightarrow$ to be expressed by means of $\sqcap, \neg$):

$$\begin{array}{ll}
\delta(A) = A & \delta(P) = P \\
\delta(C_1 \sqcap C_2) = \delta(C_1) \wedge \delta(C_2) & \delta(\neg C) = \neg\delta(C) \\
\delta(\exists R.C) =< \delta(R) > \delta(C) & \delta(\forall R.C) = [\delta(R)]\delta(C) \\
\delta(R_1 \sqcup R_2) = \delta(R_1) \cup \delta(R_2) & \delta(R_1 \circ R_2) = \delta(R_1); \delta(R_2) \\
\delta(R^*) = \delta(R)^* & \delta(id(C)) = \delta(C)? \\
\delta(R^-) = \delta(R)^-. &
\end{array}$$

From δ one can easily obtain a mapping δ^+ from $\mathcal{CI}$-TBoxes to $\mathcal{DI}$-formulae. Namely, if $\mathcal{K} = \{K_1, \cdots, K_n\}$ is a TBox in $\mathcal{CI}$, and $P_1, \ldots, P_m$ are all atomic roles appearing in $\mathcal{K}$, then

$$\begin{array}{l}
\delta^+(\mathcal{K}) = [(P_1 \cup \cdots \cup P_m \cup P_1^- \cdots \cup P_m^-)^*]\, \delta^+(\{K_1\}) \wedge \cdots \wedge \delta^+(\{K_n\}), \\
\delta^+(\{C_1 \sqsubseteq C_2\}) = (\delta(C_1) \Rightarrow \delta(C_2)).
\end{array}$$

Observe that $\delta^+(\mathcal{K})$ exploits the power of program constructs (union, converse, and transitive closure) and the "connected model property"[2] of PDLs in order to represent inclusion assertions of DLs. Based on this correspondence, we can state the following: if $\mathcal{K}$ is a TBox, then $\mathcal{K} \models C_1 \sqsubseteq C_2$ (where atomic concepts and roles in C_1, C_2 are also in $\mathcal{K}$) iff the $\mathcal{DI}$-formula

$$\delta^+(\mathcal{K}) \wedge \delta(C_1) \wedge \delta(\neg C_2)$$

is unsatisfiable. Note that the size of the above formula is polynomial with respect to the size of $\mathcal{K}, C_1$ and C_2.

By virtue of δ and δ^+, respectively, both satisfiability of $\mathcal{CI}$ concepts, and logical implication for $\mathcal{CI}$-TBoxes, can be (polynomially) reduced to satisfiability of $\mathcal{DI}$-formulae. Being satisfiability for $\mathcal{DI}$ an EXPTIME-complete problem,

[2] That is, if a formula has a model, it has a model which is connected.

so are satisfiability of $\mathcal{CI}$-concepts and logical implication for $\mathcal{CI}$-TBoxes. It is straightforward to extend the correspondence, and hence both δ and δ^+, to other DLs and PDLs.

In the rest of this section, we introduce several notions and notations that will be used in the sequel.

The *Fisher-Ladner closure* ([10]) of a $\mathcal{DI}$-formula Φ, denoted $CL(\Phi)$, is the least set F such that $\Phi \in F$ and such that (we assume, without loss of generality, $\vee, \Rightarrow, [\cdot]$ to be expressed by means of $\neg, \wedge, <\cdot>$, and the converse operator to be applied to atomic programs only[3]):

$$\begin{array}{ll}
\phi_1 \wedge \phi_2 \in F & \Rightarrow \phi_1, \phi_2 \in F, \\
\neg\phi \in F & \Rightarrow \phi \in F, \\
<r>\phi \in F & \Rightarrow \phi \in F, \\
<r_1;r_2>\phi \in F & \Rightarrow <r_1><r_2>\phi \in F, \\
<r_1 \cup r_2>\phi \in F & \Rightarrow <r_1>\phi, <r_2>\phi \in F, \\
<r^*>\phi \in F & \Rightarrow <r><r^*>\phi \in F, \\
<\phi'?>\phi \in F & \Rightarrow \phi' \in F.
\end{array}$$

The notion of Fisher-Ladner closure can be easily extended to formulae of other PDLs.

A *path* in a structure M (sometimes called *trajectory*) is a sequence $(s_0, \ldots, s_q)$ of states of M, such that for each $i = 1, \ldots, q$, $(s_{i-1}, s_i) \in \mathcal{R}_a$ for some $a = P \mid P^-$. The length of $(s_0, \ldots, s_q)$ is q. We inductively define the set of paths $Paths(r)$ of a program r in a structure M, as follows (we assume, without loss of generality, that in r all occurrences of the converse operator are moved all way in):

$$\begin{array}{ll}
Paths(a) & = \mathcal{R}_a \ (a = P \mid P^-), \\
Paths(r_1 \cup r_2) & = Paths(r_1) \cup Paths(r_2), \\
Paths(r_1; r_2) & = \{(s_0, \ldots, s_u, \ldots, s_q) \mid (s_0, \ldots, s_u) \in Paths(r_1) \\
& \quad \text{ and } (s_u, \ldots, s_q) \in Paths(r_2)\}, \\
Paths(r^*) & = \{(s) \mid s \in \mathcal{S}\} \cup (\bigcup_{i>0} Paths(r^i)), \\
Paths(\phi'?) & = \{(s) \mid M, s \models \phi'\}.
\end{array}$$

We say that a path (s_0) in M *satisfies* a formula ϕ which is not of the form $<r>\phi'$ if $M, s_0 \models \phi$. We say that a path $(s_0, \ldots, s_q)$ in M *satisfies* a formula ϕ of the form $<r_1> \cdots <r_l>\phi'$, where ϕ' is not of the form $<r'>\phi''$, if $M, s_q \models \phi'$ and $(s_0, \ldots s_q) \subseteq Paths(r_1; \cdots; r_l)$.

Finally, if a denotes the atomic program P (resp. the inverse of an atomic program P^-), then we write a^- to denote P^- (resp. P).

3 Functional Restrictions

In this section, we consider an extension of $\mathcal{CI}$, called $\mathcal{CIF}$, which is obtained from $\mathcal{CI}$ by adding the concept construct $(\leq 1\, a)$, where $a = P \mid P^-$. The meaning of the construct in an interpretation $\mathcal{I}$ is the following:

[3] We recall that the following equations hold: $(r_1;r_2)^- = r_1^-;r_2^-$, $(r_1 \cup r_2)^- = r_1^- \cup r_2^-$, $(r_1^*)^- = (r_1^-)^*$, $(\phi?)^- = (\phi?)$.

$(\leq 1\ a)^{\mathcal{I}} = \{d \in \Delta^{\mathcal{I}} \mid$ there exists *at most one* d' such that $(d, d') \in a^{\mathcal{I}}\}$.

The corresponding PDL will be called $\mathcal{DIF}$, and is obtained from $\mathcal{DI}$ by adding the same construct $(\leq 1\ a)$, where, again, $a = P \mid P^-$, whose meaning in $\mathcal{DIF}$ can be immediately derived by the semantics of $\mathcal{CIF}$. Observe that the construct $(\leq 1\ a)$ allows the notion of local determinism for both atomic programs and their converse to be represented in PDL. With this construct, we can denote states from which the running of an atomic program or the converse of an atomic program is deterministic, i.e., it leads to at most one state. It is easy to see that this possibility allows one to impose the so-called global determinism too, i.e., that a given atomic program, or the converse of an atomic program, is (globally) deterministic. Therefore, $\mathcal{DIF}$ subsumes the logic studied in [30], called Converse Deterministic PDL where atomic programs, *not their converse*, are (globally) deterministic.

From the point of view of DLs, as mentioned in the Introduction, the fact that in the $(\leq 1\ a)$ construct, a can be either an atomic role or the inverse of an atomic role, greatly enhances the expressive power of the logic, and makes $\mathcal{CIF}$ one of the most expressive DLs among those studied in the literature.

The decidability and the complexity of both satisfiability of $\mathcal{CIF}$-concepts and logical implication for $\mathcal{CIF}$-TBox, can be derived by exploiting the correspondence between $\mathcal{CIF}$ and $\mathcal{DIF}$. This is realized through the mappings δ and δ^+ described in Section 2, suitably extended in order to deal with functional restrictions.

Note however that the decidability and the complexity of satisfiability in $\mathcal{DIF}$ are to be established, yet. We establish them below by showing an encoding of $\mathcal{DIF}$-formulae in $\mathcal{DI}$. More precisely we show that, for any $\mathcal{DIF}$-formula Φ, there is a $\mathcal{DI}$-formula, denoted $\gamma(\Phi)$, whose size is polynomial with respect to the size of Φ, and such that Φ is satisfiable iff $\gamma(\Phi)$ is satisfiable. Since satisfiability in $\mathcal{DI}$ is EXPTIME-complete, this ensures us that satisfiability in $\mathcal{DIF}$, and therefore both satisfiability of $\mathcal{CIF}$-concepts and logical implication for $\mathcal{CIF}$-TBoxes, are EXPTIME-complete too.[4] In what follows, we assume, without loss of generality, that $\mathcal{DIF}$-formula Φ is in negation normal form (i.e., negation is pushed inside as much as possible). We define the *$\mathcal{DI}$-counterpart* $\gamma(\Phi)$ of the $\mathcal{DIF}$-formula Φ as the conjunction of two formulae, $\gamma(\Phi) = \gamma_1(\Phi) \wedge \gamma_2(\Phi)$, where:

- $\gamma_1(\Phi)$ is obtained from Φ by replacing each $(\leq 1\ a)$ with a new propositional letter $A_{(\leq\ 1\ a)}$, and each $\neg(\leq 1\ a)$ with $(< a > H_{(\leq\ 1\ a)}) \wedge (< a > \neg H_{(\leq\ 1\ a)})$, where $H_{(\leq\ 1\ a)}$ is again a new propositional letter.
- $\gamma_2(\Phi) = [(P_1 \cup \cdots \cup P_m \cup P_1^- \cdots \cup P_m^-)^*]\gamma_2^1 \wedge \cdots \wedge \gamma_2^q$, where $P_1, \ldots, P_m$ are all atomic roles appearing in Φ, and with one conjunct γ_2^i of the form

$$((A_{(\leq\ 1\ a)} \wedge < a > \phi) \Rightarrow [a]\phi)$$

for every $A_{(\leq\ 1\ a)}$ occurring in $\gamma_1(\Phi)$ and every $\phi \in CL(\gamma_1(\Phi))$.

[4] Indeed $\gamma(\delta^+(\mathcal{K}) \wedge \delta(C_1) \wedge \delta(\neg C_2))$ is the $\mathcal{DIF}$-formula corresponding to the implication problem $\mathcal{K} \models C_1 \sqsubseteq C_2$ for $\mathcal{CIF}$-TBoxes.

Intuitively $\gamma_2(\Phi)$ constrains the models M of $\gamma(\Phi)$ so that: for every state s of M, if $A_{(\leq 1\ a)}$ holds in s, and there is an a-transition from s to t_1 and an a-transition from s to t_2, then t_1 and t_2 are equivalent wrt the formulae in $CL(\gamma_1(\Phi))$. We show that this allow us to actually collapse t_1 and t_2 into a single state.

To prove that a $\mathcal{DIF}$-formula is satisfiable iff its $\mathcal{DI}$-counterpart is, we proceed as follows. Given a model $M = (\mathcal{S}, \{R_r\}, \Pi)$ of $\gamma(\Phi)$, we build a tree-like structure $M^t = (\mathcal{S}^t, \{R_r^t\}, \Pi^t)$ such that $M^t, root \models \gamma(\Phi)$ ($root \in \mathcal{S}^t$ is the root of the tree-structure), and the local determinism requirements are satisfied. From such M^t, a model $M^t_{\mathcal{F}}$ of Φ can easily be derived. In order to construct M^t we make use of the following notion: For each state s in M, we call by $ES(s)$ the smallest set of states in M such that

- $s \in ES(s)$, and
- if $s' \in ES(s)$, then for every s'' such that $(s', s'') \in \mathcal{R}_{a;A_{(\leq 1\ a^-)}?;a^-}$, $ES(s'') \subseteq ES(s)$.

The set $ES(s)$ is the set of states of M that are going to be collapsed into a single state of M^t. Note that, by $\gamma_2(\Phi)$, all the states in $ES(s)$ satisfy the same formulae in $CL(\gamma_1(\Phi))$. The construction of M^t is done in three stages.

Stage 1. Let $< a_1 > \psi_1, \ldots, < a_h > \psi_h$ be all the formulas of the form $< a > \phi'$ included in $CL(\Phi)$.[5] We consider an infinite h-ary tree $\mathcal{T}$ whose root is *root* and such that every node x has h children $child_i(x)$, one for each formula $< a_i > \psi_i$. We write $father(x)$ to denote the father of a node x in $\mathcal{T}$. We define two partial mappings m and l: m maps nodes of $\mathcal{T}$ to states of M, and l is used to label the arcs of $\mathcal{T}$ by either atomic programs, converse of atomic programs, or a special symbol 'undefined'. For the definition of m and l, we proceed level by level. Let $s \in \mathcal{S}$ be any state such that $M, s \models \gamma(\Phi)$. We put $m(root) = s$, and for all arcs $(root, child_i(root))$ corresponding to a formula $< a_i > \psi_i$ such that $M, s \models < a_i > \psi_i$ we put $l((root, child_i(root))) = a_i$. Suppose we have defined m and l up to level k, let x be a node at level $k+1$, and let $l((father(x), x)) = a_j$. Then $M, m(father(x)) \models < a_j > \psi_j$, and therefore, there exists a path $(s_o, s_1, \ldots, s_q)$, with $s_o = m(father(x))$ satisfying $< a_j > \psi_j$. Among the states in $ES(s_1)$ we choose a state t such that there exists a *minimal* path (i.e., a path with minimal length) from t satisfying ψ_j. We put $m(x) = t$ and for every $< a_i > \psi_i \in CL(\Phi)$ such that $M, t \models < a_i > \psi_i$ we put $l((x, child_i(x))) = a_i$.

Stage 2. We change the labeling l, proceeding level by level. If $M, m(root) \models A_{(\leq 1\ a)}$, then for each arc $(root, child_i(root))$ labeled a, except for one randomly chosen, we put $l((root, child_i(root))$ = 'undefined'. Assume we have modified l up to level k, and let x be a node at level $k+1$. Suppose $M, m(x) \models A_{(\leq 1\ a)}$. Then if $l((father(x), x)) = a^-$, for each arc $(x, child_i(x))$ labeled a, we put $l((x, child_i(x))$ = 'undefined', otherwise (i.e. $l((father(x), x)) \neq a^-$) we put $l((x, child_i(x))$ = 'undefined' for every arc $(x, child_i(x))$ labeled a, except for one randomly chosen.

Stage 3. For each P, let $\mathcal{R}'_P = \{(x, y) \in \mathcal{T} \mid l((x, y)) = P \text{ or } l((y, x)) = P^-\}$. We define the structure $M^t = (\mathcal{S}^t, \{\mathcal{R}_r^t\}, \Pi^t)$ as follows: $\mathcal{S}^t = \{x \in$

[5] Notice that the formulas ψ_i may be of the form $< r > \phi$, and that $\psi_i \in CL(\Phi)$.

$\mathcal{T} \mid (root, x) \in (\bigcup_P(\mathcal{R}'_P \cup \mathcal{R}'^{-}_P))^*\}$, $\mathcal{R}^t_P = \mathcal{R}'_P \cap (\mathcal{S}^t \times \mathcal{S}^t)$, and $\Pi^t(x) = \Pi(m(x))$ $(\forall x. x \in \mathcal{S}^t)$. From $\{\mathcal{R}^t_P\}$ we get all $\{\mathcal{R}^t_r\}$ as usual.

The basic property of M^t is stated in the following lemma.

Lemma 1. *Let Φ be a $\mathcal{DIF}$-formula, and let M be a model of $\gamma(\Phi)$. Then, for every formula $\phi \in CL(\gamma_1(\Phi))$ and every $x \in \mathcal{S}^t$, $M^t, x \models \phi$ iff $M, m(x) \models \phi$.*

From M^t, we can define a new structure $M^t_{\mathcal{F}} = (\mathcal{S}^t_{\mathcal{F}}, \{\mathcal{R}^t_{\mathcal{F}r}\}, \Pi^t_{\mathcal{F}})$ where, $\mathcal{S}^t_{\mathcal{F}} = \mathcal{S}^t$, $\{\mathcal{R}^t_{\mathcal{F}r}\} = \{\mathcal{R}^t_r\}$, and $\Pi^t_{\mathcal{F}}(x) = \Pi^t(x) - \{A_{(\leq 1\ a)}, H_{(\leq 1\ a)}\}$, for each $x \in \mathcal{S}^t_{\mathcal{F}}$. The structure $M^t_{\mathcal{F}}$ has the following property.

Lemma 2. *Let Φ be a $\mathcal{DIF}$-formula, and let $M^t, M^t_{\mathcal{F}}$ be obtained from a model M of $\gamma(\Phi)$ as specified above. Then $M^t, root \models \gamma_1(\Phi)$ implies $M^t_{\mathcal{F}}, root \models \Phi$.*

Considering that every model of Φ can be easily transformed in a model of $\gamma(\Phi)$ we can state the main result of this section.

Theorem 3. *A $\mathcal{DIF}$-formula Φ is satisfiable iff its $\mathcal{DI}$-counterpart $\gamma(\Phi)$ is satisfiable.*

Corollary 4. *Satisfiability in $\mathcal{DIF}$ and both satisfiability of $\mathcal{CIF}$-concepts and logical implication for $\mathcal{CIF}$-TBoxes are EXPTIME-complete problems.*

The fact that $\mathcal{DIF}$-formulae can be encoded in $\mathcal{DI}$, calls for some comments. Notice that $\mathcal{DI}$-formulae have always a finite model M (finite model property) while $\mathcal{DIF}$-formulae don't - e.g. the formula $A \wedge [(P^-)^*]((\leq 1\ P) \wedge < P^- > \neg A)$ does not have any finite model [6]. Indeed, M^t, and thus $M^t_{\mathcal{F}}$, built from a finite model M are not finite in general.

It is also interesting to observe that, since $\mathcal{DIF}$ subsumes Converse Deterministic PDL, also formulae of that logic can be encoded in $\mathcal{DI}$. This fact gives us procedures to decide satisfiability of Converse Deterministic PDL formulae that do not rely on techniques based on automata on infinite structures as those in [30].

Finally, the construction above can be easily modified/restricted to encode Deterministic PDL formulae in PDL. In fact, the original construction, used in [3] to study satisfiability of Deterministic PDL, is similar in the spirit, though not in the development, to such a restricted version of the our construction.

In concluding the section we would like to present some examples of $\mathcal{CIF}$ concepts, that demonstrate the power of this DL. The examples concern the definition of concepts denoting common data structures in computer science. The first example regards lists. A LIST can be (inductively) defined as: a NIL is a LIST, a NODE that as exactly one successor that is a LIST, is a LIST. From this definition it follows that a list is a chain of NODEs of any length, that

[6] This formula is a variant of the Converse Deterministic PDL formula $A \wedge [(P^-)^*] < P^- > \neg A$ (see [30]).

terminates with a NIL. Therefore, we can denote the class of LISTs as (we use $C_1 \doteq C_2$ as a shorthand for $C_1 \sqsubseteq C_2, C_2 \sqsubseteq C_1$):

$$List \doteq \exists(id(Node \sqcap (\leq\ 1\ succ)) \circ succ)^{*}.Nil.$$

The second example concerns (possible infinite) trees. A TREE is formed by a single NODE, that has no father (the root), whose all children are inner NODEs of a TREE, where an inner NODE of a TREE is a NODE having exactly one father, whose children are themselves all inner NODEs of a TREE. This definition implies that TREEs are formed by a NODE that has no father and such that all its descendants are NODEs having exactly one father. Note that infinite TREEs are allowed. The $\mathcal{CIF}$ concept corresponding to this definition of TREE is

$$Tree \doteq \forall child^{-}.\bot \sqcap (\forall child^{*}.(Node \sqcap (\leq\ 1\ child^{-}))).$$

As last example, we specialize the above definition of TREE, to BINARY-TREE where left and right subtrees are identified through different roles. That is, BINARY-TREEs are TREEs such that each NODE has at most one LEFT child and at most one RIGHT child. The corresponding $\mathcal{CIF}$-concept is

$$\begin{aligned} BinTree \doteq\ & (\forall left^{-}.\bot) \sqcap (\forall right^{-}.\bot) \sqcap \\ & (\forall (left \sqcup right)^{*}.(Node \sqcap (\leq\ 1\ left) \sqcap (\leq\ 1\ right) \sqcap \\ & (\leq\ 1\ left^{-}) \sqcap (\leq\ 1\ right^{-}) \sqcap ((\forall left^{-}.\bot) \sqcup (\forall right^{-}.\bot)))). \end{aligned}$$

Observe that, in order to fully capture the latter two concepts, we need to make use of functional restrictions on both atomic roles and inverse of atomic roles. To the best of our knowledge, $\mathcal{CIF}$ is the only DL allowing for a correct and precise definition of TREE and BINARY-TREE.

4 N-ary Relations

In this section we extend $\mathcal{CIF}$ by means of suitable mechanisms to aggregate individuals in *tuples*. Each tuple has an associated *arity* which is the number of individuals constituting the tuple. Tuples of the same arity n can be grouped into sets forming *n-ary relations*.

An n-ary relation is described by a *name* and n relation roles (r-roles in the following). Each r-role names a component of the relation, i.e., a component of each of its tuples. For each relation $\mathbf{R}$ the set of its r-roles is denoted by $rol(\mathbf{R})$. The cardinality of this set is greater or equal to 2, and implicitly determinates the arity of $\mathbf{R}$. We call "U-component" the component of $\mathbf{R}$ named by the r-role $U \in rol(\mathbf{R})$.

We present a DL, called $\mathcal{CIFR}$, with suitable constructs to deal with relations, having the following abstract syntax:

$$\begin{aligned} C \longrightarrow\ & \top \mid \bot \mid A \mid C_1 \sqcap C_2 \mid C_1 \sqcup C_2 \mid C_1 \Rightarrow C_2 \mid \neg C \mid \forall R.C \mid \exists R.C \mid \\ & (\leq\ 1\ P) \mid (\leq\ 1\ P^{-}) \mid (\leq\ 1\ \mathbf{R}[U]) \mid \\ & \forall \mathbf{R}[U].T_1 : C_1, \ldots, T_m : C_m \mid \exists \mathbf{R}[U].T_1 : C_1, \ldots, T_m : C_m \\ R \longrightarrow\ & P \mid \mathbf{R}[U, U'] \mid R_1 \sqcup R_2 \mid R_1 \circ R_2 \mid R^{*} \mid R^{-} \mid id(C). \end{aligned}$$

The intuitive meaning of the new constructs is explained below (the other constructs have the usual meaning).

- $\mathbf{R}[U]$ denotes the relation between individuals d and tuples of $\mathbf{R}$ that have d as U-component - i.e., the inverse of the function projecting $\mathbf{R}$ onto its U-component.
- $\mathbf{R}[U, U']$ denotes the function projecting the relation $\mathbf{R}$ onto its U, U' components.
- $(\leq 1\ \mathbf{R}[U])$ represents the individuals d that occur at most once as U-component of the relation $\mathbf{R}$.[7]
- $\forall \mathbf{R}[U].T_1 : C_1, \ldots, T_m : C_m$ represents the set of individuals x such that for each tuple r of $\mathbf{R}$ with x as U-component, the T_i-component of r belongs to the extension of C_i $(i = 1, \ldots, m)$.
- $\exists \mathbf{R}[U].T_1 : C_1, \ldots, T_m : C_m$ represents the set of individuals x such that there exists a tuple r of $\mathbf{R}$ with x as U-component and x_i $(i = 1, \ldots, m)$ as T_i-component such that x_i belongs to the extension of C_i.

The semantics of $\mathcal{CIFR}$ is given, as usual, through an interpretation $\mathcal{I} = (\Delta^{\mathcal{I}}, \cdot^{\mathcal{I}})$, now extended to interpret relations and the new constructs. In particular, if $\mathbf{R}$ is a (n-ary) relation whose set of r-roles is $rol(\mathbf{R}) = \{U_1, \ldots, U_n\}$, then $\mathbf{R}^{\mathcal{I}}$ is a set of labeled tuples of the form $< U_1 : d_1, \ldots, U_n : d_n >$ where $d_1, \ldots, d_n \in \Delta^{\mathcal{I}}$. We write $r[U]$ to denote the value associated with the U-component of the tuple r. The new constructs are interpreted as follows:

- $\mathbf{R}[U]^{\mathcal{I}} = \{(d, r) \in \Delta^{\mathcal{I}} \times \mathbf{R}^{\mathcal{I}} \mid d = r[U]\}$.
- $\mathbf{R}[U, U']^{\mathcal{I}} = \{(d, d') \in \Delta^{\mathcal{I}} \times \Delta^{\mathcal{I}} \mid \exists r \in \mathbf{R}^{\mathcal{I}}.d = r[U] \wedge d' = r[U']\}$.
- $(\leq 1\ \mathbf{R}[U])^{\mathcal{I}} = \{d \in \Delta^{\mathcal{I}} \mid$ there exists at most one $r \in \mathbf{R}^{\mathcal{I}}$ such that $r[U] = d\}$.
- $(\forall \mathbf{R}[U].T_1 : C_1, \ldots, T_m : C_m)^{\mathcal{I}} = \{d \in \Delta^{\mathcal{I}} \mid \forall r \in \mathbf{R}^{\mathcal{I}}.r[U] = d \Rightarrow (r[T_1] \in C_1^{\mathcal{I}} \wedge \cdots \wedge r[T_m] \in C_m^{\mathcal{I}})\}$.
- $(\exists \mathbf{R}[U].T_1 : C_1, \ldots, T_m : C_m)^{\mathcal{I}} = \{d \in \Delta^{\mathcal{I}} \mid \exists r \in \mathbf{R}^{\mathcal{I}}.r[U] = d \wedge r[T_1] \in C_1^{\mathcal{I}} \wedge \cdots \wedge r[T_m] \in C_m^{\mathcal{I}}\}$.

$\mathcal{CIFR}$-TBoxes are defined as a finite set of inclusion assertions $C_1 \sqsubseteq C_2$, where C_1, C_2 are $\mathcal{CIFR}$-concepts. Satisfiability of $\mathcal{CIFR}$-concepts, as well as logical implication, in $\mathcal{CIFR}$-TBoxes is defined as usual.

Let us show some examples of use of $\mathcal{CIFR}$. Consider the relation **Parents**, with $rol(\mathbf{Parents}) = \{child, father, mother\}$, denoting the set of tuples child and his/her (natural) parents (father and mother). An inclusion assertion regarding this relation can be:

$$Human \sqsubseteq \forall \mathbf{Parents}[child].father : Human, mother : Human$$

stating that both the father and the mother of a child, who is human, must be human as well (more precisely, every individual who is $Human$ is such that, if

[7] Note that $\mathcal{CIFR}$ does not include the concept construct $(\leq 1\ \mathbf{R}[U]^-)$, because, by definition, $\mathbf{R}[U]^-$ is always functional.

(s)he participates, as *child*-component, in a tuple r of the relation **Parents**, then both the *father*-component of r and the *mother*-component of r are *Human*). Note that, in order to represent the (natural) parents of a child, the relation **Parent** must be so that child has exactly one father and one mother in the relation **Parents** - that is, individuals may occur as *child*-component in at most one tuple of the relation. This fact can easily be represented in $\mathcal{CIFR}$ by asserting that:

$$\top \sqsubseteq (\leq\ 1\ \mathbf{Parents}[child]).$$

Next we investigate the decidability and the complexity of the reasoning tasks for $\mathcal{CIFR}$. For ease of exposition, we concentrate on satisfiability of $\mathcal{CIFR}$-TBoxes. In fact, it is easy to check that, satisfiability of $\mathcal{CIFR}$-TBoxes and logical implication in $\mathcal{CIFR}$-TBoxes are (linearly) reducible one into the other, and satisfiability of $\mathcal{CIFR}$-concepts is a subcase of both of them. We show that there exists a one-to-one mapping from $\mathcal{CIFR}$-TBoxes $\mathcal{K}$ to $\mathcal{CIF}$-TBoxes $\mathcal{K}'$ such that $\mathcal{K}$ is satisfiable if and only if $\mathcal{K}'$ is satisfiable. To define this mapping we make use of an auxiliary mapping t.

The mapping t is defined inductively, in the obvious way for the common constructs, and as follows, for the new ones:

$$t(\mathbf{R}[U]) = P^-_{\mathbf{R}[U]} \ (P_{\mathbf{R}[U]} \text{ is a new atomic role})$$
$$t(\mathbf{R}[U,U']) = P^-_{\mathbf{R}[U]} \circ P_{\mathbf{R}[U']}$$
$$t((\leq\ 1\ \mathbf{R}[U])) = (\leq\ 1\ P^-_{\mathbf{R}[U]})$$
$$t(\forall\mathbf{R}[U].T_1 : C_1, \ldots, T_m : C_m) = \forall P^-_{\mathbf{R}[U]}.\exists P_{\mathbf{R}[T_1]}.t(C_1) \sqcap \ldots \sqcap \exists P_{\mathbf{R}[T_m]}.t(C_m)$$
$$t(\exists\mathbf{R}[U].T_1 : C_1, \ldots, T_m : C_m) = \exists P^-_{\mathbf{R}[U]}.\exists P_{\mathbf{R}[T_1]}.t(C_1) \sqcap \ldots \sqcap \exists P_{\mathbf{R}[T_m]}.t(C_m).$$

Inclusion assertions $C_1 \sqsubseteq C_2$ are mapped to $t(C_1) \sqsubseteq t(C_2)$.

Let us call $t(\mathcal{K})$ the TBox thus obtained. From $t(\mathcal{K})$ we obtain $\mathcal{K}'$ by adding to it the following inclusion assertions:

1. $\top \sqsubseteq (\leq\ 1\ P_{\mathbf{R}[U]})$ for all roles $P_{\mathbf{R}[U]}$.
2. $\exists P_{\mathbf{R}[U]}.\top \sqsubseteq \exists P_{\mathbf{R}[U_1]}.\top \sqcap \ldots \sqcap \exists P_{\mathbf{R}[U_n]}.\top$ where $U \in rol(\mathbf{R})$ and $rol(\mathbf{R}) = \{U_1, \ldots, U_n\}$, for all roles $P_{\mathbf{R}[U]}$.

The inclusion assertions (1) constrain the models of $\mathcal{K}'$ so that each $P_{\mathbf{R}[U]}$ is (globally) functional. The inclusion assertions (2) constrain the models of $\mathcal{K}'$ so that if an individual has a link that is an instance of $P_{\mathbf{R}[U]}$ then it also has links that are instance of $P_{\mathbf{R}[U_i]}$ (for all $U_i \in rol(\mathbf{R})$). Indeed, (1) and (2) allow us to represent a n-ary relation $\mathbf{R}$ by the concept $\exists P_{\mathbf{R}[U_1]}.\top \sqcap \ldots \sqcap \exists P_{\mathbf{R}[U_n]}.\top$, i.e., the tuples of $\mathbf{R}$ are represented by instances of $\exists P_{\mathbf{R}[U_1]}.\top \sqcap \ldots \sqcap \exists P_{\mathbf{R}[U_n]}.\top$. Observe that this representation is accurate only in the models $\mathcal{I}$ of $\mathcal{K}'$ where tuples of $\mathbf{R}$ corresponds to a single individual, otherwise, in $\mathcal{I}$ there would be two individuals representing the same tuple. However, we can show that if $\mathcal{K}'$ admits a model, then it admits a model satisfying the above condition. Formally, the following lemma holds.

Lemma 5. *The $\mathcal{CIF}$-TBox $\mathcal{K}'$ obtained by the above construction is satisfiable if and only if it has a model $\mathcal{I}$ satisfying the constraint:*

$$\begin{array}{l} d, d' \in (\exists P_{\mathbf{R}[U_1]}.\top \sqcap \ldots \sqcap \exists P_{\mathbf{R}[U_n]}.\top)^{\mathcal{I}} \Rightarrow \\ \quad \neg((d, d_1) \in (P_{\mathbf{R}[U_1]})^{\mathcal{I}} \wedge (d', d_1) \in (P_{\mathbf{R}[U_1]})^{\mathcal{I}} \wedge \\ \quad \ldots \wedge (d', d_n) \in (P_{\mathbf{R}[U_n]})^{\mathcal{I}} \wedge (d', d_n) \in (P_{\mathbf{R}[U_n]})^{\mathcal{I}}) \end{array}$$

for every n-ary relation $\mathbf{R}$ *with* $rol(\mathbf{R}) = \{U_1, \ldots, U_n\}$.

Now, we are ready to state the desired result.

Theorem 6. *A $\mathcal{CIFR}$-TBox $\mathcal{K}$ is satisfiable if and only if the $\mathcal{CIF}$-TBox $\mathcal{K}'$ defined as above is satisfiable.*

Considering that $\mathcal{K}'$ is polynomially bounded to $\mathcal{K}$, the decidability and the complexity of reasoning in $\mathcal{CIFR}$ are an immediate consequence of the results in the previous section.

Corollary 7. *Satisfiability of $\mathcal{CIFR}$-TBoxes, logical implication for $\mathcal{CIFR}$-TBoxes, satisfiability of $\mathcal{CIFR}$-concepts, are EXPTIME-complete problems.*

5 Discussion and Conclusion

By exploiting the correspondence between DLs and PDLs, we have analyzed the decidability and the complexity of a very expressive DL, $\mathcal{CIF}$, which includes functional restrictions on both atomic roles and their inverse. On top of $\mathcal{CIF}$ we have been able to design constructs involving n-ary relations, thus presenting a DL, $\mathcal{CIFR}$, whose characteristics are quite unusual in the context of Frame Based Languages, and more typical of other class-based formalisms such as Semantics Data Models or Object-Oriented Data Models.

It is possible to show that our results on functional restrictions extend to full *qualified number restrictions* (generalizations of functional restrictions stating the minimum and the maximum number of links between instances of classes and instances of another concept through a specified role or relation), by which general cardinality constrains on components of relations can be expressed.

We conclude remarking that, the issues presented in this paper can be relevant also in the setting of Modal Mu-Calculus, a logic of programs which includes explicit constructs for least and greatest fixpoints of formulae (PDL is a fragment of it), that has been recently used to model, in a single framework, terminological cycles interpreted according to Least Fixpoint Semantics, Greatest Fixpoint Semantics, and Descriptive Semantics (see [23, 27, 11]).

References

1. G. Attardi and M. Simi. Consistency and completeness of omega, a logic for knowledge representation. In *Proc. of the Int. Joint Conf. on Artificial Intelligence IJCAI-81*, pages 504–510, 1981.

2. F. Baader and B. Hollunder. A terminological knowledge representation system with complete inference algorithm. In *Proc. of the Workshop on Processing Declarative Knowledge, PDK-91*, Lecture Notes in Artificial Intelligence, pages 67–86. Springer-Verlag, 1991.
3. M. Ben-Ari, J. Y. Halpern, and A. Pnueli. Deterministic propositional dynamic logic: Finite models, complexity, and completeness. *Journal of Computer and System Sciences*, 25:402–417, 1982.
4. A. Borgida and P. F. Patel-Schneider. A semantics and complete algorithm for subsumption in the CLASSIC description logic. Submitted for publication, 1993.
5. R. J. Brachman and H.J. Levesque. The tractability of subsumption in frame-based description languages. In *Proc. of the 4th Nat. Conf. on Artificial Intelligence AAAI-84*, pages 34–37, 1984.
6. R. J. Brachman, D. L. McGuinness, P. F. Patel-Schneider, L. Alperin Resnick, and A. Borgida. Living with CLASSIC: when and how to use a KL-ONE-like language. In John F. Sowa, editor, *Principles of Semantic Networks*, pages 401–456. Morgan Kaufmann, 1991.
7. R. J. Brachman and J. G. Schmolze. An overview of the KL-ONE knowledge representation system. *Cognitive Science*, 9(2):171–216, 1985.
8. M. Buchheit, F. M. Donini, and A. Schaerf. Decidable reasoning in terminological knowledge representation systems. In *Proc. of the 13th Int. Joint Conf. on Artificial Intelligence IJCAI-93*, pages 704–709, 1993.
9. D. Calvanese, M. Lenzerini, D. Nardi. A unified framework for class-based representation languages. In *Proc. Int. Conf. on Principles of Knowledge Representation and Reasoning KR-94*, 1994.
10. M. J. Fisher and R. E. Ladner. Propositional Dynamic Logic of regular programs. *Journal of Computer and System Sciences*, 18:194–211, 1979.
11. G. De Giacomo and M. Lenzerini. Concept language with number restrictions and fixpoints, and its relationship with mu-calculus. In *Proc. of the 11th Eur. Conf. on Artificial Intelligence ECAI-94*, pages 355–360, 1994.
12. F. M. Donini, B. Hollunder, M. Lenzerini, A. Marchetti Spaccamela, D. Nardi, and W. Nutt. The complexity of existential quantification in concept languages. *Artificial Intelligence*, 2–3:309–327, 1992.
13. F. M. Donini, M. Lenzerini, D. Nardi, and W. Nutt. The complexity of concept languages. In *Proc. of the 2nd Int. Conf. on Principles of Knowledge Representation and Reasoning KR-91*, pages 151–162, 1991.
14. F. M. Donini, M. Lenzerini, D. Nardi, and W. Nutt. Tractable concept languages. In *Proc. of the 12th Int. Joint Conf. on Artificial Intelligence IJCAI-91*, pages 458–463, 1991.
15. F. M. Donini, M. Lenzerini, D. Nardi, W. Nutt, and A. Schaerf. Adding epistemic operators to concept languages. In *Proc. of the 3rd Int. Conf. on Principles of Knowledge Representation and Reasoning KR-92*, 18–3:201–260, 1987.
16. R. Hull and R. King. Semantic database modeling: Survey, applications, and research Issues. *ACM Computing Surveys*, pages 342–353, 1992.
17. D. Kozen and J. Tiuryn. Logics of programs. In *Handbook of Theoretical Computer Science – Formal Models and Semantics*, pages 789–840. Elsevier, 1990.
18. H. J. Levesque and R. J. Brachman. Expressiveness and tractability in knowledge representation and reasoning. *Computational Intelligence*, 3:78–93, 1987.
19. R. MacGregor. Inside the LOOM description classifier. *SIGART Bulletin*, 2(3):88–92, 1991.

20. B. Nebel. Computational complexity of terminological reasoning in BACK. *Artificial Intelligence*, 34(3):371–383, 1988.
21. B. Nebel. Terminological reasoning is inherently intractable. *Artificial Intelligence*, 43:235–249, 1990.
22. B. Nebel. *Reasoning and revision in hybrid representation systems.* Springer-Verlag, 1990.
23. B. Nebel. Terminological cycles: Semantics and computational properties. In John F. Sowa, editor, *Principles of Semantic Networks*, pages 331–361. Morgan Kaufmann, 1991.
24. R. Parikh. Propositional dynamic logics of programs: a survey. In *Proc. of 1st Workshop on Logic of Programs*, Lecture Notes in Computer Science 125, pages 102–144. Springer-Verlag, 1981.
25. P. F. Patel-Schneider. A hybrid, decidable, logic-based knowledge representation system. *Computational Intelligence*, 3(2):64–77, 1987.
26. K. Schild. A correspondence theory for terminological logics: Preliminary report. In *Proc. of the 12th Int. Joint Conf. on Artificial Intelligence IJCAI-91*, pages 466–471, 1991.
27. K. Schild. Terminological cycles and the propositional μ-calculus. In *Proc. of the 4th Int. Conf. on Knowledge Representation and Reasoning KR-94*, 1994.
28. M. Schmidt-Schauß and G. Smolka. Attributive concept descriptions with complements. *Artificial Intelligence*, 48(1):1–26, 1991.
29. G. Schreiber, B. Wielinga, J. Breuker. *KADS: A principled approach to knowledge-based system development.* Academic Press, 1993.
30. M. Y. Vardi and P. Wolper. Automata-theoretic techniques for modal logics of programs. *Journal of Computer and System Sciences*, 32:183–221, 1986.

On the Concept of Generic Object: A Nonmonotonic Reasoning Approach and Examples

Leopoldo E. Bertossi
Pontificia Universidad Catolica de Chile
Departamento de Ciencia de la Computacion
Escuela de Ingenieria, Casilla 306, Santiago 22, Chile
e-mail: bertossi@ing.puc.cl
Raymond Reiter
Department of Computer Science
University of Toronto
Toronto, Canada M5S 1A4
and
The Canadian Institute for Advanced Research
e-mail: reiter@ai.toronto.edu

Abstract. In this paper we discuss the logical and commonsense intuitions behind the concept of generic object and the possibility of characterizing this concept by means of both classical first-order logic and circumscription. We also show that circumscription can be used in the characterization of the concept of generic point of a curve, as it appears in algebraic geometry. Finally, we show that several occurrences of the concept of generic object in computer science and mathematics can be shaped according to the same pattern of definition.

1 Introduction

We all have intuitions about the concept of generic object, more precisely, of generic element of a class O of objects. Since the notion of generic object seems to be a commonsense concept, it is natural to investigate the application of logical formalizations of commonsense reasoning, as they appear in artificial intelligence [24], to its characterization. In [5], the concept of generic mathematical object, more precisely, of generic triangle in geometry, was investigated from the point of view of circumscriptive commonsense reasoning [17, 21]. Sometimes, the concept has been defined precisely in the context of special mathematical theories. Examples are algebraic geometry [30, 31], set theory [8], and also model theory [27], as an abstraction and generalization of the concept appearing in set theory. The concept also appears in many areas of computer science.

In section 2, we list some of the intuitions behind the concept of generic object, and explore the possibility of formalizing them in classical logical terms. In section 3, we discuss the kind of properties that may be considered for genericity. In section 4, we discuss in general terms the application of circumscription to the characterization of the concept of generic object. In section 5, we investigate the concept of generic object as it appears in some mathematical theories, more specifically, in algebraic geometry, set theory, and model theory. In section 6 we present some examples of generic objects as they appear in computer science.

2 Intuitions Behind the Concept of Generic Object

Let us assume that there is a base theory Σ for the domain of discourse. In general we do not mention it explicitly in the formalizations that follow. We use two distinguished predicates, $O(\cdot)$ and $Gen(\cdot)$, which apply to objects in a specific class and to generic objects in that class, respectively. We further assume that the formula $\forall x(Gen(x) \rightarrow O(x))$ is entailed by Σ.

First Intuition: *If a generic object of a class has a property, then all objects in the class have that property.*

A natural attempt to formalize this intuition can be made through the following formula:

$$Gen(a) \wedge P(a) \rightarrow \forall y(O(y) \rightarrow P(y)). \tag{1}$$

In this sentence we have an implicit (second-order) universal quantification over P. Notice that in case we instantiate P with predicate Gen, all objects in the class would be generic (if there is a generic object a). This conclusion goes against our intuition of a generic object, which shows that in (1) we have to restrict the possibilities for P, in order not to get to trivial or counterintuitive situations. Also, any property appearing in (1) that applies to a generic object would be a logical consequence of the definition of the class of objects. A generic object is then not allowed to have accidental properties that are not part of the specification of the class. According to this result, if generic triangles were not right-angled, we would conclude that there are no right-angled triangles. To avoid this undesirable conclusion, we should reformulate the first intuition by constraining the range of possibilities for properties of generic objects: *If a generic object of a class has a property, then all the objects in the class have that property, except in the case of singular properties that do not qualify for genericity.*

Second Intuition: *A generic object has as properties only those properties that are explicit or implicit in the definition of the class of objects.*

This intuition says that a generic object has as properties only those properties that are logical consequence of the definition of the class of objects to which it belongs. If we try to formalize this intuition, we get first to the same formula (1). We have again a un quantification over properties P. Nevertheless, this formula captures only part of the intuition. If P is a property that is not a logical consequence of the definition, i.e., $\not\models O(x) \rightarrow P(x)$, then it is not true that $P(a)$. This part can be seen as a form of Closed World Assumption applied to generic objects. The problem with this formalization, apart from its nonmonotonicity, is that the truth value of a property can be undetermined with respect to the predicate O, i.e., neither $\models (O(x) \rightarrow P(x))$ nor $\models (O(x) \rightarrow \neg P(x))$ holds. In this case, we would jump to both conclusions: $\neg P(a)$ and $\neg\neg P(a)$. We have to restrict the possible P's in the formalization, instead of quantifying universally over all properties.

Third Intuition: *A generic object defines the class of objects to which it belongs.*

This assertion means that $O =_{def} \{x \mid Gen(a) \wedge P(a) \rightarrow P(x)\}$. In the formula defining the class O, there is an implicit universal quantification over P, that is, $O(x) \leftrightarrow \forall P(Gen(a) \wedge P(a) \rightarrow P(x))$. Direction "$\rightarrow$" gives formula (1). The other direction tells us that x belongs to the class O if it possesses all the properties of a generic object. It is clear that we have to restrict the properties P we are considering in the quantification: If a generic object does not have a property that is admissible for an element of the class, i.e., consistent with the definition of the class, then there would be no element with that property in the class. So, we are faced again with the problem of concentrating only on some particular properties for the objects in the class.

Fourth Intuition: *A generic object does not have any singular property.*

First of all, we have to say what is a singular property. Although the concept of singularity seems to import a high degree of commonsense and, accordingly, a treatment of the concept in similar terms to those of the concept of genericity should be attempted, we consider a singular property as an accidental, or contingent, property of objects. An accidental property is a possible but not necessary property of the objects. More precisely, P is singular if $\exists y(O(y) \wedge P(y))$ and $\exists y(O(y) \wedge \neg P(y))$ are both true. The second fact is more relevant because it tells us that property P is not inherent to the definition of class O. Then, we could attempt to formalize the concept in the form $\exists y(O(y) \wedge \neg P(y)) \rightarrow (Gen(a) \rightarrow \neg P(a))$. The contrapositive of the implication gives formula (1). As before, a universal quantification over P is implicit. Then, the problems already mentioned in the formalization of the first intuition appear again.

3 Properties that Are Relevant to Genericity

The moral that we can draw from the considerations above is that it is necessary to concentrate on some particular properties in order to formalize the concept of generic object. Actually, in all the previous attempts to formalizations, if we allow P to be an arbitrary property, then a generic object is indistinguishable, in the second-order language, from a non generic object, which is not desirable. How we restrict and choose relevant properties depends on the context.

Let us give an example illustrating the previous ideas. Assume that we have a base theory Σ, i.e. a set of sentences in a formal language $\mathcal{L}$. Σ might be the theory that describes in $\mathcal{L}$ a given model (or structure) $\mathcal{M} = \langle M, O^M, a^M, \ldots \rangle$. In $\mathcal{L}$ we have a predicate symbol O and a (tuple of) name(s) a to denote the class of objects O^M and the distinguished individual(s) a^M, respectively. For example, Σ could be the theory of a model of geometry, and a binary predicate $O(\cdot,\cdot)$ could denote the class of pairs of different and intersecting lines. A pair $a^M = (l_1, l_2)$, with $(l_1, l_2) \in O^M$, would be non generic if there is a property (formula) $\psi(x)$ such that: $\psi(a) \in \Sigma$, but $\forall \bar{x}(O(\bar{x}) \rightarrow \psi(\bar{x})) \notin \Sigma$.

In the model depicted in Figure 1, the pair of lines in (a) would be considered generic, but not so the pair in (b). Why? The reason is that in the first case we are not able to *say* anything more specific about (l_1, l_2) beyond the fact that $(l_1, l_2) \in O^M$ (i.e., anything not implied by $O(a)$). In the second case, instead, we can *say* that l_1 and l_2 form a right angle. The key in this example can be found in the word "say" above: *What we can say is restricted by the language that is available to us.*

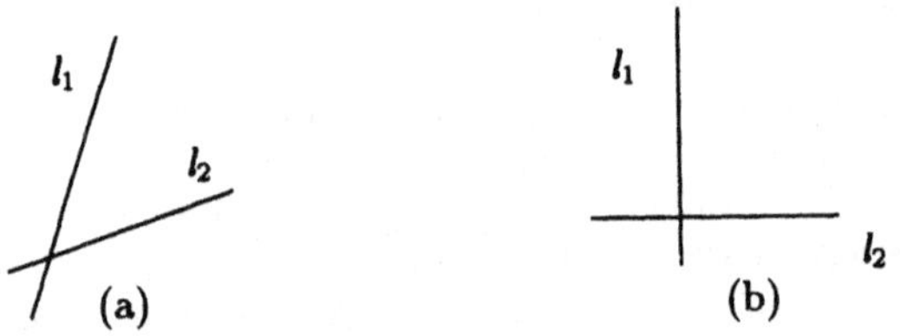

Fig. 1. Pairs of Intersecting Lines

It is so common that our geometric languages allow us to talk about right angles, that we took it for granted that it is possible to do so in $\mathcal{L}$ and, in consequence, that it is possible to singularize specific properties of (l_1, l_2). According to this reasoning, (l_1, l_2) would lose its genericity if we had a way of talking in $\mathcal{L}$ about the measures (in degrees) of the angles that the lines form. Even worse, if we were only allowed to talk about "forming a right angle", the first example would not be generic either because it has the specific property of "not intersecting at a right angle", which our language allows us to state. So, our first case is not generic either.

One could argue that we should not give the same status to the properties "intersecting at right angle" and "not intersecting at right angle". After all, the former is an immediate, positive property, but not the latter. We know that this distinction is rather tricky because it is not difficult to transform negative properties into positive ones by redefining predicates. In addition, we could be interested in properties that are more involved in their definitions, in which case their evaluation in terms of "immediateness" or "positiveness" is less natural.

Despite the problems with this approach, it still seems intuitive and attractive.

If we want to keep it in some form, it seems necessary to think at a metalevel which are the properties we consider for (non)genericity. At this point is where commonsense comes in. Asking for the genesis of the singularization of specific properties in connection with the concept of generic object is more or less the same as asking for the source of defaults or, more generally, for the source of commonsense knowledge. This is an area where much research has still to be done. Nevertheless, a way of achieving a characterization of genericity at an object level, which is at the same time general and uniform, is by using some formal system for doing commonsense reasoning. For example, a system based on circumscription.

4 What about a Nonmonotonic Approach to the Concept of Generic Object?

In the following, we try to be consistent with the ideas and objections raised in the previous sections, and to keep most of the problematic attempts to formalize the concept of generic object given there, but restricting the predicates that are relevant to genericity to a given list. We also detect if there exists some similarity to a circumscriptive approach.

Assume that we have a base theory Σ in a formal language $\mathcal{L}$ that has a predicate symbol O and a name a (or tuple of names) to denote the class of objects and a (tuple of) distinguished individual(s), respectively. Let $S_1, \ldots, S_n$ be new predicate symbols. Usually, there are definitions for these predicates in the theory. We are thinking of the S_i's as the singular properties of objects in O. We may further assume that $\Sigma \vdash \exists y(O(y) \wedge \neg S_i(y))$ for $i = 1, \ldots, n$. That is, the S_i's are accidental properties of objects.

We want a to be a generic object. Intuitively, we do not want a to have any of the singular properties $S_1, \ldots, S_n$. The approach followed in [5] —where the base theory was elementary geometry, a was a triangle, and a singular property was right-angledness— consists in circumscribing with scope [9] the singular properties in the sentence given by the base theory plus the fact $O(a)$. Accordingly, in the present case we would consider a as a generic object in O with respect to Σ if it satisfies the specification $Circum_{scoped}(\Sigma';\ W(P, x))$, that is, the scoped circumscription[1] of formula $W(P, x)$ in the theory Σ', where: (1)

[1] We consider a second-order version of scoped circumscription.
$Circum_{scoped}(A;\ W(\bar{P}, \bar{x}))$, the *scoped circumscription* of formula $W(\bar{P}, \bar{x})$ in the theory A, with variable predicate parameters $\bar{P}$ and *Scope*, is given by the conjunction of A with the second-order sentence:

$$A(\bar{P}', Scope') \wedge \forall \bar{x}[W(\bar{P}', \bar{x}) \wedge Scope'(\bar{x}) \supset W(\bar{P}, \bar{x}) \wedge Scope(\bar{x})]$$
$$\supset \forall \bar{x}[W(\bar{P}, \bar{x}) \wedge Scope(\bar{x}) \supset W(\bar{P}', \bar{x}) \wedge Scope'(\bar{x})].$$

With scoped circumscription, the scope of the domain of individuals we are reasoning about, more precisely, the restriction of predicate W to *Scope*, is kept narrow by minimization.

$P = (S_1, \ldots, S_n)$, is the tuple of singular predicates; (2) $W(P, x)$ is the formula $S_1(x) \vee \ldots \vee S_n(x)$, i.e., $W(P, x)$ says that x has some singular property; and (3) Σ' is the new theory given by the conjunction of the following sentences: (a) Σ, (b) $\exists x(O(x) \wedge \neg W(P, x))$ (there are nonsingular objects), (c) $O(a)$, and (d) $Scope(a)$.

Remark 1 Special and interesting cases are those where: 1. Conjuncts (a) and (b) in Σ' are simultaneously consistent (maybe (b) follows from (a)), and 2. Predicate *Scope* appears only in (d), and tuple a appears only in (c) and (d). ■

With our circumscriptive specification, we hope to obtain:

$$Circum_{scoped}(\Sigma';\ W(P, x)) \models \neg W(P, a), \tag{2}$$

that is, the genericity or nonsingularity of a should be a circumscriptive consequence of the theory. Under certain general circumstances this will be the case:

Proposition 1. ([9]) If $A \not\vdash \exists x(W(P, x) \wedge Scope(x))$, and the following hypothesis **H**: *"A entails a domain-closure axiom and decides the equality predicate for all ground terms of the language"* holds, then $Circum_{scoped}(A;\ W(P, x)) \vdash \forall x(Scope(x) \rightarrow \neg W(P, x))$, provided all predicates are variable. ■

From this proposition we obtain:

Proposition 2. If Σ' satisfies hypothesis **H** in Proposition 1, and also 1., 2. in Remark 1, then (2) holds.

Proof: From 1. and 2. in Remark 1, we obtain that Σ' is consistent. Then, Σ' is also consistent with $\forall x(Scope(x) \rightarrow \neg W(P, x))$, because a model $\mathcal{M} = \langle M, P^M, \ldots\rangle$ of Σ' can be modified, if necessary, in such a way that a^M, the interpretation of a in $\mathcal{M}$, is a point in O^M that does not satisfy $W(P, x)$ in $\mathcal{M}$. We may choose $Scope^M = \{a^M\}$; then, $\Sigma' \not\vdash \exists x(Scope(x) \wedge W(P, x))$. By Proposition 1, $Circum_{scoped}(\Sigma';\ W(P, x)) \models \forall x(Scope(x) \rightarrow \neg W(P, x))$; from which we obtain $Circum_{scoped}(\Sigma';\ W(P, x)) \models \neg W(P, a)$. ■

This result depends on condition **H**. Nevertheless, as pointed out in [9], this condition is sufficient, but not necessary[2]. Whether we will succeed in obtaining (2) will depend on the context.

Assume that $Circum_{scoped}(\Sigma';\ W(P, x)) \models \neg W(P, a)$. That is, $\bigwedge_1^n \neg S_i(a)$ is a circumscriptive consequence of Σ'. This fact will be written in the form $\Sigma' \vdash_{circum} \bigwedge_1^n \neg S_i(a)$. Following our convention of keeping the base theory Σ implicit, we simply write this as follows:

$$O(a) \wedge \exists y(O(y) \wedge \bigwedge_1^n \neg S_i(y)) \vdash_{circum} \bigwedge_1^n \neg S_i(a). \tag{3}$$

[2] A counterexample can be found in [5], where an intuitively expected result of this kind —the nonsigularity of a triangle a— was obtained with scoped circumscription. In that application, condition **H** was not only unnecessary, but also false.

Now, we can use "a deduction theorem" for circumscription:

Lemma 3. If $(\Sigma \wedge \varphi) \vdash_{circum} \psi$, then $\Sigma \vdash_{circum} (\varphi \rightarrow \psi)$.

Proof: Let $\mathcal{M} = \langle M, O^M, S_1^M, \ldots, S_n^M, Scope^M, a^M, \ldots \rangle$ be a model of the scoped circumscription of $Scope \cap \bigcup_1^n S_i$ in Σ with variable predicates $O, Scope, S_1, \ldots$. Then, $\mathcal{M}$ is a $(Scope \cap \bigcup_1^n S_i;\ Scope, O, S_1, \ldots)$-minimal model of Σ. Assume that $\mathcal{M} \models \varphi$. We have to show that $\mathcal{M} \models \psi$. This proof is easy because in that case $\mathcal{M}$ turns out to be a model of $\Sigma \wedge \varphi$. But it is also a $(Scope \cap \bigcup_1^n S_i;\ Scope, O, S_1, \ldots)$-minimal model since minimality arises from a preorder relation in a class of interpretations. By the hypothesis in this lemma, we have $\mathcal{M} \models \psi$. ■

Then, (3) can be reformulated in the form $O(a) \vdash_{circum} \exists y(O(y) \wedge \bigwedge_1^n \neg S_i(y)) \rightarrow \bigwedge_1^n \neg S_i(a)$, which is equivalent to $O(a) \vdash_{circum} \bigvee_1^n S_i(a) \rightarrow \forall y(O(y) \rightarrow \bigvee_1^n S_i(y))$. If we add to the base theory the definition $\forall x(S(x) \leftrightarrow \bigvee_1^n S_i(x))$, then this can be written in the form:

$$O(a) \vdash_{circum} S(a) \rightarrow \forall y(O(y) \rightarrow S(y)). \tag{4}$$

Actually, if it is possible to obtain the stronger conclusion $O(a) \vdash_{circum} S_i(a) \rightarrow \forall y(O(y) \rightarrow S_i(y))$, $i = 1, \ldots, n$; then, we could have instead of (4)[3]:

$$O(a) \vdash_{circum} \bigwedge_i^n (S_i(a) \rightarrow \forall y(O(y) \rightarrow S_i(y))). \tag{5}$$

The right-hand sides of formulas (4) and (5) can be seen as versions of formula (1), but where only a restricted universal quantification over properties of objects appears. This result means that a circumscriptive characterization of generic objects is, first, close to the initial intuitions behind the concept and, second, close to a natural, but problematic, attempt to a direct (non circumscriptive) specification in usual second-order logic.

5 Genericity in Mathematics

In 5.1, we consider the concept of generic object in algebraic geometry. In 5.2, we consider yet another intuition behind the concept of generic object and its materialization in set theory and model theory.

[3] Of course, this will not be always possible. In geometry, it is not possible to have a triangle that is neither scalene nor obtuse.

5.1 Generic Points in Algebraic Geometry

In algebraic geometry, we find the concept of generic point of a curve [30, 31]. We show, first, that the intuition behind this concept corresponds to the general intuitions we presented before and, second, that this notion can be characterized by means of circumscription.

Let $\mathbb{C}$ be the field of complex numbers, $\mathbf{K}$ an extension field, and $\mathcal{C}$ a curve in, say, $\mathbf{K}^2$, with equation $F(\bar{x}) = 0$, where $F(\bar{x}) \in \mathbb{C}[\bar{x}]$ is an irreducible polynomial with coefficients in $\mathbb{C}$. A point $\bar{a} \in \mathbf{K}^2$ is a *generic point* of $\mathcal{C}$ if: (a) $F(\bar{a}) = 0$ (i.e., $\bar{a}$ belongs to the curve $\mathcal{C}$), and (b) for all $G(\bar{x}) \in \mathbb{C}[\bar{x}]$, if $G(\bar{a}) = 0$, then for all points $\bar{p} \in \mathbf{K}^2$ that belong to the curve $\mathcal{C}$, $G(\bar{p}) = 0$.

Example:[4] Let us start from the field of complex numbers $\mathbb{C}$. We may introduce a new parameter (or indeterminate) t and pass to the extension field $\mathbf{K} = \mathbb{C}(t)$ of rational functions in t, with complex coefficients. Let $F(x, y) = x^2 + y^2 - 1$, be a polynomial in x and y with complex coefficients. This polynomial is irreducible in the ring $\mathbb{C}[x, y]$ of polynomials in x, y with complex coefficients, that is, there is no pair of polynomials in $\mathbb{C}[x, y]$ of degrees smaller than 2 that factorize $F(x, y)$. Furthermore, $F(x, y)$ defines a curve $\mathcal{C}$ on the plane $\mathbf{K}^2$, namely the set of points in $\mathbf{K}^2$ that satisfy the equation $x^2 + y^2 - 1 = 0$. For example, the point $(\frac{2t}{t^2+1}, \frac{t^2-1}{t^2+1}) \in \mathbf{K}^2$ lies on the curve. Actually, this point is a generic point of the curve. ■

We can see that this definition of generic object corresponds to our first intuition: A point on the curve $\mathcal{C}$ (our class of objects) is generic if *all its properties* are also possessed by the other points on the curve. In this case "*all its properties*" refers to the algebraic properties that are definable by polynomial equations with coefficients in the ground field $\mathbb{C}$. A circumscriptive characterization —of the form indicated in (2)— of generic points of a curve should coincide with the algebraic definition.

The formal language has the following symbols: (a) Constants: $0, 1, a_1, a_2$ ($\bar{a} = (a_1, a_2)$ stands for the generic point); (b) Predicates: C (a unary predicate for the set of complex numbers or the base field), $\mathcal{C}$ (a binary predicate for the bidimensional curve), and S (a binary predicate for a set of singular points); (c) Functions: $+$ and $\cdot$ (for the usual field operations), and $p(x, y)$ for a fixed polynomial in two variables.

We consider the following axioms: (1) Axioms for fields; (2) Axioms for fields relativized to C; (3) Axioms stating that C is an algebraic substructure of the intended field (the extension field of C), that is, the axioms: $C(0)$, $C(1)$, $\forall xy(C(x) \wedge C(y) \rightarrow C(x + y) \wedge C(x \cdot y))$; (4) $\exists x \neg C(x)$ (C is a proper subfield); (5) An infinite, but recursive, list of axioms (a schema) stating that C is algebraically closed, that is, for each natural number $n > 0$, the axiom $\forall x_0 x_1 \cdots x_n (\bigwedge_{i=0}^{n} C(x_i) \wedge x_n \neq 0 \rightarrow \exists x(C(x) \wedge x_0 + x_1 \cdot x + \cdots + x_n \cdot x^n =$

[4] John McCarthy made the authors aware of the concept of generic point in algebraic geometry; he gave us this example.

0)); (6) An infinite, but recursive, list of axioms (a schema) stating that C has characteristic zero, that is, for each natural number $n > 0$, the axiom $1 + \cdots + 1$ (n times) $\neq 0$; (7) A definition of C as an algebraic curve, i.e., the axiom $\forall x, y(C(x, y) \leftrightarrow p(x, y) = 0)$, where $p(x, y)$ is an operation symbol standing for a fixed polynomial in x, y; (8) Axioms stating that $p(x, y)$ is a polynomial with coefficients in C: $\exists x_0 \cdots x_m (\bigwedge_{i=0}^{m} C(x_i) \wedge \forall xy(p(x, y) = \sum_{i=0}^{m} x_i \cdot x^{h_i} \cdot y^{k_i}))$ ($m, h_0, \ldots, h_m, k_0, \ldots, k_m$ are fixed, we might also use some extra constants for the coefficients of the polynomial); (9) A finite list of axioms stating that $p(x, y)$ is nontrivial, i.e., $\forall z(C(z) \rightarrow \exists xy(p(x, y) \neq z))$, and irreducible, that is, there is no pair of polynomials with coefficients in C of degree smaller than the degree of $p(x, y)$ that factorize $p(x, y)$.

This is our base theory Σ. In addition, we introduce two extra binary predicates S and *Scope*. By circumscribing the singularity predicate S, we expect to obtain the genericity of a point $\bar{a}$ on the curve C. For this purpose, we will assume that the axiom $\forall x \forall y(S(x, y) \leftrightarrow \exists z(z \neq 0 \wedge C(z \cdot y) \wedge C(z \cdot y)))$ that defines predicate S is included in theory Σ. S is the set of *ground points* of the plane based on the extension field. Ground points in $\mathbf{K}^2$ can be reduced, by multiplication of its coordinates by a common non-zero factor in $\mathbf{K}$, to points with both coordinates in $\mathbb{C}$. For example, $(t + ti, t)$ is a ground point of $\mathbb{C}(t) \times \mathbb{C}(t)$. Non ground points are called *transcendental*. The point $(\frac{2t}{t^2+1}, \frac{t^2-1}{t^2+1})$ is transcendental in $\mathbb{C}(t) \times \mathbb{C}(t)$. For all practical (and theoretical) purposes, we may assume that the ground points are those of the form (c, d) with c and d in the ground field $\mathbb{C}$.

We depart from the approach in [5], in the sense that, in this case, we consider point $\bar{a}$ as variable in the scoped circumscription.

Proposition 4. *Circum*$_{scoped}(\Sigma \cup \{C(\bar{a}), Scope(\bar{a})\};\ S;\ \bar{a}) \models \neg S(\bar{a})$.

Remark 2 1. By definition of scoped circusmcription, predicate *Scope* and the parameters that appear in the definition of predicate S are variable in the circumscription. In this case, we also make explicit the fact that point $\bar{a}$ is allowed to vary in the minimization process.
2. Proposition 4 tells us that the scoped circumscription of Σ plus "$\bar{a}$ belongs to the curve" and "$\bar{a}$ is in the scope of our default reasoning" entails that $\bar{a}$ is not a singular point.
3. As was shown in [5], this circumscription is the same as circumscribing, in the classical form, the defined predicate $S \cap Scope$ in the theory obtained from $\Sigma \cup \{C(\bar{a}), Scope(\bar{a})\}$ plus the definition of predicate S.
4. Since the base theory is infinite, our circumscription is not given in terms of a finitary first-order formula[5]. Nevertheless, there is no problem in defining such

[5] See [15] for some remarks on circumscription for infinitary theories. But, if we are willing to give up a first-order base theory, we may obtain a syntactical version for the circumscription by starting from a finite theory Σ' corresponding to a categorical second-order axiomatization of the complete ordered field of real numbers. This approach would require a new unary predicate R for real numbers. We might elim-

a circumscription in semantical terms, that is, in terms of minimal models of the infinite theory. ■

Proof: (of Proposition 4) Let $\mathfrak{K} = \langle K, C^K, +^K, \ldots, C^K, S^K, Scope^K, \bar{a}^K \rangle$ be a $(S \cap Scope)$-minimal model of $\Sigma \cup \{C(\bar{a}), Scope(\bar{a})\}$ plus the definition of predicate S. Assume that $\bar{a}^K \in S^K$, that is, a_1^K, and $a_2^K \in C^K$. Then, $\bar{a}^K \in S^K \cap Scope^K$.

Since C^K correspond to an irreducible C-algebraic curve in $K \times K$, we know by results from algebraic geometry that C has a transcendental (non ground) point, say, $\bar{b} = (b_1, b_2)$. We can use this point to give a new model $\mathfrak{K}'$ for the initial theory: Leave all the interpretations as in $\mathfrak{K}$ except for $\bar{a}$ and *Scope*, which are interpreted as $\bar{b}$ and $\{\bar{b}\}$, respectively.

Now, we have: $S^{K'} \cap Scope^{K'} = \emptyset \subsetneq \{\bar{a}^K\} \subseteq S^K \cap Scope^K$. This inclusion contradicts the minimality of model $\mathfrak{K}$. ■

The proof of Proposition 4 was simple because we allowed point $\bar{a}$ to vary in the circumscription. Despite the variability of $\bar{a}$, the use of *Scope* was decisive in the proof. The same proof does not work without using this auxiliary predicate. A natural question is whether we could get by without scoped circumscription, but allowing the point to vary[6]. In [5], some heuristic reasons were given for the failure of this alternative approach. The proof above provides additional evidence in that direction. Since these are not conclusive answers, the matter should be explored further.

From Proposition 4 and the fact that a point is a generic point of a curve if and only if it is transcendental (see [30]), we obtain the genericity of point $\bar{a}$:

Corollary 5. For every $n > 0$ and formula $\varphi(x, y, x_1, \ldots, x_n)$ of the form $\sum_{i=1}^{n} x_i \cdot x^u \cdot y^v = 0$, which defines an algebraic curve in x, y with coefficients $x_1, \ldots, x_n$,

$$Circum_{scoped}(\Sigma \cup \{C(\bar{a}), Scope(\bar{a})\};\ S;\ \bar{a}) \models \forall x_1, \ldots, x_n (\varphi(\bar{a}, x_1, \ldots, x_n) \rightarrow \forall x, y (C(x, y) \rightarrow \varphi(x, y, x_1, \ldots, x_n))).$$

■

In [32], Thom shows connections between the concept of generic point in algebraic geometry and the concept of generic object in catastrophe theory and differential dynamical systems. Thus, we foresee applications of circumscription to the characterization of this central concept in two other important areas of mathematics.

5.2 Another Intuition and Forcing

Besides the intuitions behind the concept of generic object that were discussed in section 2, there is yet another one, with stronger mathematical contents:

inate schemas (5) and (6) in Σ by introducing a new constant i and the axioms: $(C(i) \wedge i \cdot i + 1 = 0)$, and $\forall x (C(x) \rightarrow \exists yz (R(y) \wedge R(z) \wedge x = y + z \cdot i)$.

[6] This question was raised to the authors by Vladimir Lifschitz. This idea was influential in letting point $\bar{a}$ vary in our scoped circumscription.

Fifth Intuition: *If it is possible to prove something using (the properties of) a generic object, then it is possible to prove the same without the object*[7].

This is the intuition behind Cohen's forcing method for establishing the independence of some axioms of set theory, but in his case the concept of generic set has a precise meaning inside the theory. Actually, his idea consists in constructing a model, say of the Zermelo-Fraenkel axioms ZF of set theory plus the negation of the axiom whose independence is to be established, by adjoining a *generic* set G to a given model $\mathcal{M}$ of ZF, and thus obtaining a new model $\mathcal{N}$ with the desired properties. The intuition that "what can be proved with the generic G can also be proved without G" is made precise and established in the so-called Cohen's Truth Lemma that tells us that a proposition is true in the extended model $\mathcal{N}$ if and only if it is forced to hold by a condition imposed on G in the original model $\mathcal{M}$.

Cohen's idea of using forcing for the construction of models of set theory was developed by Abraham Robinson [27] in the more general context of model theory [7][8]. In this model-theoretic forcing, generic objects are outside the object level language or outside the universe of the model; they are *generic sets of sentences* in the language expanded with new names for individuals. The "generic model theorem" shows that, starting from a generic set of sentences G, it is possible to construct a *generic* model $M(G)$, such that each element of the universe is denoted by one of the new constants, and the sentences satisfied by $M(G)$ are exactly those sentences that are forced by a condition in G. Another theorem establishes that $M(G)$ is a model of the base theory Σ if Σ is an inductive theory, that is, if it can be axiomatized by $\forall\exists$-sentences. A detailed treatment of model-theoretic forcing can be found in [14].

A natural question that arises is whether generic models of theories correspond to models obtained by means of formalisms for commonsense reasoning. A generic model has the "problem" of being too big, in the sense of being existentially closed, that is, it must satisfy every existential sentence that is satisfied by any algebraic superstructure. On the other hand, models obtained by circumscription are minimal in a precise sense. For example, consider the very simple inductive theory $\Sigma = \{P(a) \vee P(b)\}$. A P-minimal model of Σ is $\mathcal{M} = \langle \{a\}, \{a\}, a, a \rangle$.

[7] This intuition is not far from a common mathematical practice: When we try to prove, under given hypothesis, that all objects have a given property, we usually consider a "new" object a that does not appear explicitly mentioned in the hypothesis, and prove that a has the given property. Accordingly, we consider as proven that all objects have the property since they inherit the property from that "generic" object. This kind of reasoning is sanctioned in some formal deductive systems through a "generalization rule".

[8] This is not surprising, if we think that set-theoretical forcing reminds us of the introduction of generic points in algebraic geometry, an older subject. It is also interesting that Robinson made a generalization and abstraction of the ideas behind generic points in algebraic geometry and introduced them in model theory, before the emergence of the concept of forcing [25, 26, 28].

This model cannot be generic because it is not existentially closed: Its superstructure $\langle\{a,b\},\{a,b\},a,b\rangle$ satisfies the sentence $\exists x(x \neq a \wedge P(x))$.

6 Some Examples of Genericity in Computer Science

In this section we show how the concept of generic object appears in different areas of computer science, either explicitly or implicitly. We also try to make clear that, in essence, all those occurrences follow the same pattern of definition, namely, that a generic object having certain — well determined — properties implies that all other objects in the class have those properties.

6.1 Generic Databases

In [4], Armstrong gives a set of inference rules for functional dependencies in databases. This system of inference rules is complete and sound in the sense that it is possible to obtain all and only the functional dependencies that are logical (semantical) consequences of a given set of functional dependencies, by application of the rules. This result can be obtained [10, 19, 29] from the following preliminary result due to Armstrong: "If F is a set of functional dependencies and F^+ denotes the deductive closure of F with respect to the inference rules, then there is a relation R such that for any functional dependency φ, $\varphi \in F^+$ if and only if φ is true in R".

According to this proposition, R satisfies all the dependencies in F, and we can check whether an arbitrary functional dependency deductively follows from F by checking its validity in the relation R. R is called an Armstrong's relation for F. We have the following situation: (1) Relation R satisfies F, and (2) For every functional dependency φ: If R satisfies φ, then every relation R' that satisfies F also satisfies φ.

Thus, we can say that R is a *generic* relation in the class $\mathcal{O}$ of relations that satisfy F: All the relations in the class inherit the functional dependencies (the properties) from the generic relation.

6.2 Generic Models of Sets of Clauses

Let $L(S)$ be a first-order language, U_H the Herbrand Universe, and B_H the Herbrand Base [18]. A Herbrand interpretation for L(S) has U_H as its universe, and each n-ary function symbol $f \in S$ is interpreted by the function that sends each $(t_1,\ldots,t_n) \in U_H^n$ to the ground term $f(t_1,\ldots,t_n)(\in U_H)$. The interpretation of the predicate symbols is determined by a subset of the Herbrand Base. We have the following theorem that tells us that we can check the consistency of sets of clauses by considering only Herbrand interpretations: A set of closed clauses

$C \subseteq L(S)$ is satisfiable if and only if it is satisfied by a Herbrand interpretation for $L(S)$.

Let us consider the class $\mathcal{I}(S)$ of interpretations for language $L(S)$. Our objects, collected in a class $\mathcal{O}$, are the subsets of $\mathcal{I}(S)$. The class H of Herbrand interpretations for $L(S)$ is a particular object in the class. H is a generic object in the following sense: For each set C of closed clauses in $L(S)$, we may define the following property P_C of sets of interpretations: $P_C(A) \;:\Leftrightarrow\; \forall \mathfrak{A} \in A : \mathfrak{A} \not\models C$.

The non trivial direction of the previous theorem tells us: For all P_C: $P_C(H) \Rightarrow \forall A \in \mathcal{O} : P_C(A)$. That is, if the generic object H has any of the properties P_C, then all objects in the class $\mathcal{O}$ have that property.

6.3 Horn Clauses Admit Generic Models

We consider here only propositional formulas, say, of a language $L(P)$. According to Makowsky [20], a truth valuation σ is *generic* for a set of propositional formulas Σ iff: (1) $\sigma \models \Sigma$, and (2) for each propositional variable $p \in P$: $\sigma(p) = 1 \Leftrightarrow \Sigma \models p$. This definition means that a generic model of a set of formulas makes true only those propositional variables that are forced to be true by Σ. It is extremely economical in making propositional variables true.

Let us denote by $\mathcal{O}$ the class of valuations that satisfy Σ, and define the following properties S_p $(p \in P)$ on valuations: $S_p(\sigma) \;:\Leftrightarrow\; \sigma(p) = 1$. We can easily prove the following:

Proposition 6. A valuation σ_0 is generic for Σ iff:

(a) $\sigma_0 \in \mathcal{O}$, and
(b) for all S_p: $S_p(\sigma_0) \Rightarrow$ for all σ: $(\mathcal{O}(\sigma) \Rightarrow S_p(\sigma))$. ■

Our proposition shows that Makowsky's definition captures one of the usual intuitive notions of genericity: If the generic valuation has the distinguished properties of the class to which it belongs, then all valuations in the class have those properties. We can also easily show from results in [20] that the generic model of a set of formulas can be obtained as the parallel propositional circumscription (see [15]) of all the propositional variables in the formula. As Makowsky points out, generic models give a semantics to Reiter's Closed World Assumption [23] applied to propositional knowledge bases (see also [13]). The relationship between CWA and circumscription has been explored in [16].

6.4 Data Types and Generic Functions

There is a notion of generic function that appears in relation to the concept of parametric polymorphic function in data type specifications [6]. Polymorphic functions are functions whose operands can be of more than one type. Polymorphic types may be defined as as types whose operations are applicable to operands of more than one type. An example of a polymorphic function is the length of a list of objects, which returns integer values independently of the types of the objects. This function is not only an example of a polymorphic function, but also an example of a generic function: It works for arguments of many types, generally doing the same kind of work independently of the argument's type. Other examples of generic polymorphic functions (or procedures) are provided by the usual comparison-based sorting algorithms [22]; they work by comparison of keys, without considering the type and contents of the records.

Let us consider a fixed set of data types A and the class $\mathcal{O}$ of functions that, for some type $a \in A$, and for each $x \in a$, compute the unique value y, such that $R(x, y)$ holds. $R(\cdot, \cdot)$ is a functional relation that exists, say, in a broader setting, e.g. set theory, and supports objects of different types. For example, $R([x_1, \ldots, x_n], n)$ could be the functional relation that defines the length of a list of objects $x_1, \ldots, x_n$. These objects could be integers, reals, or objects of any type. Notice that according to our definition of the class $\mathcal{O}$, there might be more than one type associated to a function in the class, in the sense that the function computes the function values for different types. Let us denote with f_a a function that computes the function for arguments of type a.

A function $g \in \mathcal{O}$ is *generic* if the following holds: $\forall\, a \in A \,\forall\, x \in a \,\forall\, y (f_a(x) = y \Leftrightarrow g(x) = y)$.

If we denote with upper-case letters the functional relations corresponding to the functions in $\mathcal{O}$, then some facts follow from the definition of generic function: (a) $\bigcup_a F_a = G$, (b) $\forall\, a\ F_a = G \cap a$, and (c) $\forall\, x, y\ ((x, y) \notin G \Rightarrow \forall\, a\ (x, y) \notin F_a)$. In particular, item (c) expresses something of the form *"if a generic function has a property, then all functions in the class have that property"*.

6.5 Most General Unifiers Are Generic

Let us consider the class $\mathcal{O}$ of unifiers for a set of expressions E as they appear in logic programming [18]. A unifier μ is a most general unifier (an mgu) in the class if any unifier λ in the class can be represented in the form $\lambda = \mu \circ \delta$, for some assignment δ. That is, we obtain any unifier in the class by starting with the mgu μ and applying to the remaining variables the assignment δ. In this sense, the mgu is a unifier that assigns exactly those ground terms to variables in E that it is forced to assign. We have the following property for the mgu: For all variables x in E and ground term t: $\mu : x \mapsto t \Rightarrow \forall \lambda \in \mathcal{O} :\ \lambda : x \mapsto t$,

Again, this statement follows a known pattern: *"If a mgu has the property of assigning a ground term to a variable, then all the unifiers in the class have the same property"*. We can see that the notions "(most) general" and "generic" can be sensibly seen as interchangeable.

6.6 Generic Query Languages for Databases

In database theory there is the concept of generic query language [2, 3, 12]. The idea is the following: Databases usually do not show the internal representation of the data. Furthermore, when we query or update the database, the kind of computations that are performed treat data in an abstract, generic, form. On the other side, query languages usually refer only to logical properties of the data without referring to internal data representation. Database transformations should treat values as essentially uninterpreted objects [1]. More precisely [3], a query is a mapping ϕ from finite structures that are instances of an input database schema to finite structures over an output database schema. The mapping must be computable, and generic, in the sense that for each automorphism ν of the input I, $\varphi(\nu(I)) = \nu(\phi(I))$.

Again, we find the common pattern behind this notion of genericity. Results that can be obtained by means of languages that provide generic queries, or by means of computational transformations that are generic, can be obtained by means of any other (correct) query language or computational transformation, in particular, by means of semantics– or internal representation–based query languages or transformations.

6.7 Generic Points for Negation in Logic Programming

Assume that we have a normal program in logic P [18]. It is known that there are problems in the presence of a non ground negated subgoal of the form "$\leftarrow$ *not* $g(x)$".

The first-order declarative meaning of a program clause of the form "$f(x) \leftarrow$ *not* $g(x)$", containing "*not* $g(x)$" as a subgoal, can be captured by the formula "$\forall x(f(x) \leftarrow \neg g(x))$". Accordingly, if our original query is "$\leftarrow f(x)$", asking if there exists some value for x that makes f true, then we should ask if there exists a value for x that makes $\neg g(x)$ true. On its turn, this leads us to ask if *for all values of* x, $g(x)$ is true. There have been some attempts to capture in logic programming this classical meaning for negated uninstantiated subgoals. Nevertheless, in current implementations of PROLOG the procedural meaning is quite different, and is closer to "negation as failure". Actually, if we ask the query ": – *not* $g(X)$.", then: *not* $g(X)$ succeeds $\Leftrightarrow$ query $g(X)$ finitely fails.

Now, query $g(X)$ finitely fails if the control strategy makes possible to check exhaustively whether for all possible ground terms $\mathbf{t}$, $g(\mathbf{t})$ (finitely) fails. Then, the last double implication above can be put in equivalent terms: *not* $g(X)$ succeeds $\Leftrightarrow$ *for all* $\mathbf{t}$, $g(\mathbf{t})$ finitely fails.

That is: *not* $g(X)$ succeeds $\Leftrightarrow$ *for all* $\mathbf{t}$, *not* $g(\mathbf{t})$ succeeds. This equivalence reminds us of our notion of genericity: X seems to be a generic element with respect to the property "*not* $g(\cdot)$". This suggests that, in order to evaluate a negated non ground query ": – *not* $g(X)$", we might attempt to evaluate the query" : – *not* $g(\mathrm{e})$", where e is a generic element.

It seems that in [11], Ginsberg attempts to materialize this kind of approach to the evaluation of uninstantiated negated subgoals, by defining appropriate test elements **e**.

Acknowledgments: We are grateful to Tania Bedrax-Weiss, Alvaro Campos, Vladimir Lifschitz, Jorge Lobo, John McCarthy, and Pablo Straub for discussions, suggestions, and pointers to the literature; and to Gabriel Kuper, Rene Peralta, and Moshe Vardi for pointers to the literature and for making some papers available. The first author gratefully acknowledges the financial support of several grants (FONDECYT # 1930554 , DIUC, Fundacion DICTUC).

References

1. Aho, A.V. and Ullman, J.D. "Universality of Data Retrieval Languages". Proc. 6th ACM Symposium on Principles of Programming Languages (San Antonio, Texas), 1979, 110–120.
2. Abiteboul, S. and Vianu, V. "Procedural Languages for Database Queries and Updates", J. Computer and Systems Sciences, 41, 1990, 181–229.
3. Abiteboul, S. and Vianu, V. "Generic Computation and its Complexity", Proceedings STOC'91, ACM, 209–219.
4. Armstrong, W.W. "Dependency Structures of Database Relationships". Proc. IFIP 74, North Holland, 1974, 580–583.
5. Bertossi, L.E. and Reiter, R. "Circumscription and Generic Mathematical Objects". Working Notes of the 4th International Workshop on Nonmonotonic Reasoning, Vermont, May 1992, AT&T and AAAI. To appear in Fundamenta Informaticae.
6. Cardelli, L. and Wegner, P. "On Understanding Types, Data Abstraction, and Polymorphism". ACM Computing Surveys, 17, 4, 1985, 471–522.
7. Chang, C.C. and Keisler, J. "Model Theory". North-Holland, 1978.
8. Cohen, P.A. "Set Theory and the Continuum Hypothesis". Benjamin, 1966.
9. Etherington, D.W., Krauss, S. and Perlis, D. "Nonmonotonicity and the Scope of Reasoning". Artificial Intelligence 52, 1991, 221–261.
10. Fagin, R. "Functional Dependencies in a Relational Database and Propositional Logic". Revision of IBM Research Report RJ 1776, San Jose, California, 1976.
11. Ginsberg, M. "Negative Subgoals with Free Variables". Journal of Logic Programming, 11, 1991.
12. Hull, R. and Yap, C.K. "The Format Model: A Theory of Database Organization", J. ACM, 31, 3, 1984, 518–537.
13. Hodges, W. "Logical Features of Horn Clauses". Handbook of Logic in Artificial Intelligence and Logic Programming, vol. 1, 449–503, Oxford University Press, 1993.
14. Keisler, J.H. "Fundamentals of Model Theory". In *Handbook of Mathematical Logic*, Barwise, J. (ed.), North-Holland, 1977.
15. Lifschitz, V. "Circumscription". In 'Handbook of Logic in Artificial Intelligence and Logic Programming', Oxford University Press. To appear.
16. Lifschitz, V. "Closed-World Databases and Circumscription". Artificial Intelligence, 27, 1985, 229–235.

17. Lifschitz, V. "Computing Circumscription". Proc. 9th IJCAI 1985, Morgan Kaufmann Publishers, 1985, 121–127.
18. Lloyd, J.W. "Foundations of Logic Programming". Springer, 2nd edition, 1987.
19. Maier, D. "The Theory of Relational Databases". Computer Science Press, 1983.
20. Makowsky, J.A. "Why Horn Formulas Matter in Computer Science: Initial Structures and Generic Examples". Journal of Computer and Systems Sciences 34, 1987, 266–292.
21. McCarthy, J. "Circumscription – a form of non-monotonic reasoning". Artificial Intelligence, 13, 1980, 27–39.
22. Mehlhorn, K. "Algorithms and Data Structures". Springer, vol. 1, 1984.
23. Reiter, R. "On Closed World Data Bases". In 'Logic and Data Bases', Gallaire, H. and Minker, J. (eds.), Plenum Press, 1978, 55–76.
24. Reiter, R. "Nonmonotonic Reasoning". Annual Reviews in Computer Science, vol. 2, 1987.
25. Robinson, A. "On the Metamathematics of Algebra". North Holland, 1951.
26. Robinson, A. "Théorie Métamathématique des Idéaux". Gauthier-Villars, Paris, 1955.
27. Robinson, A. "Forcing in Model Theory". Symposia Mathematica, vol. 5, Academic Press, New York, 1971, 69–82.
28. Robinson, A. "Introduction to Model Theory and to the Metamathematics of Algebra". North-Holland, 1974.
29. Sagiv, Y., Delobel, C., Parker, D.S. and Fagin, R. "An Equivalence between Relational Database Dependencies and a Fragment of Propositional Logic". Journal of the ACM, 48, 3, 1981, 435–453.
30. Semple, J.G. and Kneebone, G.T. "Algebraic Curves". Oxford University Press, 1959.
31. Seidenberg, A. "Elements of the Theory of Algebraic Curves". Addison-Wesley, 1968.
32. Thom, R. "Structural Stability and Morphogenesis", W.A. Benjamin, Inc., 1975.

Autoepistemic Logic of Minimal Beliefs

Teodor C. Przymusinski

Computer Science Department
University of California at Riverside
Riverside CA 92521-0304
USA

Abstract

In recent years, various formalizations of non-monotonic reasoning and different semantics for normal and disjunctive logic programs have been proposed, including autoepistemic logic, circumscription, *CWA*, *GCWA*, *ECWA*, epistemic specifications, stable, well-founded, stationary and static semantics of normal and disjunctive logic programs.

We introduce a simple non-monotonic knowledge representation framework which isomorphically contains all of the above mentioned non-monotonic formalisms and semantics as special cases and yet is significantly more expressive than each one of these formalisms considered individually. The new formalism, called the *AutoEpistemic Logic of minimal Beliefs*, *AELB*, is obtained by augmenting Moore's autoepistemic logic, *AEL*, with an additional *minimal belief* operator, $\mathcal{B}$, which allows us to explicitly talk about minimally entailed formulae.

The existence of such a uniform framework not only results in a new powerful non-monotonic formalism but also allows us to compare and better understand mutual relationships existing between different non-monotonic formalisms and semantics and enables us to provide simpler and more natural definitions of some of them. It also naturally leads to new, even more expressive and flexible formalizations and semantics.

Keywords: Non-Monotonic Reasoning, Logics of Knowledge and Belief, Semantics of Logic Programs and Deductive Databases.

How to Use Modalities and Sorts in Prolog

Andreas Nonnengart

Max-Planck-Institut für Informatik,
Im Stadtwald, 66123 Saarbrücken, Germany,
Tel.: (+49) 681 302 5369, Fax: (+49) 681 302 5401,
Email: Andreas.Nonnengart@mpi-sb.mpg.de

Abstract. Standard logic programming languages like Prolog lack the possibility of dealing with modalities and/or sorts. A first idea how to overcome this problem (and that without changing anything on Prolog itself) would be to apply the well-known relational translation approaches from modal and sorted logic into first-order predicate logic and to feed this translation result into Prolog. This, however, leads into other problems: firstly, the transformed problem is usually of much bigger size (number of clauses) than the original one and, secondly, very often it is not even in Horn form anymore.
In this paper a translation approach is proposed which avoids both of these problems, i.e. the number of clauses after translation is exactly as big as it would have been if we simply ignored the modal operators and sort restrictions and, also, the result is in Horn form provided it was already before (modulo modal operators and sorts).

1 Introduction

For many applications in the AI community (as for instance knowledge representation and taxonomic reasoning for several agents) there is a need for modalities and/or sort declarations. On the other hand, one of the most popular AI programming languages is Prolog and it is thus evident that a combination of Prolog with modalities and sorts apparently makes sense for AI implementations.

However, it is not at all obvious how modalities and sorts could be integrated into Prolog without changing Prolog itself. A first idea might be to use standard relational translation techniques into first-order predicate logic as they are perfectly well known from the literature. With these techniques it would be possible to translate modal and sorted logic programs into classical logic before being fed to the logic programming environment.

However, soon one will have to realize that there occur quite a lot of problems with this approach. First, in the extension by modalities, the translation results in an exponential explosion of the output formula. Also, certain properties of the underlying accessibility relation, as for example symmetry, result in clauses which inevitably lead to infinite loops under the Prolog control stucture. And finally, for both modal logics and sorted logics, the resulting formula is not necessarily in Horn form even if the original formula was (if we ignored the modalities or the sort declarations respectively).

The translation approach proposed in this paper avoids these difficulties. Its theoretical basis has already been introduced in [4], although the possible application to logic programming has not been mentioned there (but in [3]) and the application to sorted logic has not been published before at all.

One of the most remarkable things about this approach is that it very often allows a significant simplification of the background theory given by the modal logic accessibility relation properties and sort declarations.

Note that this method is very closely related to the functional translation approach proposed by Ohlbach in [5], [6], [7] and others. At least the idea of replacing occurrences of certain binary predicates by suitable sets of functions is in essence identical to the approach presented there. Nevertheless, there are some quite important differences between the functional translation method and the one propose here. In the functional translation, theory clauses which stem from certain properties of the underlying accessibility relation get replaced by more or less complicated equations. These require either a pretty strong equality handling of the inference procedure or a theory unification algorithm which suits to the theory induced by the equations. Usually, however, neither of these features are available in standard logic programming languages like Prolog. Therefore, if we want to avoid changes in the logic programming language itself, we have to think of something else. Indeed, the translation approach presented in this paper does not require any changes in the programming language at all. To some extent it can be viewed as a mixed approach between relational and functional translation which tries to combine advantages from both and also tries to avoid their respective disadvantages as much as possible.

This paper is organized as follows: section 2 contains the basics of the relational translation for modal and sorted logics. Section 3 then explains how the functional simulation approach works and how it can be applied to modal and sorted logics. After that a certain saturation technique is introduced which allows to considerably simplify the background theories (for modal logics). Finally, the effect of the overall approach is demonstrated with two examples.

2 Relational Translation

Here and in the sequel we assume that the reader is already familiar with the basic principles of modal and sorted logics. Also, proofs are usually omitted in this section. In case the reader is interested in these s/he is referred to [4] and [3].

The idea of the relational translation is to make the implicit semantics of the extra features of the underlying logic explicit in the translation. I.e. in case of modal logics each predicate symbol, say P, gets replaced by a new predicate symbol, say P', which accepts one more argument, namely (a symbol denoting) the current (or actual world) (similar for function symbols). In addition, two new predicate symbols are to be introduced, R and E, which represent the accessibility relation and the domain element existence predicate respectively (in case of varying domains).

The relational translation approach for sorted logic as it is presented here is maybe not exactly as the reader might expect. For reasons which will become clear later it is not the case that for each sort symbol a unary predicate symbol is introduced. Rather a single binary predicate symbol S will be used to express elementship of domain elements in a certain sort which itself is represented by a constant.

2.1 (Serial) Modal Logics

The relational translation approach for modal logics is defined as follows:

Definition 1. The formula morphism $\lfloor \Phi \rfloor_u$ which accepts a modal logic formula and a symbol denoting a world is defined as:

$$\begin{aligned} \lfloor x \rfloor_u &= x \\ \lfloor f(t_1,\ldots,t_n) \rfloor_u &= f'(u, \lfloor t_1 \rfloor_u, \ldots, \lfloor t_n \rfloor_u) \\ \lfloor P(t_1,\ldots,t_n) \rfloor_u &= P'(u, \lfloor t_1 \rfloor_u, \ldots, \lfloor t_n \rfloor_u) \end{aligned}$$

The case for the usual connectives should be clear

$$\begin{aligned} \lfloor \forall x\, \Phi \rfloor_u &= \forall x\; E(u,x) \Rightarrow \lfloor \Phi \rfloor_u \\ \lfloor \exists x\, \Phi \rfloor_u &= \exists x\; E(u,x) \wedge \lfloor \Phi \rfloor_u \\ \lfloor \Box \Phi \rfloor_u &= \forall v\; R(u,v) \Rightarrow \lfloor \Phi \rfloor_v \\ \lfloor \Diamond \Phi \rfloor_u &= \exists v\; R(u,v) \wedge \lfloor \Phi \rfloor_v \end{aligned}$$

The initial call for the translation of an arbitrary formula Φ is thus $\lfloor \Phi \rfloor_\iota$, where ι denotes the initial (or actual) world.

Translating a given modal logic program (and query) with the help of the formula morphism from above is evidently not enough if the accessibility relation obeys certain extra properties as e.g. reflexivity, transitivity or symmetry. These properties have to be made explicit as well. Therefore there are usually some extra clauses to be added to the clause set after translation, namely those which reflect the background theory.

Theorem 2. *Relational translation is sound and complete, i.e. Φ is (modal logic) satisfiable iff $\lfloor \Phi \rfloor_\iota \wedge ML$ is (predicate logic) satisfiable, where ML denotes those formulae which stem from the additional properties of the accessibility relation of the modal logic under consideration.*

Proof. Can be found in [3].

Thus, in principle, it is possible to translate modal logic programs and queries into predicate logic and feed the result (up to the possibility of a getting non-Horn programs) into Prolog. However, as it turns out, this is not very satisfactory. Mainly there are two problems: one is that there is an exponential explosion in the size and the number of clauses because of the translation. Another, even more restrictive effect is that usually Prolog does not terminate when it is fed with these translated versions of the original modal logic program. The following example illustrates this:

Example 1. Consider the simple program: $\Diamond\Box\Diamond P$ and suppose the question is whether $\Diamond P$ holds (i.e. the query is: $\Leftarrow \Box P$). Moreover assume that the modal logic under consideration is KDB, i.e. the accessibility relation is serial and symmetric. Negation, relational translation and transformation into Horn form results in the following Horn clause set:

$$
\begin{array}{ll}
\text{From } \Diamond\Box\Diamond P & \\
R(\iota, a). & \\
R(x, b(x)) & \Leftarrow R(a, x). \\
P(b(x)) & \Leftarrow R(a, x). \\
\text{Symmetry and Seriality} & \\
R(u, v) & \Leftarrow R(v, u). \\
R(u, f(u)). & \\
\text{The query} & \\
 & \Leftarrow P(y), R(\iota, y).
\end{array}
$$

Running this as a Prolog program results in an infinite loop. In fact, a closer examination of the clause set shows that there is no possible arrangement of the given clauses such that an infinite loop can be avoided, although the set of clauses is indeed unsatisfiable.

The method presented later avoids such infinite loops and, in addition, reduces the number (and sometimes size) of clauses after translation considerably.

2.2 Sorted Logic

Definition 3. The formula morphism $\lceil \Phi \rceil$ which accepts a sorted logic formula Φ and returns a predicate logic formula is defined as:

$$
\begin{aligned}
\lceil \forall x{:}\, A\ \Phi \rceil &= \forall x\ S(a, x) \Rightarrow \lceil \Phi \rceil \\
\lceil \exists x{:}\, A\ \Phi \rceil &= \exists x\ S(a, x) \wedge \lceil \Phi \rceil
\end{aligned}
$$

$$
\begin{aligned}
\lceil A \sqsubseteq B \rceil &= \forall x\ S(a, x) \Rightarrow S(b, x) \\
\lceil e{:}\, A \rceil &= S(a, e)
\end{aligned}
$$

where A, B are arbirary sorts and Φ is an arbitrary (sorted logic) formula. For the other logical connectives just apply the usual homomorphic extension of the above.

Note that, instead of introducing unary predicate symbols for the sorts, we just add a single new (binary) predicate symbol, namely S, and for each sort a new constant such that $S(a, x)$ simply means that x belongs to the sort A. This has no effect on the soundness and completeness of the translation; it just fits better to the way we handled modalities.

Theorem 4. *Relational translation for sorted logics is sound and complete, i.e. for any sorted logic formula Φ: Φ is (sorted logic) satisfiable iff $\lceil\Phi\rceil$ is (predicate logic) satisfiable.*

Here again we have the difficulty that the translation of some sorted Horn formula is not necessarily Horn after translation (as an example consider the formula $\exists x{:}\,A\ P(x) \vee \exists y{:}\,B\ \neg Q(y)$). In the following section we will describe how this problem can be solved.

3 Functional Simulation of Binary Relations

3.1 Basic Principles

The last section provides a possibility of translating modal logic or sorted logic programs (formulae) into predicate logic by introducing new auxiliary predicates, namely R (which represents the accessibility relation for modal logics), E (which denotes the existence relation for domain elements in worlds), and S (for the "is-element-of-sort" relation for sorted logics). These are called *auxiliary* predicates since they do not occur in the original formulae; they are introduced by the translation. What is interesting about these predicates is that they occur solely in particular places inside formulae, namely either as $\forall x\ P(\alpha, x) \Rightarrow \ldots$ or as $\exists x\ P(\alpha, x) \wedge \ldots$. This fact can be utilized by the introduction of *functional simulators.* Intuitively, a functional simulator F_R for some binary predicate R is a set of functions which covers the responsibilities of this very relation.

In fact, it is not really necessary to consider F_R as a set of functions, i.e. it is nowhere really required that the elements of a functional simulator do indeed have such typical properties like: two (total) functions are identical iff they have the same result on every input. It is rather recommended to view them as nothing but a new set of elements which all can be "applied" to other kind of elements. Thus, and in order to stay in first-order syntax, instead of writing: $\exists f\ P(f(x))$ we will write: $\exists f\ P(x{:}f)$.

3.2 Application to Serial Modal Logics

In serial modal logics we get formulae of the kind $\exists x\ R(\alpha, x) \wedge \Phi(x)$ after translating modal logic programs of the form $\Diamond\Phi$. The alternative formula morphism which we are going to define next handles therefore exactly this situation.

Definition 5. The formula morphism $\lfloor\Phi\rfloor^\star_u$ which accepts a modal logic formula Φ in negation normal form[1] and a symbol u denoting a world is exactly like $\lfloor\Phi\rfloor_u$ with two exceptions, namely

$$\begin{aligned} \lfloor\Diamond\Phi\rfloor^\star_u &= \exists x_R\ \lfloor\Phi\rfloor^\star_{u:x_R} \\ \lfloor\exists x\ \Phi\rfloor^\star_u &= \exists y_E\ \lfloor\Phi\rfloor^\star_u[x/u{:}y_E] \end{aligned}$$

[1] A formula is in negation normal form if all the implication and equivalence signs are eliminated and all negations are put inwards as far as possible. It is evident that any formula can be equivalently transformed into its corresponding negation normal form.

This alternative translation method does not yet preserve unsatisfiablity (although it preserves satisfiability). Something is missing which we had from the relational translation and which we don't have in this kind of translation anymore, namely the duality of $\Box$ and $\Diamond$ and the duality of $\forall$ and $\exists$. However, we certainly need these dualities and therefore we have (in addition to the axiom schemata which we get by considering special modal logics) two additional axiom schemata, namely $\Box\Phi \Leftrightarrow \neg\Diamond\neg\Phi$ and $\forall x\ \Phi \Leftrightarrow \neg\exists x\ \neg\Phi$. We cannot deal with these axiom schemata directly, however, we can try to find their respective first-order correspondences provided they exist at all. And indeed there are such characteristic correspondences, namely: $\forall u, x_R\ R(u, u{:}x_R)$ and $\forall u, v\ R(u, v) \Rightarrow \exists x_R\ u{:}x_R = v$ for the duality of $\Box$ and $\Diamond$ and similarly $\forall u, x_E\ E(u, u{:}x_E)$ and $\forall u, v\ E(u, v) \Rightarrow \exists x_E\ u{:}x_E = v$ for the duality of $\forall$ and $\exists$.

The usefulness of this result would be questionable if we had not the following theorem which states that the conditioned equations are not necessary.

Theorem 6. *Let Φ be a (serial) modal logic formula in negation normal form and let* `Axioms` *be the set of clauses induced by the properties of the underlying accessibility relation (i.e. the first-order representation of the accessibility relation properties of ML as they are known from the relational translation). Then Φ is ML-unsatisfiable iff $\lfloor\Phi\rfloor_\iota^\star \wedge R(u, u{:}x_R) \wedge E(u, u{:}x_E) \wedge$* `Axioms` *is predicate logic unsatisfiable where* `Axioms` *are again the characteristic accessibility relation properties for the logic ML.*

Proof. (Sketch; for more details see [3] and [4])
For the "only-if"-direction of the proof it first has to be shown that a suitable functional simulator can be defined. This is quite easy by showing that any serial binary relation can be split into enumerably many functional subrelations.
In order to guarantee that the $\Box$ and $\Diamond$-operators are still duals (in spite of the "non-dual" translation) we have to find out what the first-order equivalent of the axiom schema $\Box\Phi \Leftrightarrow \neg\Diamond\neg\Phi$ is and it turns out that this results in the conjunction of the unit clause $R(u, u{:}x)$ with the conditioned equation $R(u, v) \Rightarrow \exists x\ u{:}x = v$ (and similarly with the existence-predicate E). Thus the only thing which remains to be shown is that this conditioned equation is indeed superfluous. To this end we consider this equation as being directed from left to right (this is allowed because after skolemization the right-hand-side variables form a proper subset of the left-hand-side variables) and give it a high priority in the resolution process, i.e. paramodulation steps have higher priority than resolution steps (ordered resolution). Now the negation normal form comes into play, because it can easily be seen that if the original formula to be translated is in negation normal form then there is no application of the equation possible on the translation result. Also there is no possible application of the equation on any of the elements in `Axioms`. Therefore they can be applied only to $R(u, u{:}x)$ (or $E(u, u{:}x_E)$ respectively) which results in a tautology. Thus, the conditioned equations are indeed superfluous.
For the right-to-left direction of the proof it is very easy to see that a fully relational model can be constructed from a translated model by adding $R(\alpha, \beta)$ for

each subterm $\alpha':\beta'$ such that the interpretation of α in the one model is identical to the interpretation of α' in its translation and, analogously, the interpretation of β is identical to the (translated) interpretation of $\alpha':\beta'$.

Thus, instead of using the full relational translation we are allowed to perform the alternative version which is much more compact provided we add the single unit clause $R(u, u{:}x_R)$ (and also maybe $E(u, u{:}x_E)$ if we consider varying domains) to the program clauses. What indeed has been gained is stated in the following lemma.

Lemma 7. *Let Φ be a set of modal Horn formulae in negation normal form and let Ψ be exactly like Φ if we ignore the modal operators. Then*

- $\lfloor \Phi \rfloor_\iota^\star$ *contains exactly as many clauses as* Ψ
- $\lfloor \Phi \rfloor_\iota^\star$ *is Horn iff* Ψ *is Horn*

Proof. Simple induction over the structure of Φ.

3.3 Application to Sorted Logic

Definition 3 already gives us a hint how the functional simulation approach can be applied to sorted logics as well. Here we have an auxiliary binary predicate S which can only occur in the form we require (note that in analogy to modal logics we have to assume that sorts are always non-empty). Thus we are allowed to change the relational translation accordingly to:

$$\lceil \exists x{:}\, A\ \Phi \rceil^\star = \exists y\ \lceil \Phi \rceil^\star [x/a{:}y]$$

where the other possible cases remain as in Definition 3. By the same argument as for the application to modal logics we still have to guarantee the ∀–∃–duality and we do this by adding the unit clause $S(u, u{:}x)$ (note that here as well the other direction of the duality, namely the conditioned equation, is again superfluous provided the formula to be translated is in negation normal form). Thus we have:

Theorem 8. *Let Φ be a sorted logic formula in negation normal. Then Φ is (sorted logic–)unsatisfiable iff $\lceil \Phi \rceil^\star \wedge S(u, u{:}x)$ is predicate logic unsatisfiable.*

Proof. Analogous (and even simpler) to the proof above.

What we gain from this translation is, in analogy to modal logics, a reduction in the number of generated clauses and, also, that the result is Horn if the sorted program was Horn (modulo sort declarations). We fix this by the following lemma:

Lemma 9. *Let Φ be a set of sorted Horn formulae in negation normal form and let Ψ be exactly like Φ if we ignore the sort restrictions on variables. Then*

- $\lceil \Phi \rceil^\star$ *contains exactly as many clauses as* Ψ
- $\lceil \Phi \rceil^\star$ *is Horn iff* Ψ *is Horn*

Proof. Simple induction over the structure of Φ.

4 (Partial) Saturation

What is remarkable about the alternative translation is not only that it does not increase the number of clauses or that it keeps the Horn structure of formulae but also that it results in clauses in which the auxiliary predicate symbol occurs only negatively. Thus the only positive occurrences of such an auxiliary predicate symbol are in the background theory, i.e. either in the unit clauses $R(u, u{:}\,x_R)$, $E(u, u{:}\,x_E)$ and $S(u, u{:}\,x)$ which stem from the duality schemata or in the additional `Axioms` which describe the modal logic accessibility relation properties or the subsort relationships. In particular in the case of modal logics it can be shown that this can be utilized in a way which sometimes significantly simplifies this background theory.

The idea of this approach is to consider only this background theory, knowing that it is characteristic for the logic in question and not for the theorem to be proved.

Definition 10 (Saturation). Let C be a set of clauses and let P be a designated predicate symbol which occurs positively in each of the elements of the clause set. Then we call the clause set which we can get by resolution in C and whose elements do not contain any negative occurrence of P the *saturation* of C with respect to P.

As a little example consider the clause set $\{P(a), \neg P(x) \vee P(f(x))\}$. Its saturation obviously is: $\{P(f^n(a)) \mid n \geq 0\}$.

Knowing about the saturation of a given clause set is often quite helpful as the following lemma states.

Lemma 11. *Let C be a clause set and let P be a designated predicate symbol. Moreover let $C' \subseteq C$ be exactly the subset of C which contains P positively and let C'' be the saturation of C' with respect to P. Then C is unsatisfiable iff $C \setminus C' \cup C''$ is unsatisfiable.*

Proof. Easy by definition of saturation.

The problem with the above lemma is that saturations are usually infinite. However, if we are able to find a finite alternative clause set with exactly the same saturation we can use this one instead.

Theorem 12. *Let C be a finite clause set, P be a designated predicate symbol, $C' \subseteq C$ be exactly the subset of C which contains P positively and C'' be a finite set of clauses whose elements have a positive occurrence of P and whose saturation with respect to P is identical to the saturation of C' with respect to P. Then C is unsatisfiable iff $C \setminus C' \cup C''$ is unsatisfiable.*

Proof. Follows directly from the lemma above.

Thus the idea is to extract C' and to find a hopefully simpler clause set C'' with the same saturation.

4.1 Application to serial modal logics

Recall that the translation of an arbitrary set of (Horn) clauses results in a clause set which contains no positive R- or E-literal at all. Therefore the only positive occurrences are within the background theory given by the two duality unit clauses and the accessibility relation properties (and also maybe the domain properties). It is thus evident that this background theory is a wonderful candidate for the clause subset to be extracted and that in particular because this background theory is characteristic for the logic in question and not the theorem to be proved.

Up to now the saturation of quite a lot of modal logics have been examined in order to find suitable simpler alternatives. We won't go into detail for each of these logics but simply summarize the results in a table (the interested reader is again referred to [4] and [3]).

Logic	Axioms	Alternative
KD	$R(u, u{:}x)$	$R(u, u{:}x_R)$
KT	$R(u, u{:}x)$ $R(u, u)$	$R(u, u{:}x_R)$ $R(u, u)$
KDB	$R(u, u{:}x)$ $R(u, v) \Rightarrow R(v, u)$	$R(u, u{:}x_R)$ $R(u{:}x_R, u)$
KD4	$R(u, u{:}x)$ $R(u, v) \wedge R(v, w) \Rightarrow R(u, w)$	$R(u, u{:}x_R)$ $R(u, v) \Rightarrow R(u, v{:}x_R)$
S4	$R(u, u{:}x_R)$ $R(u, u)$ $R(u, v) \wedge R(v, w) \Rightarrow R(u, w)$	$R(u, u)$ $R(u, v) \Rightarrow R(u, v{:}x_R)$
KD5	$R(u, u{:}x_R)$ $R(u, v) \wedge R(u, w) \Rightarrow R(v, w)$	$R(u, u{:}x_R)$ $R(u{:}x_R, v{:}y_R)$
KD45	$R(u, u{:}x_R)$ $R(u, v) \wedge R(v, w) \Rightarrow R(u, w)$ $R(u, v) \wedge R(u, w) \Rightarrow R(v, w)$	$R(u, v{:}x_R)$
S5	$R(u, u{:}x_R)$ $R(u, u)$ $R(u, v) \Rightarrow R(v, u)$ $R(u, v) \wedge R(u, w) \Rightarrow R(v, w)$	$R(u, v)$

As an example how these alternative are constructed we take a little closer look at the logic KD45.
First recall that for any KD45 theorem to be proved the only clauses which contain positive R-literals are those of the background theory given by

$$R(u, u{:}x_R)$$
$$R(u, v) \wedge R(v, w) \Rightarrow R(u, w)$$
$$R(u, v) \wedge R(u, w) \Rightarrow R(v, w)$$

The saturation of this clause set results in the (infinitly many) clauses

$$R(u{:}x_1 \cdots {:}x_n, u{:}y_1 \cdots {:}y_m)$$

where $n \geq 0$ and $m \geq 1$. Now note that for each world it can be assumed that it is accessible from the initial world by finitely many R-steps (generatedness of models). In our terminology this means that any u can be represented by a term of the form $\iota{:}z_1 \cdots {:}z_k$ with $k \geq 0$. Therefore we have as a special case from above that

$$R(\iota{:}x_1 \cdots {:}x_n, \iota{:}y_1 \cdots {:}y_m)$$

and thus $R(u, v{:}x)$ for any worlds u, v and functional simulator element x since $\iota{:}x_1 \cdots {:}x_n$ and $\iota{:}y_1 \cdots {:}y_{m-1}$ represent arbitrary worlds.

The table can be read as follows: given the alternative translation and a modal logic: instead of using the characteristic `Axioms` of this logic include the usually simpler alternative form and you lose neither satisfiablity nor unsatisfiablity.

Similarly, the whole approach of saturation can also be applied to varying domains. In this case one has (besides the duality axiom $E(u, u{:}x_E)$) also clauses which state for example that the domains are increasing, decreasing or identical. The respective results are tabled as well.

Increasing Domains

KD KD4 S4	$E(u, u{:}y_E)$ $E(u, x) \Rightarrow E(u{:}y_E, x)$
KD5 KD45	$E(u, u{:}y_E)$ $E(u{:}x_R, v{:}x_E)$

Decreasing Domains

KD KD4 S4	$E(u, u{:}y_E)$ $E(u{:}x_R, y) \Rightarrow E(u, y)$
KD5 KD45	$E(u, u{:}y_E)$ $E(u, v{:}x_R{:}y_E)$

5 Examples

5.1 Modal Logics

Let us consider the example 1 on page 4 which inevitabely runs into an endless loop after relational translation. Functional Simulation translates the program $\Diamond\Box\Diamond P \Rightarrow \Diamond P$ into

$$\begin{array}{l} P(u{:}f(u)) \Leftarrow R(\iota{:}a, u). \\ R(u, u{:}x). \\ R(u{:}x, u). \\ \qquad\qquad \Leftarrow P(v), R(\iota, v). \end{array}$$

Thus there are no more loops and Prolog runs without any problems.

Now let us have a look at a more complex example: the Red Hat Puzzle. Assume three people are sitting in a row and there are three red hats and two blue ones and each of these people wears a hat which s/he can't see. The first person (who sees both of the others) is asked whether s/he knows the colour of her/his hat and s/he admits that s/he doesn't know it. The same happens with the second person (who sees only the one in front of her/him). Then the third person (who doesn't see anyone of the others) knows that s/he got a red hat. Why?

What we need here is actually a multi modal logic, i.e. we need several modal operators, one for each person and one for their mutual belief. This certainly increases complexity but causes no particular problems. For more information see again [3]. For our example we consider the belief modalities $\Box_A$, $\Box_B$, and $\Box_C$ for the three people and $\Box_{\mathrm{MB}}$ for their mutual belief. As a belief theory we choose KD45, i.e. beliefs are assumed to be consistent and the agents have complete introspection. Recall that the KD45 background theory can be represented by a single unit clause for each agent, namely $R_x(u, u{:}y_x)$. Furthermore there is also a single unit clause necessary for the mutual belief $(R_{\mathrm{MB}}(\iota, u{:}y_x))$[2].
The puzzle can then be represented by the following modal logic formulae:

$$\begin{array}{l} \Box_{\mathrm{MB}}(\Box_A \mathrm{red}(B) \vee \Box_A \neg\mathrm{red}(B)) \\ \Box_{\mathrm{MB}}(\Box_A \mathrm{red}(C) \vee \Box_A \neg\mathrm{red}(C)) \\ \Box_{\mathrm{MB}}(\Box_B \mathrm{red}(C) \vee \Box_B \neg\mathrm{red}(C)) \\ \Box_{\mathrm{MB}}(\neg(\Box_A \mathrm{red}(A) \vee \Box_A \neg\mathrm{red}(A))) \\ \Box_{\mathrm{MB}}(\neg(\Box_B \mathrm{red}(B) \vee \Box_A \neg\mathrm{red}(B))) \\ \Box_{\mathrm{MB}}(\mathrm{red}(A) \vee \mathrm{red}(B) \vee \mathrm{red}(C)) \\ \Box_{\mathrm{MB}}(\Box_x red(y) \Rightarrow red(y)) \end{array}$$

[2] This might require some further explanation. We interpret a mutual belief as: each agent believes and also each agent believes that each agent believes and so on. In our terminology this means that regardless how many (agent's) steps are gone, the resulting world is still reachable from the initial world by the mutual belief relation. We do only have to consider the worlds accessible from the initial world since the mutual belief operator will only be put at the beginning of formulae, i.e. the agents do not reason about mutual belief.

where the top three of these formulae represent the facts that A and B know the colour of C and, additionally, that B knows the colour of C by saying that e.g. A knows whether B's hat is red[3]. The next two formulae state that neither A nor B know their own colour and the sixth formula represents that at least one of the three hats is indeed red. From these six statements is does not yet follow that C knows her/his colour. Note that we do not assume reflexivity for the respective accessibility relations; and indeed such an infallibility property seems to be too strong for our purpose. Nevertheless we have to be able to guarantee that the respective agents are right in their beliefs at least what the colour of the hats of the people they can see is concerned. This is stated by the seventh formula above which tells us that whenever the agent x believes that y's hat is red then s/he is right in this respect.
Note that all formula are put in a mutual belief box because the facts of the puzzle are not only true but also known to be true and known to be known to be true and so on.
The theorem to be proved then finally is:

$$\Box_C \mathrm{red}(C)$$

Unfortunately there is a non-Horn formula, however, since we know that "not red" means the same here as "blue" we can easily make it Horn and finally get as a Prolog program:

$$\begin{array}{ll}
blue(u{:}a_A(u), A) \Leftarrow R_{\mathrm{MB}}(\iota, u). & \\
blue(u{:}c_B(u), B) \Leftarrow R_{\mathrm{MB}}(\iota, u). & \\
blue(e_C, C). & \\
blue(u{:}k_x(u), y) & \Leftarrow blue(u, y), R_{\mathrm{MB}}(\iota, u). \\
blue(v, B) & \Leftarrow R_A(u, v), R_A(u, w), R_{\mathrm{MB}}(\iota, u), blue(w, B). \\
blue(v, C) & \Leftarrow R_A(u, v), R_A(u, w), R_{\mathrm{MB}}(\iota, u), blue(w, C). \\
blue(v, C) & \Leftarrow R_B(u, v), R_B(u, w), R_{\mathrm{MB}}(\iota, u), blue(w, C). \\
R_{\mathrm{MB}}(\iota, v{:}y_x). & \\
R_x(u, u{:}y_x). & \\
 & \Leftarrow blue(u{:}b_A(u), A), R_{\mathrm{MB}}(\iota, u). \\
 & \Leftarrow blue(u{:}d_B(u), B), R_{\mathrm{MB}}(\iota, u). \\
 & \Leftarrow blue(u, A), blue(u, B), blue(u, C), R_{\mathrm{MB}}(\iota, u).
\end{array}$$

We fed this to Quintus Prolog and it got easily be solved in less than a millisecond. The reader interested in more details is again referred to [3].

5.2 Sorted Logics

We applied the approach to an example which is not really an example for sorted logic programs but one which to some extent implements a matrix method solution to the world famous Steamroller problem (see Stickel's paper and Pelletier's

[3] Note that "knowing whether" is equivalent to "either knowing that or knowing that not"; and this is *not* the same as "either knowing or not knowing".

Problem Corner contribution in the Journal of Automated Reasoning, volume 2, 1986). Subgoals already proved are put into a hypothesis list which gets checked for each further subgoal (either positively with success or negatively with failure). Non-Horn clauses are dealt with by adding all possible contrapositive clauses to the clause set.

Interestingly, although there are not many existentially quantified formulae occurring in this problem, it was able to solve the puzzle in less than a millisecond with Quintus Prolog on a Spark10. We also fed the problem in its functionally simulated version to the Otter theorem prover (with Hyperresolution strategy) and it showed that the number of clauses to be generated during the inference process gets reduced by a factor of 30. This shows that, although there are no saturations possible in this sorted case, the alternative translation alone suffices for a significant time saving.

6 Summary and Conclusion

We presented a translation approach from modal or sorted logic programs into classical predicate logic programs which avoids the exponential explosion which we usually get with relational translation and also avoids the necessity to be able to deal with equations or theory unification in standard logic programming. It can be viewed as a mixed approach between relational translation and the functional translation approach proposed by Hans Jürgen Ohlbach [7], Fariñas del Cerro and Herzig [2], Enjalbert and Auffray [1] and others. It guarantees that the number of clauses generated by translation does not exceed the number of clauses we would get if we entirely ignored the modal operators (or sort restrictions). However, the clauses themselves might get bigger.
Moreover, its output is Horn if and only if the original clause set is Horn which is not for sure in the pure relational translation approach.
One of the major improvements to the purely relational or purely functional approach lies in the possibility to optimize the theory clauses which stem from the modal logic under consideration, such that they sometimes even get more or less trivial (simple unit clauses).
However, to be fair, it is not always the case that there are no more loops possible in the background theory (as the symmetry example might suggest). There still is a possibility of getting loops in the theory of S4 and KD4, although I believe that these an be avoided if the instantiation of world variables is appropriately restricted. This, however, has not been examined very deeply yet.
For all the other cases such loops are impossible because the background theory results in unit clauses.

What the application of the approach to sorted logics is concerned: usually such optimizations by saturation are not possible in sorted logics simply because there are no general facts about sorts. The best we can get (and which is also used in this paper) is the assumption that sorts are not empty. The saturation approach would come into play if there were some further general information as e.g. the intersection of two sorts form another sort. Without such extra information the advantages of the approach lies only in the fact that there are less clauses than we would have after relational translation (or relativization) and that the output is guaranteed to be Horn if the input was.

Note that this approach does by no means require to change the programming language environment. It simply transforms the original problem into an equivalent one which can then be treated by standard Prolog.

References

1. Yves Auffray and Patrice Enjalbert. Modal theorem proving: An equational viewpoint. *Journal of Logic and Computation*, 2(3):247 – 295, 1992.
2. Luis Fariñas del Cerro and Andreas Herzig. Quantified modal logic and unification theory. Rapport LSI 293, Languages et Systèmes Informatique, Université Paul Sabatier, Toulouse, 1988.
3. Andreas Nonnengart. First-order modal logic theorem proving and standard PROLOG. Technical Report MPI-I-92-228, Max-Planck-Institute for Computer Science, Saarbrücken, Germany, July 1992.
4. Andreas Nonnengart. First-order modal logic theorem proving and functional simulation. In Ruzena Bajcsy, editor, *Proceedings of the 13th IJCAI*, volume 1, pages 80 – 85. Morgan Kaufmann Publishers, 1993.
5. Hans Jürgen Ohlbach. A resolution calculus for modal logics. In Ewing Lusk and Ross Overbeek, editors, *Proc. of* 9th *International Conference on Automated Deduction, CADE-88 Argonne, IL*, volume 310 of *LNCS*, pages 500–516, Berlin, Heidelberg, New York, 1988. Springer-Verlag. extended version: SEKI Report SR-88-08, FB Informatik, Universität Kaiserslautern, 1988.
6. Hans Jürgen Ohlbach. *A Resolution Calculus for Modal Logics.* PhD thesis, University of Kaiserslautern, Germany, 1989.
7. Hans Jürgen Ohlbach. Semantics-based translation methods for modal logics. *Journal of Logic and Computation*, 1(5):691–746, 1991.

Towards Resource Handling in Logic Programming: the *PPL* Framework and its Semantics

Jean-Marie Jacquet[1]* and Luís Monteiro[2]

[1] Department of Computer Science, University of Namur, 5000 Namur, Belgium
[2] Departamento de Informática, Universidade Nova de Lisboa, 2825 Monte da Caparica, Portugal

Abstract. The *PPL* framework is proposed as a simple extension to logic programming aiming at handling resources. It is argued that the separation between logical treatments and resource handling is desirable and, to that end, resources are proposed to be manipulated by means of pre- and post-conditions associated with usual Horn clauses. The expressiveness of the resulting framework is evidenced through the coding of several applications involving objects, databases, actions and changes. Operational and declarative semantics are presented as well. The operational semantics rests on a derivation relation stating how goals and conditions are evaluated. The declarative semantics extends the classical model and fixed-point theories to take into account the evaluation of pre- and post-conditions, and in particular the non-monotonic behavior of the world of resources they induce in general. As suggested, an effort has been made to keep the work close to the classical logic programming setting. In particular, the semantics are in the main streams of logic programming semantics. However, the *PPL* framework raises new problems for which fresh solutions are proposed.

1 Introduction

The execution of many programs can be depicted as the progressive access to and update of a world of resources. For instance, the execution of a simple imperative program can be schematized as the successive executions of assignments, each one being caracterized by the three following actions: access to a state associating a value to each variable, computation of a value for the right-hand side of the assignment, update of the state in order to associate the value just computed to the left-hand side variable of the assignment. As a second example, constraint logic languages (see e.g. [27]) have a so-called store of constraints as world of resources. It is progressively accessed by means of the *tell* primitive, which adds the told constraint to the store provided consistency is preserved, and by the *ask* primitive, which tests whether a given constraint is entailed by

* Supported by the Belgian National Fund for Scientific Research as a Senior Research Assistant.

the current contents of the store. Primitives for non-monotonically modifying the store are proposed in addition in [7]. As a third example, in a new approach to parallelism in logic programming ([4, 6]), a blackboard is used as a means of communication between concurrent logic processes and is accessed by means of Linda-like primitives putting, testing and removing both passive data structures and active processes from it. Finally, to conclude this non-exhaustive list of examples, the synchronous communication mechanism introduced in [1, 5, 15, 16] can also be viewed as another instance of this general resource-based scheme. There, processes are simultaneously reduced to others and thus act as resources for their synchronizing partners.

Although resources thus appear to be central, most of the above mentioned logic languages have handled them by means of ad hoc primitives inserted into logical formulae and consequently have separated the handling of resources and the derivation of logical formulae only at the conceptual level and not at the programming one. Examples of this trend include the tell and ask primitives of [27] and the blackboard primitives of [4] which may be inserted at any place inside clause bodies and queries. From a software engineering point of view, this is quite regrettable since one better separate different issues clearly, and, consequently here, separate resource handling and logical treatments clearly. This paper proposes an alternative approach achieving this goal: on the one hand, "logical atoms" only compose Horn clauses and queries, and, on the other hand, resources are tackled in pre- and post-conditions associated with each Horn clause. Restated in programming terms, program clauses now take the form

$$H \leftarrow G \quad <C : D> \tag{1}$$

where

i) H is an (usual) atom,
ii) G is a goal, formed of (usual) atoms combined with the operators " ; " and " $\|$ ", for sequential and parallel compositions, respectively,
iii) C and D are lists of (resource-oriented) atoms separated by the symbol " , ".

Such clauses are subsequently called *pp-clauses*. They thus consist essentially of Horn clauses decorated by a pair of conditions $<C : D>$. This slight extension induces the following modificiation to the usual SLD-derivation. The condition C lists atoms (possibly duplicated) that should be present in the considered world of resources at the time of the reduction of a considered goal to the clause. If so, the atoms are removed from the resources and the (induced instance) of the body of the clause is evaluated. Then, atoms of the postconditions are added as new resources. Conditions C and D thus act as pre- and post-conditions, respectively.

The actual framework, called *PPL*, is slightly more general in two ways. On the one hand, resources may be aggregated by making use of atoms defined by means of Horn clauses. Such clauses take the usual form with body goals formed from atoms separated by the symbol ",". In particular, they do not include pre- and post-conditions. They are denoted as follows to be distinguished from the pp-clauses:

$$H \hookleftarrow G$$

The usual SLD-resolution is then used to verify the pre- and post-conditions.

On the other hand, pre-conditions are allowed to be non-destructive in the sense that their evaluation does not remove the resources they involve. Such clauses are subsequently denoted as

$$H \leftarrow G \quad <C \,|\, D> \tag{2}$$

if the whole pre-condition is non-destructive or, as

$$H \leftarrow G \quad <C_1 \,|\, C_2 : D>$$

if the C_1 part of the pre-condition is non-destructive and its C_2 part is destructive. On the point of the notation, although the above form is general and will be used for our semantic study, we shall stick, in the examples, to the notation proposed in clause (1) when the C_1 part of the pre-condition is empty and its C_2 part is not, and to the notation proposed in clause (2) when C_2 is empty. Note that with this convention, the form (2) is used when both C_1 and C_2 are empty to emphasize that no change has been operated to the resources.

Although simple, the *PPL* framework is quite expressive. This is suggested in section 2 through the coding of several applications. However, the purpose of this paper is more to identify a framework, to study semantics for it, and, on the way, to argue its declarativeness, than to create a new language. Hence, we shall not pay attention to implementation issues in the following but rather will concentratre on a semantic study. An operational semantics is first presented on the basis of a derivation relation. Then, the classical model and fixed-point theories are extended to our context. Finally, classical results of soundness and completeness are established. As will be appreciated by the reader, both the framework and the semantics have been conceived as simple extensions of logic programming remaining in its classical mainstreams.

Related work can be compared both at the language and the semantic points of views.

At the language level, it is worth stressing again that the language *PPL* clearifies the resource-handling made in other concurrent logic languages. It also shares with languages like Shared Prolog ([6]) and Multi-Prolog ([4]) the advantages induced by a communication realized at a global level and not by means of the sharing of variables: synchronization and mutual exclusion are achieved implicitly via the world of resources and do not require the coding of auxiliary merge processes and the use of commitment operators. Moreover, thanks to the decoupling of logical reductions and resource handling, one could think of combining programs developed and tested in isolation under the assumption that the suitable information will eventually be available.

The idea of pre- and post-conditions has already been exploited in Shared Prolog ([6]). The two main differences are that, on the one hand, pre- and post-conditions are introduced at the finer level of clauses instead of the larger level of theories and, on the other hand, that aggregates of resources can be manipulated as such by means of predicates defined by Horn clauses.

Resource handling is, of course, tackled by constraint languages. For simplicity, this paper has been kept in the classical lines of logic programming but a generalization to constraints is appealing and is planned for future work. However, it should be noted that such a generalization leads to higher-order constructs (as in [26]) since the state of the *PPL* computations to be generalized is composed of a substitution, describing the values computed for the variables, but also of a world of resources composed of atoms namely of first-order constructs. One of the interests of the PPL framework is that it precisely hides the higher-order constructions.

As already said, a further difference with constraint languages lies in the clear separation, both at the conceptual and programming levels, of resources handling and logical derivations. In particular, aggregating resources by means of resource-dedicated clauses is supported in *PPL* but is not supported in such a clear way in [7, 27].

Another difference with [27] is the non-monotonic behavior of the world of resources. Compared with [7], which also proposes non-monotonic stores, the main differences with our work rest in the handling of unconsumed resources and in the communication of bindings. The primitive $\mathit{update}_x(\theta)$ is proposed there to transform a given (current) store σ into the new one $\exists_x(\sigma) \sqcup \theta$ where the $\exists_x$ operator makes x a local variable in σ, thereby removing *all* the constraints on it in σ. However, in general, only part of these constraints need to be removed. Using this technique, it is thus necessary to collect all the constraints — which does not seem to be a trivial task using the proposed primitives — and to copy those constraints that should remain. In contrast, such a selected deletion is performed easily in our framework by citing the constraints that need to be present and the constraints that need to be removed in two kinds of non-destructive and destructive preconditions. In fact, these preconditions embody basic operations on mutlisets implicitly. Moreover, while checking these constraints, some unification may take place and be propagated. Again, it is not obvious to achieve this effect by using the primitives proposed in [7].

It remains that the language $\mathcal{L}$ proposed in [7] is so general that the semantics proposed for it can be used to design semantics for the *PPL* framework. However, taking profit of its specificity, we have been able to design semantics that, although they borrow from imperative techniques, still constitute reasonable extensions to the classical logic programming framework.

Linear logic provides an alternative way of handling resources (see e.g. [1, 18, 26]). Although it is certainly promising, we have preferred to follow the lines of an extension of classical logic programming. This paper proves that this is possible and does not lead to intricacies. However, the connections with linear logic will be explored in future work.

The concept of resource handling is closely related to the notions of action and change, which have recently been the subjects of many researches: see e.g. [8, 10, 11, 13, 14, 19, 20, 21, 22, 23, 25]. Our work differs from them as follows.

The language AbstrAct [25] specifies the activities of systems by action rules characterizing agents capable of performing actions that operate on a global shared data space. A feature of AbstrAct is that it distinguishes between actions to induce state transformations and deductions that can be performed in each state. Our work differs from this work in that we do not have an explicit notion of agent, relying rather on the more logic-based notion of goal, and use an aggregation technique of resources based on Horn clauses.

The works [20, 22] have introduced the so-called situation calculus. Essentially, it consists of using an argument to state that a particular resource is available in a particular situation and of using so-called frame axioms, one for each action and each resource, to solve the so-called frame problem, i.e. to express that a resource remains invariant after an action. As noted in [14], the essential problem with this solution is that the number of frame axioms rapidly increases with the number of actions and resources. This number has been reduced in [19] and has become linear with respect to the number of actions. Our proposal does not suffer from these problems since the non-consumption of resources is expressed either by default or by means of non-destructive pre-conditions.

The article [21] has proposed to use nonmonotonic inference rules and a default law of inertia to tackle the frame problem. The paper [23] has employed linear logic to describe actions and changes. The paper [10] has used extended logic programs with both classical negation and negation-as-failure for that purpose. The paper [8] has used normal logic programs with abduction. Our work differs from all of them by using a slight extension of Horn clause programs involving no negation and remaining in the (classical) mainstreams of logic programming.

The work reported in [11, 13, 14] is the closest to ours. Rephrased in our terms, it proposes to describe each action by incorporating our pre- and post-conditions as predicate arguments and the non-consumption problem by extending normal unification to cope with multisets. As an example (taken from [14]), the clause

$$action(Pre, load, Post) \leftarrow Pre =_{AC1} unloaded \wedge Post =_{AC1} loaded \quad (3)$$

states that the action of loading a gun assumes that the gun is unloaded and, if so, it moves the gun in a loaded status. The effect of the sequences of actions is expressed by means of the ternary predicate *causes*, defined by the following two clauses

$$\begin{aligned} &causes(Pre, [], Post) \leftarrow Pre =_{AC1} Post \\ &causes(Pre, [A|As], Post) \leftarrow action(Pre\_A, A, Post\_A) \\ &\qquad\qquad\qquad\qquad\qquad\qquad \wedge\, Pre\_A \;o\; Pre\_rem =_{AC1} Pre \\ &\qquad\qquad\qquad\qquad\qquad\qquad \wedge\, causes(Post\_A \;o\; Pre\_rem, As, Post). \end{aligned}$$

The frame problem is solved by using a multiset structure, defined in a way similar to lists but with the binary function symbol o , and by using a new unification, denoted $=_{AC1}$ which actually defines the operator o as associative, commutative and admitting the constant $\emptyset$ as unit element.

Our proposal first contrasts by liberating the *action* predicate from the pre- and post-conditions arguments and by avoiding extended unification. As an example, clause (3) rewrites in our framework as

$$action(load) \leftarrow \triangle \qquad <unloaded : loaded>$$

or, even, in a simpler way, as

$$load \leftarrow \triangle \qquad <unloaded : loaded>$$

where $\triangle$ denotes the empty goal. Besides the syntactic difference, we believe that separating logical derivations from resource handling has interesting consequences from a software engineering point of view. For instance, since in our framework resources are only treated when necessary by means of pre- and post-conditions and without explicit (extended) unification, the *causes* predicate can be rewritten more naturally as

$$\begin{array}{l} causes([\,]) \leftarrow \triangle \quad <\Box \mid \Box> \\ causes([A|As]) \leftarrow action(A)\,;\, causes(As) \quad <\Box \mid \Box> \end{array}$$

where $\Box$ denotes the empty (pre- and post-) condition. Another difference with our work is that we care here for the execution of the "$\wedge$" connector and provide semantics for both its sequential and parallel versions. As noted in [8], finding a way to represent parallel and non-deterministic actions is a real challenge, to which we believe to have given a solution.

Finally, this paper grew up as a continuation of our previous work [15, 16] and its related work [5, 9]. As already suggested, one difference with this work is the clear separation between resource handling and logical treatment. As an interesting consequence, resource need not be considered as active data when they are not actually. Compare for instance the stack example of section 2 and that of [15]. Moreover, the semantics are quite different from that presented in previously cited work. On the one hand, with respect to [5] and [9], one should note that, in addition to the richer context of pre- and post-conditions handled here, arbitrary mixing of sequential and parallel compositions inside goals are treated here as well as an unrestricted form of variable sharing. On the other hand, pre- and post-conditions handling and the discard for synchronous communication makes the semantics reported in this paper quite different from that of [15] and of [16].

The remainder of this paper is organized as follows. Section 2 suggests the interest of pp-clauses through the coding of examples of actions, database manipulations, and programs integrating the object and logic programming styles. Section 3 describes the basic constructs of the language and explains our terminology. Section 4 defines auxiliary concepts. Section 5 presents the operational semantics $\mathcal{O}$. Section 6 discusses the declarative semantics $Decl_{\mathrm{m}}$ and $Decl_{\mathrm{f}}$, based on model and fixed point theories, respectively. Section 7 establishes the soundness and completeness properties and consequently connects the three semantics. Finally, section 8 gives our conclusions.

2 Examples

2.1 Objects

The behavior of objects is classically represented in logic programming by the evaluation of a call to a procedure defined recursively, the successive values of the arguments representing the successive states of the object. Following this line, the treatment of a message `mess(M)` by an object `obj(S)` by means of a method `method(M)` can be schematized by one of the following pp-clauses depending upon whether the message is consumed or not:

$$obj(S) \leftarrow method(M, S, NewS)\,;\, obj(NewS) \qquad < mess(M) : \Box >$$
$$obj(S) \leftarrow method(M, S, NewS)\,;\, obj(NewS) \qquad < mess(M) \mid \Box >$$

Indeed, in the *PPL* framework, any call to $obj(S)$ results successively

i) in the evaluation of the precondition $mess(M)$, which identifies the message from the world of resources,
ii) in the reduction of the call $method(M, S, NewS)$, which computes the new state $NewS$, and in the reduction of the recursive call to $obj(NewS)$, which describes the object with its new state
iii) in the evaluation of the empty post-condition $\Box$.

Note that, depending upon whether the message has to be consumed or not, the message $mess(M)$ is consumed or not from the world of resources by the evaluation of the pre-condition.

An instance of this scheme is given by the following description of the class of stacks:

$$stack(Id, S) \leftarrow stack(Id, [X|S]) \qquad < push(Id, X) : \Box >$$
$$stack(Id, [X|S]) \leftarrow stack(Id, S) \qquad < pop(Id) : \Box >$$

Stacks are identified there by the Id argument of the ***stack*** predicate and their state, implemented as a list, moves, respectively from S, $[X|S]$ to $[X|S]$, S according to the treatment of a ***push*** or ***pop*** message.

The classical airline reservation system provides another interesting instance of the above scheme. The task consists here of simulating an airline reservation system composed of n agencies communicating with a global database about m flights. It is achieved by evaluating the *PPL* query

$$agency(Id_1) \;||\; \cdots \;||\; agency(Id_n) \;||\; air\_syst(DB\_init)$$

where $agency(Id_j)$ represents the jth agency, identified by Id_j, and where DB_init represents the initial information about the m flights. The complete description of the agencies is out of interest for our illustrative purposes. It is here sufficient to assume that some internal actions successively generate queries for the database and behave correctly according to the answers. We will just consider the message $reserve(Ag\_id, Flight\_id, Nb\_seats)$, other ones being treated in a similar way. This message consists of asking the reservation of Nb_seats in the flight

Flight_id for the agency identified by *Ag_id*. A message *reserve_ans*(*Ag_id*, *Ans*) reporting the issue of the reservation is expected as a result. According to the above scheme and using the auxiliary predicates *remaining_seats* and *update*, with obvious meanings, the treatment of the message can be coded as follows (as easily checked by the reader):

$$
\begin{array}{l}
air\_syst(DB) \leftarrow \\
\qquad make\_reservation(Ag\_id, Flight\_id, Nb\_seats, DB, New\_DB)\ ; \\
\qquad air\_syst(New\_DB) \\
\qquad\qquad < reserve(Ag\_id, Flight\_id, Nb\_seats) : \Box >
\end{array}
$$

$$
\begin{array}{l}
make\_reservation(Ag\_id, Flight\_id, Requested\_seats, DB, New\_DB) \leftarrow \\
\qquad remaining\_seats(Flight\_id, Seats)\ ; \\
\qquad Seats \geq Requested\_seats\ ; \\
\qquad update(DB, Requested\_seats, New\_DB) \\
\qquad\qquad < \Box \,|\, reserve\_ans(Ag\_id, successful\_reservation) > \\
make\_reservation(Ag\_id, Flight\_id, Requested\_seats, DB, DB) \leftarrow \\
\qquad remaining\_seats(Flight\_id, Seats)\ ; \\
\qquad Seats < Requested\_seats \\
\qquad\qquad < \Box \,|\, reserve\_ans(Ag\_id, failed\_reservation) >
\end{array}
$$

The following points are worth noting from this example. They contrast with what is usually done with concurrent logic programming languages. First, accessing the database is achieved without handling lists of messages explicitly and without using merge processes. Second, assuming the atomicity of the evaluation of the pre- and post-conditions, mutual access to the database is ensured without using commitment.

2.2 Databases

The airline reservation system has already given an example of database handling. There, the database has been manipulated as an argument of the *air_syst* predicate. However, databases can also be tackled as worlds of resources. As an illustration, let us consider the education database of [11]. Three relations are specified about students, courses, and grades:

i) the relation $enr(St, C)$ states that the student St is enrolled in course C.
ii) the relation $grd(St, C, G)$ states that the grade of the student St for course C is G
iii) the relation $pre(L, C)$ states that the courses of the list L are prerequisite courses for the course C.

The aim of the program is to code the action of registering a student St for a course C given the constraint that a grade of at least 50 should have been obtained by the student for all prerequisite courses of C. This is achieved by the following pp-clause:

$$
regis(St, C) \leftarrow satisfiable(L) \quad < list\_grd\_prec(St, C, L) \,|\, enr(St, C) >
$$

The $regis(St, C)$ performs the registration transaction. Its definition uses the precondition $list\_grd\_prec(St, C, L)$ which builds the list L of grades corresponding to the prerequisite courses for C. This list L is then tested for the fulfillment of the above constraint rule by the $satisfiable(L)$ goal. Finally, in case this rule holds, the item $enr(St, C)$ is added to the database.

It is here worth noting that resource handling and logical treatment are well separated. Indeed, on the one hand, information retrieving and update on the database, is isolated in the pre- and post-conditions of the rule. This reflects the status of the database as a resource. On the other hand, the logical treatment on these data (i.e. $satisfiable(L)$) is performed in the body of the clause.

From a programming point of view, the reader should also note the ease of programming given, on the one hand, by distinguishing destructive and non-destructive preconditions, and, on the other hand, by allowing predicates to be defined on primitive resource-atoms.

The auxiliary predicate $satisfiable(L)$ is defined by the following self-explanatory clauses.

$$satisfiable([]) \leftarrow \Delta \quad < \square \mid \square >$$
$$satisfiable([G|Gs]) \leftarrow G \geq 50 \mid\mid satisfiable(Gs) \quad < \square \mid \square >$$

The auxiliary predicate $list\_grd\_prec(St, C, L)$ is defined by first identifying the list of prerequisite courses for the course C (reduction of $pre(Lc, L)$) and then by collecting all the grades got by the student St for all these courses (reduction of $list\_grd(St, Lc, L)$).

$$list\_grd\_prec(St, C, L) \hookleftarrow pre(Lc, C), list\_grd(St, Lc, L)$$

$$list\_grd(St, [], []) \hookleftarrow \square$$
$$list\_grd(St, [C|Cs], [G|Gs]) \hookleftarrow grd(St, C, G), list\_grd(St, Cs, Gs)$$

2.3 Actions and Change

Actions and change can also be coded in the PPL framework. As a support to this claim, we now show how to code the classical Yale Shooting problem of [12]. It has as actors a gun, which is either unloaded or loaded, and a turkey, which is either alive or dead. Three actions are possible:

i) loading, which results in moving the status of the gun to loaded both if the gun was unloaded or loaded
ii) shooting, which results in unloading the gun and in killing the turkey if the gun was loaded
iii) waiting, which is a passive action.

The problem can be coded in PPL by representing the states of the turkey and of the gun by the following tokens and by taking them as resources:

i) loaded: the gun is loaded
ii) unloaded: the gun is unloaded
iii) alive: the turkey is alive

iv) dead: the turkey is dead

The loading, shooting and waiting actions, can then be described by the following pp-clauses:

$$wait \leftarrow \Delta \quad < \Box \,|\, \Box >$$
$$load \leftarrow \Delta \quad < unloaded : loaded >$$
$$load \leftarrow \Delta \quad < loaded \,|\, \Box >$$
$$shoot \leftarrow \Delta \quad < unloaded \,|\, \Box >$$
$$shoot \leftarrow \Delta \quad < loaded, alive : unloaded, dead >$$
$$shoot \leftarrow \Delta \quad < loaded, dead \,|\, \Box >$$

Other classical problems such as the fragile object problem ([28]), the murder mystery problem ([2]), and the stolen car problem ([17]) can be coded in a similar way.

3 The Language

We now turn to the formal definition of the language *PPL*. As usual in logic programming, it comprises denumerably infinite sets of *variables*, *functions* and *predicates*, subsequently referred to as *Svar*, *Sfunct* and *Spred*, respectively. They are assumed to be pairwise disjoint. The notions of term, atom, substitution, ... are defined therefrom as usual. Their sets are subsequently referred to as *Sterm*, *Satom*, *Ssubst*, ..., respectively. In contrast, two kinds of goals are subsequently distinguished. On the one hand, so-called pp-goals are formed from the atoms by combining them with the operators " ; " and " || ", for sequential and parallel compositions, respectively. They are typically denoted by the letter G. Their set is subsequent referred to by *Sppgoal*. On the other hand, so-called l-goals consist of the usual lists of atoms, separated by the symbol " , ". They are typically denoted by the letters C and D, respectively. Their set is subsequently referred to as *Slgoal*.

Clauses are extended by pre- and post-conditions and take the form:

$$H \leftarrow G \quad < C_1 \,|\, C_2 : D >$$

where H is an atom, G is a pp-goal, C_1, C_2, and D are l-goals. They are called *pp-clauses*. Our semantic study requires that resources are ground only. To fulfill this property, we shall subsequently restrict pp-clauses in such a way that all the variables of the postcondition D occur in the preconditions C_1 and C_2. Note that, although restrictive at first sight, this constraint is satisfied by the programs of section 2. Such clauses are subsequently called pp-clauses.

Besides pp-clauses, usual Horn clauses are also considered. They take the form

$$H \hookleftarrow B$$

with H an atom and B a l-goal. They are called *r-clauses*. Again for our semantic purposes, we shall assume subsequently that r-clauses defining atoms appearing in postconditions obey the following property: all their variables appear both in their heads and in their bodies.

Programs are then composed of two parts: a set of pp-clauses and a set of r-clauses. They are typically denoted as P or as (P_p, P_r) with P_p the set of pp-clauses and P_r the set of r-clauses. Their set is subsequently denoted by *Sprog*.

4 Auxiliary Concepts

It turns out that it is possible to treat the sequential and parallel composition operators in a very similar way by introducing the auxiliary notion of context. Basically, a context consists of a partially ordered structure where the place holder $\diamond$ has been inserted at some top-level places i.e. places not constrained by the previous execution of other atoms. Viewing pp-goals as partially ordered structures too, the atoms to be reduced are those that can be substituted by a place holder $\diamond$ in a context. Furthermore, the pp-goals resulting from the reductions are basically obtained by substituting the place holder by the corresponding clause bodies. The evaluation of the postconditions of the clauses under consideration need simply to be processed in addition. This is achieved by decorating the clause bodies with constructs of the form $\uparrow D$, where D stands for a postcondition. The intention is that the postcondition is evaluated as soon as its decorating goal is reduced to the empty goal. As it is necessary to decorate already decorated goals in such a way, we are naturally lead to the following notion of gpp-goals.

Definition 1. The gpp-goals are defined inductively by the following rules:

i) any pp-goal is a gpp-goal
ii) if G is a gpp-goal and if D is a l-goal then the construct $G \uparrow D$ is a gpp-goal.

The set of gpp-goals is subsequently referred to as *Sgppgoal*.

The precise definition of the contexts is as follows.

Definition 2. The *contexts* are the functions from *Sgppgoal* to *Sgppgoal* inductively defined by the following rules.

i) $\diamond$ is a context that maps any gpp-goal to itself. For any gpp-goal G, this application is subsequently referred to as $\diamond[\![G]\!]$.
ii) If c is a context and if G is a gpp-goal, then $(c\,;G)$, $(c \mid\mid G)$, $(G \mid\mid c)$ are contexts. Their applications are defined as follows: for any gpp-goal G',

$$(c\,;G)[\![G']\!] = (c[\![G']\!]\,;G)$$
$$(c \mid\mid G)[\![G']\!] = (c[\![G']\!] \mid\mid G)$$
$$(G \mid\mid c)[\![G']\!] = G \mid\mid (c[\![G']\!])$$

iii) If c is a context and if D is a l-goal, then $(c \uparrow D)$ is a context. Its application is defined as follows : for any gpp-goal G',

$$(c \uparrow D)[\![G']\!] = (c[\![G']\!] \uparrow D).$$

In the above rules, we further state that the structure (*Sgppgoal*, "; ", " $\mid\mid$ ", "Δ") is a bimonoid. Moreover, in the following, we will simplify the gpp-goals resulting from the application of contexts accordingly.

5 Operational Semantics

The operational semantics O of the language *PPL* is defined in terms of derivation relations, themselves defined by means of rules of the form

$$\frac{\textit{Assumptions}}{\textit{Conclusion}} \quad \textit{if} \quad \textit{Conditions},$$

asserting the *Conclusion* whenever the *Assumptions* and *Conditions* hold. Note that *Assumptions* and *Conditions* may be absent from some rules, in which case they are considered to be trivially satisfied.

Two auxiliary derivation relations need first to be defined. They formalize the evaluation of pre- and post-conditions. They are attached the following meaning. The relation $rr \Downarrow\!\vdash C\ [rr^*]\ [\theta]$ expresses the property that, assuming a given program and given a world of resources rr, the precondition C has a successful derivation producing the substitution θ and using the resources of rr^*. Similarly, the relation $\Uparrow\!\vdash D\ [rr^*]\ [\theta]$ expresses the property that, assuming a given program, the post-condition D has a successful derivation producing the substitution θ and the resources of rr^*.

The definitions are as follows. As usual, the above notations are used instead of the relational ones with the aim of clarity. Moreover, the notation $\mathcal{M}(E)$ is used to denote the set of multisets with elements from E. Usual set operations are also extended to multisets.

Definition 3 (Precondition). Define $\Downarrow\!\vdash$ as the smallest relation of $\mathcal{M}(\textit{Satom}) \times \textit{Slgoal} \times \mathcal{M}(\textit{Satom}) \times \textit{Ssubst}$ that satisfies the following rules (C_1) to (C_3).

$$(\mathrm{C}_1) \qquad \frac{}{rr\ \Downarrow\!\vdash\ \Box\ [\emptyset]\ [\epsilon]}$$

$$(\mathrm{C}_2) \qquad \frac{rr\ \Downarrow\!\vdash\ G\theta\ [rr^*]\ [\sigma]}{rr\ \Downarrow\!\vdash\ c, G\ [rr^* \cup \{c\theta\}]\ [\theta\sigma]}$$

$$\text{if} \left\{ \begin{array}{l} c\theta \in rr \\ c \text{ not defined in } P_r \end{array} \right\}$$

$$(\mathrm{C}_3) \qquad \frac{rr\ \Downarrow\!\vdash\ (B, G)\theta\ [rr^*]\ [\sigma]}{rr\ \Downarrow\!\vdash\ c, G\ [rr^*]\ [\theta\sigma]}$$

$$\text{if} \left\{ \begin{array}{l} (H \hookleftarrow B) \in P_r \\ H \text{ and } c \text{ unify with mgu } \theta \end{array} \right\}$$

Definition 4 (Postcondition). Define $\Uparrow\!\vdash$ as the smallest relation of $\textit{Slgoal} \times \mathcal{M}(\textit{Satom}) \times \textit{Ssubst}$ that satisfies the following rules (D_1) to (D_3).

$$(D_1) \qquad \frac{}{\Uparrow\!\vdash \ \Box \ [\emptyset] \ [\epsilon]}$$

$$(D_2) \qquad \frac{\Uparrow\!\vdash \ G \ [rr^*] \ [\sigma]}{\Uparrow\!\vdash \ d, G \ [rr^* \cup \{d\}] \ [\sigma]}$$

$$\text{if } \{ \ d \text{ is not defined in } P_r \ \}$$

$$(D_3) \qquad \frac{\Uparrow\!\vdash \ (B, G)\theta \ [rr^*] \ [\sigma]}{\Uparrow\!\vdash \ d, G \ [rr^*] \ [\theta\sigma]}$$

$$\text{if } \left\{ \begin{array}{l} (H \hookleftarrow B) \in P_r \\ H \text{ and } d \text{ unify with mgu } \theta \end{array} \right\}$$

A word on the above rules is in order. Rules (C_1) and (D_1) tackle the empty pre- and post-conditions. For any world of resources, they state that the empty pre- and post-conditions are derivable with the empty substitution ϵ as computed answer substitution and with no resources used or to be added. The remaining rules tackle the reduction of a resource-atom. If it is defined by a clause of P_r, i.e. if there is a clause of P_r whose head has the same predicate name and arity, then rules (C_3) and (D_3) state that a unifiable clause should be employed and that the induced instance should be reduced. Otherwise, on the one hand, with respect to the pre-conditions, an instance of the resource-atom should be in the world of resources and, if so, should be considered as used (rule (C_2)). On the other hand, as far as the post-conditions are concerned, the resource-atom is to be added to the world of resources (rule (D_2)).

The following properties illustrate the usefulness of the restrictions on pp-clauses and r-clauses.

Proposition 5.

1) *For any l-goal C and any multiset rr composed of ground atoms only, if the relation $rr \ \Downarrow\!\vdash \ C \ [rr^*] \ [\theta]$ holds, then rr^* is composed of ground atoms only and $C\theta$ is ground. In particular, for any pp-clause $H \leftarrow B \quad < C_1 \mid C_2 : D >$, if the above relation holds both for C_1 and C_2 and for such a multiset rr, then $D\theta$ is ground as well.*
2) *For any ground l-goal D, if the relation $\Uparrow\!\vdash \ D \ [rr] \ [\theta]$ holds, then rr is composed only of ground atoms.*

We are now in a position to define the main derivation relation. It is written as $rr \ \vdash G \ [\theta]$ and expresses the property that, assuming a program and given the world of resources rr, the gpp-goal G has a successful derivation with θ as computed answer substitution. It is defined as follows.

Definition 6. Define $\vdash$ as the smallest relation of $\mathcal{M}(Satom) \times Sgppgoal \times Ssubst$ that satisfies the following rules (E_1), (E_2), and (A).

$$(\mathrm{E}_1) \qquad \frac{}{rr \ \vdash \ \Delta \ [\epsilon]}$$

$$(\mathrm{E}_2) \qquad \frac{rr \cup rr^* \ \vdash \ (c[\![\Delta]\!])\theta \ [\sigma]}{rr \ \vdash \ c[\![\Delta \uparrow D]\!] \ [\theta\sigma]}$$

$$\text{if } \{ \Uparrow\!\vdash \ D \ [rr^*] \ [\theta] \}$$

$$(\mathrm{A}) \qquad \frac{rr \backslash rr_2 \ \vdash \ (c[\![B \uparrow D]\!])\theta\mu\nu \ [\sigma]}{rr \ \vdash \ c[\![A]\!] \ [\theta\mu\nu\sigma]}$$

$$\text{if } \left\{ \begin{array}{l} (H \leftarrow B \quad < C_1 \,|\, C_2 : D >) \in P_p \\ H \text{ unifies with } A \text{ with mgu } \theta \\ rr \ \Downarrow\!\vdash \ C_1\theta \ [rr_1] \ [\mu] \\ rr \ \Downarrow\!\vdash \ C_2\theta\mu \ [rr_2] \ [\nu] \\ (rr_1 \cup rr_2) \subseteq rr \end{array} \right\}$$

The above rules essentially rephrase the intuitive explanation already given. Rule (E_1) states that the empty goal is derivable for any world of resources with the empty substitution as answer substitution. Rule (E_2) expresses the reduction of a postcondition. Finally, rule (A) explains the reduction of an atom: it consists of finding a unifiable clause, of evaluating the (induced instance of the) pre-conditions, of reducing the (induced instances of the) clause body and finally the post-condition.

The conciseness and expressiveness of the contexts are worth stressing. Thanks to them, it is not necessary to specify rules for the sequential and parallel composition of atoms inside gpp-goals. Moreover, pp-goals as well as their extensions incorporating postconditions, the gpp-goals, are selected for reduction in a uniform manner.

An operational semantics can be derived from the derivation relation, as follows.

Definition 7. Define the operational semantics $O : \mathit{Sprog} \rightarrow \mathit{Sppgoal} \rightarrow \mathit{Ssubst}$ as the following function: for any $P \in \mathit{Sprog}$, any $G \in \mathit{Sppgoal}$, $O(P)(G) = \{\theta : \emptyset \ \vdash \ G \ [\theta]\}$.

6 Declarative Semantics

One of the features of a logic programming language is that its semantics can be understood in two complementary ways, inherited from logic. On the one hand, the operational semantics, based on proof theory, describes the method of executing programs. On the other hand, the declarative semantics, based on model theory, explains the meaning of programs in terms of the set of their logical consequences. In our opinion, any claim that a given language is a logic programming language must be substantiated by providing suitable logic-based semantic characterizations. The operational semantics of the language *PPL* has been described in the previous section. We now turn to the discussion of the declarative semantics.

6.1 Model Theory

Our first task is to find an appropriate notion of interpretation for *PPL*. Classically, an interpretation of Horn clause programs consists of a subset of the Herbrand base, intended to record all the facts that are considered to be true under the interpretation. In *PPL* which facts are true actually depend on the considered world of resources. Moreover, because of the pre- and post-conditions, this world may have a non-monotonic behavior and, because of the parallel composition inside pp-goals it may be influenced by concurrent evaluations. Hence, following previous work [24, 3], an interpretation should define the truth of an atom not in absolute terms nor with respect to a given world of resources but with respect to traces reporting the successive states of this world. We are thus naturally lead to the following definitions.

Convention 8. *For any set S, the notation ground(S) is used to denote the set of ground instances of elements of S.*

Definition 9.

1) The set *Strace* is defined as the set of finite sequences of pairs of the form (rr, rr^*) with rr, rr^* multisets of ground atoms. Such sequences are subsequently called *traces* and are typically denoted as $RR_1.RR_2.\cdots.RR_n$, with $RR_1, RR_2, \ldots, RR_N \in \mathcal{M}(ground(Satom)) \times \mathcal{M}(ground(Satom))$. They are said to start in rr if rr is the initial multiset RR_1 of their first pair. The empty trace is denoted by λ.
2) A trace $t = (rr_1, rr_2).(rr_3, rr_4).\cdots.(rr_n, rr_{n+1})$ is *continuous* iff the output resources of every pair is the input resources of the following pair in t, if any, i.e. iff the following property holds: for any even $i \in 1, \ldots, n-1$, $rr_i = rr_{i+1}$. The set of continuous traces is subsequently denoted as *Sctrace*.
3) The Herbrand base B_{H} is defined as the set *Strace* $\times$ *ground(Satom)*. An Herbrand interpretation is defined as any subset of B_{H}.

Our next task is to define the notion of truth. To that end, we first need an auxiliary notion capturing the evaluation of pre- and post-conditions. Note that, as specified in Definitions 3 and 4, they essentially consist of using clauses of P_r until non-defined atoms are reached. This is formalized in the following relation $c \lhd C$, with c a ground lgoal and C the set of non-defined ground atoms reached by the reduction.

Definition 10. For any program (P_p, P_r), we define $\lhd$ as the smallest relation of $ground(Slgoal) \times \mathcal{P}(ground(Satom))$ that satisfies the following properties: for any $c, c_1, \ldots, c_n \in ground(Satom)$, any $\overline{c} \in ground(Slgoal)$, any $C, C_1, \ldots, C_n \in \mathcal{M}(ground(Satom))$,

i) $c \lhd \{c\}$ if c is not defined in P_r
ii) $c \lhd C$ if there is a ground instance $c \hookleftarrow \overline{c}$ of a clause of P_r such that $\overline{c} \lhd C$
iii) $(c_1, \ldots, c_n) \lhd \cup_{i=1}^{n} C_i$ if $c_i \lhd C_i$, for every $i = 1, \ldots, n$.

The satisfaction relation is defined at the level of ground constructs as follows.

Definition 11. Given two traces t_1 and t_2, let $t_1 \oplus t_2$ and $t_1 \otimes t_2$ denote the concatenation and merge of the traces. Then $\models$ is defined as the smallest relation of $Strace \times \mathcal{P}(B_{\mathrm{H}}) \times (ground(Sggoal) \cup ground(Sppclause))$ that verifies the following properties: for any interpretation I, any trace t, any ground atom a, h, any ground pp-goals b, g_1, g_2, any ground l-goals c_1, c_2, d,

i) ground atom: $t \models_I a$ iff $(t, a) \in I$
ii) ground empty pp-goal: $\lambda \models_I \Delta$
iii) ground sequential pp-goal: $t \models_I g_1 ; g_2$ iff there are $t_1, t_2 \in Strace$ such that $t_1 \models_I g_1$, $t_2 \models_I g_2$, and $t = t_1 \oplus t_2$
iv) ground parallel pp-goal: $t \models_I g_1 \parallel g_2$ iff there are $t_1, t_2 \in Strace$ such that $t_1 \models_I g_1$, $t_2 \models_I g_2$, and $t \in t_1 \otimes t_2$
v) ground pp-clause: $t \models_I h \leftarrow b \quad <c_1 \,|\, c_2 : d>$ iff whenever there are $rr_1, rr_2, rr_3, rr_4, C_1, C_2, D \in \mathcal{M}(ground(Satom))$, $u \in Strace$ such that the following properties hold:
   1. $t = (rr_1, rr_2) \oplus u \oplus (rr_3, rr_4)$,
   2. $c_1 \lhd C_1$
   3. $c_2 \lhd C_2$
   4. $(C_1 \cup C_2) \subseteq rr_1$
   5. $rr_2 = rr_1 \setminus C_2$
   6. $u \models_I b$
   7. $d \lhd D$
   8. $rr_4 = rr_3 \cup D$

   then $t \models_I h$ holds as well.

The notion of model, central for the declarative semantics, can now be defined.

Definition 12. An interpretation I is a *model* of a *PPL* program P iff, for any trace t and any ground instance $h \leftarrow b \quad <c_1 \,|\, c_2 : d>$ of a clause of P, the property $t \models_I h \leftarrow b \quad <c_1 \,|\, c_2 : d>$ holds.

It is easy to verify that, for any program, the intersection of a family of its models is again a model of the considered program. Moreover, the Herbrand base B_{H} is also a model of any program. It follows that any program has a least model, which is identical to the intersection of all its models.

Proposition 13. *For any program P, the intersection $\cap_{I \in \mathcal{I}} I$ of all its models is again a model. It is called the least model M_{P} of P.*

The notion of model is defined in a slightly different way for pp-goals than for programs. Although surprising at first sight, this difference is justified by the quite different nature of the two objects. On the one hand, clauses essentially consist of implications and any implication is considered to hold iff its consequent part holds whenever its antecedent part does. This is reflected in rule 5 of definition 11 and in the universal quantification over traces of definition 12. On the other hand, calling the operational semantics as support for our intuition, pp-goals should not manifestly hold for any trace but for particular ones. Moreover, these ones should be continuous and should start with the empty world of resources. This distinction of quantification is in the essence of the following definition.

Definition 14. An interpretation I is a model of the pp-goal G iff there is a continuous trace t starting in $\emptyset$ and such that for any ground instance G_0 of G the relation $t \models_I G_0$ is satisfied. This is subsequently denoted by $\models_I G$.

We are now in a position to define the notion of logical consequence.

Definition 15. The pp-goal G is a *logical consequence* of the program P iff any model of P is also a model of G. This is subsequently denoted by $P \models G$.

The next notion to introduce in the classical study of logic program is that of success set. For a given program and a given pp-goal, it can be defined as the set of substitutions that make the pp-goal a logical consequence of the program. However, this involves checking all the models of P. Fortunately, just the least model need actually to be checked. This reduction is an easy consequence of the property that if I and J are interpretations such that $I \subseteq J$ and if $\models_I G$ holds then so does $\models_J G$.

We can then define the declarative semantics as follows.

Definition 16. Define the model declarative semantics $Decl_m : Sprog \to Sppgoal \to Ssubst$ as the following function: for any $P \in Sprog$, any $G \in Sppgoal$, $Decl_m(P)(G) = \{\theta : P \models G\theta\} = \{\theta : \models_{M_P} G\theta\}$.

6.2 Fixed-point Theory

It turns out that the least model M_P can also be characterized as the least fixed point of a continuous transformation $T_P : \mathcal{P}(B_H) \to \mathcal{P}(B_H)$ called, as usual, the immediate consequence operator.

Definition 17. Given a program $P = (P_p, P_r)$, the immediate consequence operator $T_P : \mathcal{P}(B_H) \to \mathcal{P}(B_H)$ is defined as the following function: for any interpretation $I \subseteq B_H$,

$$\begin{aligned} T_P(I) = \{\, (t,a) \in B_H :\ & a \leftarrow b \ <c_1 \mid c_2 : d> \in ground(P_p), \\ & t = (rr_1, rr_2) \oplus u \oplus (rr_3, rr_4), \\ & c_1 \lhd C_1 \,,\ c_2 \lhd C_2 \,,\ u \models_I b \,,\ d \lhd D \,, \\ & (C_1 \cup C_2) \subseteq rr_1, rr_2 = rr_1 \setminus C_2, rr_4 = rr_3 \cup D \\ & rr_1, rr_2, rr_3, rr_4, C_1, C_2 \in \mathcal{M}(ground(Satom)), \\ & u \in Strace \qquad \} \end{aligned}$$

It is easy to check that the mapping T_P is well-defined, that is that for any interpretation I it returns an interpretation. Moreover, by endowing B_H by the set inclusion as order, B_H is turned into a complete lattice with the empty set and B_H as bottom and top elements, respectively. The mapping T_P can then be shown to be continuous and it can be proved that an interpretation I is a model of the program P iff it is a prefixed-point of T_P i.e. is such that $T_P(I) \leq I$. It follows from Tarki's lemma that T_P has a least fixed point which therefore is also the least model of P. Furthermore, this model can be computed by the standard iterative procedure. All these observations constitute the contents of the next proposition.

Proposition 18. *For any program P, the mapping T_P has a least fixed point, noted $lfp(T_P)$, which verifies the following equalities: $lfp(T_P) = T_P \uparrow \omega = M_P$*

The fixed-point semantics $Decl_f$, defined in terms of the least fixed point $lfp(T_P)$ can then be defined as follows.

Definition 19. Define the fixed-point declarative semantics $Decl_f : Sprog \rightarrow Sppgoal \rightarrow Ssubst$ as the following function: for any $P \in Sprog$, any $G \in Sppgoal$, $Decl_f(P)(G) = \{\theta : \models_{lfp(T_P)} G\theta \}$.

The equivalence between the declarative semantics $Decl_m$ and $Decl_f$ follows directly from proposition 18.

Proposition 20. *The semantics $Decl_m$ and $Decl_f$ are identical.*

7 Soundness and Completeness Properties

Soundness and completeness relating the operational and declarative semantics can also be proved. They are claimed in the following proposition.

Proposition 21. *For any program P and any pp-goal G,*

i) if $\emptyset \vdash G\ [\theta]$ holds for some substitution θ, then so does $P \models G\theta$;
ii) if $P \models G\gamma$ holds for some substitution γ, then so does $\emptyset \vdash G\ [\theta]$ for a substitution θ such that $G\theta \leq G\gamma$.

In particular, if $\alpha : \mathcal{P}(Ssubst) \rightarrow \mathcal{P}(Ssubst)$ is defined by $\alpha(\Theta) = \{\theta\gamma_{|S} : \theta \in \Theta, \gamma \in Subst, dom(\theta) \subseteq S\}$, for any $\Theta \in \mathcal{P}(SSubst)$, the equalities

$$Decl_m(P)(G) = Decl_f(P)(G) = \alpha(O(P)(G))$$

hold, for any program P and pp-goal G.

8 Conclusions

The paper has presented an extension of logic programming aiming at tackling resource handling, as well as operational and declarative semantics for it. From the language point of view, the extension mainly consists of adding pre- and post-conditions acting on resource-atom to Horn clauses. Roughly speaking, these conditions state which resource should be present to let any call be reduced by the considered clause and which atoms should be added at the end of this reduction, if successful. The extended framework has been shown to be well-suited for coding applications involving objects, databases, changes and actions.

The operational and declarative semantics have extended the classical notions of success set, model and immediate consequence operator to the new framework. An effort has been made to design these semantics as simple as possible as well as in the classical lines of logic programming semantics. However, the pre- and post-conditions have raised new problems for which fresh solutions have been proposed. Among these problems are the non-monotonic behavior of resources and the context dependency of reduction. They have been handled by means of traces. Also of interest is the notion of context which allows to treat in a uniform manner both sequential and parallel composition operators. All these semantics have been related in the paper, as established by propositions 20 and 21.

Acknowledgments

The first author likes to thank the members of the C.W.I. concurrency group, headed by J.W. de Bakker, for their weekly intensive discussions, as well as B. Le Charlier, for his interest in his work. He also likes to thank the Belgian National Fund for support through a Senior Research Assistantship. The second authors wishes to thank Junta Nacional de Investigação Científica e Tecnológica for financial support.

References

1. J.-M. Andreoli and R. Pareschi. Linear Objects: Logical Processes with Built-in Inheritance. In D.H.D. Warren and P. Szeredi, editors, *Proc. 7th Int. Conf. on Logic Programming*, pages 495–510, Jerusalem, Israel, 1990. The MIT Press.
2. A. Baker. Nonmonotonic Reasoning in the Framework of the Situation Calculus. *Artificial Intelligence*, 49:5–23, 1991.
3. K. De Bosschere and J.-M. Jacquet. Comparative Semantics of *μlog*. In D. Etiemble and J.-C. Syre, editors, *Proceedings of the PARLE'92 Conference*, volume 605 of *Lecture Notes in Computer Science*, pages 911–926, Paris, 1992. Springer-Verlag.
4. K. De Bosschere and J.-M. Jacquet. Multi-Prolog: Definition, Operational Semantics, and Implementation. In D.S. Warren, editor, *Proceedings of the ICLP'93 Conference*, pages 299–314, Budapest, Hungary, 1993. The MIT Press.

5. A. Brogi. And-Parallelism without Shared Variables. In D.H.D. Warren and P. Szeredi, editors, *Proc. 7th Int. Conf. on Logic Programming*, pages 306–321, Jerusalem, Israel, 1990. The MIT Press.
6. A. Brogi and P. Ciancarini. The Concurrent Language Shared Prolog. *ACM Transactions on Programming Languages and Systems*, 13(1):99–123, January 1991.
7. F.S. de Boer, J. Kok, C. Palamidessi, and J.J.M.M. Rutten. Non-Monotonic Concurrent Constraint Programming. In D. Miller, editor, *Proc. Int. Symp. on Logic Programming*, pages 315–334, Vancouver, Canada, 1993.
8. P.M. Dung. Representing Actions in Logic Programming and its Applications in Database Updates. In D.S. Warren, editor, *Proc. 10th Int. Conf. on Logic Programming*, pages 222–238, Budapest, Hungary, June 1993. The MIT Press.
9. M. Falaschi, G. Levi, and C. Palamidessi. A Synchronization Logic: Axiomatics and Formal Semantics of Generalized Horn Clauses. *Information and Control*, 60:36–69, 1984.
10. M. Gelfond and V. Lifschitz. Representing Actions in Extended Logic Programming. In K.R. Apt, editor, *Proc. Joint International Conference and Symposium on Logic Programming*, pages 559–573, Washington, USA, November 1992. The MIT Press.
11. G. Große, S. Hölldobler, J. Schneeberger, U. Sigmund, and M. Tielscher. Equational Logic Programming, Actions, and Change. In K.R. Apt, editor, *Proc. Joint International Conference and Symposium on Logic Programming*, pages 177–191, Washington, USA, November 1992. The MIT Press.
12. S. Hanks and D. Mac Dermott. Nonmonotonic Logic and Temporal Projection. *Artificial Intelligence*, 33(3):379–412, 1987.
13. S. Hölldobler and J. Schneeberger. A New Deductive Approach to Planning. *New Generation Computing*, 8:225–244, 1990.
14. S. Hölldobler and M. Thielscher. Actions and Specificity. In D. Miller, editor, *Proc. Int. Symp. on Logic Programming*, pages 164–180, Vancouver, Canada, October 1993. The MIT Press.
15. J.-M. Jacquet and L. Monteiro. Extended Horn Clauses: the Framework and its Semantics. In J.C.M. Baeten and J.F. Groote, editors, *Proc. 2nd Int. Conf. on Concurrency Theory (Concur'91)*, volume 527 of *Lecture Notes in Computer Science*, pages 281–297, Amsterdam, The Netherlands, 1991. Springer-Verlag.
16. J.-M. Jacquet and L. Monteiro. Communicating Clauses: the Framework and its Semantics. In K.R. Apt, editor, *Proc. Joint International Conference and Symposium on Logic Programming*, Series in Logic Programming, pages 98–112, Washington, USA, November 1992. The MIT Press.
17. H. Kautz. The Logic of Persistence. In *Proc. AAAI*, pages 401–405, 1986.
18. N. Kobayashi and A. Yonezawa. ACL — A Concurrent Linear Logic Programming Paradigm. In D. Miller, editor, *Proc. Int. Symp. on Logic Programming*, pages 295–314, Vancouver, Canada, 1993.
19. R. Kowalski. *Logic for Problem Solving*. North Holland, New York, 1979.
20. J. Mac Carthy. Situations and Actions and Causal Laws. Standford Artificial Intelligence Project Memo 2, Stanford University, Palo Alto, CA, USA, 1963.
21. J. Mac Carthy. Applications of Circumscription to Formalizing Commonsense Knowledge. *Artificial Intelligence*, 28:89–116, 1986.
22. J. Mac Carthy and P.J. Hayes. Some Philosophical Problems from the Standpoint of Artificial Intelligence. *Machine Intelligence*, 4:463–502, 1969.

23. M. Masseron, C. Tollu, and J. Vauzeilles. Generating Plans in Linear Logic. In *Foundations of Software Technology and Theoretical Computer Science*, volume 472 of *Lecture Notes in Computer Science*, pages 63–75. Springer-Verlag, 1990.
24. L. Monteiro. Distributed Logic, A Theory of Distributed Programming in Logic. Research report, Departamento de Informática, Universidade de Lisboa, 2885 Monte da Caparica, Lisbon, Portugal, 1986.
25. A. Porto and P. Rosado. The AbstrAct Scheme for Concurrent Programming. In E. Lamma and P. Mello, editors, *Extensions of Logic Programming*, volume 660 of *Lecture Notes in Artificial Intelligence*, pages 216–241, Berlin, 1993. Springer-Verlag.
26. V. Saraswat and P. Lincoln. Higher-Order, Linear, Concurrent Constraint Programming. Research report, Xerox Palo Research Center, Palo Alto, CA, USA, 1992.
27. V.A. Saraswat. *Concurrent Constraint Programming Languages.* The MIT Press, 1993.
28. L. Schubert. Monotonic Solution for the Frame Problem in the Situation Calculus: an Efficient Method for Worlds with Fully Specified Actions. In H.E. Kyburg, R. Loui, and G. Carlson, editors, *Knowledge Representation and Defeasible Reasoning*, pages 23–67. Kluwer, 1990.

Extending Horn Clause Theories by Reflection Principles

Stefania Costantini[1], Pierangelo Dell'Acqua[2*] and Gaetano A. Lanzarone[1]

[1] Università degli Studi di Milano, Dip. di Scienze dell'Informazione
via Comelico 39/41, I-20135 Milano, Italy
[2] Uppsala University, Computing Science Dept.
Box 311, S-751 05 Uppsala, Sweden

Abstract. In this paper, we introduce logical reflection as a principled way to empower the representation and reasoning capabilities of logic programming systems. In particular, *reflection principles* take the role of axiom schemata of a particular form that, once added to a given logic program (the basic theory, or the initial axioms), enlarge the set of consequences sanctioned by those initial axioms. The main advantage of this approach is that it is much easier to write a basic theory and then to augment it with condensed axiom schemata, than it is to write a corresponding large (or even infinite) set of axioms in the first place. Moreover, the well-established semantic properties of Horn clauses, carry over to Horn clauses with reflection. In fact, the semantics of Reflective SLD Resolution and the semantics of the Reflective Least Herbrand Model are obtained by making slight variations to, respectively, the procedural and the declarative semantics classically defined for Horn clauses. We present a complete formalization of this concept of reflection, that should constitute a simple way of understanding reflective programs; and a description of how reflection allows one to treat uniformly different application areas. To support this claim, the following three case studies will be discussed: metalevel reasoning; reasoning with multiple communicating theories (agents); and analogical reasoning. For each of these areas, the choice of a suitable reflection principle is shown, which tries to capture the specificity of the problem domain.

1 Introduction

Logic programming languages have proved to be powerful means for knowledge representation and automated reasoning. They are generally based on clausal form of first-order logic and on resolution inference rule, with special emphasis on Horn clauses and SLD resolution. These languages have well-behaved, convenient, computational properties, and well-developed, clean model and proof theories.

Limitations of these kinds of formal systems have also been recognized, however, and several directions of research are active to overcome them [11].

* P. Dell'Acqua has been supported financially by both Univ. degli Studi di Milano and Uppsala Univ.

In this paper, we introduce logical reflection as a principled way to further empower the representation and reasoning capabilities of logic programming systems. What we are referring to here are reflection principles inspired by those introduced in logic by Feferman [9], where a reflection principle is understood as "a description of a procedure for adding to any set of axioms A certain new axioms whose validity follow from the validity of the axioms A and which formally express within the language of A evident consequences of the assumption that all the theorems of A are valid".

In the context of logic programming, we use reflection as a simple yet powerful way to extend a basic theory. Reflection principles take the role of axiom schemata of a particular form that, once added to a given logic program (the basic theory, or the initial axioms), enlarge the set of consequences sanctioned by those initial axioms.

One advantage of this procedure, most evident yet far from secondary, is that it is much easier to write a basic theory and then to augment it with condensed axiom schemata, than it is to write a corresponding large (or even infinite) set of axioms in the first place. Greater expressivity and naturalness of representation may thus be attained by the logic programmer, and the logic program results more compact and manageable.

In the following sections, we show how it is possible to fully embed reflection principles in the context of Horn clauses and SLD resolution. With these additions, which notationally can be kept small, Horn clauses become a richer language.

As a second, important advantage of this approach, the well-established semantic properties of Horn clauses, mentioned above, carry over to Horn clauses with reflection. In fact, the semantics of Reflective SLD Resolution and the semantics of the Reflective Least Herbrand Model are obtained by making slight variations to, respectively, the procedural and the declarative semantics classically defined for Horn clauses [12].

Though this is a novel approach to reflective logic programming, it should not be seen as a tentative proposal. Rather, it is the outcome of the systematization and generalization of results accumulated in several years of study and experimentation on using reflection principles in the logic programming framework. In this paper we present in fact a complete formalization of a concept of reflection, that should constitute a simple way of understanding reflective programs; and a description of how reflection allows one to treat uniformly different application areas. The applications of reflection that we have previously studied (and reported in detail in previous papers [5, 6, 7, 8]) become instances of the new formalization (modulo a suitable reformulation). Thus we are able to present them as case studies, that show how reflective logic programming can constitute a uniform framework for several problem domains. In particular we will discuss how to model by means of reflection: 1) metalevel reasoning; 2) reasoning with multiple communicating theories (agents); and 3) analogical reasoning.

In the first two areas, the language of Horn clauses is provided with an encoding facility, to give names to language expressions, and a substitution facility, taking into account such named expressions. Reflection principles take advantage of these facilities to yield reflective extensions of a basic theory. In any case, it turns out that the reflection principles and the semantics of the reflective Horn clause programs embedding them are independent of the peculiarities of the chosen encoding and substitution facilities. The former can thus be very simple or quite sophisticated, depending on the demands of the problem area. The latter can be realized by means of whatever rewriting system is deemed more appropriate and efficient in the considered domain.

For each of the areas previously mentioned, the choice of a suitable reflection principle is shown, which tries to capture the specificity of the problem domain. Given a basic theory expressing a particular problem in that domain, its extension determined by that underlying reflection principle contains the consequences intended by the choice of that principle, but not entailed by the basic theory alone.

Thus, this use of reflection is different in essence from previous use of reflection rules in logic programming. Following [3], metaprogramming in logic programming [1, 4] has often been based on so-called metainterpreters. These are meta-level clauses which explicitly represent the derivability relation of the object level language, and are added to the clauses representing the problem domain. Language and metalanguage are amalgamated and interact by means of linking rules, which reflect meta-level conclusions to the object level and vice versa. The amalgamation, however, is conservative, in the sense that no new theorems are provable in the amalgamated theory that were not already provable in either the language or the metalanguage.

On the contrary, the goal of our conception and use of reflection principles is precisely to make the set of theorems provable from the basic theory augmented with reflective axioms differ from the set of theorems provable from the basic theory alone. This capability allows one to model different forms of reasoning within the same formal framework.

How this is made possible is described in the rest of the paper. The next section formally defines our conception of reflection principles, and specifies their integration in the declarative and procedural semantics of the Horn-clause language (giving the necessary definitions and properties of the formalism). The following three sections show how basic theories are enlarged to include consequences deriving from metaknowledge (section 3), from knowledge communicated among agents (section 4), and from analogies (section 5). Differently from Feferman's, the reflection principles used for such purposes sanction consequences which are intended but not necessarily valid. In conclusion, section 6 briefly mentions implementation issues of the proposed approach, and ends with some concluding remarks.

2 Reflective Semantics

In this section we show how to model several extensions to the language of Horn clauses with no modification of the standard semantics. This will be done by implicitly augmenting a program with a finite or denumerable set of new axioms, characterized by a program transformation schema called a Reflection Principle. In the following, let P be a definite program and B_P its Herbrand base. Let also "terms" and "atoms" be terms and atoms in the language of P. It is useful to assume that the variables appearing in P are *standardized apart* [12].

Definition 1 *A transformation schema AX is a mapping from Horn clauses to Horn clauses. We indicate the application of AX by* $\Leftarrow_{AX}$. □

The following general definitions can be given.

Definition 2 *Let* $I \subseteq B_P$ *be an interpretation for P. Let AX be a transformation schema, and AX′ the (finite or denumerable) set of ground instances of the clauses obtained by applying AX to all the clauses of P. I is an AX-Model for P iff it is a model for* $P' = P \cup AX'$. □

Definition 3 *Let AX be a transformation schema.* $A \in B_P$ *is an AX-Logical Consequence of P iff for every interpretation* $I \subseteq B_P$, *I is an AX-Model for P implies that I is a model for A.* □

AX-Models are clearly models in the usual sense. Therefore, for definite programs there exists the Least AX-Herbrand Model. It is in general not minimal as a Herbrand model of the original program, but it is minimal with respect to the set of consequences which can be drawn from the program and the additional axioms. These new axioms allow extensions to be performed to the language of Horn clauses by leaving the underlying logic unchanged, and rather modifying the program. The drawback is that the resulting program P' will have in general a denumerable number of clauses, which is allowed in principle but difficult to manage in practice.

Below we define a specific formulation of transformation schema, that will be called a *Reflection Principle.* The peculiarity of this formulation is that it makes it possible to define a variant of the declarative and procedural semantics of the Horn-clause language which is completely equivalent to the standard one, but allows the (potentially denumerable) set of new axioms to be coped with finitely.

Definition 4 *Let A be an atom and let B be a (possibly empty) conjunction of atoms. A transposition* Σ *is a relation which, given A and B, returns a new atom A′ and a (possibly empty) new conjunction B′. A transposition is total but not necessarily injective on A and B. We write* $\Sigma(A, B) = (A', B')$. □

For the sake of simplicity we also indicate with Σ any procedure able to compute the relation.

Definition 5 (*Reflection Principle*) *Let* A, A' *be atoms and let* B, B' *be (possibly empty) conjunctions of atoms. Let* $\Sigma^{\mathcal{R}}$ *be a transposition. A Reflection principle* $\mathcal{R}$ *is a transformation schema defined as follows:*

$$(A' \leftarrow B') \Leftarrow_{\mathcal{R}} (A \leftarrow B)$$

iff $A \leftarrow B$ *is a clause in P and* $\Sigma^{\mathcal{R}}(A, B) = (A', B')$. □

Intuitively, a reflection principle maps a rule of P over a new rule, given the transposition between their conclusions and conditions. The new rules generated by a reflection principle are called *reflection axioms*.

As shown in the following sections, several extensions to the Horn-clause language and various forms of knowledge representation can be modeled by means of reflection principles.

Definitions 2 and 3 can now be rephrased to *Reflective Model* and *Reflective Logical Consequence*. Since P is a definite program, there exists the *Least Reflective Model* $LRM_P^{\mathcal{R}}$ which entails the consequences of P, the additional consequences drawn by means of the reflection axioms, and the further consequences obtained from both.

The denomination "reflection principle" is borrowed from the work of Feferman [9], where (though in a quite different context) a reflection principle is defined to be

> *"a description of a procedure for adding to any set of axioms A certain new axioms whose validity follow from the validity of the axioms A ..."*.

In fact, in the definition 2.5, the transformation schema does not generate entirely arbitrary consequences, but rather a transposition of the original ones. The difference is that the validity of the reflection axioms does not *formally* follows from the validity of the given program, but it rather follows *conceptually*, according to the intended meaning of the extension. The definition of $\Sigma^{\mathcal{R}}$ embeds the particular knowledge representation features underlying the extension, and is a key point of the approach. It is important to notice that $\Sigma^{\mathcal{R}}$ can be any suitable formal system for the application at hand. In particular, $\Sigma^{\mathcal{R}}$ might be a metaprogram in any metalogic language.

It is now easy to define an extended function $RT_P^{\mathcal{R}}$, which gives $LRM_P^{\mathcal{R}}$ a constructive characterization, as well as an extended resolution principle. They will both be composed of the normal case, plus a second case to model the reflection principle.

Definition 6 *Let $I \subseteq B_P$. The mapping $RT_P^{\mathcal{R}} : 2^{B_P} \rightarrow 2^{B_P}$ is defined as follows (let $n, k \geq 0$):*

$$
\begin{aligned}
RT_P^{\mathcal{R}}(I) = \{\; & A \in B_P : A \leftarrow B_1, \ldots, B_n \text{ is a clause in } P,\ \{B_1, \ldots, B_n\} \subseteq I\} \cup \\
\{\; & A' \in B_P : A \leftarrow B_1, \ldots, B_n \text{ is a clause in } P, \\
& \quad \Sigma^{\mathcal{R}}(A, (B_1, \ldots, B_n)) = (A', (D_1, \ldots, D_k)), \\
& \quad \{D_1, \ldots, D_k\} \subseteq I \qquad \} \quad \square
\end{aligned}
$$

The mapping $RT_P^{\mathcal{R}}$ is clearly monotonic, and the following results hold.

Theorem 1 *The mapping $RT_P^{\mathcal{R}}$ is continuous.* $\square$

Theorem 2 *Let $I \subseteq B_P$. I is a Reflective Model for P iff $RT_P^{\mathcal{R}}(I) \subseteq I$.*

Proof. According to Definitions 2 and 3, I is a Reflective Model for P iff (let $n, k \geq 0$) one of the following conditions holds:

i. for every clause $A \leftarrow B_1, \ldots, B_n$ in P,
$\{B_1, \ldots, B_n\} \subseteq I$ implies $A \in I$, since I is a model of each axiom of P.

ii. for every clause $A \leftarrow B_1, \ldots, B_n$ in P,
$\Sigma^{\mathcal{R}}(A, (B_1, \ldots, B_n)) = (A', (D_1, \ldots, D_k))$ and $\{D_1, \ldots, D_k\} \subseteq I$ implies $A' \in I$, since I is a model of the reflection axioms defined by the reflection principle

iff $RT_P(I) \subseteq I$. □

Therefore we can apply the result by Van Emden and Kowalski [12, Th 6.5] thus providing a fixpoint characterization of the Least Reflective Herbrand Model $LRM_P^{\mathcal{R}}$.

Theorem 3 $LRM_P^{\mathcal{R}} = RT_P^{\mathcal{R}} \uparrow \omega = \bigcup_{n<\omega} RT_P^{\mathcal{R}} \uparrow n$ (*where* $RT_P^{\mathcal{R}} \uparrow 0 = \emptyset$). □

Definition 7 *Reflective $SLD^{\mathcal{R}}$-Resolution ($RSLD^{\mathcal{R}}$-Resolution)*
Let G be a definite goal $\leftarrow A_1, \ldots, A_i, \ldots, A_m$ and C a clause. Let A_i be the selected atom in G. G' is derived from G and C using mgu θ iff C is a clause $A \leftarrow B_1, \ldots, B_n$ in P, and one of the following conditions holds (let $n, k \geq 0$).

*i. θ is the mgu of A_i and A,
G' is $\leftarrow (A_1, \ldots, A_{i-1}, B_1, \ldots, B_n, A_{i+1}, \ldots, A_m)\theta$.*

*ii. $\Sigma^{\mathcal{R}}(A, (B_1, \ldots, B_n)) = (A', (D_1, \ldots, D_k))$,
A_i is ground, and
θ is the mgu of A_i and A',
G' is $\leftarrow (A_1, \ldots, A_{i-1}, D_1, \ldots, D_k, A_{i+1}, \ldots, A_m)\theta$.* □

While the definitions of RSLD derivation and refutation, success set, answer, and computed answer are exactly as usual, the notion of correct answer has to be defined appropriately.

Definition 8 *Let G be a definite goal $\leftarrow A_1, \ldots, A_m$ and θ an answer for $P \cup \{G\}$. We say that θ is a correct answer for $P \cup \{G\}$ if $\forall((A_1 \wedge \ldots \wedge A_m)\theta)$ is a Reflective Logical Consequence of P.*

Theorem 4 (*Soundness of $RSLD^{\mathcal{R}}$-Resolution*) *Every computed answer for $P \cup \{G\}$ is a correct answer for $P \cup \{G\}$.* □

Theorem 5 (*Completeness of $RSLD^{\mathcal{R}}$-Resolution*) *For every correct answer θ for $P \cup \{G\}$, there exists a computed answer λ for $P \cup \{G\}$ and a substitution γ such that $\theta = \lambda\gamma$.* □

We do not report here the proofs of the above results because they are quite long and (similarly to the proof of Theorem 2) they are straightforward extensions of the corresponding proofs in [12].

Clearly, soundness and completeness of $RSLD^{\mathcal{R}}$-Resolution depend on appropriate properties of the procedural mechanism associated with $\Sigma^{\mathcal{R}}$.

The requirement that the result of transposition be applied to ground atoms only (i.e., in point (ii) of Definition 7 A_i is required to be ground) can be avoided if unification is replaced by a more general constraint-satisfaction device. The problem is extensively discussed in [2] where a suitable solution is proposed, so as reflection can be applied with no restrictions.

It is important to notice that introducing more than one reflection principle is completely straighforward: simply, this will correspond to adding more cases to function RT and to resolution.

The semantics we have just presented is applicable for introducing reflection not only in the plain Horn clause language, but possibly in all its variations and extensions which are mainly based on variations and extensions of the classical semantics of [12].

In the following sections we show how different forms of knowledge representation and reasoning can be accomodated in the proposed framework, by defining suitable reflection principles.

3 Meta-level Reasoning

A relevant application of the formalization presented in the previous Section is the language Reflective Prolog (RP) [5, 6], a metalogic programming and knowledge representation language which allows the definition and use of both knowledge and metaknowledge in a uniform way in the same program.

RP has three basic features. First, a full referentiation/dereferentiation (*naming*) mechanism which allows the representation of metaknowledge in the form of *metalevel clauses*. In the following, the name of a term/atom/clause A is simply indicated by $\uparrow A$, wherever we do not want to go into syntactic details. Second, the possibility of specifying *metaevaluation clauses*, which are metalevel clauses defining the distinguished predicate *solve*, that allow to declaratively extend the meaning of the other predicates (forming the so-called *base level*). Last, a form of logical reflection that makes this extension effective.

In terms of Definition 5, RP declarative and procedural semantics is based on two reflection rules: *reflection up* and *reflection down*. Reflection up makes available (reflected up) to the metaevaluation level every conclusion drawn at the base level. It is based on the following transposition and transformation schema:

$$\Sigma^{\mathcal{R}_{up}}(A, B) = (solve(\uparrow A), B)$$
$$(solve(\uparrow A) \leftarrow B) \Leftarrow_{\mathcal{R}_{up}} (A \leftarrow B).$$

Reflection down makes available (reflected down) to the base level every conclusion drawn at the metaevaluation level. This means that *solve* clauses play the role of additional clauses for those predicates whose (referenced) instances match its argument. It is based on the following transposition and transformation schema:

$$\Sigma^{\mathcal{R}_{down}}(solve(\uparrow A), B) = (A, B)$$
$$(A \leftarrow B) \Leftarrow_{\mathcal{R}_{down}} (solve(\uparrow A) \leftarrow B).$$

Notice that the formulation of the reflection rules does not depend on the particular naming convention adopted. Both RT_P and resolution will have three cases. Procedurally, clauses with conclusion *solve*($\uparrow A$) may be used to resolve a goal A, and vice versa clauses with conclusion A to resolve *solve*($\uparrow A$).

Many examples of use of the reflective RT_P and resolution of RP are given in [5], and we will give here only a brief illustration of the application of the reflection rules in *RP*. To this aim, it is necessary to summarize the naming mechanism of *RP*, which is based on a *naming relation NR*. Please notice that the following is a very simplified version of a much more rich, complex and precise name theory. For detail, the reader may refer to [2].

In the language of an *RP* program P there are *object variables*, that we indicate with the usual syntax (e.g. X) and *metavariables*, that here we conventionally indicate by a single quote (e.g. Y'), though this is not the concrete syntax used in *RP*. Let us assume that for every symbol k in the alphabet of P, we add a new symbol k' to the alphabet itself. Let us also assume that $(k', k) \in NR$, which means that we take k' to be the name of k. If k and k' are variables, say X and X', we mean that any term denoted by X' is taken to be the name of any term denoted by X. For every predicate symbol q in P, we add the metavariable Q' to the alphabet of P, and the couple (Q', q) to NR, meaning that Q' stands for the name q' of q. Unification is extended according to this naming convention. In practice, by means of the extended unification any metavariable in the predicate position can act as the name of any predicate symbol q. This point is relevant to understand the following *RP* program P.

Example 1 *Let P be:*

(1) $solve(P'(X', Y'))$:- $symmetric(P), solve(P'(Y', X'))$
(2) $symmetric(p')$
(3) $p(a, b)$
(4) $q(b)$:- $p(b, a)$.

By applying $\mathcal{R}_{down}$ to clause (1), since p is the only binary predicate we get:
(1′) $p(X, Y)$:- $symmetric(p'), solve(p'(Y', X'))$.
By applying $\mathcal{R}_{up}$ to clauses (3) and (4), we get:
(3′) $solve(p'(a', b'))$
(4′) $solve(q'(b'))$:- $p(b, a)$.

From (1′), (2) and (3′) we conclude $p(b, a)$ and then, from (4), $q(b)$. Both consequences do not follow from the program without the reflection principles. By means of the reflection principles, clause (1) becomes an axiomatization of symmetry, which can be applied whenever necessary. □

In section 6 we will briefly discuss how any extension based on reflection can be easily and soundly implemented in RP.

4 Communication-Based Reasoning

The ability to represent agents and multi-agent co-operation is central to many AI applications. In the context of communication-based reasoning, the interaction among agents is based on communication acts. In particular, every agent can make questions to other agents in order to solve a given problem.

Within the logic programming paradigm, an approach to communication-based reasoning has been proposed in [8]. The main idea of this approach is to represent agents and communication acts by means of theories and reflection principles, respectively. Thus, theories formalize knowledge of agents, while a reflection principle characterizes a possible interaction among agents.

Every agent has associated with it a theory represented by a finite set of clauses prefixed with the corresponding theory symbol. In the following, to specify that a clause $A \leftarrow B_1, \ldots, B_n$ belongs to a theory T, we use the notation $T{:}A \leftarrow T{:}B_1, \ldots, T{:}B_n$. A program is then defined as a set of theories. Like in the previous section, $\uparrow A$ stands for the name of the syntactic expression A, whatever naming convention is adopted.

Communication acts are formalized by means of inter-theory reflection axioms based on the distinguished binary predicate symbols *tell* and *told*, and on the following reflection principle:

$$(T{:}told(\uparrow S, \uparrow A) \leftarrow S{:}B) \Leftarrow_{\mathcal{R}_{com}} (S{:}tell(\uparrow T, \uparrow A) \leftarrow S{:}B).$$

Its intuitive meaning is that every time an atom of the form $tell(\uparrow T, \uparrow A)$ can be derived from a theory S (which means that agent S wants to communicate proposition A to agent T), the atom $told(\uparrow S, \uparrow A)$ is consequently derived also in the theory T (which means that proposition A becomes available to agent T). The reflection principle above uses the following transposition:

$$\Sigma^{\mathcal{R}_{com}}(S{:}tell(\uparrow T, \uparrow A), S{:}B) = (T{:}told(\uparrow S, \uparrow A), S{:}B).$$

The objective of this formalization is that each agent can specify, by means of clauses defining the predicate *tell*, the modalities of interaction with the other agents. These modalities can thus vary with respect to different agents or different conditions.

The extended function $RT_P^{\mathcal{R}_{com}}$ and $RSLD^{\mathcal{R}_{com}}$-Resolution have both two cases. They are shown by means of a simple example.

Example 2 *Let P be the program composed of two theories, a and b, containing the following clauses.*

(1) $a{:}tell(\uparrow b, \uparrow ciao) \text{:-}\ friend(\uparrow b)$
(2) $a{:}friend(\uparrow b)$
(3) $b{:}hate(\uparrow a)$

The transposition associated with both P and the reflection principle above is

$$\Sigma^{\mathcal{R}_{com}}(a{:}tell(\uparrow b, \uparrow ciao), a{:}friend(\uparrow b)) = (b{:}told(\uparrow a, \uparrow ciao), a{:}friend(\uparrow b)).$$

To calculate the model of P, the steps (according to definition 2.6) are the following.

$RT_P^{\mathcal{R}_{com}} \uparrow 0 = \emptyset$

$RT_P^{\mathcal{R}_{com}} \uparrow 1 = \{b{:}hate(\uparrow a), a{:}friend(\uparrow b)\}$

$RT_P^{\mathcal{R}_{com}} \uparrow 2 = RT_P^{\mathcal{R}_{com}} \uparrow 1 \cup \{a{:}tell(\uparrow b, \uparrow ciao)\} \cup \{b{:}told(\uparrow a, \uparrow ciao)\}$

$RT_P^{\mathcal{R}_{com}} \uparrow 3 = RT_P^{\mathcal{R}_{com}} \uparrow 2$

Therefore, $LRM_P^{\mathcal{R}_{com}} = RT_P^{\mathcal{R}_{com}} \uparrow 2$.
The goal $\leftarrow b{:}told(\uparrow a, X)$ can be proved with the following steps.

$\leftarrow b{:}told(\uparrow a, X)$
by case (ii) of Definition 2.7 with $\theta = \{X/\uparrow ciao\}$ we get
$\leftarrow a{:}friend(\uparrow b)$
which succeeds by fact (2).

5 Plausible Reasoning

Plausible reasoning is a suitable realm of application of reflection principles. In fact, most forms of plausible reasoning reinterpret available premises to draw plausible conclusions.

As a significant example, a knowledge representation principle which has been shown [7] to be easily formalizable in terms of reflection is *replacement based analogy* [13]. Analogy is based on the assumption that if two situations are similar in some respect, then they may be similar in other respects as well. Thus, an analogy is a mapping of knowledge from a known "source" domain into a novel "target" domain. Analogy can be applied to problem-solving, planning, proving, etc., on the basis of the following kind of inference: knowing that from premises A conclusion B follows, and that A' corresponds to A, analogically conclude B'. In particular, replacement-based analogy defines analogy as a replacement of the source object with the target object as stated in the following principle, due to Winston [13].

Assume that the premises $\beta_1, \ldots, \beta_n$ logically imply α in the source domain. Assume also that analogous premises $\gamma_1, \ldots, \gamma_n$ hold in the target domain. Then we conclude the atom α' in the target which is analogous to α.

In logic programming, given a program P viewed as divided into two subprograms P_s and P_t (which play the role of the source and the target domain respectively), analogy can be procedurally performed by transforming rules in P_s into analogous rules in P_t. The analogous rules can be computed by means of *partial identity* between terms of the two domains, like in [10], or by means of *predicate analogies* and *terms correspondence* like in [7], as exemplified below.

Example 3 *Consider the following programs P_s and P_t.*

P_s *kills*(X,Y):- *hates*(X,Y), *has_weapon*(X). $(*)$
hates(john, george).
has_weapon(john).

P_t *hates(anne, joe).*
despises(X, X):- depressed(X).
has_weapon(anne).
has_weapon(bill).
depressed(bill).

Let us assume that predicates with the same name are in analogy by default. Let us also assume an explicit analogy between predicates ***despises*** and ***hates***. Clearly, the goal ?- *kills(anne, joe)* is not provable in P_t. It is however provable by analogy by taking rule (*) in P_s, and assuming the term correspondence $\{(john, anne), (george, joe)\}$. That is, taken the instance of clause (*) in P_s:
kills(john,george):- hates(john,george),has_weapon(john).
we can apply analogy by simulating the corresponding clause of P_t:
kills(anne,joe):- hates(anne,joe),has_weapon(anne).
Similarly we are able to prove by analogy the goal ?- *kills(bill, bill)*. This is obtained by allowing the correspondence $\{(john, bill), (george, bill)\}$. In a correspondence, an element of the source domain can correspond to only one element of the target domain; on the contrary, two elements of the source domain, like the constants *john* and *george* in the example, may correspond to the same element of the target domain. □

Reflective semantics of this kind of analogical reasoning is defined in full detail in [7]. We report here only the essential points, in order to show the connection to the formalization that we propose in this paper.

Let a program P be divided into two subprograms, P_s and P_t, as mentioned above. Let $U_{P_s}, B_{P_s}, pred(P_s)$ ($U_{P_t}, B_{P_t}, pred(P_t)$) be the Herbrand universe, the Herbrand base and the set of predicate symbols of P_s (P_t). In the following, atoms and conjunctions of atoms in P_s (P_t) will be indicated with A_s, B_s respectively (A_t, B_t).

Definition 9 *A set of predicate analogies S is a finite, possibly empty set of couples (ps, pt), where $ps \in pred(P_s)$ and $pt \in pred(P_t)$.* □

Definition 10 *A correspondence σ is a finite (possibly empty) set of the form $\{t_1//s_1, \ldots, t_n//s_n\}$ where each t_i and s_i is a term, and for every i and j, such that $1 \leq i, j \leq n$, t_i is distinct from t_j and t_i is not a subterm of t_j.* □

Clearly, substitutions are particular cases of correspondences. A correspondence can be *applied* to a term or atom A in the obvious way, by substituting every occurrence of t_i with s_i. The result of the application is indicated by $A\sigma$. Correspondences can be composed with substitutions, giving a new correspondence as a result. In particular, given a substitution θ, the correspondence $\gamma = \sigma\theta$ is obtained by substituting every variable in σ with its assigment in θ, if any.

Semantics of this kind of analogical reasoning may be expressed in terms of the following *analogical reflection principle* $\mathcal{RA}$. Given a set S of predicate analogies, we have

$$\Sigma^{\mathcal{RA}}(A_s, B_s) = (A_t, B_t)$$

iff $A_s = ps(t_1,\ldots,t_k)$, $A_t = pt(f_1,\ldots,f_k)$, $(ps,pt) \in S$, and there exists a correspondence σ between $t_1,\ldots,t_k$ and $f_1,\ldots,f_k$ such that $B_t = B_s\sigma$.

By applying the previous definitions, the mapping $RT_P^{\mathcal{RA}}$ is the following.

$$
\begin{aligned}
&RT_P^{\mathcal{RA}}(I) = \\
&\quad \{\, A_t \in B_{P_t} : A_t \leftarrow B_1,\ldots,B_n \text{ is a clause in } P_t,\ \{B_1,\ldots,B_n\} \subseteq I \} \cup \\
&\quad \{\, A_t \in B_{P_t} : A_s \leftarrow B_1,\ldots,B_n \text{ is a clause in } P_s, \\
&\qquad\qquad A_s = ps(t_1,\ldots,t_k), A_t = pt(f_1,\ldots,f_k), (ps,pt) \in S, \\
&\qquad\qquad \text{there exists a correspondence } \sigma \text{ between } t_1,\ldots,t_k \text{ and} \\
&\qquad\qquad f_1,\ldots,f_k \text{ such that } \{B_1\sigma,\ldots,B_n\sigma\} \subseteq I \qquad \} \quad \square
\end{aligned}
$$

The mapping $RT_P^{\mathcal{RA}}$ characterizes the consequences of P_t with respect to the clauses of P_t itself, and to the clauses of P_s which allow analogical consequences to be derived. We show below the computation of the Least Reflective Herbrand Model of the program in Example 5.1. For the sake of clarity, every analogical conclusion is written in boldface. We also indicate the (ground instance of the) clause in P_s which allows the analogical derivation, and the related correspondence.

$RT_P^{\mathcal{RA}} \uparrow 0 = \emptyset$
$RT_P^{\mathcal{RA}} \uparrow 1 = \{hates(anne,joe), has\_weapon(anne),$
$has\_weapon(bill), depressed(bill)\}$
$RT_P^{\mathcal{RA}} \uparrow 2 = RT_P^{\mathcal{RA}} \uparrow 1 \cup \{\mathbf{despises(bill,bill)}\} \cup \{\mathbf{kills(anne,joe)}\}$
clause and correspondence:
$kills(john, george) \leftarrow hates(john, george), has\_weapon(john)$
$\sigma = \{john//anne, george//joe\}$
$RT_P^{\mathcal{RA}} \uparrow 3 = RT_P^{\mathcal{RA}} \uparrow 2 \cup \{\mathbf{kills(bill,bill)}\}$
clause and correspondence:
$kills(john, george) \leftarrow hates(john, george), has\_weapon(john)$
(recall that the predicates despises and hates are in analogy)
$\sigma = \{john//bill, george//bill\}$
$RT_P^{\mathcal{RA}} \uparrow 4 = RT_P^{\mathcal{RA}} \uparrow 3$

6 Implementation Issues and Concluding Remarks

Though a reflection principle is, semantically, an axiom schema, from the implementation point of view it can be seen as a metatheoretical statement, based on the treatment of predicate symbols and atoms as first-class objects. Thus Reflective Prolog, which provides these features in a clean and easily usable form, has proved a good language for the implementation of RSLD-resolution (definition 2.7). Below we show how an arbitrary reflection principle can be represented by means of RP reflection rules (introduced in section 3).

Let *theory_clause* be the declarative equivalent of the Prolog predefined predicate *clause* [5]. A reflection principle $\mathcal{R}$ characterized by a transposition $\Sigma(AA, BB, A, B)$ can be implemented by the following RP metaevaluation clause, to be added to the given program (recall that X' is the syntax for metavariables).

$$solve(A') : -theory\_clause(AA', BB'), Sigma(AA', BB', A', B'), solve(B').$$

(together with some heuristic guidance on how to retrieve a clause appropriate to the correspondence). According to the definitions given in section 3, this rule in RP is automatically applied whenever A fails in the given program. In general, the reflection principle $\mathcal{R}$ is automatically applied on failure of goals by SLD-Resolution.

We sketch now an informal proof of soundness of the implementation w.r.t. the given semantics. RP's Resolution has been proved correct and complete w.r.t. the Least Reflective Herbrand model of an RP program [6].

Let us assume that the clause above is a sound representation of a reflection principle (the main point is that the procedures Σ correctly implement the transposition). On this assumption we can conclude that the Least Reflective Herbrand model of the given program with respect to $\mathcal{R}$ is contained in the Least Reflective Herbrand model of the resulting RP program, and thus the correctness and completeness results formally proved for the latter carry over to the former.

In conclusion, in this paper we have reported on the possibility of using logical reflection principles for knowledge representation and reasoning within the logic programming approach. We believe that, while some particular aspect of the formalism we have adopted may be a matter of taste, a general merit may be ascribed to the extended logic programming framework that we propose. It amounts to the idea of not having to define a separate logic for every reasoning mechanism; rather, to allow a specific reasoning form to be expressed in the framework, where it both becomes executable (procedural semantics) and, at the same time, acquires a precise meaning (declarative semantics).

As mentioned in the introduction, we are since long experimenting with the usability of this approach. In addition to the three case studies presented in this paper, we have also coped with paradigms such as non-monotonic reasoning [6], explanation-based generalization, and object-oriented knowledge representation (forthcoming papers).

References

1. H. D. Abramson, M. H. Rogers (eds). *Meta-Programming in Logic Programming.* The MIT Press, 1989.
2. J. Barklund, S. Costantini, P. Dell'Acqua, and G. A. Lanzarone. *Reflection through Constraint Satisfaction.* Submitted paper.

3. K. A. Bowen, R. A. Kowalski. *Amalgamating Language and Metalanguage.* in: K. L. Clark and S.-Å Tarnlund (eds.), Logic Programming, Academic Press, 1982.
4. M. Bruynooghe (ed.). *Proceedings of the 2nd Workshop on Metaprogramming in Logic.* Leuven (Belgium), April 4-6 1990.
5. S. Costantini, G. A. Lanzarone. *A Metalogic Programming Approach: Language, Semantics and Applications.* Journal of Experimental and Theoretical Artificial Intelligence 1(1993). Extended Abstract in: G. Levi and A. Martelli (eds.) *Logic Programming,* Proceedings of the Sixth International Conference, The Mit Press, 1989.
6. S. Costantini, G. A. Lanzarone. *Metalevel Negation in Non-Monotonic Reasoning.* To appear on the Journal of Methods of Logic in Computer Science. Extended Abstract in: Proceedings of the Workshop on Logic Programming and Non-Monotonic Reasoning, Austin, TX, November 1-2, 1990.
7. S. Costantini, G. A. Lanzarone, and L. Sbarbaro. *A Formal Definition and a Sound Implementation of Analogical Reasoning in Logic Programming.* To appear on the Annals of Mathematics and Artificial Intelligence.
8. S. Costantini, P. Dell'Acqua, and G. A. Lanzarone. *Reflective Agents in Metalogic Programming.* in: A. Pettorossi (ed.) *Meta-Programming in Logic. Lecture Notes in Computer Science 649,* Springer-Verlag,1992.
9. S. Feferman. *Transfinite Recursive Progressions of Axiomatic Theories.* Journal of Symbolic Logic vol. 27, n. 3, 1962.
10. M. Haraguchi, S. Arikawa. *Reasoning by Analogy as a Partial Identity between Models.* in: K. P. Jantke (ed.), Analogical and Inductive Inference. Lecture Notes in Computer Science n. 265, Springer-Verlag, 1987.
11. R. A. Kowalski. *Problems and Promises of Computational Logic.* in: J. W. Lloyd (ed.), Computational Logic: Symposium Proceedings. Springer-Verlag 1990.
12. J. W. Lloyd. *Foundations of Logic Programming* (Second, Extended Edition). Springer-Verlag, Berlin, 1987.
13. P. Winston. *Learning and Reasoning by Analogy.* Communication of the Association for Computing Machinery, 23 (12), December 1980.

Springer-Verlag and the Environment

We at Springer-Verlag firmly believe that an international science publisher has a special obligation to the environment, and our corporate policies consistently reflect this conviction.

We also expect our business partners – paper mills, printers, packaging manufacturers, etc. – to commit themselves to using environmentally friendly materials and production processes.

The paper in this book is made from low- or no-chlorine pulp and is acid free, in conformance with international standards for paper permanency.

Lecture Notes in Artificial Intelligence (LNAI)

Vol. 638: A. F. Rocha, Neural Nets. A Theory for Brains and Machines. XV, 393 pages. 1992.

Vol. 642: K. P. Jantke (Ed.), Analogical and Inductive Inference. Proceedings, 1992. VIII, 319 pages. 1992.

Vol. 659: G. Brewka, K. P. Jantke, P. H. Schmitt (Eds.), Nonmonotonic and Inductive Logic. Proceedings, 1991. VIII, 332 pages. 1993.

Vol. 660: E. Lamma, P. Mello (Eds.), Extensions of Logic Programming. Proceedings, 1992. VIII, 417 pages. 1993.

Vol. 667: P. B. Brazdil (Ed.), Machine Learning: ECML – 93. Proceedings, 1993. XII, 471 pages. 1993.

Vol. 671: H. J. Ohlbach (Ed.), GWAI-92: Advances in Artificial Intelligence. Proceedings, 1992. XI, 397 pages. 1993.

Vol. 679: C. Fermüller, A. Leitsch, T. Tammet, N. Zamov, Resolution Methods for the Decision Problem. VIII, 205 pages. 1993.

Vol. 681: H. Wansing, The Logic of Information Structures. IX, 163 pages. 1993.

Vol. 689: J. Komorowski, Z. W. Raś (Eds.), Methodologies for Intelligent Systems. Proceedings, 1993. XI, 653 pages. 1993.

Vol. 695: E. P. Klement, W. Slany (Eds.), Fuzzy Logic in Artificial Intelligence. Proceedings, 1993. VIII, 192 pages. 1993.

Vol. 698: A. Voronkov (Ed.), Logic Programming and Automated Reasoning. Proceedings, 1993. XIII, 386 pages. 1993.

Vol. 699: G.W. Mineau, B. Moulin, J.F. Sowa (Eds.), Conceptual Graphs for Knowledge Representation. Proceedings, 1993. IX, 451 pages. 1993.

Vol. 723: N. Aussenac, G. Boy, B. Gaines, M. Linster, J.-G. Ganascia, Y. Kodratoff (Eds.), Knowledge Acquisition for Knowledge-Based Systems. Proceedings, 1993. XIII, 446 pages. 1993.

Vol. 727: M. Filgueiras, L. Damas (Eds.), Progress in Artificial Intelligence. Proceedings, 1993. X, 362 pages. 1993.

Vol. 728: P. Torasso (Ed.), Advances in Artificial Intelligence. Proceedings, 1993. XI, 336 pages. 1993.

Vol. 743: S. Doshita, K. Furukawa, K. P. Jantke, T. Nishida (Eds.), Algorithmic Learning Theory. Proceedings, 1992. X, 260 pages. 1993.

Vol. 744: K. P. Jantke, T. Yokomori, S. Kobayashi, E. Tomita (Eds.), Algorithmic Learning Theory. Proceedings, 1993. XI, 423 pages. 1993.

Vol. 745: V. Roberto (Ed.), Intelligent Perceptual Systems. VIII, 378 pages. 1993.

Vol. 746: A. S. Tanguiane, Artificial Perception and Music Recognition. XV, 210 pages. 1993.

Vol. 754: H. D. Pfeiffer, T. E. Nagle (Eds.), Conceptual Structures: Theory and Implementation. Proceedings, 1992. IX, 327 pages. 1993.

Vol. 764: G. Wagner, Vivid Logic. XII, 148 pages. 1994.

Vol. 766: P. R. Van Loocke, The Dynamics of Concepts. XI, 340 pages. 1994.

Vol. 770: P. Haddawy, Representing Plans Under Uncertainty. X, 129 pages. 1994.

Vol. 784: F. Bergadano, L. De Raedt (Eds.), Machine Learning: ECML-94. Proceedings, 1994. XI, 439 pages. 1994.

Vol. 795: W. A. Hunt, Jr., FM8501: A Verified Microprocessor. XIII, 333 pages. 1994.

Vol. 798: R. Dyckhoff (Ed.), Extensions of Logic Programming. Proceedings, 1993. VIII, 360 pages. 1994.

Vol. 799: M. P. Singh, Multiagent Systems: Intentions, Know-How, and Communications. XXIII, 168 pages. 1994.

Vol. 804: D. Hernández, Qualitative Representation of Spatial Knowledge. IX, 202 pages. 1994.

Vol. 808: M. Masuch, L. Pólos (Eds.), Knowledge Representation and Reasoning Under Uncertainty. VII, 237 pages. 1994.

Vol. 810: G. Lakemeyer, B. Nebel (Eds.), Foundations of Knowledge Representation and Reasoning. VIII, 355 pages. 1994.

Vol. 814: A. Bundy (Ed.), Automated Deduction — CADE-12. Proceedings, 1994. XVI, 848 pages. 1994.

Vol. 822: F. Pfenning (Ed.), Logic Programming and Automated Reasoning. Proceedings, 1994. X, 345 pages. 1994.

Vol. 827: D. M. Gabbay, H. J. Ohlbach (Eds.), Temporal Logic. Proceedings, 1994. XI, 546 pages. 1994.

Vol. 830: C. Castelfranchi, E. Werner (Eds.), Artificial Social Systems. Proceedings, 1992. XVIII, 337 pages. 1994.

Vol. 833: D. Driankov, P. W. Eklund, A. Ralescu (Eds.), Fuzzy Logic and Fuzzy Control. Proceedings, 1991. XII, 157 pages. 1994.

Vol. 835: W. M. Tepfenhart, J. P. Dick, J. F. Sowa (Eds.), Conceptual Structures: Current Practices. Proceedings, 1994. VIII, 331 pages. 1994.

Vol. 837: S. Wess, K.-D. Althoff, M. M. Richter (Eds.), Topics in Case-Based Reasoning. Proceedings, 1993. IX, 471 pages. 1994.

Vol. 838: C. MacNish, D. Pearce, L. M. Pereira (Eds.), Logics in Artificial Intelligence. Proceedings, 1994. IX, 413 pages. 1994.

Lecture Notes in Computer Science

Vol. 803: J. W. de Bakker, W.-P. de Roever, G. Rozenberg (Eds.), A Decade of Concurrency. Proceedings, 1993. VII, 683 pages. 1994.

Vol. 804: D. Hernández, Qualitative Representation of Spatial Knowledge. IX, 202 pages. 1994. (Subseries LNAI).

Vol. 805: M. Cosnard, A. Ferreira, J. Peters (Eds.), Parallel and Distributed Computing. Proceedings, 1994. X, 280 pages. 1994.

Vol. 806: H. Barendregt, T. Nipkow (Eds.), Types for Proofs and Programs. VIII, 383 pages. 1994.

Vol. 807: M. Crochemore, D. Gusfield (Eds.), Combinatorial Pattern Matching. Proceedings, 1994. VIII, 326 pages. 1994.

Vol. 808: M. Masuch, L. Pólos (Eds.), Knowledge Representation and Reasoning Under Uncertainty. VII, 237 pages. 1994. (Subseries LNAI).

Vol. 809: R. Anderson (Ed.), Fast Software Encryption. Proceedings, 1993. IX, 223 pages. 1994.

Vol. 810: G. Lakemeyer, B. Nebel (Eds.), Foundations of Knowledge Representation and Reasoning. VIII, 355 pages. 1994. (Subseries LNAI).

Vol. 811: G. Wijers, S. Brinkkemper, T. Wasserman (Eds.), Advanced Information Systems Engineering. Proceedings, 1994. XI, 420 pages. 1994.

Vol. 812: J. Karhumäki, H. Maurer, G. Rozenberg (Eds.), Results and Trends in Theoretical Computer Science. Proceedings, 1994. X, 445 pages. 1994.

Vol. 813: A. Nerode, Y. N. Matiyasevich (Eds.), Logical Foundations of Computer Science. Proceedings, 1994. IX, 392 pages. 1994.

Vol. 814: A. Bundy (Eds.), Automated Deduction—CADE-12. Proceedings, 1994. XVI, 848 pages. 1994. (Subseries LNAI).

Vol. 815: R. Valette (Eds.), Application and Theory of Petri Nets 1994. Proceedings. IX, 587 pages. 1994.

Vol. 816: J. Heering, K. Meinke, B. Möller, T. Nipkow (Eds.), Higher-Order Algebra, Logic, and Term Rewriting. Proceedings, 1993. VII, 344 pages. 1994.

Vol. 817: C. Halatsis, D. Maritsas, G. Philokyprou, S. Theodoridis (Eds.), PARLE '94. Parallel Architectures and Languages Europe. Proceedings, 1994. XV, 837 pages. 1994.

Vol. 818: D. L. Dill (Ed.), Computer Aided Verification. Proceedings, 1994. IX, 480 pages. 1994.

Vol. 819: W. Litwin, T. Risch (Eds.), Applications of Databases. Proceedings, 1994. XII, 471 pages. 1994.

Vol. 820: S. Abiteboul, E. Shamir (Eds.), Automata, Languages and Programming. Proceedings, 1994. XIII, 644 pages. 1994.

Vol. 821: M. Tokoro, R. Pareschi (Eds.), Object-Oriented Programming. Proceedings, 1994. XI, 535 pages. 1994.

Vol. 822: F. Pfenning (Ed.), Logic Programming and Automated Reasoning. Proceedings, 1994. X, 345 pages. 1994. (Subseries LNAI).

Vol. 823: R. A. Elmasri, V. Kouramajian, B. Thalheim (Eds.), Entity-Relationship Approach — ER '93. Proceedings, 1993. X, 531 pages. 1994.

Vol. 824: E. M. Schmidt, S. Skyum (Eds.), Algorithm Theory - SWAT '94. Proceedings. IX, 383 pages. 1994.

Vol. 825: J. L. Mundy, A. Zisserman, D. Forsyth (Eds.), Applications of Invariance in Computer Vision. Proceedings, 1993. IX, 510 pages.

Vol. 826: D. S. Bowers (Ed.), Directions in Databases. Proceedings, 1994. X, 234 pages. 1994.

Vol. 827: D. M. Gabbay, H. J. Ohlbach (Eds.), Temporal Logic. Proceedings, 1994. XI, 546 pages. 1994. (Subseries LNAI).

Vol. 828: L. C. Paulson, Isabelle. XVII, 321 pages. 1994.

Vol. 829: A. Chmora, S. B. Wicker (Eds.), Error Control, Cryptology, and Speech Compression. Proceedings, 1993. VIII, 121 pages. 1994.

Vol. 830: C. Castelfranchi, E. Werner (Eds.), Artificial Social Systems. Proceedings, 1992. XVIII, 337 pages. 1994. (Subseries LNAI).

Vol. 831: V. Bouchitté, M. Morvan (Eds.), Orders, Algorithms, and Applications. Proceedings, 1994. IX, 204 pages. 1994.

Vol. 832: E. Börger, Y. Gurevich, K. Meinke (Eds.), Computer Science Logic. Proceedings, 1993. VIII, 336 pages. 1994.

Vol. 833: D. Driankov, P. W. Eklund, A. Ralescu (Eds.), Fuzzy Logic and Fuzzy Control. Proceedings, 1991. XII, 157 pages. 1994. (Subseries LNAI).

Vol. 834: D.-Z. Du, X.-S. Zhang (Eds.), Algorithms and Computation. Proceedings, 1994. XIII, 687 pages. 1994.

Vol. 835: W. M. Tepfenhart, J. P. Dick, J. F. Sowa (Eds.), Conceptual Structures: Current Practices. Proceedings, 1994. VIII, 331 pages. 1994. (Subseries LNAI).

Vol. 836: B. Jonsson, J. Parrow (Eds.), CONCUR '94: Concurrency Theory. Proceedings, 1994. IX, 529 pages. 1994.

Vol. 837: S. Wess, K.-D. Althoff, M. M. Richter (Eds.), Topics in Case-Based Reasoning. Proceedings, 1993. IX, 471 pages. 1994. (Subseries LNAI).

Vol. 838: C. MacNish, D. Pearce, L. M. Pereira (Eds.), Logics in Artificial Intelligence. Proceedings, 1994. IX, 413 pages. 1994. (Subseries LNAI).